Lecture Notes in Computer Science 3376

Commenced Publication in 1973
Founding and Former Series Editors:
Gerhard Goos, Juris Hartmanis, and Jan van Leeuwen

Editorial Board

David Hutchison
  Lancaster University, UK
Takeo Kanade
  Carnegie Mellon University, Pittsburgh, PA, USA
Josef Kittler
  University of Surrey, Guildford, UK
Jon M. Kleinberg
  Cornell University, Ithaca, NY, USA
Friedemann Mattern
  ETH Zurich, Switzerland
John C. Mitchell
  Stanford University, CA, USA
Moni Naor
  Weizmann Institute of Science, Rehovot, Israel
Oscar Nierstrasz
  University of Bern, Switzerland
C. Pandu Rangan
  Indian Institute of Technology, Madras, India
Bernhard Steffen
  University of Dortmund, Germany
Madhu Sudan
  Massachusetts Institute of Technology, MA, USA
Demetri Terzopoulos
  New York University, NY, USA
Doug Tygar
  University of California, Berkeley, CA, USA
Moshe Y. Vardi
  Rice University, Houston, TX, USA
Gerhard Weikum
  Max-Planck Institute of Computer Science, Saarbruecken, Germany

Alfred Menezes (Ed.)

Topics in Cryptology – CT-RSA 2005

The Cryptographers' Track at the RSA Conference 2005
San Francisco, CA, USA, February 14-18, 2005
Proceedings

Volume Editor

Alfred Menezes
University of Waterloo
Department of Combinatorics and Optimization
Waterloo, Ontario, N2L 3G1, Canada
E-mail: ajmeneze@uwaterloo.ca

Library of Congress Control Number: 2004117506

CR Subject Classification (1998): E.3, G.2.1, D.4.6, K.6.5, K.4.4, F.2.1-2, C.2, J.1

ISSN 0302-9743
ISBN 3-540-24399-2 Springer Berlin Heidelberg New York

This work is subject to copyright. All rights are reserved, whether the whole or part of the material is concerned, specifically the rights of translation, reprinting, re-use of illustrations, recitation, broadcasting, reproduction on microfilms or in any other way, and storage in data banks. Duplication of this publication or parts thereof is permitted only under the provisions of the German Copyright Law of September 9, 1965, in its current version, and permission for use must always be obtained from Springer. Violations are liable to prosecution under the German Copyright Law.

Springer is a part of Springer Science+Business Media

springeronline.com

© Springer-Verlag Berlin Heidelberg 2005
Printed in Germany

Typesetting: Camera-ready by author, data conversion by Olgun Computergrafik
Printed on acid-free paper   SPIN: 11377726   06/3142   5 4 3 2 1 0

Preface

The RSA Conference is attended by over 10,000 security professionals each year. The Cryptographers' Track (CT-RSA), one of several parallel tracks at the conference, provides an excellent opportunity for cryptographers to showcase their research to a wide audience. CT-RSA 2005 was the fifth year of the Cryptographers' Track.

The selection process for the CT-RSA program is the same as for other cryptography research conferences. This year, the program committee selected 23 papers from 74 submissions (two of which were later withdrawn) that covered all aspects of cryptography. The program also included two invited talks by Cynthia Dwork and Moti Yung. These proceedings contain the revised versions of the selected papers. The revisions were not checked, and so the authors (and not the committee) bear full responsibility for the contents of their papers.

I am very grateful to the program committee for their very conscientious efforts to review each paper fairly and thoroughly. The initial review stage was followed by a tremendous amount of discussion which contributed to our high confidence in our judgements. Thanks also to the many external reviewers whose names are listed in the following pages. My apologies to those whose names were inadvertently omitted from this list.

Thanks to Eddie Ng for maintaining the submission server and the Web review system. The submission software was written by Chanathip Namprempre, and the Web review software by Wim Moreau and Joris Claessens. Thanks to Alfred Hofmann and his colleagues at Springer for the timely production of these proceedings. Finally, it is my pleasure to acknowledge Ari Juels and Mike Szydlo of RSA Laboratories for their assistance and cooperation during the past seven months.

October 2004   Alfred Menezes

RSA Cryptographers' Track 2005
February 14–18, 2005, San Francisco, CA, USA

The RSA Conference 2005 was organized by RSA Security Inc. and its partner organizations around the world. The Cryptographers' Track was organized by RSA Laboratories.

Program Chair

Alfred Menezes, University of Waterloo, Canada

Program Committee

Masayuki Abe	NTT Laboratories, Japan
Paulo Barreto	Scopus Tecnologia, Brazil
Alex Biryukov	K.U.Leuven, Belgium
John-Sebastien Coron	Gemplus, France
Steven Galbraith	Royal Holloway, University of London, UK
Amir Herzberg	Bar-Ilan University, Israel
Yuval Ishai	Technion, Israel
Stanislaw Jarecki	UC Irvine, USA
Lars Knudsen	Technical University of Denmark
Kaoru Kurosawa	Ibaraki University, Japan
Tanja Lange	Ruhr-Universität, Bochum, Germany
Helger Lipmaa	Helsinki University of Technology, Finland
Philip MacKenzie	DoCoMo, USA
Tal Malkin	Columbia University, USA
Wenbo Mao	HP Laboratories, UK
Ilya Mironov	Microsoft Research, USA
Josef Pieprzyk	Macquarie University, Australia
Palash Sarkar	Indian Statistical Institute, India
Jessica Staddon	Palo Alto Research Center, USA
Rene Struik	Certicom, Canada
Michael Szydlo	RSA Laboratories, USA
Tsuyoshi Takagi	TU Darmstadt, Germany

Steering Committee

Marc Joye	Gemplus, France
Tatsuaki Okamoto	NTT, Japan
Bart Preneel	K.U.Leuven, Belgium
Ron Rivest	MIT, USA
Moti Yung	Columbia University, USA

External Reviewers

Toru Akishita
Alexandr Andoni
Roberto Avanzi
Sara Bitan
Alexandra Boldyreva
Reinier Bröker
Daniel Brown
Bertrand Byramjee
Christophe De Canniere
Dario Catalano
Liqun Chen
Joe Cho
Carlos Cid
Mathieu Ciet
Scott Contini
Claus Diem
Yevgeniy Dodis
Eiichiro Fujisaki
Juan Garay
Craig Gentry
Philippe Golle
Shai Halevi
Darrel Hankerson
Heng Swee Huay

David Hwang
Kouichi Itoh
Tetsu Iwata
Antoine Joux
Masanobu Katagi
Jonathan Katz
Jeff King
Lea Kissner
Yuichi Komano
Hugo Krawczyk
Caroline Kudla
Joseph Lano
Kerstin Lemke
John Linn
Anna Lysyanskaya
Dahlia Malkhi
Daniele Micciancio
Anton Mityagin
Atsuko Miyaji
David Molnar
Michael Mueller
Jorge Nakahara
Wakaha Ogata
Kazuo Ohata

Yasuhiro Ohtaki
Akira Otsuka
Pascal Paillier
Zulfikar Ramzan
Leo Reyzin
Matt Robshaw
Markku-Juhani Saarinen
Taiichi Saito
Akashi Satoh
Kai Schramm
Daniel Schepers
Igor Shparlinski
Nigel Smart
Angelos Stavrou
Ron Steinfeld
Makoto Sugita
Matti Tommiska
Eran Tromer
Huaxiong Wang
Michael Wiener
Kai Wirt
Christopher Wolf
Shoko Yonezawa
Yunlei Zhao

Table of Contents

Invited Talks

Sub-linear Queries Statistical Databases: Privacy with Power 1
 Cynthia Dwork

Malicious Cryptography: Kleptographic Aspects 7
 Adam Young and Moti Yung

Cryptanalysis

Resistance of SNOW 2.0 Against Algebraic Attacks 19
 Olivier Billet and Henri Gilbert

A Study of the Security of Unbalanced Oil
and Vinegar Signature Schemes....................................... 29
 An Braeken, Christopher Wolf, and Bart Preneel

Hold Your Sessions: An Attack on Java Session-Id Generation 44
 Zvi Gutterman and Dahlia Malkhi

Update on SHA-1 58
 Vincent Rijmen and Elisabeth Oswald

A Fast Correlation Attack on the Shrinking Generator 72
 Bin Zhang, Hongjun Wu, Dengguo Feng, and Feng Bao

Public-Key Encryption

Improved Efficiency for CCA-Secure Cryptosystems Built
Using Identity-Based Encryption 87
 Dan Boneh and Jonathan Katz

A Generic Conversion with Optimal Redundancy 104
 Yang Cui, Kazukuni Kobara, and Hideki Imai

Choosing Parameter Sets for NTRUEncrypt with NAEP and SVES-3 118
 Nick Howgrave-Graham, Joseph H. Silverman, and William Whyte

Signature Schemes

Foundations of Group Signatures: The Case of Dynamic Groups 136
 Mihir Bellare, Haixia Shi, and Chong Zhang

Time-Selective Convertible Undeniable Signatures...................... 154
 Fabien Laguillaumie and Damien Vergnaud

Design Principles

On Tolerant Cryptographic Constructions 172
  Amir Herzberg

Password-Based Protocols

Simple Password-Based Encrypted Key Exchange Protocols 191
  Michel Abdalla and David Pointcheval

Hard Bits of the Discrete Log with Applications
to Password Authentication 209
  Philip Mackenzie and Sarvar Patel

Proofs for Two-Server Password Authentication 227
  Michael Szydlo and Burton Kaliski

Design and Analysis of Password-Based Key Derivation Functions 245
  Frances F. Yao and Yiqun Lisa Yin

Pairings

A New Two-Party Identity-Based Authenticated Key Agreement 262
  Noel McCullagh and Paulo S.L.M. Barreto

Accumulators from Bilinear Pairings and Applications 275
  Lan Nguyen

Computing the Tate Pairing 293
  Michael Scott

Fast and Proven Secure Blind Identity-Based Signcryption from Pairings .. 305
  Tsz Hon Yuen and Victor K. Wei

Efficient and Secure Implementation

A Systematic Evaluation of Compact Hardware Implementations
for the Rijndael S-Box 323
  Nele Mentens, Lejla Batina, Bart Preneel, and Ingrid Verbauwhede

CryptoGraphics: Secret Key Cryptography Using Graphics Cards 334
  Debra L. Cook, John Ioannidis, Angelos D. Keromytis, and Jake Luck

Side-Channel Leakage of Masked CMOS Gates 351
  Stefan Mangard, Thomas Popp, and Berndt M. Gammel

New Minimal Weight Representations for Left-to-Right Window Methods . 366
  James A. Muir and Douglas R. Stinson

Author Index 385

Sub-linear Queries Statistical Databases: Privacy with Power

Cynthia Dwork

Microsoft Research
dwork@microsoft.com

Abstract. We consider a statistical database in which a trusted administrator introduces noise to the query responses with the goal of maintaining privacy of individual database entries. In such a database, a query consists of a pair (S, f) where S is a set of rows in the database and f is a function mapping database rows to $\{0, 1\}$. The true response is $\sum_{r \in S} f(DB_r)$, a noisy version of which is released. Results in [3, 4] show that a strong form of privacy can be maintained using a surprisingly small amount of noise, provided the total number of queries is sublinear in the number n of database rows. We call this a sub-linear queries (SuLQ) database. The assumption of sublinearity becomes reasonable as databases grow increasingly large.

The SuLQ primitive – query and noisy reply – gives rise to a calculus of noisy computation. After reviewing some results of [4] on multi-attribute SuLQ, we illustrate the power of the SuLQ primitive with three examples [2]: principal component analysis, k means clustering, and learning in the statistical queries learning model.

1 Introduction

Consider a statistical database in which a trusted administrator introduces noise to the query responses with the goal of maintaining privacy of individual database entries. For concreteness, let the database consist of some number n of rows $DB_1, \ldots, DB_n$, where each row is a d-tuple of Boolean values. A query consists of a pair (S, f) where $S \subseteq [n]$ is a set of rows in the database and $f : \{0, 1\}^d \to \{0, 1\}$ is a function mapping database rows to $\{0, 1\}$. The true response to the query is $\sum_{r \in S} f(DB_r)$, a noisy version of which is released. That is, the administrator algorithm chooses a random quantity in some range and releases the sum of the true response and the random quantity.

Such databases were studied extensively in the early 1980's (see [1] for an excellent survey of results on these and other techniques for statistical disclosure control), with mixed results. However, results in [3, 4] show that a strong form of privacy can be maintained using a surprisingly small amount of noise – a random quantity whose standard deviation is of order $o(\sqrt{n})$ – provided the total number of queries is sublinear in the number n of database rows.

This is significant for the following reason. If we think of each row as a sample from some underlying probability distribution and we wish to gather statistics on a properties P that occur with possibly small but still constant probability in the population, then the sampling error in our population of size n will be of order $\Omega(\sqrt{n})$. Thus, the noise that is added for the sake of protecting privacy is significantly smaller than the sampling error. In other words, providing privacy need not interfere with accuracy, so long as the number of statistical queries is not too large. The assumption of sublinearity is reasonable as databases grow increasingly large.

The basic SuLQ primitive – noisy sums of arbitrary Boolean functions applied to each row in a set $S \subseteq [n]$ of rows – is powerful: statistics for any d-ary predicate can be very accurately obtained simply by querying the database. It is natural to ask, "Which more complex computations can be expressed using few (in n) SuLQ queries?" We have found this class to be quite rich.

Here, we review the results of [4] on multi-attribute SuLQ databases (Section 3) and then give three examples of the power of the SuLQ primitive (Section 4): principal component analysis, k means clustering, and learning in the statistical queries learning model. The treatment here is informal and without proofs. Rigorous treatment of these and other, related, results, is given in [4, 2].

2 Definitions

We model a database as an $n \times d$ binary matrix $DB = \{DB_{i,j}\}$. Intuitively, the columns in DB correspond to Boolean attributes $\alpha_1, \ldots, \alpha_d$, and the rows in DB correspond to individuals, where $DB_{i,j} = 1$ iff attribute α_j holds for individual i.

Let $\mathcal{D}$ be a distribution on $\{0,1\}^d$. We say that a database $DB = \{DB_{i,j}\}$ is chosen according to distribution $\mathcal{D}$ if every row in DB is chosen according to $\mathcal{D}$, independently of the other rows (in other words, DB is chosen according to $\mathcal{D}^n$). To capture partial information that the adversary may have obtained about individuals prior to interacting with the database, this requirement is relaxed in the privacy analysis, allowing each row i to be chosen from a (possibly) different distribution $\mathcal{D}_i$. In that case we say that the database is chosen according to $\mathcal{D}_1 \times \cdots \times \mathcal{D}_n$.

For a Boolean function $f : \{0,1\}^d \to \{0,1\}$ we let $p_0^{i,f}$ be the *a priori* probability that $f(DB_{i,1}, \ldots, DB_{i,d}) = 1$ and $p_T^{i,f}$ be the *a posteriori* probability that $f(DB_{i,1}, \ldots, DB_{i,d}) = 1$, given the answers to T queries, *as well as* all the values in all the rows of DB other than i: $DB_{i'}$ for all $i' \neq i$.

We define the monotonically-increasing 1-1 mapping $\text{conf} : (0,1) \to \mathbb{R}$ as follows:
$$\text{conf}(p) = \log \frac{p}{1-p}.$$

Note that a small additive change in conf implies a small additive change in p [1].

---

[1] The converse does not hold: conf grows logarithmically in p for $p \approx 0$ and logarithmically in $1/(1-p)$ for $p \approx 1$.

Let $\text{conf}_0^{i,f} = \log \frac{p_0^{i,f}}{1-p_0^{i,f}}$ and $\text{conf}_T^{i,f} = \log \frac{p_T^{i,f}}{1-p_T^{i,f}}$. We write our privacy requirements in terms of the random variables $\Delta\text{conf}^{i,f}$ defined as[2]:

$$\Delta\text{conf}^{i,f} = |\text{conf}_T^{i,f} - \text{conf}_0^{i,f}|.$$

Definition 1 ((δ, T)-Privacy). *A database access mechanism is (δ, T)-private if for every distribution $\mathcal{D}$ on $\{0,1\}^d$, for every row index i, for every function $f : \{0,1\}^d \to \{0,1\}$, and for every adversary $\mathcal{A}$ making at most T queries*

$$\Pr[\Delta\text{conf}^{i,f} > \delta] \le \text{neg}(n),$$

where $\text{neg}(n)$ grows more slowly than the inverse of any polynomial in n. The probability is taken over the choice of each row in DB according to $\mathcal{D}$, and the randomness of the adversary as well as the database access mechanism.

The definition of (δ, T)-privacy speaks of the probability that any single function experiences a change in confidence. The next definitions speak about sets of functions that together experience little change in confidence.

A *target set* F is a set of d-ary Boolean functions (one can think of the functions in F as being selected by an adversary; they represent information the adversary may wish to learn about someone). A target set F is δ-*safe* if $\Delta\text{conf}^{i,f} \le \delta$ for all $i \in [n]$ and $f \in F$. Let F be a target set of size polynomial in n. Definition 1 implies that under a (δ, T)-private database mechanism, F is δ-safe with probability $1 - \text{neg}(n)$.

Claim. [4] Consider a (δ, T)-private database with $d = O(\log n)$ attributes. Let F be the target set containing all the 2^{2^d} Boolean functions over the d attributes. Then, $\Pr[F \text{ is } 2\delta\text{-safe}] = 1 - \text{neg}(n)$.

3 Multi-attribute SuLQ Databases

The multi-attribute SuLQ database of [4] is easy to describe. Let $T = T(n) = O(n^c)$, $c < 1$, and define $R = (T(n)/\delta^2) \cdot \log^\mu n$ for some $\mu > 0$ (taking $\mu = 6$ will work).

SuLQ Database Algorithm $\mathcal{A}$
Input: a query (S, g).

1. Let $a_{S,g} = \sum_{i \in S} g(DB_i)$.
2. Generate a perturbation value: Let $(e_1, \ldots, e_R) \in_R \{0,1\}^R$ and
   $\mathcal{E} \leftarrow \sum_{i=1}^R e_i - R/2$.
3. Return $\tilde{a}_{S,g} = a_{S,g} + \mathcal{E}$.

---

[2] Our choice of defining privacy in terms of $\Delta\text{conf}^{i,f}$ is somewhat arbitrary, one could rewrite our definitions (and analysis) in terms of the a priori and a posteriori probabilities. Note however that limiting $\Delta\text{conf}^{i,f}$ in Definition 1 is a stronger requirement than just limiting $|p_T^{i,f} - p_0^{i,f}|$.

Note that $\mathcal{E}$ is a binomial random variable with $\mathbf{E}[\mathcal{E}] = 0$ and standard deviation $\sqrt{R}$. The analysis ignores the case where $\mathcal{E}$ largely deviates from zero, as the probability of such an event is extremely small: $\Pr[|\mathcal{E}| > \sqrt{R}\log^2 n] = \mathsf{neg}(n)$. In particular, this implies that the SuLQ database algorithm $\mathcal{A}$ is within $\tilde{O}(\sqrt{T(n)})$ perturbation, meaning that for every query (S, f)

$$\Pr[|\mathcal{A}(S,f) - a_{S,f}| \leq \mathcal{E}] = 1 - \mathsf{neg}(n).$$

The probability is taken over the randomness of the database algorithm $\mathcal{A}$.

Theorem 1. *[4] Let $T(n) = O(n^c)$ and $\delta = 1/O(n^{c'})$ for $0 < c < 1$ and $0 \leq c' < c/2$. Then the SuLQ algorithm $\mathcal{A}$ is $(\delta, T(n))$-private within $\tilde{O}(\sqrt{T(n)}/\delta)$ perturbation.*

Note that whenever $\sqrt{T(n)}/\delta < \sqrt{n}$, restricting the adversary to $T(n)$ queries allows privacy with perturbation magnitude less than $\sqrt{n}$.

Let $i \in [n]$ and $f : \{0,1\}^d \to \{0,1\}$. The proof analyzes the *a posteriori* probability p_ℓ that $f(DB_i) = 1$ given the answers to the first ℓ queries $(\tilde{a}_1, \ldots, \tilde{a}_\ell)$ and $DB^{\{-i\}}$ (where $DB^{\{-i\}}$ denotes the entire database except for the ith row). Let $\mathsf{conf}_\ell = \log_2 p_\ell/(1-p_\ell)$. Note that $\mathsf{conf}_T = \mathsf{conf}_T^{i,f}$, and (due to the independence of rows in DB) $\mathsf{conf}_0 = \mathsf{conf}_0^{i,f}$. Following [3], a random walk on the real line is defined, with $\mathsf{step}_\ell = \mathsf{conf}_\ell - \mathsf{conf}_{\ell-1}$. The proof argues that (with high probability) $T(n)$ steps of the random walk do not suffice to reach distance δ.

4 Computation with the SuLQ Primitive

The basic SuLQ operation – query and noisy reply – can be viewed as a noisy computational primitive which may be used to compute other functions of the database than statistical queries. In this section we describe three examples of the power of the primitive. In this setting, the inputs are reals drawn from the unit d-dimensional cube, and the noise is distributed according to a normal variable $N(0, R)$, where $R = R(n)$ is roughly of order $T(n) \log n \log T(n)$. The privacy analysis in the proof of Theorem 1 must be extended accordingly. A rigorous treatment of this work appears in [2].

4.1 Principal Component Analysis

Principal component analysis [6] is an extremely valuable tool in the (frequent) case in which high-dimensional data lies primarily in a low-dimensional subspace.

The input consists of n points in $\mathcal{U}^d$ (the d-dimensional cube of side length 1) and an integer $k \leq d$. The output will be the k largest eigenvalues of the $d \times d$ covariance matrix (defined below), and their corresponding eigenvectors.

For $1 \leq i \leq d$, we let $\mu_i = E_{r \in [n]}[p_r(i)]$, where $p_r(i)$ denotes the ith coordinate of the input point described by row r. We let the $d \times d$ covariance matrix C be defined by $C = \{c_{ij}\}$, where

$$c_{ij} = E_{r \in [n]}[p_r(i) p_r(j)] - \mu_i \mu_j.$$

PCA is known to be remarkably stable under random noise – so much so, that it is often used with the express intention of *removing* noise.

SuLQ Computation of PCA

1. (d queries) For $0 \leq i \leq d$, let $m_i = SuLQ(F(x) := x(i))/n$. By this we mean that $F(x)$ selects the ith coordinate of each row, so the query sums all the ith coordinates (getting a noisy version of this sum), and the algorithm divides this noisy sum by n. This gives an approximation to μ_i in the pure PCA algorithm described above.
2. (Roughly $d^2/2$ queries) Let $c_{ij} = SuLQ(F(x) = x(i)x(j))/n - m_i m_j$. That is, we first obtain a noisy average of the product of the ith and jth coordinates, and then subtract the product of the estimates of μ_i and μ_j.

Given (an approximation to) the covariance matrix C, the k largest eigenvalues and corresponding eigenvectors can be computed directly, without further queries.

We remark that, using the techniques of [4] for vertically partitioned databases, this computation can be carried out even if each column of the database is stored in a separate, independent, SuLQ database.

4.2 k Means

An instance of the k means computation is a set of n points in $\mathcal{U}^d$, together with some number k of initial candidate "means" in $\mathcal{U}^d$. The output will consist of k points in $\mathcal{U}^d$ (the "means"), together with the fraction of points in the database associated with each mean. We next describe the basic step of the k means algorithm.

Basic Step of k Means Algorithm

1. (k queries): For each mean m_i, $1 \leq i \leq k$, count the number of points closer to this mean than to every other mean. This yields cluster sizes. This is approximated via the queries, for $1 \leq i \leq k$,

   $\text{Size}_i = SuLQ(F(x) := 1$ if m_i is the closest mean to x, and 0 otherwise).

2. (kd queries): for each mean m_i, $1 \leq i \leq k$, and coordinate j, $1 \leq j \leq d$, compute the sum, over all points in the cluster associated with m_i, of the value of the jth coordinate. Divide by the size of the cluster.
   (a) $\text{Sum}_{ij} = SuLQ(F(x) := x(j)$ if m_i is the closest center to x, and 0 otherwise).
   (b) $m_{ij} = \text{Sum}_{ij}/\text{Size}_i$

The basic step is iterated until some maximum number of queries have been issued. (In practice, this usually converges after a small number of basic steps.) If any cluster size is below a threshhold (say, $\sqrt{T(n)}$), then output an exception.

For clusters that are of size $\Omega(n)$, one step of the (pure) k means computation differs from one step of the SuLQ-based k means computation by a quantity that is roughly Gaussian with mean zero and variance $\tilde{O}(\sqrt{R}/n)$.

4.3 Capturing the Statistical Queries Learning Model

The Statistical Queries Learning model was proposed by Kearns [5]. In this model the goal is to learn a *concept* $c : \{0,1\}^d \to \{0,1\}$. There is a distribution D on strings in $\{0,1\}^d$, and the learning algorithm has access to an oracle, $\text{stat}_{c,D}$, described next.

On query (f, τ), where $f = f(x, \ell)$ is any boolean function over inputs $x \in D$ and label $\ell \in \{0,1\}$, and $\tau = 1/poly(d)$ is an error tolerance, the oracle replies with a noisy estimate of the probability that $f(x, c(x)) = 1$ for a randomly selected element from D; the answer is guaranteed to be correct within additive tolerance τ. Many (but not all, see [5]) concept classes that are PAC learnable can also be learned in the statistical queries learning model.

To fit the statistical queries learning model into our setting, we require that one of the attributes be the value of c applied to the other data in the row, so that a typical row looks like $DB_r = (x, c(x))$. By definition, on input (f, S) the SuLQ database responds with a noisy version of $\sum_{r \in S} f(DB_r)$. Taking $S = [n]$, we have that so long as the noise added by the SuLQ database is within the tolerance τ, the response (divided by n) is a "valid" response of the $\text{stat}_{c,D}$ oracle. In other words, to simulate the query $\text{stat}_{c,D}(f, \tau)$ we compute $\text{SuLQ}(F(x) := f(x))/n$ a total of $\tilde{O}(R/\tau^2 n^2)$ times and return the average of these values.

With high probability the answer obtained will be within tolerance τ. Also, recall that $\tau = 1/poly(d)$; if $d = n^{o(1)}$ then repetition is not necessary.

References

1. N.R. Adam and J.C. Wortmann, Security-Control Methods for Statistical Databases: A Comparative Study, *ACM Computing Surveys 21*(4), pp. 515–556, 1989.
2. A. Blum, C. Dwork, F. McSherry, and K. Nissim, On the Power of SuLQ Databases, *manuscript in preparation*, 2004.
3. I. Dinur and K. Nissim, Revealing information while preserving privacy, *Proceedings of the Twenty-Second ACM SIGACT-SIGMOD-SIGART Symposium on Principles of Database Systems*, pp. 202-210, 2003.
4. C. Dwork and N. Nissim, Privacy-Preserving Datamining on Vertically Partitioned Databases, *Proceedings of CRYPTO 2004*
5. M. Kearns, Efficient Noise-Tolerant Learning from Statistical Queries, *JACM 45*(6), pp. 983–1006, 1998. See also *Proc. 25th ACM STOC*, pp. 392–401, 1993
6. M. J. O'Connel, Search Program for Significant Variables, *Comp. Phys. Comm. 8*, 1974.

Malicious Cryptography: Kleptographic Aspects

Adam Young[1] and Moti Yung[2]

[1] Cigital Labs
ayoung@cigital.com
[2] Dept. of Computer Science, Columbia University
moti@cs.columbia.edu

Abstract. In the last few years we have concentrated our research efforts on new threats to the computing infrastructure that are the result of combining malicious software (malware) technology with modern cryptography. At some point during our investigation we ended up asking ourselves the following question: what if the malware (i.e., Trojan horse) resides within a cryptographic system itself? This led us to realize that in certain scenarios of black box cryptography (namely, when the code is inaccessible to scrutiny as in the case of tamper proof cryptosystems or when no one cares enough to scrutinize the code) there are attacks that employ cryptography itself against cryptographic systems in such a way that the attack possesses unique properties (i.e., special advantages that attackers have such as granting the attacker exclusive access to crucial information where the exclusive access privelege holds even if the Trojan is reverse-engineered). We called the art of designing this set of attacks "kleptography." In this paper we demonstrate the power of kleptography by illustrating a carefully designed attack against RSA key generation.

Keywords: RSA, Rabin, public key cryptography, SETUP, kleptography, random oracle, security threats, attacks, malicious cryptography.

1 Introduction

Robust backdoor attacks against cryptosystems have received the attention of the cryptographic research community, but to this day have not influenced industry standards and as a result the industry is not as prepared for them as it could be. As more governments and corporations deploy public key cryptosystems their susceptibility to backdoor attacks grows due to the pervasiveness of the technology as well as the potential payoff for carrying out such an attack.

In this work we discuss what we call kleptographic attacks, which are attacks on black box cryptography. One may assume that this applies only to tamper proof devices. However, it is rarely that code (even when made available) is scrutinized. For example, Nguyen in Eurocrypt 2004 analyzed an open source digital signature scheme. He demonstrated a very significant implementation error, whereby obtaining a single signature one can recover the key [3].

In this paper we present a revised (more general) definition of an attack based on embedding the attacker's public key inside someone else's implementation of a public-key cryptosystem. This will grant the attacker an exclusive advantage that enables the subversion of the user's cryptosystem. This type of attack employs cryptography against another cryptosystem's implementation and we call this kleptography. We demonstrate a kleptographic attcak on the RSA key generation algorithm and survey how to prove that the attack works.

What is interesting is that the attacker employs modern cryptographic tools in the attack, and the attack works due to modern tools developed in what some call the "provable security" sub-field of modern cryptographic research. From the perspective of research methodologies, what we try to encourage by our example is for cryptographers and other security professionals to devote some of their time to researching new attack scenarios and possibilities. We have devoted some of our time to investigate the feasibility of attacks that we call "malicious cryptography" (see [6]) and kleptographic attacks were discovered as part of our general effort in investigating the merger of strong cryptographic methods with malware technology.

2 SETUP Attacks

A number of backdoor attacks against RSA [5] key generation (and Rabin [4]) have been presented that exploit secretly embedded trapdoors [7-9]. Also, attacks have been presented that emphasize speed [1]. This latter attack is intended to work even when Lenstra's composite generation method is used [2] whereas the former three will not. However, all of these backdoor attacks fail when half of the bits of the composite are chosen pseudorandomly using a seed [7] (this drives the need for improved public key standards, and forms a major motivation for the present work). It should be noted that [1] does not constitute a SETUP attack since it assumes that a secret key remains hidden even after reverse-engineering.

We adapt the notion of a strong SETUP [8] to two games. For clarity this definition is tailored after RSA key generation (as opposed to being more general). The threat model involves three parties: the designer, the eavesdropper, and the inquirer.

The designer is a malicious attacker and builds the SETUP attack into some subset of all of the black-box key generation devices that are deployed. The goal of the designer is to learn the RSA private key of a user who generates a key pair using a device contained in this subset when the designer only has access to the RSA public keys. Before the games start, the eavesdropper and inquirer are given access to the SETUP algorithm in its entirety[1]. However, in the games they play they are not given access to the internals of the particular devices that are used (they cannot reverse-engineer them).

Assumptions: The eavesdropper and inquirer are assumed to be probabilistic poly-time algorithms. It is assumed that the RSA key generation algorithm is deployed in tamper-proof black-box devices. It is traditional to supply an RSA

---

[1] e.g., found in practice via the costly process of reverse-engineering one of the devices.

key generation algorithm with 1^k where k is the security parameter. This tells the generator what security parameter is to be used and assures that running times can be derived based on the size of the input. For simplicity we assume that the generator takes no input and that the security parameter is fixed. It is straightforward to relax this assumption.

Let D be a device that contains the SETUP attack.

Game 1: The inquirer is given oracle access to two devices A and B. So, the inquirer obtains RSA key pairs from the devices. With 50% probability A has a SETUP attack in it. A has a SETUP attack in it iff B does not. The inquirer wins if he determines whether or not A has the SETUP attack in it with probability significantly greater than $1/2$.

Property 1: (indistinguishability) The inquirer fails Game 2 with overwhelming probability.

Game 2: The eavesdropper may query D but is only given the public keys that result, not the corresponding private keys. He wins if he can learn one of the corresponding private keys.

Property 2: (confidentiality) The eavesdropper fails Game 1 with overwhelming probability.

Property 3: (completeness) Let (y, x) be a public/private key generated using D. With overwhelming probability the designer computes x on input y.

In a SETUP attack, the designer uses his or her own private key in conjunction with y to recover x. In practice the designer may learn y by obtaining it from a Certificate Authority.

Property 4: (uniformity) The SETUP attack is the same in every black-box cryptographic device.

When property 4 holds it need not be the case that each device have a unique identifier ID. This is important in a binary distribution in which all of the instances of the "device" will necessarily be identical. In hardware implementations it would simplify the manufacturing process.

Definition 1. *If a backdoor RSA key generation algorithm satisfies properties 1, 2, 3, and 4 then it is a* **strong SETUP**.

3   SETUP Attack Against RSA Key Generation

The notion of a SETUP attack was presented at Crypto '96 [7] and was later improved slightly [8]. To illustrate the notion of a SETUP attack, a particular attack on RSA key generation was presented. The SETUP attack on RSA keys from Crypto '96 generates the primes p and q from a skewed distribution. This skewed distribution was later corrected while allowing e to remain fixed[2] [9]. A backdoor attack on RSA was also presented by Crépeau and Slakmon [1]. They showed that if the device is free to choose the RSA exponent e (which is often not the case in practice), the primes p and q of a given size can be generated uniformly at random in the attack. Crépeau and Slakmon also give an attack similar to PAP in which e is fixed. Crépeau and Slakmon [1] noted the skewed distribution in the original SETUP attack as well.

3.1 Notation and Building Blocks

Let $L(x/P)$ denote the Legendre symbol of x with respect to the prime P. Also, let $J(x/N)$ denote the Jacobi symbol of x with respect to the odd integer N.

The attack on RSA key generation makes use of the probabilistic bias removal method (PBRM). This algorithm is given below [8].

$PBRM(R, S, x)$:
input: R and S with $S > R > \frac{S}{2}$ and x contained in $\{0, 1, 2, ..., R-1\}$
output: e contained in $\{-1, 1\}$ and x' contained in $\{0, 1, 2, ..., S-1\}$
1. set $e = 1$ and set $x' = 0$
2. choose a bit b randomly
3. if $x < S - R$ and $b = 1$ then set $x' = x$
4. if $x < S - R$ and $b = 0$ then set $x' = S - 1 - x$
5. if $x \geq S - R$ and $b = 1$ then set $x' = x$
6. if $x \geq S - R$ and $b = 0$ then set $e = -1$
7. output e and x' and halt

Recall that a random oracle $R(\cdot)$ takes as input a bit string that is finite in length and returns an infinitely long bit string. Let $H(s, i, v)$ denote a function that invokes the oracle and returns the v bits of $R(s)$ that start at the ith bit position, where $i \geq 0$. For example, if $R(110101) = 01001011110101...$ then,

$$H(110101, 0, 3) = 010$$

and

$$H(110101, 1, 4) = 1001$$

and so on.

The following is a subroutine that is assumed to be available.

$RandomBitString1()$:
input: none
output: random $W/2$-bit string
1. generate a random $W/2$-bit string str
2. output str and halt

Finally, the algorithm below is regarded as the "honest" key generation algorithm.

---
[2] For example, with $e = 2^{16} + 1$ as in many fielded cryptosystems.

*GenPrivatePrimes*1():
input: none
output: $W/2$-bit primes p and q such that $p \neq q$ and $|pq| = W$
1. for $j = 0$ to ∞ do:
2. $p = RandomBitString1()$ /* at this point p is a random string */
3.   if $p \geq 2^{W/2-1} + 1$ and p is prime then break
4. for $j = 0$ to ∞ do:
5. $q = RandomBitString1()$
6.   if $q \geq 2^{W/2-1} + 1$ and q is prime then break
7. if $|pq| < W$ or $p = q$ then goto step 1
8. if $p > q$ then interchange the values p and q
9. set $S = (p, q)$
10. output S, zeroize all values in memory, and halt

3.2 The SETUP Attack

When an honest algorithm *GenPrivatePrimes*1 is implemented in the device, the device may be regarded as an honest cryptosystem C. The advanced attack on composite key generation is specified by *GenPrivatePrimes*2 that is given below. This algorithm is the infected version of *GenPrivatePrimes*1 and when implemented in a device it effectively serves as the device C' in a SETUP attack.

The algorithm *GenPrivatePrimes*2 contains the attacker's public key N where $|N| = W/2$ bits, and $N = PQ$ with P and Q being distinct primes. The primes P and Q are kept private by the attacker. The attacker's public key is half the size of p times q, where p and q are the primes that are computed by the algorithm.

In hardware implementations each device contains a unique $W/2$-bit identifier ID. The IDs for the devices are chosen randomly, subject to the constraint that they all be unique. In binary distributions the value ID can be fixed. Thus, it will be the same in each copy of the key generation binary. In this case the security argument applies to all invocations of all copies of the binary as a whole.

The variable i is stored in non-volatile memory and is a counter for the number of compromised keys that the device created. It starts at $i = 0$. The variable j is not stored in non-volatile memory. The attack makes use of the four constants (e_0, e_1, e_2, e_3) that must be computed by the attacker and placed within the device. These quantities can be chosen randomly, for instance. They must adhere to the requirements listed in Table 1.

It may appear at first glance that the backdoor attack below is needlessly complicated. However, the reason for the added complexity becomes clear when the indistinguishability and confidentiality properties are proven. This algorithm effectively leaks a Rabin ciphertext in the upper order bits of pq and uses the Rabin plaintext to derive the prime p using a random oracle.

Note that due to the use of the probabilistic bias removal method, this algorithm is not going to have the same expected running time as the honest algorithm *GenPrivatePrimes*1(). The ultimate goal in the attack is to make it produce outputs that are indistinguishable from the outputs of an honest

Table 1. Constants used in key generation attack.

Constant	Properties
e_0	$e_0 \in \mathbb{Z}_N^*$ and $L(e_0/P) = +1$ and $L(e_0/Q) = +1$
e_1	$e_2 \in \mathbb{Z}_N^*$ and $L(e_2/P) = -1$ and $L(e_2/Q) = +1$
e_2	$e_1 \in \mathbb{Z}_N^*$ and $L(e_1/P) = -1$ and $L(e_1/Q) = -1$
e_3	$e_3 \in \mathbb{Z}_N^*$ and $L(e_3/P) = +1$ and $L(e_3/Q) = -1$

implementation. It is easiest to utilize the Las Vegas key generation algorithm in which the only possible type of output is (p,q) (i.e., "failure" is not an allowable output).

The value Θ is a constant that is used in the attack to place a limit on the number of keys that are attacked. It is a restriction that simplifies the algorithm that the attacker uses to recover the private keys of other users.

GenPrivatePrimes2():
input: none
output: $W/2$-bit primes p and q such that $p \neq q$ and $|pq| = W$
1. if $i > \Theta$ then output *GenPrivatePrimes1()* and halt
2. update i in non-volatile memory to be $i = i + 1$
3. let I be the $|\Theta|$-bit representation of i
4. for $j = 0$ to ∞ do:
5.   choose x randomly from $\{0, 1, 2, ..., N-1\}$
6.   set $c_0 = x$
7.   if $gcd(x, N) = 1$ then
8.   choose bit b randomly and choose u randomly from $\mathbb{Z}_N^*$
9.   if $J(x/N) = +1$ then set $c_0 = e_0^b e_2^{1-b} u^2 \bmod N$
10.   if $J(x/N) = -1$ then set $c_0 = e_1^b e_3^{1-b} u^2 \bmod N$
11.   compute $(e, c_1) = PBRM(N, 2^{W/2}, c_0)$
12.   if $e = -1$ then continue
13.   if $u > -u \bmod N$ then set $u = -u \bmod N$ /* for faster decr. */
14.   let T_0 be the $W/2$-bit representation of u
15.   for $k = 0$ to ∞ do:
16.   compute $p = H(T_0 || ID || I || j, \frac{kW}{2}, \frac{W}{2})$
17.   if $p \geq 2^{W/2-1} + 1$ and p is prime then break
18.   if $p < 2^{W/2-1} + 1$ or if p is not prime then continue
19. $c_2 = RandomBitString1()$
20.   compute $n' = (c_1 \;||\; c_2)$
21.   solve for the quotient q and the remainder r in $n' = pq + r$
22.   if q is not a $W/2$-bit integer or if $q < 2^{W/2-1} + 1$ then continue
23.   if q is not prime then continue
24.   if $|pq| < W$ or if $p = q$ then continue
25.   if $p > q$ then interchange the values p and q
26.   set $S = (p, q)$ and break
27. output S, zeroize everything in memory except i, and halt

It is assumed that the user, or the device that contains this algorithm, will multiply p by q to obtain the public key $n = pq$. Making n publicly available is perilous since with overwhelming probability p can easily be recovered by the attacker. Note that c_1 will be displayed verbatim in the upper order bits of $n = n' - r = pq$ unless the subtraction of r from n' causes a borrow bit to be taken from the $W/2$ most significant bits of n'. The attacker can always add this bit back in to recover c_1.

Suppose that the attacker, who is either the malicious manufacturer or the hacker that installed the Trojan horse, obtains the public key $n = pq$. The attacker is in a position to recover p using the factors (P, Q) of the Rabin public key N. The factoring algorithm attempts to compute the two smallest ambivalent roots of a perfect square modulo N. Let t be a quadratic residue modulo N. Recall that a_0 and a_1 are ambivalent square roots of t modulo N if $a_0^2 \equiv a_1^2 \equiv t \bmod N$, $a_0 \neq a_1$, and $a_0 \neq -a_1 \bmod N$. The values a_0 and a_1 are the two smallest ambivalent roots if they are ambivalent, $a_0 < -a_0 \bmod N$, and $a_1 < -a_1 \bmod N$. The Rabin decryption algorithm can be used to compute the two smallest ambivalent roots of a perfect square t, that is, the two smallest ambivalent roots of a Rabin ciphertext.

For each possible combination of ID, i, j, and k the attacker computes the algorithm $FactorTheComposite$ given below. Since the key generation device can only be invoked a reasonable number of times, and since there is a reasonable number of compromised devices in existence, this recovery process is tractable.

$FactorTheComposite(n, P, Q, ID, i, j, k)$:
input: positive integers i, j, k with $1 \leq i \leq \Theta$
      distinct primes P and Q
      n which is the product of distinct primes p and q
      Also, $|n|$ must be even and $|p| = |q| = |PQ| = |ID| = |n|/2$
output: $failure$ or a non-trivial factor of n
1. compute $N = PQ$
2. let I be the Θ-bit representation of i
3. $W = |n|$
4. set U_0 equal to the $W/2$ most significant bits of n
5. compute $U_1 = U_0 + 1$
6. if $U_0 \geq N$ then set $U_0 = 2^{W/2} - 1 - U_0$ /* undo the PBRM */
7. if $U_1 \geq N$ then set $U_1 = 2^{W/2} - 1 - U_1$ /* undo the PBRM */
8. for $z = 0$ to 1 do:
9.   if U_z is contained in $\mathbb{Z}_N^*$ then
10.   for $\ell = 0$ to 3 do: /* try to find a square root */
11.   compute $W_\ell = U_z e_\ell^{-1} \bmod N$
12.   if $L(W_\ell/P) = +1$ and $L(W_\ell/Q) = +1$ then
13.   let a_0, a_1 be the two smallest ambivalent roots of W_ℓ
14.   let A_0 be the $W/2$-bit representation of a_0
15.   let A_1 be the $W/2$-bit representation of a_1
16.   for $b = 0$ to 1 do:
17.   compute $p_b = H(A_b||ID||I||j, \frac{kW}{2}, \frac{W}{2})$

18.   if p_0 is a non-trivial divisor of n then
19.   output p_0 and halt
20.   if p_1 is a non-trivial divisor of n then
21.   output p_1 and halt
22. output $failure$ and halt

The quantity $U_0 + 1$ is computed since a borrow bit may have been taken from the lowest order bit of c_1 when the public key $n = n' - r$ is computed.

4 Security of the Attack

In this section we argue the success of the attack and how it holds unique properties.

The attack is indistinguishable to all adversaries that are polynomially bounded in computational power[3]. Let C denote an honest device that implements the algorithm $GenPrivatePrimes1()$ and let C' denote a dishonest device that implements $GenPrivatePrimes2()$. A key observation is that the primes p and q that are output by the dishonest device are chosen from the same set and same probability distribution as the primes p and q that are output by the honest device. So, it can be shown that p and q in the dishonest device C' are chosen from the same set and from the same probability distribution as p and q in the honest device C[4].

In a nutshell confidentiality is proven by showing that if an efficient algorithm exists that violates the confidentiality property then either $W/2$-bit composites PQ can be factored or W-bit composites pq can be factored. This reduction is not a randomized reduction, yet it goes a long way to show the security of this attack.

The proof of confidentiality is by contradiction. Suppose for the sake of contradiction that a computationally bounded algorithm A exists that violates the confidentiality property. For a randomly chosen input, algorithm A will return a non-trivial factor of n with non-negligible probability. The adversary could thus use algorithm A to break the confidentiality of the system. Algorithm A factors n when it *feels* so inclined, but must do so a non-negligible portion of the time.

It is important to first set the stage for the proof. The adversary that we are dealing with is trying to break a public key pq where p and q were computed by the cryptotrojan. Hence, pq was created using a call to the random oracle R. It is conceivable that an algorithm A that breaks the confidentiality will make oracle calls as well to break pq. Perhaps A will even make some of the *same* oracle calls as the cryptotrojan. However, in the proof we cannot assume this. All we can assume is that A makes at most a polynomial[5] number of calls to the oracle and we are free to "trap" each one of these calls and take the arguments.

---

[3] Polynomial in $W/2$, the security parameter of the attacker's Rabin modulus N.
[4] The key to this being true is that n' is a random W-bit string and so it can have a leading zero. So, $|pq|$ can be less than W bits, the same as in the operation in the honest device before p and q are output.
[5] Polynomial in $W/2$.

Consider the following algorithm $SolveFactoring(N, n)$ that uses A as an oracle to solve the factoring problem.

$SolveFactoring(N, n)$:
input: N which is the product of distinct primes P and Q
       n which is the product of distinct primes p and q
      Also, $|n|$ must be even and $|p| = |q| = |N| = |n|/2$
output: $failure$, or a non-trivial factor of N or n
1. compute $W = 2|N|$
2. for $k = 0$ to 3 do:
3.   do:
4.   choose e_k randomly from $\mathbb{Z}_N^*$
5.   while $J(e_k/N) \neq (-1)^k$
6. choose ID to be a random $W/2$-bit string
7. choose i randomly from $\{1, 2, ..., \Theta\}$
8. choose bit b_0 randomly
9. if $b_0 = 0$ then
10.   compute $p = A(n, ID, i, N, e_0, e_1, e_2, e_3)$
11.   if $p < 2$ or $p \geq n$ then output $failure$ and halt
12.   if $n \bmod p = 0$ then output p and halt /* factor found */
13.   output $failure$ and halt
14. output $CaptureOracleArgument(ID, i, N, e_0, e_1, e_2, e_3)$ and halt

$CaptureOracleArgument(ID, i, N, e_0, e_1, e_2, e_3)$:
1. compute $W = 2|N|$
2. let I be the Θ-bit representation of i
3. for $j = 0$ to ∞ do: /* try to find an input that A expects */
4.   choose x randomly from $\{0, 1, 2, ..., N-1\}$
5.   set $c_0 = x$
6.   if $gcd(x, N) = 1$ then
7.   choose bit b_1 randomly and choose u_1 randomly from $\mathbb{Z}_N^*$
8.   if $J(x/N) = +1$ then set $c_0 = e_0^{b_1} e_2^{1-b_1} u_1^2 \bmod N$
9.   if $J(x/N) = -1$ then set $c_0 = e_1^{b_1} e_3^{1-b_1} u_1^2 \bmod N$
10.   compute $(e, c_1) = PBRM(N, 2^{W/2}, c_0)$
11.   if $e = -1$ then continue
12.   if $u_1 > -u_1 \bmod N$ then set $u_1 = -u_1 \bmod N$
13.   let T_0 be the $W/2$-bit representation of u_1
14.   for $k = 0$ to ∞ do:
15.   compute $p = H(T_0||ID||I||j, \frac{kW}{2}, \frac{W}{2})$
16.   if $p \geq 2^{W/2-1} + 1$ and p is prime then break
17.   if $p < 2^{W/2-1} + 1$ or if p is not prime then continue
18. $c_2 = RandomBitString1()$
19.   compute $n' = (c_1 \;||\; c_2)$
20.   solve for the quotient q and the remainder r in $n' = pq + r$
21.   if q is not a $W/2$-bit integer or if $q < 2^{W/2-1} + 1$ then continue
22.   if q is not prime then continue

23. if $|pq| < W$ or if $p = q$ then continue
24. simulate $A(pq, ID, i, N, e_0, e_1, e_2, e_3)$, watch calls to R, and
    store the $W/2$-most significant bits of each call in list ω
25. remove all elements from ω that are not contained in $\mathbb{Z}_N^*$
26. let L be the number of elements in ω
27. if $L = 0$ then output $failure$ and halt
28. choose α randomly from $\{0, 1, 2, ..., L-1\}$
29. let β be the αth element in ω
30. if $\beta \equiv \pm u_1 \bmod N$ then output $failure$ and halt
31. if $\beta^2 \bmod N \neq u_1^2 \bmod N$ then output $failure$ and halt
32. compute $P = gcd(u_1 + \beta, N)$
33. if $N \bmod P = 0$ then output P and halt
34. compute $P = gcd(u_1 - \beta, N)$
35. output P and halt

Note that with non-negligible probability A will not balk due to the choice of ID and i. Also, with non-negligible probability e_0, e_1, e_2, and e_3 will conform to the requirements in the cryptotrojan attack. So, when $b_0 = 0$ these four arguments to A will conform to what A expects with non-negligible probability. Now consider the call to A when $b_0 = 1$. Observe that the value pq is chosen from the same set and probability distribution as in the cryptotrojan attack. So, when $b_0 = 1$ the arguments to A will conform to what A expects with non-negligible probability. It may be assumed that A balks whenever e_0, e_1, e_2, and e_3 are not appropriately chosen without ruining the efficiency of $SolveFactoring$. So, for the remainder of the proof we will assume that these four values are as defined in the cryptotrojan attack.

Let u_2 be the square root of $u_1^2 \bmod n$ such that $u_2 \neq u_1$ and $u_2 < -u_2 \bmod n$. Also, let T_1 and T_2 be u_1 and u_2 padded with leading zeros as necessary such that $|T_1| = |T_2| = W/2$ bits, respectively. Denote by E the event that in a given invocation algorithm A calls the random oracle R at least once with either T_1 or T_2 as the $W/2$ most significant bits. Clearly only one of the two following possibilities hold:

1. Event E occurs with negligible probability.
2. Event E occurs with non-negligible probability.

Consider case (1). Algorithm A can detect that n was not generated by the cryptotrojan by appropriately supplying T_1 or T_2 to the random oracle. Once verified, A can balk and not output a factor of n. But in case (1) this can only occur at most a negligible fraction of the time since changing even a single bit in the value supplied to the oracle elicits an independently random response. By assumption, A returns a non-trivial factor of n a non-negligible fraction of the time. Since the difference between a non-negligible number and negligible number is a non-negligible number it follows that A factors n without relying on the random oracle. So, in case (1) the call to A in which $b_0 = 0$ will lead to a non-trivial factor of n with non-negligible probability.

Now consider case (2). Since E occurs with non-negligible probability it follows that A may in fact be computing non-trivial factors of composites n by making oracle calls and constructing the factors in a straightforward fashion. However, whether or not this is the case is immaterial. Since A makes at most a polynomial number of calls[6] to R the value for L cannot be too large. Since with non-negligible probability A passes either T_1 or T_2 as the $W/2$ most significant bits to R and since L cannot be too large it follows that β and u_1 will be ambivalent roots with non-negligible probability. Algorithm A has no way of knowing which of the two smallest ambivalent roots *SolveFactoring* chose in constructing the upper order bits of pq. Algorithm A, which may be quite uncooperative, can do no better than guess at which one it was, and it could in fact have been either. Hence, *SolveFactoring* returns a non-trivial factor of N with non-negligible probability in this case.

It has been shown that in either case, the existence of A contradicts the factoring assumption. So, the original assumption that adversary A exists is wrong. This proves that the attack satisfies Property 2 of a SETUP attack.

Immediately following the test for $p = q$ in C and in C' it is possible to check that $gcd(e,(p-1)(q-1)) = 1$ and restart the entire algorithm if this does not hold. This handles the generation of RSA primes by taking into account the public RSA exponent e. This preserves the indistinguishability of the output of C' with respect to C.

5 Conclusion

Attacks on cryptosystems can occur from many different angles: a specification may be incorrect which requires provable security as a minimum requirement – preferably based on a complexity theoretic assumption and if not than on some idealization (e.g., assuming a random oracle like the idealization of unstructured one-way hash functions). However, implementations can have problems of their own. Here a deliberate attack by someone who constructs the cryptosystem (e.g., a vendor) has been demonstrated. This attack is not unique to the RSA cryptosystem and is but one of many possible attacks. However, it serves to demonstrate the overall approach. At a minimum, the message that we try to convey is that the scrutiny of code and implementations is crucial to the overall security of the cryptographic infrastructure, and if practitioners exercise scrutiny then we should be aware that we may need to completely trust each individual implementation to be correct in ways that may not be efficiently black-box testable (as our attack has demonstrated).

References

1. C. Crépeau, A. Slakmon. Simple Backdoors for RSA Key Generation. In *The Cryptographers' Track at the RSA Conference – CT-RSA '03*, pages 403–416, 2003.
2. A. K. Lenstra. Generating RSA Moduli with a Predetermined Portion. In *Advances in Cryptology – Asiacrypt '98*, pages 1–10, 1998.

---

[6] Polynomial in W.

3. P. Q. Nguyen. Can We Trust Cryptographic Software? Cryptographic Flaws in GNU Privacy Guard v1.2.3. In *Advances in Cryptology - Eurocrypt '04*, pages 555–570, 2004.
4. M. Rabin. Digitalized signatures and public-key functions as intractable as factorization. TR-212, MIT Laboratory for Computer Science, January 1979.
5. R. Rivest, A. Shamir, L. Adleman. A method for obtaining Digital Signatures and Public-Key Cryptosystems. In *Communications of the ACM*, volume 21, n. 2, pages 120–126, 1978.
6. A. Young, M. Yung. "Malicious Cryptography: Exposing Cryptovirology," Wiley Publishing Inc., Feb. 2004.
7. A. Young, M. Yung. The Dark Side of Black-Box Cryptography, or: Should we trust Capstone? In *Advances in Cryptology - Crypto '96*, pages 89–103, 1996.
8. A. Young, M. Yung. Kleptography: Using Cryptography Against Cryptography. In *Advances in Cryptology - Eurocrypt '97*, pages 62–74, 1997.
9. A. Young. Kleptography: Using Cryptography Against Cryptography. PhD Thesis, Columbia University, 2002.

Resistance of SNOW 2.0 Against Algebraic Attacks

Olivier Billet and Henri Gilbert

France Télécom R&D,
38–40, rue du Général Leclerc,
92794 Issy les Moulineaux Cedex 9 – France
{olivier.billet,henri.gilbert}@francetelecom.com

Abstract. SNOW 2.0, a software oriented stream cipher proposed by T. Johansson and P. Ekdahl in 2002 as an enhanced version of the NESSIE finalist SNOW 1.0, is usually considered as one of the strongest stream ciphers designed so far. This paper investigates the resistance of SNOW 2.0 against algebraic attacks. This is motivated by the fact that the main source of non-linearity in SNOW 2.0 comes from a permutation build upon the AES S-box, which inputs and outputs are well known to be related by numerous quadratic equations. We show that a slightly modified version of SNOW 2.0 is susceptible to an algebraic attack with time complexity about 2^{50}, and which requires no more than 1000 words of output. We then explore various ways to extend this attack to the actual stream cipher.

Keywords: SNOW 2.0, stream ciphers, algebraic attacks.

1 Introduction

SNOW 2.0 [9] is a software oriented stream cipher proposed by T. Johansson and P. Ekdahl in 2002 as a replacement of an earlier version named SNOW 1.0 [8]. SNOW 2.0 is generally considered as one of the strongest stream cipher designs currently available, together with ciphers like the Shrinking Generator [3], SCREAM [12], and carefully initialized versions of RC4 [11]. SNOW 1.0 was one of the finalists of the European project NESSIE. One of the main reasons for the rejection of SNOW 1.0 from the NESSIE portfolio of recommended cryptographic primitives – which eventually lacked a stream cipher design – was the discovery of a statistical distinguisher with time complexity 2^{95} due to Coppersmith *et al.* [2]. A key recovery attack of expected complexity 2^{224} against SNOW 1.0 was also found H. Hawkes and G. Rose [13]. Both attacks require a known key stream length of 2^{95}. Those attacks motivated the introduction of a new version of SNOW, SNOW 2.0 [9], which eliminated at the same time some other minor flaws. The most characteristic features of SNOW 2.0 are

- an LFSR defined over a large field, $\mathrm{GF}(2^{32})$ with a new feedback polynomial as to avoid the flaws detected in the previous design, SNOW 1.0;

– a finite state machine involving two non-linearly updated memory registers of size 32 bits. The non-linearity results from two modular additions, and a 32 bit to 32 bit function S based on the well-known and highly studied AES S-box [14].

The best attack against SNOW 2.0 so far is a distinguishing attack of complexity 2^{225} due to D. Watanabe, A. Biryukov, C. de Cannière [16], and requires 2^{225} key stream words. It consists in an enhanced variant of the linear masking method [2] which exploits the feedback polynomial of the LFSR over $GF(2^{32})$ instead of requiring low weight multiples with $GF(2)$ coefficients, as in the original attack.

This paper investigate the resistance of SNOW 2.0 against algebraic attacks. Although the relevance of such attacks in the context of block ciphers – like AES, for instance – remains unclear, it has been proved to be of interest in the context of regularly clocked stream ciphers [5,6,1]. Considering that SNOW 2.0 is a regularly clocked stream cipher which non-linearity mainly rests on the AES S-box, it seems natural to probe its resistance against algebraic attacks.

We first establish that if the function S based on the AES S-box was the only source of non-linearity, SNOW 2.0 would be vulnerable to a very efficient algebraic attack. More precisely, we consider the close variant of SNOW 2.0 obtained by replacing the two modular additions by additions over $GF(2^{32})$, leaving the other parts (LFSR, S function based on AES S-box...) unchanged. We explain how to recover the initial state of the LFSR using a linearization attack of complexity about 2^{50}, requiring no more than 1000 clocks of key stream. We then examine the consequences of this result for the actual stream cipher, and show that the knowledge of a small key stream sequence (slightly more than 17 key stream outputs) allows the attacker to write a rather large – still, overdetermined and sparse – system of quadratic equations. Solving of such sparse quadratic systems and its complexity are not yet fully understood, but there is a growing research effort on the subject, due in large part to its potential application to the AES block cipher standard [15,7].

The paper is organized as follows. Section 2 provides a brief description of the SNOW 2.0 stream cipher. Section 3 describe the algebraic attack against a slightly modified SNOW 2.0, while Sec. 4 analyzes different means to extend this attack to the actual stream cipher.

2 Description of SNOW

The stream cipher SNOW 2.0 is made of a linear feedback shift register (LFSR) with sixteen 32 bit words and a finite state machine (FSM) with two 32 bit memory registers. SNOW 2.0 mixes additions over $GF(2^{32})$ hereafter denoted by '$\oplus$', together with additions modulo 2^{32} denoted by '$\boxplus$'.

2.1 The Linear Feedback Shift Register

The linear feedback shift register (LFSR) is defined over $GF(2^{32})$, which allows good performance for software implementations. It is made of sixteen 32 bit words, thus exhibiting 512 bit internal state size. The field of definition can be further described as $\mathrm{GF}(2^{32}) = \mathrm{GF}(2)(\alpha, \beta)$, where β is a root of the $\mathrm{GF}(2)[x]$ polynomial $x^8 + x^7 + x^5 + x^3 + 1$, and α is a root of the $\mathrm{GF}(2^8)[x]$ polynomial $x^4 + \beta^{23}x^3 + \beta^{245}x^2 + \beta^{48}x + \beta^{239}$. The feedback polynomial is then defined by

$$\alpha x^{16} + x^{14} + \alpha^{-1} x^5 + 1 \, .$$

This choice of a tower extension to describe $\mathrm{GF}(2^{32})$ is justified by the simple expression of the feedback polynomial in this context: it only consists of byte shifts/xors, since each word can be expressed on the base $\{\alpha^3, \alpha^2, \alpha, 1\}$.

In the following, the word that the LFSR outputs at clock t is denoted by s^t.

2.2 Finite State Machine

The other part of the stream cipher is a finite state machine (FSM), which contains two 32 bit memory registers r_1 and r_2. This FSM is intended to produce the non-linear part of the stream cipher. To this end, it contains a non-linear 32 bit to 32 bit non-linear bijection denoted by $\mathcal{S}$, based on the AES S-box [14], and defined as follows. If we decompose the register r_1 at clock t on the base $\{\alpha^3, \alpha^2, \alpha, 1\}$ as explained in the above section as $r_1^t = a_1^t \alpha^3 + b_1^t \alpha^2 + c_1^t \alpha + d_1^t$, and similarly the register r_2 at next clock as $r_2^{t+1} = a_2^{t+1} \alpha^3 + b_2^{t+1} \alpha^2 + c_2^{t+1} \alpha + d_2^{t+1}$, the rule $r_2^{t+1} = \mathcal{S}(r_1^t)$ to update r_2 from r_1 can be defined as

$$\begin{bmatrix} a_2^{t+1} \\ b_2^{t+1} \\ c_2^{t+1} \\ d_2^{t+1} \end{bmatrix} = \begin{bmatrix} X & X+1 & 1 & 1 \\ X+1 & 1 & 1 & X \\ 1 & 1 & X+1 & X \\ 1 & X+1 & X & 1 \end{bmatrix} \times \begin{bmatrix} S(a_1^t) \\ S(b_1^t) \\ S(c_1^t) \\ S(d_1^t) \end{bmatrix}$$

where S represents the AES S-box, the matrix is the one MixColumn step in AES when its four input bytes are considered as elements of the $\mathrm{GF}(2^8)$ definition of the AES, i.e. $\mathrm{GF}(2)[X]/(X^8 + X^4 + X^3 + X + 1)$. This completes the definition of the non-linear function $\mathcal{S}$.

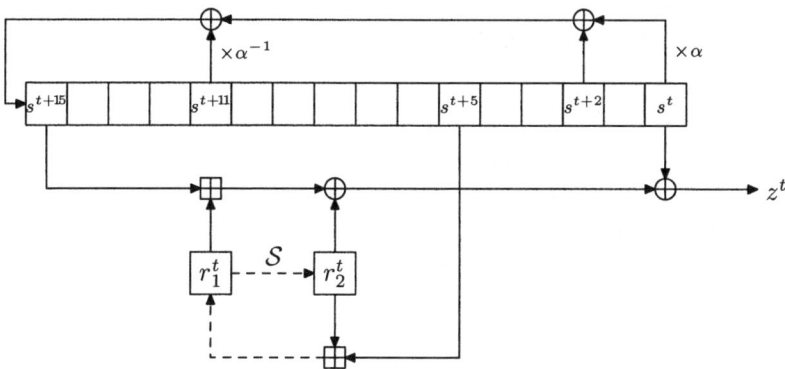

Fig. 1. SNOW 2.0.

Now the rule to update the register r_1 from r_2 is given by $r_1^{t+1} = r_2^t \boxplus s^{t+5}$. The output of the FSM at clock t, which we denote by F^t, is finally defined by $F^t = (r_1^t \boxplus s^{t+15}) \oplus r_2^t$. Let us summarize the behavior of the FSM below

$$\begin{cases} r_2^{t+1} \stackrel{\text{def}}{=} \mathcal{S}(r_1^t) \ , \\ r_1^{t+1} \stackrel{\text{def}}{=} r_2^t \boxplus s^{t+5} \ , \\ \quad F^t \stackrel{\text{def}}{=} (r_1^t \boxplus s^{t+15}) \oplus r_2^t \ . \end{cases}$$

2.3 Output of the Stream Cipher

The output of SNOW 2.0 is a classical example of linear masking, that is the output of the LFSR is xored with the output of the (non-linear) FSM. Thus the key stream output at clock t, which we henceforth denote by z^t, is defined by $z^t = s^t \oplus F^t$, or equivalently by

$$z^t = (s^{t+15} \boxplus r_1^t) \oplus r_2^t \oplus s^t \ .$$

2.4 Key Initialization

The stream cipher SNOW 2.0 can be used with 128 bit or 256 bit keys. For the key initialization, the LFSR is loaded with the secret key K, a publicly known initialization vector IV, and the two memory registers are set to zero. The cipher is then clocked 32 times in a special mode where no key stream is produced, and the FSM output is injected in the feedback value

$$s^{t+16} \stackrel{\text{def}}{=} \alpha^{-1} s^{t+11} \oplus s^{t+2} \oplus \alpha s^t \oplus F^t \ .$$

The cipher is then switched into the normal mode described in 2.3, but the first output of the keystream is discarded.

3 Attack on a Modified Version of SNOW 2.0

We now describe the algebraic attack against the close variant of SNOW 2.0 where modular additions '$\boxplus$' are replaced with xors '$\oplus$' in its description, while everything else remains identical.

3.1 Deriving the System

Let us construct a system of equation in the LFSR's initial state variables alone, and solve it. In order to do so, we need to eliminate the memory from the set of equations. This is done by looking at the key stream generation and the update rule for the register r_1. Indeed, combining those relations

$$\begin{cases} z^t = s^{t+15} \oplus r_1^t \oplus s^t \oplus r_2^t \ , \\ r_1^t = r_2^{t-1} \oplus s^{t+4} \ , \end{cases}$$

which can be further reduced into

$$r_2^t = r_2^{t-1} \oplus z^t \oplus s^{t+15} \oplus s^{t+4} \oplus s^t,$$

we get an expression of the register r_2, for any clock t, which only involves the key stream, the LFSR initial state variables $s^0, \ldots, s^{15}$, and the initial state r_2^0 of the register r_2. Put it in equation, for each clock t, there are known binary coefficients ϵ_t^i such that

$$r_2^t = r_2^0 \bigoplus_{i=0}^{t} z^i \bigoplus_{j=0}^{15} \epsilon_t^j s^j.$$

Let us assign $t = 0$ to the clock of the first key stream output. One easily checks that the register r_1, updated against the rule $r_1^{t+1} = r_2^t \oplus s^{t+5}$, benefits from the same property. (Note that the initial state of the register r_1 can be derived from the knowledge of r_2^0 and the relation $r_1^0 = r_2^0 \oplus s^0 \oplus s^{15} \oplus z^0$.) In other words, we got rid of the memory, since for any clock $t > 0$, it can be expressed linearly in terms of the initial state variables and the initial memory value r_2^0.

The property that the knowledge of the key stream allows to track the linear functions of $r_2^0, s^0, \ldots, s^{15}$ contained in r_1 and r_2 may be visualized on Fig. 3. (Note that a similar property involving non-linear expressions, also holds for the actual stream cipher.) Now we need to derive some equations involving the initial LFSR state variable, and r_2^0. Those are obtained from the second update rule, namely $r_2^{t+1} = \mathcal{S}(r_1^t)$. Since the non-linear function $\mathcal{S}$ maps the four bytes of r_1^t to the four bytes of r_2^t via the AES S-box, and then mixes the resulting bytes linearly at the bit level, we are able to write down 156 linearly independent quadratic equations relating the bits of $r_1^t = r_2^t \oplus s^t \oplus s^{t+15} \oplus z^t$ and the bits of r_2^{t+1}.

To explain why, it is suffices to recall the well known property of the AES S-box: there are linearly independent quadratic equations involving the S-box input and output bits. To see why, just write $S = A \circ I$, where A denotes the $GF(2)$-affine mapping, and I maps zero to zero and equals the inversion over $GF(2^8)$ everywhere else. Then if $w = S(u) = A \circ I(u)$ and $v = I(u)$, we get

$$uv = 1, \quad u^2 v = u, \quad uv^2 = v, \quad uv^4 = v^3, \quad u^4 v = u^3,$$

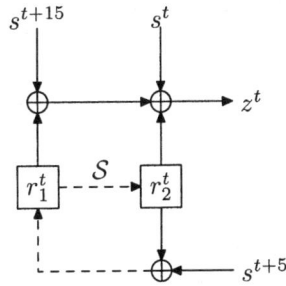
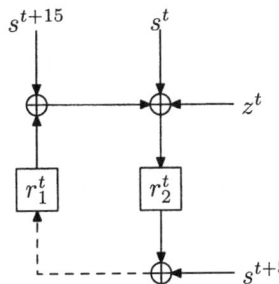

Fig. 2. A variant of SNOW 2.0.   **Fig. 3.** Tracking memory registers r_1 and r_2.

the first equation being true for all bits, except the least significant one, because $I(0) = 0$. And since $x \mapsto x^2$ is GF(2)-linear, we deduce that the bits of u and v are related by $5 \times 8 - 1 = 39$ quadratic equations. Now this property obviously remains true after the application of A.

Going back to $\mathcal{S}$, we are now able to write $4 \times 39 = 156$ quadratic equations relating r_1^t and r_2^{t+1}. Remember here that both registers are linear functions of the LFSR initial state variables $s^0, \ldots, s^{15}$ and r_2^0, for any $t > 0$.

3.2 Recovering the Initial State and the Key

The problem of recovering the initial state of the LFSR can be directly translated into that of solving the system of quadratic equations constructed in the previous section. However, two distinct strategies can be devised.

First one may wish to entirely linearize the system. The number of monomials involved are, in the worst case, all monomials of degree up to 2 involving a total of $512 + 32$ variables over GF(2). There are $N = \sum_{k=0}^{2} \binom{544}{k}$ such variables, which is slightly more than 2^{17}. To be able to linearize the system, we thus need to get about $N/156 < 951$ key stream words, that is we need to get less than 1000 consecutive output words of the stream cipher – under the usual assumption that the small number of linear dependencies occurring before a full rank system is obtained do not much affect the required number of outputs. A very conservative estimation of the time complexity to solve the system is the cube of the number of variables, that is about 2^{51}.

One could also want to solve the system of quadratic equations as soon as it is overdefined and without requiring it to be linearized, since there exists algorithms especially designed for this task [7, 10]. We note that in such case, only slightly more than 17 key stream output words are needed for the system to be overdefined. In this case however, the complexity to solve the system is not fully understood in the current state of the art, and is expected to be notably higher than for solving a linearized system.

Once the initial state $s^0, \ldots, s^{15}$, and r_2^0 have been recovered using the above linearization method, r_1^0 is given by the relation $r_1^0 = r_2^0 \oplus s^0 \oplus s^{15} \oplus z^0$, and so the entire state of the cipher at clock $t = 0$ is known. In order to derive the secret key K – and thus be able to predict the key stream sequence for other IVs – it suffices to run the cipher backward, one clock in the normal operation mode, then 32 clocks in the special feedback mode. It is easy to see that the state transitions of the SNOW 2.0 in both normal and special modes are invertible. Therefore, we are able to get the LFSR state at the initialization time, wich gives, from the knowledge of IV, the value of the secret key K.

4 Implications for SNOW 2.0

In this section, we seek for an extension of our attack described in Sec. 3 to the actual SNOW 2.0 stream cipher. We mainly identified two possible methods to take into account the extra source of non-linearity introduced by the modular additions of the FSM. The first one is to guess the carries' values for a small number of consecutive clocks. The other one consists in introducing new variables for the carries, and building a system of quadratic equations involving the LFSR initial state variables, the FSM initial memory variables and the extra carry variables. As will be shown in the sequel, the first method appears to require an impractical amount of guessing, while the second one seems more promising at first glance from a cryptanalytic point of view.

In the following, the carry corresponding to the addition $s^{t+15} \boxplus r_1^t$ of the FSM will be denoted by c_1^t, while the carry corresponding to the addition $s^{t+5} \boxplus r_2^t$ of the FSM will be denoted by c_2^t. Hence,

$$\begin{aligned} s^{t+15} \boxplus r_1^t &= s^{t+15} \oplus r_1^t \oplus c_1^t, \\ s^{t+5} \boxplus r_2^t &= s^{t+5} \oplus r_2^t \oplus c_2^t. \end{aligned} \tag{1}$$

As in previous section we denote by $t = 0$ the clock of the first observable output of SNOW 2.0 and call initial state the state of the LFSR at $t = 0$.

4.1 Guessing the Carries

This method strives to take benefit of the specificities of the carry bits' distribution occurring in modular additions. According to Eq. 1, we can track affine functions of r_2^0, s^0, ..., s^{15} contained in the memory registers r_1 and r_2 in the same way as done in Sec. 3 – and then, apply the attack therein described – just by guessing the values of the carries c_1^t and c_2^t for about 16 consecutive clocks. The single difference with Sec. 3 is that the expressions of r_1^t and r_2^t now involve constant terms from the guessed carry values. However, due to the very particular distribution of carry bits, the cost of one guess is far less than 2^{32}. Actually, it can be shown that the most probable carry – i.e. with no carry at all during the addition – has one chance over $\left(\frac{3}{4}\right)^{31}$ to occur. Indeed, this will happen when any two matching bits are not simultaneously 1, which represents three possibilities out of four. Thus a rough estimation for an upper bound on the probability to make a right guess for the carries c_1 and c_2 during 16 consecutive clocks is $\left(\frac{3}{4}\right)^{31 \times 2 \times 16}$. As it is much less than 2^{-256}, this approach seems impractical.

4.2 Quadratic System with Carry Variables

This second method consists in building a system of quadratic equations describing the actual SNOW 2.0 stream cipher. To this end, it suggests to introduce new $GF(2)$ variables for the carry bits of the two modular additions '$\boxplus$' at each clock. This results in gathering quadratic equations during slightly more than 17 clocks – for the system to be overdetermined – and trying to solve the corresponding system.

Deriving the set of quadratic equations goes along the lines of the method exposed in Sec. 3. Indeed, just inserting the carries due to modular additions gives

$$\begin{cases} z^t = c_1^t \oplus s^{t+15} \oplus r_1^t \oplus s^t \oplus r_2^t, \\ r_1^t = c_2^{t-1} \oplus r_2^{t-1} \oplus s^{t+4}, \end{cases}$$

which is this time reduced into

$$r_2^t = r_2^{t-1} \oplus z^t \oplus s^{t+15} \oplus s^{t+4} \oplus s^t \oplus c_1^t \oplus c_2^{t-1}.$$

Eventually, we come to the fact that the memory registers can be expressed at any clock $t > 0$ as a linear combination of the initial LFSR state variables, the initial value r_2^0 of the register r_2, and all the carry bits occurring between clock 0 and clock t. The $(i+1)$th carry bit in the modular addition of two 32 bit words x and y can be defined as the majority of the ith bits of x, y, and ith carry bit. For each clock t, Eq. 1 thus implies

$$c_{1,[0]}^t = 0$$
$$c_{1,[1]}^t = s_{[0]}^{t+15} \oplus r_{1,[0]}^t$$
$$0 < i < 32, \quad c_{1,[i+1]}^t = s_{[i]}^{t+15} r_{1,[i]}^t \oplus s_{[i]}^{t+15} c_{1,[i]}^t \oplus c_{1,[i]}^t r_{1,[i]}^t,$$

as well as

$$c_{2,[0]}^t = 0$$
$$c_{2,[1]}^t = s_{[0]}^{t+5} \oplus r_{2,[0]}^t$$
$$0 < i < 32, \quad c_{2,[i+1]}^t = s_{[i]}^{t+5} r_{2,[i]}^t \oplus s_{[i]}^{t+5} c_{2,[i]}^t \oplus c_{2,[i]}^t r_{2,[i]}^t,$$

where $x_{[i]}$ denotes the ith bit of the 32 bit word x.

Of course, we have to add to these the quadratic equations holding between the registers r_1 and r_2. As stated above in Sec. 3, there are 156 such equations at each clock t, but this time involving the LFSR initial state variables, the variables for the bits of r_2^0, and all the carries' bits.

Let us now count the number of variables that appear in the system after n consecutive clocks. There are the 512 variables from the LFSR initial state, the 32 variables from r_2^0, plus for each clock, 62 carry bit variables. Hence, a total of $544 + 62n$ variables. On the other hand, there are $156n$ equations coming from the relation S, and $62n$ equations from the definition of each carry bit, all at most quadratic, which amounts to a total of $218n$ equations. Hence the minimum value $n = 17$ for the system to be overdetermined, gives a total of 3706 equations with 1598 variables. For larger value of n, the system is more overdefined, but the equations to variables ratio is asymptotically bounded above by $\frac{7}{2}$.

The above system has been derived as to minimize the number of variables, not to maximize its sparsity. One can easily see that only the equations defining the carry bits are extremely sparse. Alternatively, one might write an equivalent, still overdefined and much more sparse system, by introducing the auxiliary variables r_1^t and r_2^t and their related linear equations, at the expense of increasing the number of variables. The system would have $544 + 126n$ variables, $282n$ quadratic or linear equations, and about the same sparsity as the equations on the AES block cipher. Its intractability remains, as in the case of the AES, an open question.

5 Conclusion

We exposed in this paper a very efficient attack against a close variant of the stream cipher SNOW 2.0. Various ways to extend this attack to the actual SNOW 2.0 design were also tried. The key search problem for the actual SNOW 2.0 was shown to be reducible to the solving of an overdetermined system of quadratic equations, the complexity of which remains unknown nowadays.

References

1. J. Y. Cho and J. Pieprzyk. Algebraic attacks on SOBER-t32 and SOBER-128. In W. Meier and B. K. Roy, editors, *Fast Software Encrytion – FSE 2004*, Lecture Notes in Computer Science. Springer-Verlag, 2004.
2. D. Coppersmith, S. Halevi, and C. S. Jutla. Cryptanalysis of stream ciphers with linear masking. In M. Yung, editor, *Advances in Cryptology – CRYPTO 2002*, Lecture Notes in Computer Science, pages 515–532. Springer-Verlag, 2002.
3. D. Coppersmith, H. Krawczyk, and Y. Mansour. The shrinking generator. In D. R. Stinson, editor, *Advances in Cryptology – CRYPTO '93*, Lecture Notes in Computer Science, pages 22–39. Springer-Verlag, 1993.
4. N. T. Courtois. Algebraic attacks on combiners with memory and several outputs. Cryptology ePrint Archive, Report 2003/125, 2003. http://eprint.iacr.org/.
5. N. T. Courtois. Fast algebraic attacks on stream ciphers with linear feedback. In D. Boneh, editor, *Advances in Cryptology – CRYPTO 2003*, volume 2729 of *Lecture Notes in Computer Science*, pages 176–194. Springer-Verlag, 2003.
6. N. T. Courtois and W. Meier. Algebraic attacks on stream ciphers with linear feedback. In E. Biham, editor, *Advances in Cryptology – EUROCRYPT 2003*, volume 2656 of *Lecture Notes in Computer Science*, pages 345–359. Springer-Verlag, 2003.
7. N. T. Courtois and J. Pieprzyk. Cryptanalysis of block ciphers with overdefined systems of equations. In Y. Zheng, editor, *Advances in Cryptology – ASIACRYPT 2002*, volume 2501 of *Lecture Notes in Computer Science*, pages 267–287. Springer-Verlag, 2002.
8. P. Ekdahl and T. Johansson. SNOW – a new stream cipher. Submission can be downloaded at http://www.cryptonessie.org, 2000.
9. P. Ekdahl and T. Johansson. A new version of the stream cipher SNOW. In K. Nyberg and H. M. Heys, editors, *Selected Areas in Cryptography – SAC 2002*, Lecture Notes in Computer Science, pages 47–61. Springer-Verlag, 2002.
10. J.-C. Faugère. A New Efficient Algorithm for Computing Groebner Bases without Reduction to Zero (F5). In *Proceedings of ISSAC*, pages 75–83. ACM Press, 2002.
11. S. R. Fluhrer, I. Mantin, and A. Shamir. Weaknesses in the key scheduling algorithm of rc4. In S. Vaudenay and A. M. Youssef, editors, *Selected Areas in Cryptography – SAC 2001*, Lecture Notes in Computer Science, pages 1–24. Springer-Verlag, 2001.
12. S. Halevi, D. Coppersmith, and C. S. Jutla. SCREAM: A software-efficient stream cipher. In J. Daemen and V. Rijmen, editors, *Fast Software Encrytion – FSE 2002*, Lecture Notes in Computer Science, pages 195–209. Springer-Verlag, 2002.
13. P. Hawkes and G. G. Rose. Guess-and-determine attacks on snow. In K. Nyberg and H. M. Heys, editors, *Selected Areas in Cryptography – SAC 2002*, Lecture Notes in Computer Science, pages 37–46. Springer-Verlag, 2002.

14. National Institute of Standards and Technology. Advanced encryption standard. FIPS publication 197, 2001.
    http://csrc.nist.gov/publications/fips/fips197/fips-197.pdf.
15. M. J. B. Robshaw and S. Murphy. Essential Algebraic Structure within the AES. In M. Yung, editor, *Advances in Cryptology – CRYPTO 2002*, volume 2442 of *Lecture Notes in Computer Science*, pages 1–16. Springer-Verlag, 2002.
16. D. Watanabe, A. Biryukov, and C. de Cannière. A distinguishing attack on SNOW 2.0 with linear masking method. In M. Matsui and R. Zuccherato, editors, *Selected Areas in Cryptography – SAC 2003*, Lecture Notes in Computer Science, pages 222–233. Springer-Verlag, 2003.

A Study of the Security of Unbalanced Oil and Vinegar Signature Schemes

An Braeken, Christopher Wolf, and Bart Preneel

K.U.Leuven, ESAT-COSIC,
Kasteelpark Arenberg 10, B-3001 Leuven-Heverlee, Belgium
{An.Braeken,Christopher.Wolf,Bart.Preneel}@esat.kuleuven.ac.be
http://www.esat.kuleuven.ac.be/cosic/

Abstract. The Unbalanced Oil and Vinegar scheme (UOV) is a signature scheme based on multivariate quadratic equations. It uses m equations and n variables. A total of v of these are called "vinegar variables". In this paper, we study its security from several points of view. First, we are able to demonstrate that the constant part of the affine transformation does not contribute to the security of UOV and should therefore be omitted. Second, we show that the case $n \geq 2m$ is particularly vulnerable to Gröbner basis attacks. This is a new result for UOV over fields of odd characteristic. In addition, we investigate a modification proposed by the authors of UOV, namely to chose coefficients from a small subfield. This leads to a smaller public key. But due to the smaller key-space, this modification is insecure and should therefore be avoided. Finally, we demonstrate a new attack which works well for the case of small v. It extends the affine approximation attack from Youssef and Gong against the Imai-Matsumoto Scheme B for odd characteristic and applies it against UOV. This way, we point out serious vulnerabilities in UOV which have to be taken into account when constructing signature schemes based on UOV.

1 Introduction

1.1 Public Key Cryptography in General

Public key cryptography is used in e-commerce systems for authentication (electronic signatures) and secure communication (encryption). In terms of key distribution, public key cryptography has significant advantages over secret key cryptography. Moreover, efficient signature schemes cannot be obtained by secret key schemes. The security of widely used public key algorithms relies on the difficulty of a small set of problems from algebraic number theory. The RSA scheme relies on the difficulty of factoring large integers, while the difficulty of solving discrete logarithms provides the basis for the ElGamal and Elliptic Curve schemes [18]. Given that the security of these public key schemes rely on such a small number of problems that are *currently* considered hard, research on new schemes that are based on other classes of problems is worthwhile. Such work provides a greater diversity and avoids the risk that the information society joints all its "crypto eggs" in one basket.

In addition, important results on the potential weaknesses of existing public key schemes are emerging. Techniques for factorisation and solving discrete logarithm continually improve. Polynomial time quantum algorithms can be used to solve both problems [25]; fortunately, quantum computers with more than 7 bits are not yet available and it seems unlikely that quantum computers with 100 bits will be available within the next 10–15 years. Nevertheless, this stresses the importance of research into new algorithms for asymmetric encryption and signature schemes that may not be vulnerable to quantum computers.

1.2 Multivariate Cryptography

One way to achieve more variety in asymmetric cryptology are schemes based on the problem of solving $\mathcal{M}$ultivariate $\mathcal{Q}$uadratic equations ($\mathcal{MQ}$-problem), e.g., see [17, 21, 22, 3, 12, 19, 4, 28, 11]. These schemes use the fact that the $\mathcal{MQ}$-problem, i.e., finding a solution $x \in \mathbb{F}^n$ for a given system of m polynomial equations in n variables each

$$\begin{cases} y_1 = p_1(x_1, \ldots, x_n) \\ y_2 = p_2(x_1, \ldots, x_n) \\ \vdots \\ y_m = p_m(x_1, \ldots, x_n), \end{cases}$$

for given $y_1, \ldots, y_m \in \mathbb{F}$ and unknown $x_1, \ldots, x_n$ is difficult, namely $\mathcal{NP}$-complete (cf [9, p. 251] and [24, App.] for a detailed proof)). In the above system of equations, the polynomials p_i have the form

$$p_i(x_1, \ldots, x_n) := \sum_{1 \leq j \leq k \leq n} \gamma_{i,j,k} x_j x_k + \sum_{j=1}^{n} \beta_{i,j} x_j + \alpha_i,$$

for $1 \leq i \leq m; 1 \leq j \leq k \leq n$ and $\alpha_i, \beta_{i,j}, \gamma_{i,j,k} \in \mathbb{F}$ (constant, linear, and quadratic terms). This polynomial-vector $\mathcal{P} := (p_1, \ldots, p_m)$ forms the public key of these systems. Moreover, the private key consists of the triple $(S, \mathcal{P}', T)$ where $S \in \mathrm{AGL}_n(\mathbb{F}), T \in \mathrm{AGL}_m(\mathbb{F})$ are affine transformations and $\mathcal{P}' \in \mathcal{MQ}_{n,m}$ is a polynomial-vector $\mathcal{P}' := (p'_1, \ldots, p'_m)$ with m components; each component is a polynomial in n variables $x'_1, \ldots, x'_n$. Throughout this paper, we will denote components of this private vector $\mathcal{P}'$ by a prime '. In contrast to the public polynomial vector $\mathcal{P} \in \mathcal{MQ}_{n,m}$, the private polynomial vector $\mathcal{P}'$ does allow an efficient computation of $x'_1, \ldots, x'_n$ for given $y'_1, \ldots, y'_m$. At least for secure $\mathcal{MQ}$-schemes, this is not the case if the public key $\mathcal{P}$ alone is given. The main difference between $\mathcal{MQ}$-schemes lies in their special construction of the central equations $\mathcal{P}'$ and consequently the trapdoor they embed into a specific class of $\mathcal{MQ}$-problems. We refer to [13] for an overview of the different proposed schemes. Note that most of them are already broken e.g., [5, 8, 10, 15, 20, 27]. We describe in this paper some new results on the cryptanalysis of the Unbalanced

Oil and Vinegar scheme which is still considered to be secure for certain choices of parameters.

1.3 Outline and Achievement

We start with an explanation of the Unbalanced Oil and Vinegar scheme (UOV). Second, we outline in Sect. 3.1 why the constant part of the initial affine transformation can be omitted as it does not contribute to the overall security of UOV. In Sect. 3.2, we give a short description of the Shamir and Kipnis attack against the (balanced) oil and vinegar scheme together with its extension on the unbalanced case. Then we show how this attack breaks the scheme proposed in [13, Sect. 14, ex. 4]. Moreover, we show that the case $n \geq 2m$ is particularly vulnerable to Gröbner basis attacks (Sect. 3.3). This way, we improve a result of Courtois *et al.* who were able to defeat the cases $n \geq 4m$ [2] – and to some extent also $n \geq 3m$. However, for their most efficient attack to work, they need an even characteristic. The attacks demonstrated in this paper do not have this restriction. Finally, we extend the attack from Youssef and Gong [29] against the Scheme B from Imai and Matsumoto [16] against Unbalanced Oil and Vinegar scheme – both for even and odd characteristic in Sect. 3.4. The algorithm presented in [29] only works for the even case. We conclude with Section 4.

2 Oil and Vinegar Signature Schemes

In 1997, Jacques Patarin suggested a scheme called "Oil and Vinegar" for public key cryptography [23]. This scheme uses multivariate quadratic polynomial equations over small finite fields as public key and similar polynomials as the private keys.

In Oil and Vinegar Schemes, the trapdoor is achieved by a special structure of multivariate quadratic polynomials p'_i. Let $o \in \mathbb{N}$ be the number of oil variables and $v \in \mathbb{N}$ the number of vinegar variables. We have $n = o + v$. Moreover, we have $m = o$ and $o = v$ (or also $n = 2m$) for the case of Oil and Vinegar Schemes[1]. The private polynomials p'_i for $1 \leq i \leq m$ can be represented by

$$p'_i(x'_1, \ldots, x'_n) := x'_1 Lin'_{i,1}(x'_1, \ldots, x'_n) + \ldots + x'_v Lin'_{i,v}(x'_1, \ldots, x'_n) +$$
$$+ Af'_i(x'_1, \ldots, x'_n)$$
$$= \sum_{\substack{1 \leq j \leq v \\ 1 \leq k \leq n}} \gamma'_{i,j,k} x'_j x'_k + \sum_{1 \leq k \leq n} \beta'_{i,k} x'_k + \alpha'_i,$$

for $Lin'_{i,j}$ linear, Af'_i affine or – more general – for $1 \leq i \leq m, 1 \leq j \leq v$ and $1 \leq k \leq n$ and $\alpha'_i, \beta'_{i,k}, \gamma'_{i,j,k} \in \mathbb{F}$. Here the vinegar variables $x'_1, \ldots, x'_v$ may be quadratically combined while oil variables $x'_{v+1}, \ldots, x'_n$ do not mix with oil variables.

---
[1] The above notation clearly has some redundancies. The problem in this context is that different papers about these schemes use very different notation. With the above settings, we use a kind of "generalised notation" which suits most of them.

The trapdoor consists of an affine transformation $S \in \mathrm{AGL}_n(\mathbb{F})$ that mixes the oil and vinegar variables, i.e., $(x'_1, \ldots, x'_n) = S(x_1, \ldots, x_n)$ leads to an affine relation between the public variables x_i and the private variables x'_i. In order to obtain a solution for such a system, the legitimate user fixes all vinegar variables to random values. This way, he obtains a (random) linear equation in the oil variables which can be solved with ordinary Gauss elimination.

Generally speaking, the (unbalanced) oil and vinegar scheme is designed for a signature scheme. It is not suitable for encryption because of the parameter v, which should be chosen too high for an appropriate security level. To sign a message $M \in \mathbb{F}^m$, we perform the following steps:

1. Assign random variables $a_1, \ldots, a_v$ to all the vinegar variables.
2. After substituting the random values, the system $M = \mathcal{P}'(a)$ becomes linear. Solve this linear system for the remaining m variables $a_1, \ldots, a_o$ of a by Gaussian elimination. If the linear system is singular, return to the first step and try with new random vinegar variables.
3. Map the solution a to the signature x by $x = S^{-1}(a)$.

Verifying the signature $x \in \mathbb{F}^n$ is just the evaluation of x by the public system $\mathcal{P}$. An attacker wants to forge signature on a given message $M = (M_1, \ldots, M_m)$, needs to solve the system:

$$M_1 = p_1(x_1, \ldots, x_n)$$
$$\vdots$$
$$M_m = p_m(x_1, \ldots, x_n)$$

In general, this is an $\mathcal{MQ}$-problem and therefore difficult to solve.

As the original Oil and Vinegar scheme was broken in [14], Kipnis et al. extended it to the so-called "Unbalanced Oil and Vinegar" signature scheme [12] (see also the extended version [13]). For an Unbalanced Oil and Vinegar Scheme (UOV), we have $v > o$ (or equivalently $n > 2m$). According to [12, 13], this case is considered to be secure if the number of vinegar variables is not too "close" to the number of oil variables. In symbols: $v \not\approx o$.

3 Cryptanalysis

3.1 Attacking the Constant Part of UOV

We first show that the affine transformation S in the oil and vinegar scheme should be replaced by a linear transformation.

Consider the affine transformation $S \in \mathrm{AGL}_n(\mathbb{F})$, which can be uniquely represented by an invertible matrix $M_S \in \mathbb{F}^{n \times n}$ and a vector $m_s \in \mathbb{F}^n$, i.e., $S(x) = M_S x + m_s$ for all $x \in \mathbb{F}^n$. Moreover, we can uniquely rewrite S as $S(x) = (x' + m_s) \circ (M_S x)$ where x' denotes the output of $M_S x$ and $\circ$ represents the composition of functions. We now express the public key $\mathcal{P}$ as a composition of the private key $(\mathcal{P}', S)$:

$$\begin{aligned}
\mathcal{P} &= \mathcal{P}' \circ S \\
&= \mathcal{P}' \circ [(x' + m_s) \circ (M_S x)] \\
&= [\mathcal{P}' \circ (x' + m_s)] \circ (M_S x) \\
&= \mathcal{P}'' \circ (M_S x)
\end{aligned}$$

for some system of equations $\mathcal{P}''$. As $(x' + m_s)$ is a transformation of degree 1, it does not change the overall degree of $\mathcal{P}''$, i.e., as $\mathcal{P}'$ consists of equations of degree 2 at most, so will $\mathcal{P}''$. In addition, due to its construction, $(M_S, \mathcal{P}'')$ forms a private key for the public key $\mathcal{P}$. Moreover, the private key equations $\mathcal{P}'$ were random equations. The transformation $(x' + m_s)$ does not change the internal structure of $\mathcal{P}'$.

Therefore, we can conclude that the use of an affine instead of a linear transformation does not enhance the overall security of the (unbalanced) oil and vinegar schemes. In fact, we can draw a similar conclusion for all such systems – as long as it is possible to replace the equation $\mathcal{P}'$ by an equation of similar shape. This is always the case if $\mathcal{P}'$ allows a constant, non-zero term and also non-zero linear terms. The corresponding observation for HFE has been made by Toli [26].

3.2 The Kipnis and Shamir Attack

After this initial observation, we move on to the attack of Kipnis and Shamir against the *Balanced* Oil and Vinegar scheme. The main idea in this attack is to separate the oil and the vinegar variables, which enables the attacker to access an isomorphic copy of the private key. This way, an attacker can forge arbitrary signatures. The attack is very efficient for all $v \leq m$. We describe the attack here for $v = m$ and thus $2m = n$.

We take only the quadratic terms of the private $\mathcal{P}'$ and the public $\mathcal{P}$ equations into account. In odd characteristic, we can uniquely represent the private key equations (resp. public key equations) by $x^t P_i' x$ (resp. $x'^t P_i x'$) for $0 \leq i \leq m$, where P_i' and P_i are symmetric matrices (here ^{t} denotes transposition). For even characteristic, the unique symmetric matrices $P_i' + P_i'^t$ and $P_i + P_i^t$ where P_i' and P_i are upper-triangular matrices belonging to $\mathbb{F}^{m \times m}$ are considered. For simplicity, we denote these matrices again by P_i' and P_i.

Note that because of the special structure of the private equations $\mathcal{P}'$, the matrices P_i' for $1 \leq i \leq m$ have the form:

$$P_i' = \begin{pmatrix} 0 & A_i \\ B_i & C_i \end{pmatrix},$$

where $0, A_i, B_i, C_i$ are submatrices of dimension $m \times m$. Because $\mathcal{P} = \mathcal{P}' \circ S$, we obtain

$$P_i = M_S \begin{pmatrix} 0 & A_i \\ B_i & C_i \end{pmatrix} M_S^T.$$

It is clear that each P'_i maps the subspace $x_{m+1} = \cdots = x_{2m}$ (oil subspace) to the subspace $x_1 = \cdots = x_m = 0$ (vinegar subspace). If P'_j is invertible, we can then conclude that each $P'_i {P'_j}^{-1}$ maps the oil subspace to itself. Consequently the image of the oil subspace under S, called the subspace O, is a common eigenspace for each $P_i P_j^{-1}$ with $1 \leq i < j \leq m$. In [14, Sect. 4], Shamir and Kipnis describe two very efficient algorithms for computing the common eigenspace O of a set of transformations. Picking a subspace V for which $O + V = \mathbb{F}^m$ allows us to separate the oil and the vinegar variables. This way, we obtain an isomorphic copy of the private key $(\mathcal{P}, S)$.

In [12, Sect. 4], an extension based on a probabilistic approach of the previous attack is described which also works for $v > m$ (or $n > 2m$) with complexity $O(q^{v-m-1} m^4) = O(q^{n-2m-1} m^4)$.

Application Against the Parameters from [13, Sect. 14]: In order to avoid the birthday paradox, [12, Sect. 8] describes a modification of UOV which fixes the linear terms of the public equations depending on the message M. This way, it is no longer possible to obtain a collision for different messages $M_1 \neq M_2$ and the same public key, as this public key now also depends on the message M. We consider this construction to be secure and therefore refer to [12, Sect. 8] for a detailed description. However, its application in [13, Sect. 14], Example 4 is flawed. In order to derive a smaller public key, the authors use the trick of restricted coefficients (cf [13, Sect. 10]). In a nutshell, all coefficients in the affine transformation S and the system of private polynomials $\mathcal{P}'$ are not chosen from the field $\mathbb{F}$ but from a strictly smaller subfield $\tilde{\mathbb{F}}$. This way, the public key $\mathcal{P}$ will only have coefficients from $\tilde{\mathbb{F}}$ as $\mathcal{P} = \mathcal{P}' \circ S$ and subfields are closed under addition and multiplication. Thus, we derive a public key which is a factor of $(\log |\tilde{\mathbb{F}}|/\log |\mathbb{F}|)$ smaller than the original key.

In Example 4, the authors of [13] propose $\mathbb{F} = \mathrm{GF}(16)$, $\tilde{\mathbb{F}} = \mathrm{GF}(2)$, $m = 16$, $v = 32/48$ and obtain a public key with 2.2kB/4 kB – this is 4 times smaller than without this trick. However, we can apply the attack from the [12, Sect. 4] (see above) against the UOV system over $\tilde{\mathbb{F}} = \mathrm{GF}(2)$. This is possible as the Kipnis-Shamir attack does not take linear terms into account but only quadratic terms. The crucial point is that the linear terms are from $\mathrm{GF}(16)$ while the quadratic terms are from a subfield isomorphic to $\mathrm{GF}(2)$. As soon as we derived an isomorphic copy of the private key $(\mathcal{P}, S)$ over $\mathrm{GF}(2)$, we can translate it to $\mathrm{GF}(16)$ and are now in the same position as a legitimate user. In particular, we can do all computations necessary to translate the linear parts of the public key (over $\mathrm{GF}(16)$) to the corresponding private key (now, also over $\mathrm{GF}(16)$). As we have $q = 2$, the attack complexity is $2^{32-16-1}.16^4 = 2^{32}$ or $2^{48-16-1}.16^4 = 2^{47}$ and therefore far less than the claimed security level of 2^{64}.

Remark: Although the algorithms from [2] achieve a lower running time, they are not applicable in this case: they are only able to solve a given instance of an $\mathcal{MQ}$-problem. For this attack, we need the fact that we actually derive a valid private key of the UOV-system.

3.3 Attacks Using Gröbner Basis Algorithms

The article of Daum, Felke, and Courtois [5] outlines a way of attacking HFE with Gröbner Basis algorithms. The attack works for $m < n$, i.e., less equations than variables. The idea is to add $n-m$ linear equations. This way, the number of variables can be reduced to m. On the other hand, a system with n variables and m equations is expected to have q^{n-m} solutions on average. Therefore, adding a total of $n-m$ linear equations will lead to one solution on average. Repeating this experiment a few times (e.g., 6, cf Fig. 1), we will find at least one solution.

In our experiments, we fixed $n-m$ variables to random values from $\mathbb{F}$ instead of adding $n-m$ linear equations. From a mathematical point of view, both ideas are equivalent, as the transformation S already gives a random system of degree 1 equations. In a first step, we computed the average number of tries for a series of experiments where n takes values from 10 to 24, and v goes from 1 to $n-1$. Figure 1 shows that we need only a few tries for a given system of equations until we find a solution. In more than 60% of the cases, we obtain a solution with the first random fixing of variables, after that the number of necessary tries converts quickly to zero.

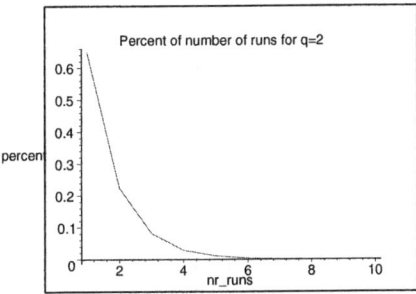

Fig. 1. Occurrence of number of runs.

In a second step, we investigated the time complexity of the attack for fixed m and varying v. From experiments, we could conclude that the time complexity increases exponentially with increasing v. This fact can be understood intuitively by the observation that for increasing v, the scheme becomes more random, which makes it more difficult to solve. However, as the number of solutions increases by q^v, i.e., exponentially, the probability of finding one out of these q^v solution becomes higher, too.

In particular, we investigated the logarithmic time complexity (T) for varying the number of equations m for the two values $v = 2m$, $v = 3m$ in characteristics $q = 2, q = 3$ and $q = 16$. The corresponding graphs can be shown in figures 2, 3, and 4. In Table 1, we computed the line that approximately fitted the points from our experiments for the extended Gröbner attack on UOV.

From these experiments, we conclude that the number m of equations should be higher than 38 for characteristic 2 and higher than 24 for characteristic 3 both

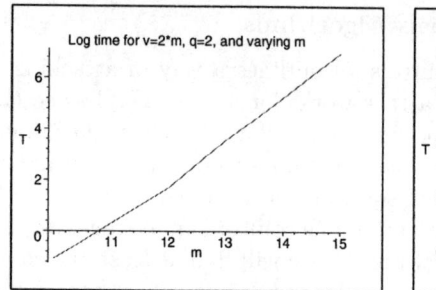
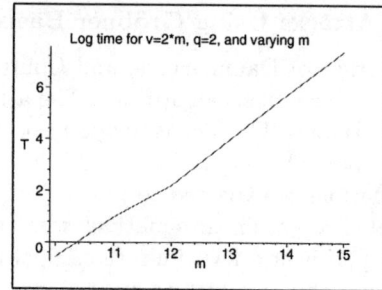

Fig. 2. Graphs for logarithmic time in function of m with $v = 2m$, resp. $3m$, and $q = 2$.

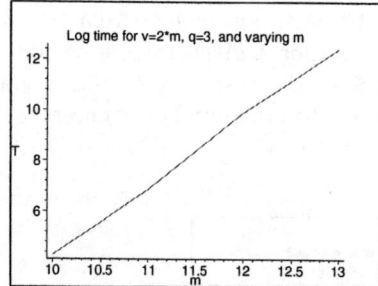
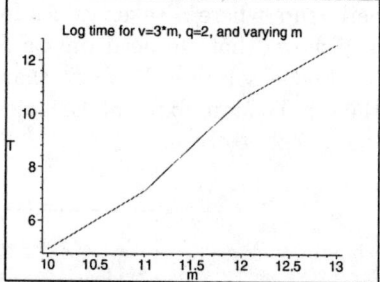

Fig. 3. Graphs for logarithmic time in function of m with $v = 2m$, resp. $3m$, and $q = 3$.

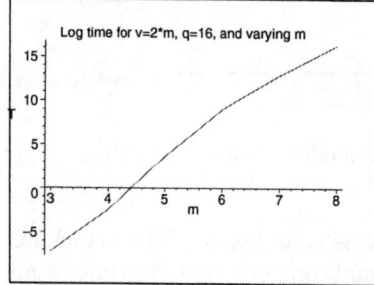
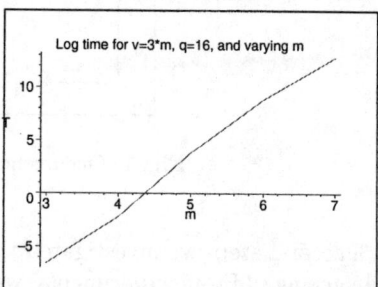

Fig. 4. Graphs for logarithmic time in function of m with $v = 2m$, resp. $3m$, and $q = 16$.

for $n \geq 2m$ and $n \geq 3m$ in order to obtain a security level greater than 2^{64}. In this paper, we do not predict the behaviour of the curve for $q = 16$ as the graph does not clearly convert to a straight line. To see its behaviour for $m > 8$ – and therefore, to make predictions, we would need to run more experiments. Unfortunately, the current computational power available does not permit this.

These lowerbounds on the minimum number of equations are much higher than the bounds proposed in [13] and later in [2]. All experiments in this sec-

Table 1. Equations representing the time complexity of the extended Gröbner Attack.

v	q	Equation	Base
$v = 2m$	$q = 2$	-17.53+1.62m	3.07
$v = 3m$	$q = 2$	-16.66+1.60m	3.03
$v = 2m$	$q = 3$	-23.17+2.74m	6.68
$v = 3m$	$q = 3$	-21.85+2.67m	6.36
$v = 2m$	$q = 16$	-21.14+4.82m	28.20
$v = 3m$	$q = 16$	-21.89+5.03m	32.63

tion were carried out with MAGMA and used its implementation of Faugere's algorithm F_4 [6]. Given the fact that his algorithm F_5 [7] has a far better running time, we expect the attack to be even more efficient with this method. Unfortunately, we do not have access to an actual implementation of it.

3.4 Exploiting the Existence of Affine Subspaces

This attack extends the attack of Youssef and Gong [29] against the Imai and Matsumoto Scheme B [16]. It exploits the fact that a cryptosystem can be approximated by several affine equations. The original attack was designed for fields of even characteristic. The attack described in this section is generalised to all characteristics.

In a nutshell, the attack assembles several points belonging to the same affine subspace W. Having w points $x_1, \ldots, x_w \in \mathbb{F}^n$ for which UOV is affine, a function $F(x) = Ax + b$ can be used to describe the output of UOV. To launch the attack, we first compute the corresponding $y_i = UOV(x_i)$ for $1 \leq i \leq w$ and $y_i \in \mathbb{F}^m$. With this knowledge, we can determine for any given y' if it belongs to the subspace W and – if this is the case – compute a vector $a \in \mathbb{F}^w$ with $y' = \sum_{i=1}^{w} a_i y_i$. As the subspace W is affine, we can then determine the corresponding $x' \in \mathbb{F}^n$ as $\sum_{i=1}^{w} a_i x_i$. In the following section, we will present several ways of computing the points x_i, i.e., to determine one or several subspaces W.

For UOV, there exist approx. q^v subspaces of dimension $o = m$ on which UOV is affine. Moreover, all these subspaces are disjunct. If we can find $(o+1)$ linearly independent points of the same subspace, we completely broke the scheme for this subspace. If we find fewer, e.g., w points, we have at least covered q^w points of the corresponding subspace W. Repeating the search for $(o + 1)$ points q^v times, we break the whole scheme. Note that it is sufficient for the signature forgery of a given $y \in \mathbb{F}^m$ if we know **one** subspace W for which $y \in W$. Therefore, we do not need to know all q^v subspaces but only a small number for forging any given signature $x \in \mathbb{F}^n$ with high probability.

In order to search for points which are in the same subspace, we use the following observation: if the 3 points $R_1, R_2, R_3 \in \mathbb{F}^n$ are in the same affine subspace with respect to UOV, the following condition has to be satisfied:

$$UOV(R_1) - UOV(R_2) - UOV(R_3) + UOV(-R_1 + R_2 + R_3) = 0. \quad (1)$$

```
Input:   point R₁, public key 𝒫 of UOV
Output:  A pair (R₁, R₂) of points which belong to the same affine subspace
repeat
    pass ← 0
    trials ← 0
    R₂ ← Random(𝔽ⁿ)
    δₓ ← −R₁ + R₂
    repeat
        trials ← trials + 1
        R₃ ← Random(𝔽ⁿ)
        R₄ ← δₓ + R₃
        δ_y ← UOV(R₁) − UOV(R₂) − UOV(R₃) + UOV(R₄)
        if (δ_y = 0) then pass ← pass + 1
    until (pass > threshold) or (trials > qᵛ · threshold)
until (pass > threshold) or (trials > qᵛ · threshold)
OUTPUT (R₁, R₂)
```

Fig. 5. Algorithm to find a pair of points in the same affine subspace for which UOV is affine.

Using this property, we can determine points of the same affine subspace repeating the heuristic algorithm described in Figure 5 several times. The corresponding algorithm for even characteristic has been described in [29].

Repeating this algorithm often enough for a fixed point R_1, we obtain $(o+1)$ linearly independent points of one affine subspace. The complexity of the algorithm will be roughly $O(q^{2v})$, according to the probability that R_1, R_2 and R_3 belong to the same affine subspace.

This attack can be improved using the relation

$$UOV(R_1) + UOV(R_2) - UOV(R_1 + R_2) = b \qquad (2)$$

for some fixed $b \in \mathbb{F}^m$. As soon as we find a triple $(R_1, R_2, R_3) \in (\mathbb{F}^n)^3$ of points which yield $\delta_y = 0$ in Algorithm 5, we use (2) to check if all of them yield the same constant b. If this is the case, we can conclude with probability q^{-2m} that all three points belong to the same subspace. At this point, we can change to another algorithm: instead of checking triples, we now check pairs. If the pair (R_1, R') yields the constant b, we found a new candidate belonging to the same subspace as R_1. Using the other points found so far, we can increase the probability that R' is genuine further by q^{-m} with each point we try. We summarise this algorithm:

1. Find a triple $(R_1, R_2, R_3) \in (\mathbb{F}^n)^3$ which satisfies (1).
2. Using this triple and (2), determine the value of the constant $b \in \mathbb{F}^m$.
3. Use (2) to find more points $R' \in \mathbb{F}^n$ in the same subspace.
4. As soon as $(o+1)$ points $R \in \mathbb{F}^n$ are known, determine the value of the matrix A by Gaussian elimination.

The running time of this algorithm is $O(q^{2v} + (n-v)q^v)$ on average as we chose the points R_2 and R_3 independently from the point R_1 in the first step and R' also independently from R_1. The overall running time to find a total of $(o+1)$ points in the same subspace becomes therefore $O(q^{2v})$ as $O(oq^v)$ is negligible in comparison to $O(q^{2v})$.

We are able to speed up Algorithm 5 from Section 3.4 if we can spend some memory and also have $m > v$, i.e., we do have "enough" equations in relation to the dimension v of the affine subspaces to be found. This is certainly not true for UOV – here we have typically $m < v$ or even $m < 2v$ (see above). However, for other multivariate quadratic systems, this condition may hold. In particular, it is the case for System B of Matsumoto-Imai, cf [29]. We therefore present two ways of speeding up Algorithm 5. We explain it for the example of UOV to simplify the discussion but want to stress that it also works against System B or any other multivariate quadratic system which has affine approximations of small dimension.

Triple-Algorithm If we can spend $O(kq^{2v})$ of memory for some small k (e.g., $10 \leq k \leq 20$), we can achieve a time/memory-tradeoff for finding **all** subspaces in UOV by using the following technique. In the precomputation phase, we evaluate random pairs $(R_1, R_2) \in_R \mathbb{F}^n \times \mathbb{F}^n$ using (2). The probability for each of these pairs to have points in the same affine subspace is q^{-v} (birthday paradox). Moreover, we know that two points in the same subspace will yield the same constant $b \in \mathbb{F}^m$. On the other hand, two points which are not in the same subspace will yield a random value $v \in \mathbb{F}^m$. The probability for each of these values to occur is q^{-m} with $m > v$. As we were dealing with a total of kq^{2v} pairs, we do not expect two random values $v_1, v_2 \in \mathbb{F}^n$ to occur more often than, say, $\frac{k}{2}$ times. Therefore, all values occurring more often than $\frac{k}{2}$ are constants b with very high probability. Checking the points in the corresponding pairs using (1), we can even distinguish pairs of different subspaces which yield the same constant b. After this precomputation step, we can check for each point $R' \in \mathbb{F}^n$ to which of the q^v subspaces it belongs, using $O(q^v)$ computations on average. After $O(oq^v)$ trials, we have $(o+1)$ points for each subspace and can therefore determine the matrix $A \in \mathbb{F}^{n \times n}$ and the vector b for the affine equation $F(x) = Ax + b$. The above algorithm can be summarised as follows:

1. Use Equation 2 on kq^{2v} random pairs $(R_1, R_2) \in_R \mathbb{F}^n \times \mathbb{F}^n$ and store triples $(b, R_1, R_2) \in \mathbb{F}^m \times (\mathbb{F}^n)^2$
2. Check for each value $b_i \in \mathbb{F}^m$ how often it occurs in the stored list
3. For values b_i which occur at least $\frac{k}{2}$ times, use (1) to check whether the corresponding triples belong to the same affine subspace.
4. Use (2) to determine more points $R' \in \mathbb{F}^n$ for each of these subspaces.

The overall running time of this algorithm is $O(q^{2v})$. However, the drawback is that we need an amount of memory that grows exponentially with $2v$. Therefore, it seems to be advisable to use the following algorithm $O(q^v)$ times instead. This leads to the same overall running time but requires less memory, namely only $O(q^v)$.

Pair-Algorithm Using a similar idea, we can also reduce the running time for finding the corresponding subspace W for **one** given point $R_1 \in \mathbb{F}^n$. However, we need $O(kq^v)$ memory for some small k, e.g., $10 \leq k \leq 20$. In this setting, we evaluate pairs (R_1, R_2) for randomly chosen $R_2 \in_R \mathbb{F}^n$ and store the corresponding triples $(b, R_1, R_2) \in \mathbb{F}^m \times (\mathbb{F}^n)^2$. With a similar argument as for the previous algorithm, we expect a random distribution for the values $b_i \in \mathbb{F}^m$ – except if the pair (R_1, R_2) for given R_1, R_2 is in the same vector space W. This event occurs with probability q^{-v}. Therefore, we can assume that the correct value b will occur k times on average and with very high probability at least $\frac{k}{2}$ times. As soon as we have found this value b, we can look for more values R' which satisfy (2). The overall running time of this algorithm is $O(kq^v)$ for the first step and $O(oq^v)$ for the second step, i.e., $O(q^v)$ in total. However, the drawback is that we need an amount of memory that grows exponentionally with v.

Both speed-ups do no longer work for $v, m = \frac{n}{2}$ as the "gap" between q^{-v} and q^{-m} no longer exists. Therefore, we cannot distinguish anymore between values b and random values.

The advantage of the affine approximation attack against UOV is that we know exactly the structure of these affine subspaces. In addition, all these affine subspaces are disjunct. This was not the case for System B from Matsumoto-Imai [16]. Theoretical predictions were therefore more difficult.

4 Conclusions

In this paper, we studied the security of the public key signature scheme "Unbalanced Oil and Vinegar" which has been proposed by Kipnis, Patarin, and Goubin in [12] and extended in [13]. We studied its resistance against a modified Gröbner basis attack and concluded that the case $2m < v < 4m$ is particularly vulnerable. In addition, we demonstrated that the choice of parameters in [13, Sect. 14] for Example 4 is insecure under an attack from the previous paper [12]. Moreover, we implemented and simulated an attack using Gröbner bases against the other parameter sets described in [13, Sect. 14]. We conclude that they allow a security-level of 2^{64}, as claimed in the paper. However, as we did not have access to the algorithm F_5 [7], we recommend to be cautious as this algorithm is expected to have a rather small running time, therefore, its effect on UOV should be studied more carefully.

In addition, we showed that the constant part of the affine transformation S does not contribute to the overall security of UOV – at least not for attacks which recover the private key.

Finally, we described a new attack against cryptosystems which have small affine subspaces and applied it against UOV. In particular, parameters with q^v small are shown to be very vulnerable against this type of attack. The attack is very elegant and the occurrence of affine subspaces is a very natural property. We therefore expect it to be efficient against other multivariable cryptographic schemes which have a high number of affine subspaces.

Acknowledgements

We want to thank Jacques Patarin (University of Versailles, France) for fruitful discussions about UOV and pointing out the algorithm from Meier and Tacier [2] to us. In addition, we want to thank Willi Meier (FH Aargau, Swiss) for answering our questions about the Meier-Tacier algorithm.

This work was supported in part by the Concerted Research Action (GOA) GOA Mefisto 2000/06, GOA Ambiorix 2005/11 of the Flemish Government and the European Commission through the IST Programme under Contract IST-2002-507932 ECRYPT. The first author was mainly supported by the FWO.

Disclaimer

The information in this document reflects only the author's views, is provided as is and no guarantee or warranty is given that the information is fit for any particular purpose. The user thereof uses the information at its sole risk and liability.

References

1. Computational Algebra Group, University of Sydney. *The MAGMA Computational Algebra System for Algebra, Number Theory and Geometry.* http://magma.maths.usyd.edu.au/magma/.
2. Nicolas Courtois, Louis Goubin, Willi Meier, and Jean-Daniel Tacier. Solving underdefined systems of multivariate quadratic equations. In *Public Key Cryptography – PKC 2002*, volume 2274 of *Lecture Notes in Computer Science*, pages 211–227. David Naccache and Pascal Paillier, editors, Springer, 2002.
3. Nicolas Courtois, Louis Goubin, and Jacques Patarin. *Quartz: Primitive specification (second revised version)*, October 2001. https://www.cosic.esat.kuleuven.ac.be/nessie/workshop/submissions/quartz v21-b . zip, 18 pages.
4. Nicolas Courtois, Louis Goubin, and Jacques Patarin. $SFlash^{v3}$, *a fast asymmetric signature scheme – Revised Specificatoin of SFlash, version 3.0*, October 17th 2003. ePrint Report 2003/211, http://eprint.iacr.org/, 14 pages.
5. Nicolas T. Courtois, Magnus Daum, and Patrick Felke. On the security of HFE, HFEv- and Quartz. In *Public Key Cryptography – PKC 2003*, volume 2567 of *Lecture Notes in Computer Science*, pages 337–350. Y. Desmedt, editor, Springer, 2002. http://eprint.iacr.org/2002/138.
6. Jean-Charles Faugère. A new efficient algorithm for computing Gröbner bases (F_4). *Journal of Pure and Applied Algebra*, 139:61–88, June 1999.
7. Jean-Charles Faugère. A new efficient algorithm for computing Gröbner bases without reduction to zero (F_5). In *International Symposium on Symbolic and Algebraic Computation – ISSAC 2002*, pages 75–83. ACM Press, July 2002.
8. Jean-Charles Faugère and Antoine Joux. Algebraic cryptanalysis of Hidden Field Equations (HFE) using gröbner bases. In *Advances in Cryptology – CRYPTO 2003*, volume 2729 of *Lecture Notes in Computer Science*, pages 44–60. Dan Boneh, editor, Springer, 2003.

9. Michael R. Garay and David S. Johnson. *Computers and Intractability – A Guide to the Theory of NP-Completeness*. W.H. Freeman and Company, 1979. ISBN 0-7167-1044-7 or 0-7167-1045-5.
10. Louis Goubin and Nicolas T. Courtois. Cryptanalysis of the TTM cryptosystem. In *Advances in Cryptology – ASIACRYPT 2000*, volume 1976 of *Lecture Notes in Computer Science*, pages 44–57. Tatsuaki Okamoto, editor, Springer, 2000.
11. Masao Kasahara and Ryuichi Sakai. A construction of public key cryptosystem for realizing ciphtertext of size 100 bit and digital signature scheme. *IEICE Trans. Fundamentals*, E87-A(1):102–109, January 2004. Electronic version: http://search.ieice.org/2004/files/e000a01.htm#e87-a,1,102.
12. Aviad Kipnis, Jacques Patarin, and Louis Goubin. Unbalanced oil and vinegar signature schemes. In *Advances in Cryptology – EUROCRYPT 1999*, volume 1592 of *Lecture Notes in Computer Science*, pages 206–222. Jacques Stern, editor, Springer, 1999. Extended version: [13].
13. Aviad Kipnis, Jacques Patarin, and Louis Goubin. Unbalanced oil and vinegar signature schemes – extended version, 2003. 17 pages, citeseer/231623.html, 2003-06-11, based on [12].
14. Aviad Kipnis and Adi Shamir. Cryptanalysis of the oil and vinegar signature scheme. In *Advances in Cryptology – CRYPTO 1998*, volume 1462 of *Lecture Notes in Computer Science*, pages 257–266. Hugo Krawczyk, editor, Springer, 1998.
15. Aviad Kipnis and Adi Shamir. Cryptanalysis of the HFE public key cryptosystem. In *Advances in Cryptology – CRYPTO 1999*, volume 1666 of *Lecture Notes in Computer Science*, pages 19–30. Michael Wiener, editor, Springer, 1999. http://www.minrank.org/hfesubreg.ps or http://citeseer.nj.nec.com/kipnis99cryptanalysis.html.
16. Tsutomu Matsumoto and Hideki Imai. Algebraic methods for constructing asymmetric cryptosystems. In *Algebraic Algorithms and Error-Correcting Codes, 3rd International Conference, AAECC-3, Grenoble, France, July 15-19, 1985, Proceedings*, volume 229 of *Lecture Notes in Computer Science*, pages 108–119. Jacques Calmet, editor, Springer, 1985.
17. Tsutomu Matsumoto and Hideki Imai. Public quadratic polynomial-tuples for efficient signature verification and message-encryption. In *Advances in Cryptology – EUROCRYPT 1988*, volume 330 of *Lecture Notes in Computer Science*, pages 419–545. Christoph G. Günther, editor, Springer, 1988.
18. Alfred J. Menezes, Paul C. van Oorschot, and Scott A. Vanstone. *Handbook of Applied Cryptography*. CRC Press, 1996. ISBN 0-8493-8523-7, online-version: http://www.cacr.math.uwaterloo.ca/hac/.
19. T. Moh. A public key system with signature and master key function. *Communications in Algebra*, 27(5):2207–2222, 1999. electronic version at http://citeseer/moh99public.html.
20. Jacques Patarin. Cryptanalysis of the Matsumoto and Imai public key scheme of Eurocrypt'88. In *Advances in Cryptology – CRYPTO 1995*, volume 963 of *Lecture Notes in Computer Science*, pages 248–261. Don Coppersmith, editor, Springer, 1995.
21. Jacques Patarin. Asymmetric cryptography with a hidden monomial. In *Advances in Cryptology – CRYPTO 1996*, volume 1109 of *Lecture Notes in Computer Science*, pages 45–60. Neal Koblitz, editor, Springer, 1996.

22. Jacques Patarin. Hidden Field Equations (HFE) and Isomorphisms of Polynomials (IP): two new families of asymmetric algorithms. In *Advances in Cryptology – EUROCRYPT 1996*, volume 1070 of *Lecture Notes in Computer Science*, pages 33–48. Ueli Maurer, editor, Springer, 1996. Extended Version: http://www.minrank.org/hfe.pdf.
23. Jacques Patarin. The oil and vinegar signature scheme. presented at the Dagstuhl Workshop on Cryptography, September 1997. transparencies.
24. Jacques Patarin and Louis Goubin. Trapdoor one-way permutations and multivariate polynomials. In *International Conference on Information Security and Cryptology 1997*, volume 1334 of *Lecture Notes in Computer Science*, pages 356–368. International Communications and Information Security Association, Springer, 1997. Extended Version: http://citeseer.nj.nec.com/patarin97trapdoor.html.
25. Peter W. Shor. Polynomial-time algorithms for prime factorization and discrete logarithms on a quantum computer. *SIAM Journal on Computing*, 26(5):1484–1509, October 1997.
26. Ilia Toli. Cryptanalysis of HFE, June 2003. arXiv preprint server, http://arxiv.org/abs/cs.CR/0305034, 7 pages.
27. Christopher Wolf, An Braeken, and Bart Preneel. Efficient cryptanalysis of RSE(2)PKC and RSSE(2)PKC. In *Conference on Security in Communication Networks – SCN 2004*, Lecture Notes in Computer Science, September 8–10 2004. 14 pages.
28. Bo-Yin Yang and Jiun-Ming Chen. Rank attacks and defence in Tame-like multivariate PKC's. Cryptology ePrint Archive, Report 2004/061, 23rd March 2004. http://eprint.iacr.org/, 21 pages.
29. Amr M. Youssef and Guang Gong. Cryptanalysis of Imai and Matsumoto scheme B asymmetric cryptosystem. In *Progress in Cryptology – INDOCRYPT 2001*, volume 2247 of *Lecture Notes in Computer Science*, pages 214–222. C. Pandu Rangan and Cunsheng Ding, editors, Springer, 2001.

Hold Your Sessions:
An Attack on Java Session-Id Generation

Zvi Gutterman and Dahlia Malkhi

School of Engineering and Computer Science,
The Hebrew University of Jerusalem,
Jerusalem 91904, Israel
{zvikag,dahlia}@cs.huji.ac.il

Abstract. HTTP session-id's take an important role in almost any web site today. This paper presents a cryptanalysis of Java Servlet 128-bit session-id's and an efficient practical prediction algorithm. Using this attack an adversary may impersonate a legitimate client. Through the analysis we also present a novel, general space-time tradeoff for secure pseudo random number generator attacks.

Keywords: pseudo random number generators, space-time tradeoff, HTTP, web security

1 Introduction

At the root of many security protocols, one finds a secret seed which is supposedly generated at random. Unfortunately, truly random bits are hard to come by, and as a consequence, often security hinges on shaky, low entropy sources. In this paper, we reveal such a weakness in an important e-commerce building block, the Java Servlets engine.

Servlets generate a session-id token which consists of 128 hashed bits and must be unpredictable. Nevertheless, this paper demonstrates that this is not the case, and in fact it is feasible to hijack client sessions, using a few legitimately-obtained session-id's and moderate computing resources.

Beyond the practical implication to the thousands [16] of servers using Servlets, this paper has an important role in describing an attack on a pseudo-random-number-generator (PRNG) based security algorithm and in demonstrating a nontrivial reverse engineering procedure. Both can be used beyond the Servlets attack described henceforth.

Web server communication with clients (browsers) often requires state. This enables a server to "remember" the client's already visited pages, language preferences, "shopping basket" and any other session or multi-session parameters. As HTTP [9] is stateless, these sites need a way to maintain state over a stateless protocol. Section 2 describes various alternatives for implementing state over HTTP. However, the common ground of all these schemes is a token traversing between the server and the client, the session-id.

The session-id is supported by all server-side frameworks, be it ASP, ASP.net, PHP, Delphi, Java or old CGI programming. Session-id's are essentially a random value, whose security hinges solely on the difficulty of predicting valid session id's. HTTP session hijacking is the act where an adversary is able to conduct a session with the web server and pretend to be the session originator. In most cases, the session-id's are the only means of recognizing a subscribing client returning to a site. Therefore, guessing the unique session-id of a client suffices to act on its behalf.

Driven by this single point of security, we set out to investigate the security of session-id's deployments, and as our first target, we have analyzed the generation of session-id's by Apache Tomcat. Apache [2] is an open-source software projects community. The Apache web server is the foundation's main project. According to Netcraft [16] web study of more than 48,000,000 web servers, the Apache web server is used by more than 67% of the servers and hence the most popular web server for almost a decade.

At the time of this writing (April 2004), sites such as www.nationalcar.com, www.reuters.com, www.kodak.com and ieeexplore.ieee.org are using Java Servlets on their production web sites.

In many of these sites, the procedure for an actual credit-card purchase requires a secure TLS [8] sessions, separated from the "browsing and selection" session. However, this is not always the case. For example, Amazon's patented [10] "one-click" checkout option permits subscribing customers to perform a purchase transaction within their normal browsing session. In this case, the server uses a client's credit-card details already stored at the server, and debits it based solely on their session-id identification.

In either of these scenarios, an attacker that can guess a valid client id can easily hijack the client's session. At the very least, it can obtain client profile data such as personal preferences. In the case of a subscriber to a sensitive service such as Amazon's "one-click", it can order merchandize on behalf of a hijacked client.

Briefly, our study of the generation of Java Servlets' session-id's reveals the following procedure. A session-id is obtained by taking an *MD5* hash over 128-bits generated using one of Java's pseudo-random number generators (PRNG). Therefore, two attacks can be ruled out right away. First, a brute force search of valid session-id's on a space of 2^{128} is clearly infeasible. Second, various attacks on PRNGs, e.g., Boyar's [6] attack on linear congruential generators, fail because PRNG values are hidden from an observer by the *MD5* hashing.

Nevertheless, we are able to mount two concrete attacks. We first show a general space-time attack on any PRNG whose internal state is reasonably small, e.g., 2^{64}–2^{80}. Our attack is resilient to any further transformation of the PRNG values, such as the above *MD5* hashing. Using this attack, we are able to guess session-id's of those Servlets that use the java.util.Random package, whose internal PRNG state is 64-bits. Beyond that, our generic PRNG attack is the first to use space-time tradeoffs, and may be of independent interest.

Our second attack is on the seed-generation algorithm of Java Servlets. Using intricate reverse engineering, we show a feasible bound for the seed's entropy. Consequently, we are able to guess valid session-id's even when Servlets are using the java.security.SecureRandom secure PRNG (whose internal state is 160 bits).

The paper is organized as follows. In Section 2 we describe the HTTP state mechanisms. In Section 3 we describe and analyze the Tomcat session-id generation algorithm. Java hashCode() study is presented in Section 4. In section 5 we present our attacks on the session-id. We conclude in Section 6.

2 Stateful Web Browsing

HTTP is a client/server protocol designed for a light-weight and quick delivery of content from servers to clients. HTTP is stateless, in that a server responds to a client's request with a hypertext page and then breaks down the connection. Any additional request from the same client requires the client to build a new, seemingly unrelated connection with the server. Statelessness is part of what makes HTTP efficient and fast to implement.

However, a typical client/server interaction entails repeated interaction. For example, often a web page contains links to images and multi-media objects. Obtaining each one of these is done in a separate TCP/IP connection to the server, but they appear to be part of a single prolonged interaction. The new HTTP standard [9] (HTTP 1.1) is already in place, allowing multiple retrievals instead of a single one. Nevertheless, it is not meant to keep connections up through an involved client/server interaction, which could span multiple screens and forms. And it does not address clients returning to the same site after days have passed.

Cookies [13] change this situation. Introduced originally by Netscape and thereafter adopted widely and as part of HTTP 1.1, cookies were designed with the intention of solving the vexing problem of keeping long-lived relationships between web servers and their clients. Cookies extend the HTTP protocol by allowing a server to hand a client certain information to keep. The client's browser automatically hands the server this information, the cookie, the next time it connects to the same site. Cookies are used by servers to store a variety of information, from client membership identification to complete shopping basket contents. They greatly enhance the web browsing experience, allowing a client to be recognized by a server, accumulate shopping selections, and so on.

An analog mechanism to cookies is URL rewriting. In this framework, instead of sending a fixed web page to the client the web server encodes the session information as part of the page, e.g., within embedded URL links. URL rewriting requires less from the client side, but as far as this paper is concerned is the same session mechanism and our attack is equally applicable to it.

From a privacy point of view, it should be noted that the cookies mechanism and likewise, URL re-writing were designed to prevent leakage of information between sites, in that a cookie is returned only to the site that originally sent it. In this way, a server may only obtain information that it already had about a user. Unfortunately, there are examples of cookie-abuse, e.g., the infamous doubleclick.com site, that collects client clicking-profile through its advertisements on partner sites.

This work, however, is concerned with a different weakness of cookies, and more generally, with stateful web browsing. True, recognizing a returning client through cookies alleviates the need to tediously re-type a user name and a password upon each connection establishment to a site. Unfortunately, it also poses a web-identity theft potential: If one can guess a valid cookie, one can impersonate another client. As simple as that.

There is hardly a limit to what an attacker may obtain through such identity theft: She may be able to learn private user data, such as names and addresses. She could collect clients' profile information, such as preferences and shopping history. She could penetrate access protected sites. In a particularly vicious attack, using Amazon's "one-click" option, she might be able to order merchandize on behalf of impersonated customers. Essentially, there are limitless hazards.

3 Tomcat Session-Id Generation Algorithm

In this section, we describe our study of Tomcat 5 [1], the Apache Java implementation for Servlet 2.4 [19] and JSP 2.0 [14] specifications. We study version 5.0.18, which was released on January 2004. Our full study involves additional, and more challenging reverse-engineering of relevant modules of the JVM which are written in native-code. This part is deferred to the next section.

The remainder of this section describes the Tomcat session-id generation scheme, which includes two parts. One is a session-id allocation used during the set up of each new session. The second is an initialization phase that is executed once when the server comes up. We hint about potential weaknesses as we go along. The description omits unimportant implementation details such as irrelevant Java **class** names.

3.1 Allocation

We begin by examining the algorithm for generating new sessions-id's during the set up of new sessions. Session-id's are allocated within method generateSessionId(), and consists of 16 bytes, or equivalently, 128 bits.

Inside generateSessionId(), the allocation consists of the following steps:

1. Method getRandomBytes fills a sixteen bytes array. If /dev/urandom exists the bytes are read from it. If not, a Java pseudo-random number generator (PRNG) is invoked. Method getRandom() is invoked to obtain a handle either to Java.Security.SecureRandom or Java.Util.Random. Figures 1,2 presents these functions.
2. The 16 bytes obtained from getRandomBytes are mixed using a digest function which is *MD5* [18] by default.
3. The result is the 128-bit session-id. For convenience, it is converted into 32 ASCII characters, where each 4 bits are mapped to a matching character between '0' ... 'F'.

$$x_n := \begin{cases} \text{initial seed} & n = 0 \\ (25,214,903,917 \times x_{n-1} + 11) \bmod (2^{48} - 1) & n > 0 \end{cases}$$

Fig. 1. java.util.Random. x_n holds the PRNG next output.

$$x_n := SHA1(s_n) \qquad n = 0$$

$$s_n := \begin{cases} \text{initial seed} & n = 0 \\ (x_{n-1} + s_{n-1} + 1) \bmod 2^{160} & n > 1 \end{cases}$$

Fig. 2. java.security.SecureRandom. x_n holds the PRNG next output and s_n is the internal state.

3.2 Initialization

Given that generateSessionId() employs a Java PRNG for allocating session-id's, the next thing to investigate is how it is initialized inside the Tomcat package. We had initially hoped to find a simple weakness, e.g., initialization by a hard-wired constant, which would render session-id's easily predictable. Such weakness were found frequently in the past, e.g, [12].

That is not the case here, and the seeding of the PRNG within Tomcat is an intricate, thoughtful process, consisting of the following steps.

1. Set C = System.currentTimeMillis(). This is 64 bit field measuring the time since January 1, 1970 in milliseconds.
2. Set *Entropy* = toString(org.apache.catalina.session.ManagerBase.java). The value of *Entropy* is equal to the Java String org.apache.catalina.session. ManagerBase.java@X. The prefix of the term is always the same, and the part following the @ sign is variable. Section 4 describes our study of the X value and how we can predict it.
3. Set $Seed = f(C, Entropy)$. The function f is depicted in Figure 3. It takes the *Entropy* and spreads it byte by byte (letter by letter), with 8 bytes per row (or 64 bits per row). It computes a xor of all the rows, xor'ed also with C, yielding a 64-bit value.
4. *Seed* is used for initializing the PRNG.

Despite the intricate seeding process above, this is the important part where our attack will take place. As we show below, we can indeed predict with reasonable effort the *Seed* value. As all other steps are deterministic and known from the server code, once we find the *Seed* we can predict each session-id value. This will be later presented in Section 5.

4 Java Object.toString() Algorithm

The Java Object.toString() function is used by the initialization algorithm presented in Section 3 for generating the PRNG seed. In this section, we take a

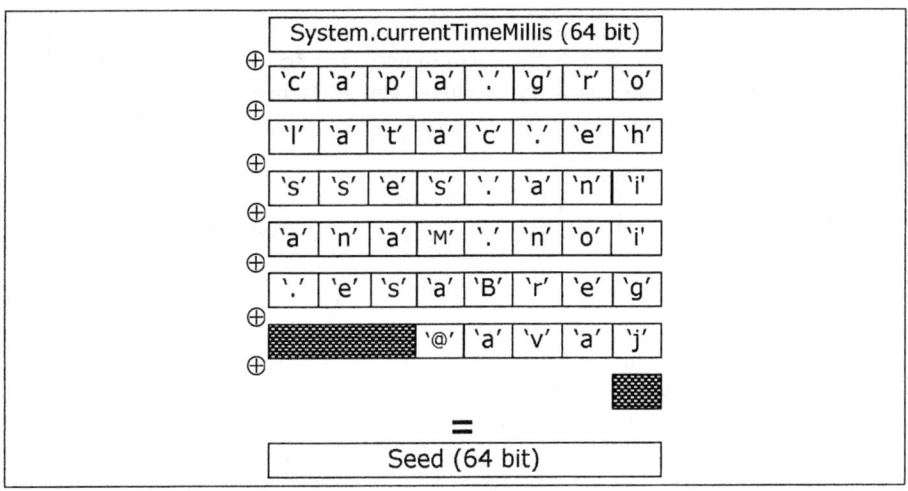

Fig. 3. The function $f()$ employed to generate a seed out of C, the time in milliseconds, and *Entropy*, a string containing org.apache.catalina.session.ManagerBase.java@X, where X is the hashCode 32 bit field marked as scattered area.

close look at Object.toString(), and show that this value is actually a very low entropy source.

The Java Object method toString returns the value
getClass().getName()+"@"+Integer.toHexString(hashCode()); Hence, the returned string has a fixed prefix, which is the class name, followed by the @ sign and a 32 bit field which is the result of the method hashCode.

The function java.lang.Object.hashCode() is a native one, which requires each Java virtual machine implementor to bring its own implementation.

According to the Java documentation the hashCode method must have the following properties.

1. Whenever it is invoked on the same object more than once during an execution of a Java application, the hashCode method must return the same integer (32 bit).
2. If two objects are equal they return the same hashCode
3. it is not required that two object which are not equal return distinct values of hashCode.
4. As much as is reasonably practical, the hashCode method defined by class Object does return distinct integers for distinct objects (this is typically implemented by converting the internal address of the object into an integer, but this implementation technique is not required by the Java programming language).

It is important to note here that reading the Java documentation may lead the reader (and maybe also the Tomcat implementor) to think that the hashCode is hard to predict.

However, this is not always the case. In particular, the Microsoft Windows platform [15] does not follow the recommendation to use the pointer address space in generating the hashCode. Instead the JVM uses a linear congruential generator (LCG) to get the different hash codes. Using the IDA-Pro [7] disassembler we get

$$hashCode(object) := \begin{cases} (a \times x_n + b) \mod m & \text{first object access} \\ \text{hash is given from a history table} & \text{otherwise} \end{cases} \quad (1)$$

We can now predict the hashCode value using the LCG values. What we need to know is the server boot sequence where our object will be called. This information should usually be available for an attacker, which in most cases can deploy the same server and verify the class loading sequence. Even when this procedure is hard to perform an adversary can narrow the valid range into 256 possible values with only few trials. This brings the Java hashCode into 8 entropy bits or less, which is far lower entropy than the presumed 32 bits and will take part in our general attack scheme.

5 Attacks

We remind the reader that the goal of an attacker in our settings is to predict legitimate session-id's that are allocated to clients, and impersonate these clients over HTTP connections with servers.

We describe two attacks. The first one is a generic attack on any PRNG whose internal state is feasibly small, e.g., 2^{64}–2^{80}. The second is an attack on the seeding procedure of Java Servlets.

5.1 Space-Time Tradeoffs for PRNG Attacks

A space-time tradeoff attack is the notion of using a large space of pre-computed values in order to reduce the time of an online attack. Ours is the first general space-time tradeoff on secure PRNG based protocols. In the following, we first present the general attack and then tailor it for the session-id's case. Our attack is a direct adaptation of a space-time tradeoff attack on stream-ciphers, recently demonstrated by Biryokuv and Shamir in [4]. For completeness, we first introduce space-time tradeoffs for block and stream ciphers.

Background. A block cipher space-time attack lets an adversary tune the values of memory M and online attack time T for a given key space K of size $N = |K|$. Hellman [11] introduced this method with a $TM^2 = N^2$ tradeoff.

Hellman's space-time tradeoff block cipher attack is made of two parts. We first conduct a pre-computation stage to set the memory tables, with computation cost $P = N$. The second stage includes the online attack. Given a cipher-text y the online stage returns the key $k \in K$ such that $y = E_k(p)$, where E is the encryption function and p is a pre-chosen plain text.

The pre-computation includes building several tables of chains as follows. For the first element in the chain, we first randomly select a key $k^0 \in K$. The second chain element is $k^1 = R(E_{k^0}(p))$, where $R(y) \in K$ is an arbitrary reduction function which maps a cipher text block to a valid key value. The reduction function can be simple truncation, or a selection of $|k|$ bits from the cipher text y, but as explained below, it is important that R is uniformly distributed over K.

A chain of length t contains repeated invocations of $R(E_{k^i}(p))$. We mark SP and EP as the start and end points. The resulting length t chain, with reduction function $R()$ is as follows:

$$SP := k^0 \longrightarrow k^1 := R(E_{k^0}(p)) \longrightarrow \ldots \longrightarrow EP := k^{t-1} := R(E_{k^{t-2}}(p)) \quad (2)$$

The goal is to cover K with the different chains and with low or no collisions at all. Each chain starts with a different SP, and we assume that the application of $R(E_k())$ over the initial random starting points is like a random selection of elements from K.

We can repeat the chain building procedure and make m such chains. In order to complete our attack these chains must cover K. However, some collisions will occur, i.e., a chain will occasionally reach a key that already appears in a previous chain. Once such a collision occurs, the remainder of the chain, which is computed in a deterministic way, will repeat the same, already computed chain. Furthermore, when existing chains cover as little as N/t out of the N elements, the probability for collision in the next t elements is a non negligible constant.

Hellman suggested to solve the collision problem using r different reduction functions $R_1, \ldots, R_r$. Each reduction function is chosen as a different selection of $|k|$ bits out of the cipher text y. For each reduction function we build a table with m chains, each of length t, such that $mt = N/t$ (the point beyond which producing additional chains is wasteful). The different reduction functions ensure that even when an element occurs in two different tables, the next element in the chains will be different in the two tables, hence the total number of collisions is low.

The assignment of m, t and r such that $mtr \geq N$ solves both our collision and coverage concerns. In the rare occasion that during our pre-computation two chains end with the same EP we select the longer chain.

An additional important technique which can improve the table lookup performance is due to Rivest. Instead of stopping after t steps we can stop at a *Distinguished Point* which is a point with some easy to verify property, e.g., all its $\log_2 t$ first bits are zero. As $R_i(E_k(p))$ is distributed uniformly, the average chain length will be t. In this way, instead of looking up each key value in the pre-computed endpoints, we will only need to look for values which are *Distinguished Points*.

Here, care should be taken to avoid loops. When building a chain, there is a small probability of a loop, in which case we may never reach a distinguished point. In this rare event we just keep any such loop chain. The additional computational and storage complexities are negligible.

The second part of the space-time attack is the online attack. At this stage we assume that r tables with m different chains, each of length t were computed and stored. Each such chain is stored as a pair of SP and EP.

Given a cipher text y' we can now find a key k' such that $E_{k'}(p) = y'$ as follows. The idea is to find the chain in which y' appears, and then find y''s predecessor in the chain, which is k'. We locate the chain by setting $k^0 = R_i(y)$, and then repeatedly applying $R_i(E_{k^j}(p))$ with the r different reduction functions. Once getting to a distinguished point we look it up in the i-th table. If matched, we found the chain represented as SP,EP. We can now repeat the $R_i(E_{k^j}(p))$ invocation starting from SP, until we find k' such that $y' = E_{k'}(p)$.

Neglecting logarithmic factors, we can conclude Hellman's space-time attack for block ciphers with online cost $T = tr$ (though only r expensive table lookups), space $M = mr$, pre-computation $P = trm = N$. Together, these yield $TM^2 = N^2$.

Hellman's attack can be quite practical. In fact, Oechslin demonstrates in [17] a very feasible implementation of Hellman's space-time attack for breaking Windows passwords. That work is based on the fact that the key space is rather small, 2^{37}, and on the fact that Windows password encryption uses the password to encrypt a fixed known plain text.

That said, Hellman's method has two main drawbacks. The first is the pre-computation cost, which is equal to the entire key space size N. The second is that it is a chosen plain-text attack. All the table values were computed using a chosen plain text and are relevant only for attacking that plain text cipher.

Recently, Biryokuv and Shamir [4] extended space-time attacks for stream ciphers. A stream cipher works as a state machine that is initialized with a secret key and outputs a keystream sequence that contains bits from the internal state of the machine. Encryption consists of xor-ing the keystream bits with the plain text. Once we find a correct state of the stream cipher machine, not necessarily the initial key or the first state but **any** state, the remainder of the stream cipher output is predictable. Hence, the search space K is no longer the initial key space but rather the internal stream cipher state. That is, given a state s of the stream cipher, the next keystream $k(s)$ (of some pre-determined length) produced by the stream cipher is determined. Now, given a known plain-text p' and its cipher-text c', we can determine whether $k(s)$ is the key producing the cipher and conclude that the stream-cipher's internal state is s. This may be done for any known plain-text, not a specifically chosen plain-text p as before.

Hellman's attack framework presented above is used in a similar way here with one important change. The chain step maps an internal state of the stream cipher into the appropriate keystream it generates, and from the keystrem is reduces back using a reduction function to an internal state. The rest of the parameters – N, m, t, r and the distinguished points can be used in the same way.

When working on stream ciphers, Biryukov and Shamir explain how the two main drawbacks for block ciphers are solved. Cipher stream encryption is used as a one time pad for the plain text. Therefore, given any exposed plain-text, we recover the keystream with which it is encrypted. This keystream is the same for a given internal state of the streamcipher, regardless of the plaintext it encrypts. Given an exposed cipher text, we first (trivially) find the keystream that encrypts it, and then we attempt to recover the stream-cipher's internal state that results in this keystream. Hence, this is a **known** plain text attack and not a chosen plain text attack as in the block cipher case. The distinction is huge, since we can use a one-time preparation stage for all future attacks on the stream-cipher.

We can also use this fact to reduce the search space using multiple known plain-texts. Let us denote the number of exposed cipher texts given to the adversary by D. Since every exposed cipher text (equivalently, every keystream) corresponds to some unknown internal state of the stream cipher, we can find one of the keystreams with good probability if we cover only N/D of the states space. Thus, if an adversary can expose D cipher texts, it is enough to pre-compute only N/D of the states space. We therefore set $r = t/D$ instead of $r = t$, and compute only r different tables.

The space-time tradeoff for stream ciphers can now be written as time $T = Dtr = t^2$ (as in Hellman's attack), space $M = mt/D$, where $mt^2 = N$ which is better than before, and likewise the pre-computation $P = N/D$ is lower. We get a tradeoff of $TM^2D^2 = N^2$, which is much better than the block-cipher tradeoff of $TM^2 = N^2$.

Session-Id's Space-Time Tradeoffs Attacks on pseudo random generators can be addressed in a similar way to stream ciphers, thus we attack the PRNG internal state using a space-time attack. Below, we demonstrate the attack using the specific example of the Tomcat session-id generation algorithm. However, the same principles can be applied for other uses of the bits produced by a PRNG.

We can describe a PRNG as a state machine with states $x_1, x_2, x_3, \ldots$. In any state x_n, some bits are made available as output, and then the PRNG shifts to state x_{n+1}. Consequently, there is a deterministic sequence of bits $b_1, b_2, b_3, \ldots$ produced by the PRNG from any particular state x_n onward. For example, in java.util.Random(), the bits produced by the LCG state x_n are x_n itself. We denote $f(x_n)$ the deterministic 128-bit sequence produced by the PRNG from state x_n. The Tomcat session-id is generated as follows:

$$y := session\_id := MD5(f(x_n)) \qquad (3)$$

Although the *MD5* transformation (or any other transformation, for that sake) effectively masks the values of the PRNG, we do not need to break *MD5* in order to predict session-id's. The session-id generation algorithm is deterministic and has no additional entropy sources along the algorithm. In this sense, our PRNG algorithm is similar to the stream-cipher where the encryption is based on the internal state cipher. Once we break any session-id value and reverse it to its state value x_n we can generate the entire series of next values.

Assume for the sake of demonstration that states are 64 bit values. The space-time attack we employ targets the "key space" K of PRNG internal states. Thus, $N = |K| = 2^{64}$.

We denote the transformation of Equation 3 by F. Given a value y, our goal is to find x such that $x = F^{-1}(y)$. We do this with a time-space tradeoff as follows. The start-point of chains are m randomly selected values k representing states of the PRNG. The chaining step from k_i to k_{i+1} is the transformation F followed by reduction functions R_j, $j = 1..r$. We use for R_j a truncation and a simple xor in order to reduce the 128 bits F values into a 64 bits internal PRNG states: $R_i(y_{0-127}) := y_{0-63} \oplus i$ where $i \in \{1 \ldots r\}$. As before, we maintain r tables, each containing m chains, and each terminating with a distinguished end-point (e.g., whose lowest $\log_2 t$ bits are zero). For each chain, we store only the start and the end points.

Suppose we are able to obtain D distinct valid session-id's. In practice, collecting session-id's from a working web-server is easy, and even a large number of sessions requested by the same client over a short time frame may not raise suspicion. Note that, these session-id's need not be consecutive, which is important in the framework of current distributed clients accessing a web server.

Our attack is then mounted as follows: For each of the D known session-id's y, and for $j = 1..r$, apply $R_j(F())$ repeatedly until a distinguished point is reached, and search for it in the j'th pre-made table. If found, then go back to the start point, and reach the state x_i such that $F(x_i) = y$. From state x_i onward, the session-id's generated by this server are predictable.

Letting $r = t/D$ as in the stream cipher attack, we obtain a tradeoff of $P = N/D$ pre-computation time, space $M = mt/D$ where $mt^2 = N$, and on-line computation time $T = t^2$. This yields $TM^2D^2 = N^2$.

For concrete numbers, we assume that it is possible to obtain $D = 1000$ valid session id's without raising suspicion. We put $t = 2^{22}$. Then our space of $N = 2^{64}$ PRNG states can be broken with storage $M = 2^{64-22-10} = 2^{32}$, and an on-line computation time $T = t^2 = 2^{44}$, both very feasible today with a moderately powerful workstation.

5.2 The Seed Attack

Some installations of Java Servlets use the java.security.SecureRandom PRNG, rather than java.util.Random. As outlined in Section 4 above, SecureRandom has an internal state of 160 bits. Hence, the general PRNG attack we described so far is not feasible against it. Here, we attack the protocol using another weakness, a low-entropy seed.

According to the description in Section 3, the space of seeds for the PRNG is determined by combining the range of possible clock readings in milliseconds (counted from 1970), and a value set by the method hashCode(). A day has about 2^{26} milliseconds and a year has about 2^{35}. Hence, the entropy of this value is between 26 to 35, depending on how accurately we can estimate a server's uptime. As for the value of hashCode(), Our reverse engineering of this method constrains it to within a small set of values, typically less than 128 different ones. Thus, the effective total range size of seeds is bounded between 2^{33} and 2^{42}. Certainly this is a space that can be searched exhaustively with a moderate computation power, especially if the uptime of a server is estimated relatively accurately.

While this is a weakness of the session-id generation algorithm, in itself it does not lead to a practical attack. The difficulty is in verifying the correctness of a guessed seed. The naive way is to involve the server. That is, one can guess a seed value, generate one or several "session-id's" originating with the seed value, and attempt to "hijack" a customer session with this session id. As this procedure involves an interaction with the server for each guessed value, even for for a space of 2^{32} values it is very time consuming. Moreover, it would be very easy to detect that such an attack is going on at the server side. The server can protect itself against repeated connection attempts from the same domain over a short period of time by slowing down its response or refusing recurring attempts, and thus thwart the entire attack.

Our strategy is therefore to mount an almost entirely off-line attack as follows:

1. Get a valid session-id by connecting to the attacked web server. Mark this valid session-id as Sid.
2. Set T as an upper limit for the server uptime, since the last reboot. The value is in milliseconds.
3. Set $hash\_min, hash\_max$ as the lower and upper limit on the JVM hashCode(). Mark $\Delta_{hash} = hash\_max - hash\_min$.
4. Set sid_min, sid_max as the minimal and maximal number of valid session-id's assigned so far by the attacked server. Mark $\Delta_{sid} = sid\_max - sid\_min$.
5. Generate all the possible session-id's using all the possible $T \times (hash\_max - hash\_min)$ seeds, and for each potential seed, producing $(sid\_max - sid\_min)$ session-id's. Compare Sid against this space, until a valid seed is revealed.

The above ignores the variability that different architectures and JVM versions may have in generating hashCode() values. If that is not known by the attacker, this should incur a multiplicative factor over the range of possible hashCode() values.

In the above attack, the size of the potential sessions-id's space is 2^E, where the exponent E is given by the following sum:

$$E = \log_2(T) + \log_2(\Delta_{hash}) + \log_2(\Delta_{sid}) \qquad (4)$$

If we take fairly conservative values, a server up-time of a month, hash values range 128, and valid session-id range 32,000 we get $E = 29 + 7 + 15 = 51$. This is certainly a searchable space.

6 Conclusions

This paper presents a practical attack on one of today's main E-commerce building blocks, the session-id. Our attack shows that the presumably secure 128 bits can be broken using 2^{64} or less computation steps. Our attack can be mounted using limited computing resources, and has the same communication fingerprint of a legitimate user accessing the attacked web server. Hence, it is difficult for a server to detect and stop such an ongoing attack.

We implemented the attack and tested it under distilled environment conditions. In our case, we set up a Tomcat server and obtained session-id's from it. We staged our attack on the same machine, so any uncertainty about Java versions and platforms was completely alleviated. Given the session-id's we obtained, we were able to predict the PRNG sequence within a day of CPU time. We did not try our attack on working servers to avoid legal complications.

Beyond the attack on session-id generation, we present a general scheme with a space-time tradeoff for attacking pseudo random number generators. To the best of our knowledge, this is the first space-time tradeoff for PRNG attacks. The attack may have important ramifications on presumably secure uses of PRNGs, such as BlumBlumShub [5], and emphasizes the need for deploying these with a large internal state.

This paper proves again a common cryptographers' knowledge. The complexity of a security scheme does not make it secure; nor is it made secure by using building blocks such as one way functions and secure pseudo random number generators.

It is important to note that Tomcat bring web server administrator the option to harden the session-id generation. The simple option is to add secret entropy to the seed. Other options require either using a different random number generator or a different session-id scheme.

The Tomcat web server is an open-source project. As such, it is an easy target for analysis, through both dynamic and static reverse engineering. The equivalent "binary only" attack requires more sisyphean work, usually through the low level assembly code. In a sense, this is the Achilles' heel for the security aspects of open source code. We believe that this is true only for the short term. In the long term, an open source project can benefit from a large audience testing its security, while closed projects might wrongly be presumed secure just because their study is complex. One such example is the GSM encryption scheme, which was considered secure for long, but was recently proven not so [3].

Acknowledgments

The authors wish to thank Tzachy Reinman and Yaron Sella for reviewing early drafts of the paper.

References

1. Apache Software Foundation (ASF). Apache jakarta tomcat. http://jakarta.apache.org/Tomcat.
2. Apache Software Foundation (ASF). Apache web server. http://www.apache.org.
3. E. Barkan, E. Biham, and N. Keller. Instant ciphertext-only cryptanalysis of gsm encrypted communication. In *Proceedings of CRYPTO'2003, LNCS 2729*, pages 600–616, 2003.
4. A. Biryukov and A. Shamir. Cryptanalytic time/memory/data tradeoffs for stream ciphers. In *Lecture Notes in Computer Science 1976, proceedings of ASIACRYPT'2000*, pages 1–13, 2000.
5. L. Blum, M. Blum, and M. Shub. A simple unpredictable pseudo-random number generator. 15:364–383, 1986.

6. J. Boyar. Inferring sequences produced by a linear congruential generator missing low-order bits. *Journal of Cryptology*, 1(3):177–184, 1989.
7. Datarescue. Ida: The interactive disassembler. http://www.datarescue.com/idabase/.
8. T. Dierks and C. Allen. The TLS protocol version 1.0. RFC 2246, Internet Engineering Task Force, January 1999.
9. R. Fielding, J. Gettys, J. C. Mogul, H. Frystyk, L. Masinter, P. J. Leach, and T. Berners-Lee. Hypertext transfer protocol – HTTP/1.1. RFC 2616, Internet Engineering Task Force, June 1999.
10. Hartman. Method and system for placing a purchase order via a communications network, September 1999. U. S. patent 5,960,411.
11. M. E. Hellman. A cryptanalytic time-memory trade off. *IEEE Trans. Inform. Theory*, IT-26:401–406, 1980.
12. M. Heuse. Websphere cookie and session-id predictability, 10 2001. http://www.securiteam.com/windowsntfocus/6Q0020K2UU.html.
13. D. Kristol and L. Montulli. HTTP state management mechanism. RFC 2965, Internet Engineering Task Force, October 2000.
14. M. Roth. JSR 152: JavaServer PagesTM 2.0 Specification, November 2003. http://jcp.org/aboutJava/communityprocess/final/jsr152/index.html.
15. Sun Microsystems. The java virtual machine version 1.4.2. http://java.sun.com/j2se/1.4.2/index.jsp.
16. Netcraft. Market share for top servers across all domains august 1995 - march 2004. http://news.netcraft.com/archives/web_server_survey.html.
17. P. Oechslin. Making a faster crytanalytical time-memory trade-off. In *Advances in Cryptology - CRYPTO 2003*, volume 2729 of *Lecture Notes in Computer Science*, Santa Barbara, California, USA, August 2003. 23rd Annual International Cryptology Conference, Springer. ISBN 3-540-40674-3.
18. R. Rivest. The MD5 message-digest algorithm. RFC 1321, Internet Engineering Task Force, April 1992.
19. Y. Yoshida. JSR-000154 JavaTM Servlet 2.4 Specification (Final Release), November 2003. http://jcp.org/aboutJava/communityprocess/final/jsr154/index.html.

Update on SHA-1*

Vincent Rijmen[1,2] and Elisabeth Oswald[1]

[1] IAIK, Graz University of Technology,
Inffeldgasse 16a, A-8010 Graz, Austria
{vincent.rijmen,elisabeth.oswald}@iaik.tugraz.at
[2] Cryptomathic A/S,
Jægergårdsgade 118, DK-8000 Århus C, Denmark

Abstract. We report on the experiments we performed in order to assess the security of SHA-1 against the attack by Chabaud and Joux [5]. We present some ideas for optimizations of the attack and some properties of the message expansion routine. Finally, we show that for a reduced version of SHA-1, with 53 rounds instead of 80, it is possible to find collisions in less than 2^{80} operations.

Keywords: hash functions, cryptanalysis

1 Introduction

In [5], Chabaud and Joux presented a method to find collisions for the original Secure Hash Standard (here denoted by SHA-0). We present here the results of our attempts to apply their attack to SHA-1, as well as some extensions to the approach described in [5]. For a good understanding of our results, it is recommended to study [5] very carefully. Space restrictions do not permit us to copy all the important details of the original attack.

In the case of SHA-0, the message expansion shows a certain weakness, which allows to reduce the search space for difference patterns to a size which makes exhaustive search possible. This weakness has been fixed in SHA-1 and consequently, it was necessary to design and implement more intelligent searching algorithms.

Furthermore, we investigated the use of alternative linear approximations for the non-linear functions. We also optimized the equation solving step, which allows to solve larger systems of equations. Finally, we analyzed in much detail the complexity of an attack on a version of SHA-1 reduced to 53 rounds.

The appendix lists some expanded message words with very low weight, which we can't use in an attack.

In parallel to our research, Biham and Chen [1] improved the complexity of the Chabaud-Joux attack on SHA-0. The same authors announce forthcoming results on SHA-1 in [2]. Saarinen describes attacks on block ciphers based on SHA-1 in [6].

---
* This research was supported financially by the A-SIT, Austria and by the BSI, Germany.

2 SHA-0 and SHA-1

The SHA family of hash functions is described in [4]. Briefly, the hash functions consist of two phases: a message expansion and a state update function. These are explained in more detail in the following. SHA-0 and SHA-1 share the same state update, but SHA-0 has a simpler message expansion. Both SHA-0 and SHA-1 consist of 80 rounds. Because we will mainly study reduced versions here, we make the number of rounds variable, and denote it by R.

2.1 Message Expansion

In SHA-1, the message expansion is defined as follows. The input is a 512-bit message, denoted by a row vector m. The message is also represented by 16 32-bit words, denoted by M_t, with $t = 0, 1, \ldots, 15$.

In the message expansion, this input is expanded linearly into R 32-bit words W_t, also denoted as the $32R$-bit expanded message word w. The words W_t are defined as follows.

$$W_t = M_t, t = 0, \ldots 15 \quad (1)$$
$$W_t = (W_{t-3} \oplus W_{t-8} \oplus W_{t-14} \oplus W_{t-16}) \lll 1, t > 15 \quad (2)$$

The message expansion of SHA-0 is very similar, but uses:

$$W_t = W_{t-3} \oplus W_{t-8} \oplus W_{t-14} \oplus W_{t-16} \; . \quad (3)$$

Consequently, a bit at a certain position i in one of the words of w only depends on the bits at corresponding positions in the words of m.

2.2 State Update Transformation

The state update transformation starts from a (fixed) initial state for 5 32-bit registers and updates them in R steps, using one word W_t in every step. Figure 1 illustrates one step of the state update function. The function f depends on the round number: rounds 1 to 20 use the *IF-function*, rounds 41 to 60 use the *MAJ-function*.

$$f_{\text{if}}(X, Y, Z) = XY \oplus \overline{X}Z \quad (4)$$
$$f_{\text{maj}}(X, Y, Z) = XY \oplus XZ \oplus YZ \quad (5)$$

The remaining rounds use 3-input XOR. A round constant K_t is added in every round. There are four different constants; one for rounds 1 to 20, one for rounds 21 to 40, one for rounds 41 to 60 and one for rounds 61 to 80. After the last application of the state update transformation, the initial register values are XOR-ed to the final values, and the result is outputted.

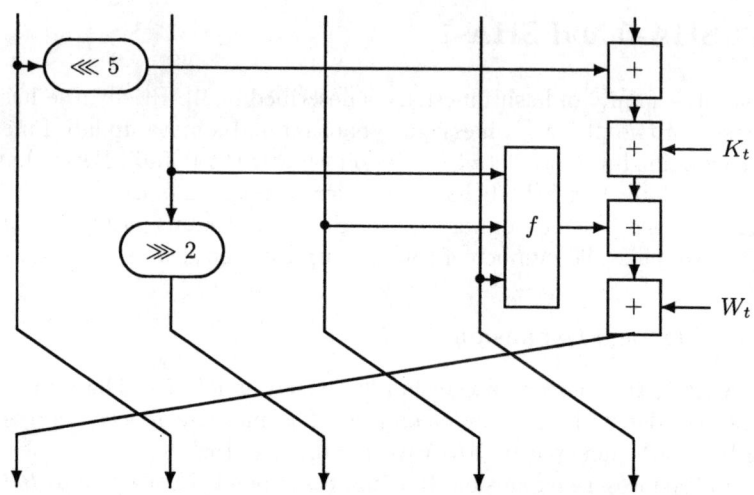

Fig. 1. One step of the state update function of SHA-1.

3 The Basic Attack Strategy

The attack on SHA-0 can be summarized as follows [5].

1. Firstly, a linear approximation of SHA-0 is constructed.
2. Secondly, collisions for the linear approximation are determined.
3. Thirdly, a collision for the real SHA-0 is searched among the collisions for the linear approximation.

We now discuss each of these steps and point out the differences between SHA-0 and SHA-1 relevant to the attack. The next sections also go into more detail for step 2 and step 3.

3.1 Determining a Linear Approximation

In this step, we replace all non-linear components by linear approximations. For our purposes, a linear function λ is a 'good' approximation for a non-linear function γ if the relation

$$\gamma(x \oplus \delta) \oplus \gamma(x) = \lambda(x \oplus \delta) \oplus \lambda(x) = \lambda(\delta) \tag{6}$$

holds for relatively many values of x and δ. The complexity of the third step of the attack is influenced by the quality of the approximation.

Both in SHA-0 and SHA-1, there are 3 non-linear components. Firstly, there is the addition modulo 2^{32}. This component is approximated by a bitwise exclusive-or of the inputs, i.e. the carry is ignored.

Next are the functions $f_{\text{if}}(X, Y, Z)$ and $f_{\text{maj}}(X, Y, Z)$. The functions operate bitwise, hence the approximation should also be a bitwise function. The authors of [5] approximate both functions by a bitwise exclusive-or of the 3 inputs. Since the f-function used in half of the rounds is exactly given by this exclusive-or, this approximation results in 80 iterations of the same round. On the other hand, the quality of this approximation seems sub-optimal. We discuss some ideas for improvements in Section 4.

3.2 Finding Collisions for the Linear Approximation

Finding a collision for a linear approximation of SHA-1 is not difficult. Whether two messages will produce a collision or not, doesn't depend on the value of the messages, but depends on the value of their difference only. Collision-producing values of the difference can be found as the solutions of an under-determined set of linear equations.

The difficulty lies in the additional constraints imposed by the third step of the attack. We want to find a collision that minimizes the work of the third step. The explanation of the third step will show that the most important requirement is that the weight of the difference should be small.

The problem of finding a collision-producing difference with small weight can be translated to the problem of finding a codeword with small weight in a linear code.

Since the message expansion is linear, there is a $512 \times 32R$ matrix E such that $w = mE$. The message expansion starts with a copy of the message, cf. (1). Hence, there is a $512 \times 32(R-16)$ matrix F such that E can be written as:

$$E_{512 \times 32R} = \begin{bmatrix} I_{512 \times 512} & F_{512 \times 32(R-16)} \end{bmatrix} . \tag{7}$$

For the linearized state update transformation, we can construct a $32R \times 160$ matrix A that produces the output vector o from the expanded message w.

$$o = wA = mEA \tag{8}$$

A message word m corresponds to a collision-producing difference if and only if $o = 0$. Hence, the set of collision-producing differences is a linear code with check matrix $H_{160 \times 512} = (EA)^t$. The dimension of the code is $k = 512 - 160 = 352$. The length of the code is $n = 512$. Later on, it becomes more useful to look at the expanded words w, which are code words of a code with $k = 352$, $n = 32R$ and check matrix

$$H'_{(32R-352) \times 32R} = \begin{bmatrix} A^t|_{160 \times 32R} \\ F^t|_{(32R-512) \times 512} \; I_{(32R-512) \times (32R-512)} \end{bmatrix} . \tag{9}$$

In [5] an additional condition is imposed on the differences. The differences are constructed as the sum of a *perturbation word* w_p and 5 *correction words* $w_{c,i}$, $i = 1, 2, 3, 4, 5$. The authors require that w_p and the 5 $w_{c,i}$ are all codewords. It can be seen that this is a sufficient, but not a necessary condition for the sum to be a codeword as well.

This restriction of the search space corresponds to adding 160 rows to the check matrix of the code. It allows to zoom in quickly on the optimal solutions in the case of SHA-0. In the case of SHA-1, our experiments indicate that this restriction on the search space leads to suboptimal results (cf. Table 4). Hence, we can't apply the perturbation-correction technique. A small side-effect is that the weights we quote in this paper, are always weights for the full codeword, which makes it difficult to compare them to the results in [5], where the authors quote the weights of the perturbation words only.

3.3 Finding the Collisions for the Real SHA-1

We now have a difference that produces collisions in the linear approximation of SHA-1. In the non-linear components of SHA-1, the propagation of the differences may be the same as in the linear case, or it may be different, depending on the value of the message bits.

In order to find a collision, we want to find the values of the inputs such that the difference propagation in SHA-1 corresponds to the difference propagation in the linear approximation. This condition results in the equations that we have to solve in order to find the collision.

For example, consider the addition of a register A_t and a word W_t of the expanded message. Let the result be denoted by B_t. Assume now that in both inputs, the differences are equal to zero except for one bit, at position i. It is clear that in the linear approximation of the addition, the result has difference zero. In B_t, the result of the real addition, the difference will be zero for the bits with positions $0, 1, \ldots, i$. The difference at positions above i will be zero if and only if the carry into position $i+1$ has difference zero. Writing down the equations for the carry results in the following requirement on the value of the operands: the bits at position i in A_t and W_t should be of opposite value.

If we look at the equations generated during the attack, we typically get groups of several equations involving the same state bit(s). It is then often possible to rework some of the equations and obtain linear equations involving bits of the expanded message words only. Linear equations in the expanded message words can easily be translated into conditions on the message words. They can easily be solved.

Besides the additions, also the approximations of f_{if} and f_{maj} result in conditions. These conditions can be derived from Table 1. If the functions are approximated by 3-input XOR, then the first 3 and the 7th equation have to be copied from the table (output difference equal to 1 is desired), while the 4th, 5th and 6th have to be inverted (output difference equal to 0 is desired). The conditions result in equations involving bits from one, two or three registers. Some special attention should go to the 4th equation for f_{if}, respectively the 7th equation for f_{maj}. For instance, when f_{if} is approximated by 3-input XOR, this input difference is problematic. It was this observation that led us to the idea of considering other linear approximations. This is discussed in Section 4.

Table 1. Conditions to have output difference equal to 1.

$f_{\text{if}}(x,y,z)$	
δ	equation
001	$x = 0$
010	$x = 1$
100	$y + z = 1$
011	always
101	$x + y + z = 1$
110	$x + y + z = 0$
111	$y + z = 0$

$f_{\text{maj}}(x,y,z)$	
δ	equation
001	$x + y = 1$
010	$x + z = 1$
100	$y + z = 1$
011	$y + z = 0$
101	$x + z = 0$
110	$x + y = 0$
111	always

4 Other Linear Approximations

4.1 Motivation

Approximation of both f_{if} and f_{maj} by 3-input XOR has as main advantage that the resulting approximation for SHA-1 has 80 identical rounds. There appear to be also a couple of disadvantages related to this choice:

1. For $\delta = 011$, the output bit of the linear approximation flips with probability 0, but the output bit of f_{if} flips with probability 1. Hence there is no input pair that produces the same behavior in f_{if} and the approximation.
2. 3-Input XOR has good diffusion properties (avalanche effect). Other linear functions on the 3 inputs have worse diffusion properties.

The first property complicates the search for a suitable collision in the linear approximation of SHA-1, because the situation where $\delta = 011$, has to be avoided. The restriction of the search space corresponds to the addition of non-linear conditions. The second property makes it more difficult to prevent a difference from expanding; more equations have to be added. Therefore we considered also other linear approximations, which produce different equations, that can also be generated by copying and inverting equations from Table 1.

Table 2 lists for both f-functions and the 7 possible linear functions with 3 inputs the probability that the output bit flips for the given input difference. By definition, the output flip probability for a linear function is either 0 or 1; it can be computed by simply evaluating $f(\delta)$.

The quality of the approximations differs only for the values $\delta = 011$ and $\delta = 111$. For these values of δ, the output bit of f_{if}, respectively f_{maj}, flips with probability 1. The first disadvantage explained above can be avoided by selecting a linear approximation that does not have output bit flip probability equal to 0 when the non-linear function has output flip probability equal to 1. The second criterion for the quality of an approximation is the diffusion: an approximation with bad diffusion is more likely to result in low-weight collision-producing differences.

For the function f_{if}, the approximations y, z, $x \oplus y$ and $x \oplus z$ appear to be better choices. For the function f_{maj}, the approximations x, y and z have equally good flip probabilities as $x \oplus y \oplus z$, and less avalanche.

Table 2. The probability that the output bit changes value when the input bits are changed according to the input difference, for the two f-functions and for all the linear approximations.

δ	output flip probability						
	f_{if}	f_{maj}	linear functions				
	$xy \oplus \bar{x}z$	$xy \oplus xz \oplus yz$	$x\ y\ z$	$x \oplus y$	$x \oplus z$	$y \oplus z$	$x \oplus y \oplus z$
000	0	0	0 0 0	0	0	0	0
001	1/2	1/2	0 0 1	0	1	1	1
010	1/2	1/2	0 1 0	1	0	1	1
011	1	1/2	0 1 1	1	1	0	0
100	1/2	1/2	1 0 0	1	1	0	1
101	1/2	1/2	1 0 1	1	0	1	0
110	1/2	1/2	1 1 0	0	1	1	0
111	1/2	1	1 1 1	0	0	0	1

4.2 Results

Despite the expected improvements in the search for low-weight codewords, the results obtained with alternative linear approximations turn out to be inferior. We tried out replacing the approximations by any of the other linear functions with the same or better flip probabilities. For versions with more than 25 rounds, we never obtained better results than with the original approximation.

We can think of two possible explanations. The use of alternative approximations for the Boolean functions results in an approximation for SHA-1 that has two different round transformations (at least), since the 40 rounds using the 3-input XOR clearly can't be approximated by another linear function. Hence we obtain a linear code with less regularity.

As we explain in Section 5, we use heuristic algorithms to search for low-weight codewords. A first explanation would be that the decrease in regularity causes an increase in the minimum distance of the code. But this almost implies that the round transformation of SHA-1 would show some kind of weakness, which results in a lower minimum distance of the corresponding linear code. An alternative explanation is that the decrease in regularity makes the heuristic search algorithms perform worse: the low-weight words are still there, but we can't find them. In that case, perhaps better search algorithms can be found.

5 Searching for Low-Weight Codewords

There is no fast, deterministic algorithm known that can find low-weight codewords in arbitrary linear codes. Different approaches are possible:

1. Exhaustive search.
2. Apply heuristic techniques that can be used to find low-weight codewords in random linear codes, e.g. [3].
3. Exploit the structure of the code and obtain an analytical solution.

In the case of SHA-0, it is possible to define a restricted search space that is very likely to contain the best codewords. Since the restricted search space has dimension 2^{16}, exhaustive search is possible. For SHA-1, it seems impossible to define a search space small enough to allow an exhaustive approach.

It seems that the dimensions we are dealing with here, are still out of reach for the algorithms discussed in [3]. Secondly, as follows from Section 4, we are clearly not in the situation of a purely random code.

The best strategy seems to combine the second and the third approach. For instance, we know that the codewords are produced by an LFSR. If $w = (W_0, W_1, \ldots)$ is a codeword, then also $(\text{rot}(W_0), \text{rot}(W_1), \ldots)$ is a codeword. More specific knowledge of the LFSR allowed us to define other strategies resulting in words with a weight probably very close to the minimal weight for $R < 50$. For larger values of R, the problem is still open.

5.1 Our Search Algorithm

Our heuristic search algorithm is based on an observation that resulted from experiments on SHA-1 versions with $R \leq 25$. First, we introduce the following notation. Let the bitwise OR operation be denoted by $\vee$, then we define the following shorthand notation.

$$W_\vee = \bigvee_{i=0}^{R-1} W_i \qquad (10)$$

Observation 1. *For codewords $w = (W_0, W_1, \ldots, W_{R-1})$ with low Hamming weight, the Hamming weight of $W_\vee$ is low. In other words: codewords with low Hamming weight have the property that the non-zero bits usually occur at the same positions in all the words W_i.*

The observation was derived from experiments, but we believe it is also in agreement with intuition. Differences introduced in the state have to be compensated for and eventually canceled. This requires that the differences in the expanded message words occur in 'bands'. Algorithm 1 uses the observation to perform an accelerated search. The results obtained with the algorithm are shown in Table 3. Further restriction of the search space is possible by using Observation 2.

Observation 2. *The non-zero bits in $W_\vee$ occur at consecutive positions, or 'almost' consecutive positions.*

By 'occur in almost consecutive positions,' we mean that there are at most two runs of ones, separated by a run of one or two zeroes. Motivated by this observation, we remove the inner for-loop of Algorithm 1, which results in an important speedup. Algorithm 2 uses a parameter u, which denotes the sum of the lengths of the runs of ones and the one or two zeroes in between.

By starting with the large values of u and moving to the lower ones, we save on the operations needed to compute the new check matrix of the code and its rank. The algorithm starts an exhaustive search when the dimension of the code

Table 3. Minimal weights of collision-producing codewords for reduced versions of SHA-1. Produced with Algorithm 1.

R	Hwt($W_\vee$)	Hwt(w)
20	3	18
25	6	34
30	6	38
31	6	38
32	6	38
33	6	38
34	6	38
35	8	76
36	8	76
37	8	80
38	8	100
39	8	112
40	10	128

gets below a parameter D. Algorithm 2 was executed for various numbers of rounds. In the cases where there were several values for u that resulted in a code with less than 2^D codewords, it was always the case that the codeword with the lowest weight was among those with the smallest u. The results are presented in Table 4.

In order to optimize the complexity of the attack, the first 15 message words are pre-computed such that the conditions in the first 15 rounds are satisfied. Hence, the weight in the first 15 rounds is not relevant. Therefore, the results in Table 4 do not take into account the first 15 rounds (neither in Hwt(w), nor in u). The results indicate that a shortcut attack finding a collision is feasible for versions reduced to 35–40 rounds. In Section 6, we examine the complexity of the attack for a version reduced to 53 rounds.

5.2 Observations on the Message Expansion

We applied Algorithm 2 also to the naked message expansion: dropping the condition to have a collision and simply searching for low-weight expanded message words. The results are given in Table 5. Since the words are obtained with a heuristic algorithm, it is not proven that these are the best words. Indeed, in Appendix A, we give three 80-rounds expanded message words with Hamming weight 44 (including the first 15 rounds), obtained by other means. Nevertheless, the values in the table give some indication about the 'penalty' in additional weight coming from the requirement to produce a collision.

For all the words listed in this paper, it can be observed that the rounds that contribute the most to the total weight, are situated at the beginning and at the end. Furthermore, the rounds with the lowest weight aren't situated exactly in the middle, but slightly more towards the end of the word. This can be explained by the fact that the diffusion of the message expansion goes slower in the backwards direction.

Algorithm 1

Input: H /* check matrix of the code */
n /* length of the code */

For $h = 1$ to 32 do
  For all values of W_V with Hamming weight h do
    Copy H to H_e
    Extend H_e with $R \times (32 - h)$ rows corresponding to the conditions on
    the bits of $W_0, W_1, \ldots$ at the positions where $W_V = 0$
    If $\text{rank}(H_e) < n$ then
      Perform an exhaustive search for low-weight codewords
      Output the word with lowest weight and exit

Algorithm 2

Input: H /* check matrix of the code */
n /* length of the code */
D /* exhaustive search space bound */

For $u = 32$ to 1 do
  Extend H with R rows corresponding to the conditions on
  the bits of $W_0, W_1, \ldots$ at position $h - 1$
  If $n - \text{rank}(H) < D$ then
    Perform an exhaustive search for low-weight codewords
    Output the word with lowest weight

Table 4. Hamming weights of the codewords with smallest weights, for reduced versions of SHA-1. Produced with Algorithm 2. The weight of the first 15 rounds is not taken into account. The fourth and the fifth column list the results obtained when the search space is restricted to the *perturbation-correction* codewords used with success on SHA-0 in [5]. The second and the third column give the results for an unrestricted search space.

	full search space		restricted space	
R	u	$\text{Hwt}(w)$	u	$\text{Hwt}(w)$
35	6	35	16	127
40	8	67	17	178
45	8	81	17	215
50	8	83	18	258
51	8	86		
53	8	95		
54	10	145		
55	10	157		
60	12	167		
65	12	226		
70	12	276		
75	13	278		
80	14	333	21	552

Table 5. Weight of low-weight codewords for the message expansion of SHA-1. The weight includes the weight of the first 15 rounds. Produced with Algorithm 2.

R	u	Hwt(m)
50	3	20
60	3	31
70	4	41
80	5	51

The diffusion of the message expansion (2) is determined by the feedback polynomial and the rotation. If we visualize the expanded message as a $32 \times R$ rectangle, then the 'influence region' of a bit occurring at position i in round t can be visualized as two triangles meeting at the point (t, i). This is illustrated in Figure 2. In the forward direction, the upper line of the triangle has a slope of 3 rounds per bit. The lower line has a slope of 16 rounds per bit. In the backwards direction, the influence region is bounded by the horizontal line and a line with a slope of 16 rounds per bit. This region expands much slower.

Consequently, a good strategy to find a low-weight codeword is to place a word with weight 1 somewhere in the middle, add 7 or 8 zero words before and after, and compute the other rounds backwards and forwards. Since the diffusion backwards goes slower, it is better to compute more rounds backwards than forwards.

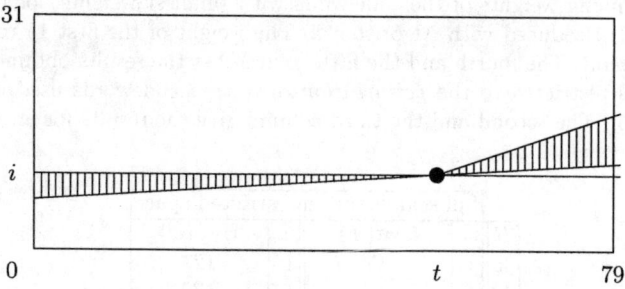

Fig. 2. Diffusion in the SHA-1 message expansion. A bit at position i in round t influences only bits in the shaded triangles.

6 Experiments with a 53-Round Characteristic

Table 4 shows that the weight of our best 54-round codeword is significantly larger than the weight of our best 53-round codeword. Therefore, we decided to use a version of SHA-1 reduced to 53 rounds to apply the third step of the attack: generating (and possibly solving) the equations. The codeword we used, is listed in Table 6.

During the generation of equations, a new problem became apparent. In the original attack on SHA-0, the modular additions have input differences in 0, 1

Table 6. Low-weight codeword for SHA-1 reduced to 53 rounds.

00000000	80000030	00000020
00000000	20000001	80000001
00000000	C0000012	C0000002
40000000	60000041	40000040
00000008	40000032	40000002
40000002	20000003	80000002
90000040	C0000042	80000040
50000011	E0000042	80000002
10000068	E0000002	80000000
E0000002	00000002	80000000
F0000022	00000040	80000000
70000051	80000001	00000000
10000010	80000060	00000000
60000041	80000001	
C0000022	40000042	
80000003	C0000040	
E0000052	40000042	
C0000040	00000000	
E0000052	80000040	
20000003	00000003	

or 2 of the input words. However, our codeword results also in situations where 3 input words have a non-zero difference at the same position. In this situation, a carry to the next position can't be avoided. Hence, there is no input that can behave the same in the linear approximation and in the real SHA-1. An exception to this rule is of course formed by the most significant bit position, where the carry simply overflows. It turns out that we are lucky enough that all the situations with 3 non-zero input differences occur at the same bit position, hence we can choose a rotated version of the codeword where all cases happen at the most significant bit position.

In order to achieve the best results, we decided to avoid the 'IF-rounds' of SHA-1, by defining the start of our reduced version at round 21. Doing this, we get a total of 166 equations. From these equations, we can isolate 62 linear equations in bits of the message words only, leaving 104 equations. 33 equations apply to the first 15 rounds, and can be solved explicitly during the pre-computation phase, leaving 71 equations for the main step. Using the most naive methods for solving non-linear Boolean equations, these equations can be solved with a complexity that is close to but below that of 2^{71} hash function evaluations. Hence, in principle it is possible to find collisions for the reduced version, faster than by the birthday paradox.

7 Conclusions

In this paper we presented the results from our attempts to extend the Chabaud-Joux attack to SHA-1. The application to SHA-1 results in several complications, which were not obvious from the start. We proposed several strategies for optimization of the attack and examined their effectiveness.

As a result, we have described a theoretical shortcut attack on a version of SHA-1 reduced to 53 rounds. The shortcut attack becomes feasible for SHA-1 reduced to 35–40 rounds. It is also clear that we are still far from even a theoretical attack on the full SHA-1.

Furthermore, we presented several observations that came out of our experiments. We hope that they might be of use for other cryptographers trying to break SHA-1.

Acknowledgements

The authors wish to thank Carlos Cajal for assistance with the programming, and Antoon Bosselaers for helpful discussions.

References

1. Eli Biham, Rafi Chen, "Near-Collisions of SHA-0," *Advances in Cryptology – Crypto '04, LNCS*, M. Franklin, Ed., Springer-Verlag, to appear.
2. Eli Biham, Rafi Chen, "Near-Collisions of SHA-0," *Cryptology ePrint Archive, Report 2004/146*, 2004, version of June 22, 2004, http://eprint.iacr.org/.
3. Anne Canteaut, Florent Chabaud, "A new algorithm for finding minimum-weight words in a linear code: application to McEliece's cryptosystem and to narrow-sense BCH codes of length 511," *IEEE Transactions on Information Theory*, Vol. 44, No. 1, January 1998.
4. *Federal Information Processing Standard 180-2, Secure Hash Standard*, August 1, 2002.
5. Florent Chabaud, Antoine Joux, "Differential Collisions in SHA-0," *Advances in Cryptology – Crypto '98, LNCS 1462*, H. Krawczyk, Ed., Springer-Verlag, 1998, pp. 56–71.
6. Markku-Juhani O. Saarinen, "Cryptanalysis of Block Ciphers Based on SHA-1 and MD5," *Fast Software Encryption 2003, LNCS 2887*, T. Johansson, Ed., Springer-Verlag, 2003, pp. 36–44.

A Some Low-Weight Codewords

Below is a codeword for 80 rounds, with weight 51. It is the word corresponding to the entry in Table 5. The weight includes the weight of the first 15 rounds. The ordering is from top to bottom, and then from left to right.

10000000	40000000	40000000	00000000
20000000	40000000	20000000	40000000
00000000	40000000	00000000	00000000
30000000	00000000	00000000	00000000
40000000	00000000	40000000	00000000
41000000	20000000	60000000	00000000
40000000	00000000	00000000	00000000
40000000	50000000	40000000	00000000
10000000	40000000	40000000	00000000
40000000	50000000	00000000	00000000
00000000	00000000	00000000	00000000
30000000	40000000	40000000	00000000
00000000	00000000	00000000	00000000
21000000	60000000	00000000	00000000
40000000	00000000	00000000	00000000
40000000	40000000	40000000	00000000
50000000	40000000	00000000	00000000
20000000	20000000	40000000	80000000
00000000	00000000	00000000	00000000
30000000	40000000	40000000	00000000

The absolutely smallest weight we found for the 80-round message expansion, is 44. We found 3 such codewords, given below. The 3 codewords have a large amount of W_i in common. The first codeword starts top left and ends at the bottom of the fourth column, the second codeword is shifted over 4 W_i's, the third one over a further two. The words have $W_\vee$ = C0000FF, C0001FF, C0003FF and $\text{Hwt}(W_\vee)$ = 10, 11, 12.

80000000	80000000	00000001	00000002	00000050
00000000	00000000	00000000	00000000	00000100
C0000000	00000001	00000001	00000000	00000010
00000001	00000001	00000000	00000004	00000000
00000001	80000000	00000001	00000000	00000210
00000001	00000000	00000000	00000000	00000000
00000000	00000000	00000000	00000008	
00000000	00000001	00000000	00000000	
80000000	80000001	00000000	00000004	
00000000	00000000	00000000	00000010	
40000001	00000001	00000000	00000000	
00000001	00000001	00000000	00000000	
40000001	00000000	00000000	00000020	
00000000	00000000	00000000	00000000	
00000001	00000001	00000000	00000014	
00000000	00000000	00000000	00000040	
80000001	00000000	00000000	0000000C	
00000000	00000000	00000000	00000000	
00000001	00000001	00000000	00000080	
00000001	00000000	00000000	00000010	

A Fast Correlation Attack on the Shrinking Generator*

Bin Zhang[1,2], Hongjun Wu[1], Dengguo Feng[2], and Feng Bao[1]

[1] Institute for Infocomm Research, Singapore
[2] State Key Laboratory of Information Security,
Graduate School of the Chinese Academy of Sciences,
Beijing 100039, P.R. China
zhangbin@mails.gscas.ac.cn
{hongjun,baofeng}@i2r.a-star.edu.sg

Abstract. In this paper we demonstrate a fast correlation attack on the shrinking generator with known connections. Our attack is applicable to arbitrary weight feedback polynomial of the generating LFSR and comparisons with other known attacks show that our attack offers good trade-offs between required keystream length, success probability and complexity. Our result confirms Golić's conjecture that the shrinking generator may be vulnerable to fast correlation attacks without exhaustively searching through all possible initial states of some LFSR is correct.

Keywords: Fast correlation attack, Shrinking generator, Linear feedback shift register.

1 Introduction

The shrinking generator (SG) is a well-known keystream generator proposed in [4] at Crypto'93. It consists of two LFSR's, say LFSR A and LFSR S. Both LFSRs are regularly clocked and the output bit of the generating LFSR A is taken iff the current output bit of the control LFSR S is 1. This generator obtains a kind of implicit non-linearity from the shrinking process, i.e. the exact positions of the remaining bits in the generated keystream become uncertain. It is proved that the generated keystream has many merits in cryptographic sense such as a long period, a desirably high linear complexity and good statistical properties. It is recommended in [4] that both the initial states of the two LFSR's and the feedback polynomials of theirs be secret key. As in [5], we stress here that our analysis is also based on the known feedback polynomials assumption.

So far, several attacks against the shrinking generator have been proposed. A simple divide-and-conquer attack is proposed in [4] requiring an exhaustive search through all possible initial states and feedback polynomials of LFSR S. A

---
* Supported by National Natural Science Foundation of China (Grant No. 60273027), National Key Foundation Research 973 project (Grant No. G1999035802) and National Science Fund for Distinguished Young Scholars (Grant No. 60025205).

correlation attack is proposed in [8] and is experimentally analyzed in [19] which takes an exhaustive search through all initial states and all possible feedback polynomials of LFSR A. At Asiacrypt'98, T. Johansson [12] presented a reduced complexity correlation attack based on searching for specific subsequences of the keystream, whose complexity and required keystream length are both exponential in the length of LFSR A. In 2001, a probabilistic correlation analysis [6] based on a recursive computation of the posterior probabilities of individual bits of LFSR A was conducted by J. D. Golić, which revealed the possibility of implementing certain fast correlation attack on the shrinking generator. A novel distinguishing attack on the shrinking generator is proposed in [5]. According to the facts that an arbitrary weight feedback polynomial of degree L is known to have a weight 4 multiple of degree $O(2^{L/3})$ and $10000 = 2^{13.2877} = 2^{L/3}$ [7, 20], that distinguisher is applicable to arbitrary shrunken LFSR's of length around 40. Very recently, an improved linear consistency attack is presented in [17] which is an completely exhaustive search through all initial states of LFSR S.

In [6], it was conjectured that the shrinking generator *may* be vulnerable to fast correlation attacks that would not require an exhaustive search through all possible initial states of LFSRs. In this paper we try to answer this question definitely even for LFSR A of length 61, as suggested in [9]. We show that given a length of only 140000 keystream bits, the initial state of LFSR A with arbitrary weight feedback polynomial of degree 61 can be recovered with success probability higher than 99% and complexity 2^{56}, which is a good trade-off between these parameters.

This paper is organized as follows. In Section 2 we present a general description of our attack. Deep analysis of our attack is made in Section 3. Experiments results together with comparisons with other attacks on the shrinking generator are provided in Section 4. Finally, conclusions are given in Section 5.

2 A General Description of Our Attack

We first present a general description of our attack. Denote the output sequence of LFSR A by $a = a_0, a_1, \cdots$ and the output sequence of LFSR S by $s = s_0, s_1, \cdots$. The output keystream of (SG) is $z = z_0, z_1, \cdots$. Our attack is composed of two phases: first, correlation analysis phase which results in a sequence $\hat{a} = \hat{a}_0, \hat{a}_1, \cdots$ associated with sequence a by the relation $P(\hat{a}_i = a_i) = \frac{1}{2} + \varepsilon$ with $\varepsilon > 0$; second, fast correlation attack phase which aims at recovering the secret initial state of LFSR A. Here we adopt the BSC (binary symmetric channel) model for fast correlation attack, as shown in Figure 1.

Our main idea is to regard the sequence $\hat{a}$ as the noisy version of sequence a through the binary symmetric channel representing the noise introduced by the shrinking generator, i.e. $1 - p = P(\hat{a}_i = a_i)$, given p as the crossover probability in the BSC. W.l.o.g assume $p < 0.5$. Our aim is to restore sequence a from $\hat{a}$ by efficient fast correlation attack techniques. Note that several new efficient fast correlation attacks on stream ciphers are proposed recently, [2, 3, 15, 16], enabling us to construct an efficient fast correlation attack on the shrinking generator, which is impossible by traditional techniques. In this paper, we follow

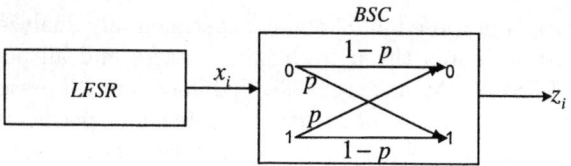

Fig. 1. Model for fast correlation attack.

the method in [3] to mount our attack on the shrinking generator. In nature, our correlation analysis has nothing to do with the decoding algorithm which means other decoding techniques may also be applied, as discussed in Section 4.

The original idea of correlation analysis phase goes back to [21]. We made crucial improvements to the initial method. For simplicity, assume that both the LFSR sequences generated by LFSR A and LFSR S are purely random (a sequence of independent uniformly distributed random variables is called purely random). Consider the probability that z_k equals a_r in the (SG). It is obvious that $k \leq r$. If we regard the event that $s_i = 1$ as success, then the event that z_k equals a_r is equivalent to the event that the kth success of sequence s occurs at the rth trial which obeys the Pascal Distribution. Thus the probability that z_k equals a_r is:

$$P(z_k = a_r) = \binom{r}{k}(\frac{1}{2})^{r+1}. \tag{1}$$

On the other hand, if a_r appears in the keystream z, the following equation holds:

$$a_r = z_{\sum_{i=0}^{r-1} s_i}. \tag{2}$$

When r grows large, the distribution of the sum $\sum_{i=0}^{r-1} s_i$ approximates the Normal Distribution, i.e.

$$\frac{\sum_{i=0}^{r-1} s_i - r/2}{\sqrt{r/4}} \mapsto N(0,1). \tag{3}$$

Let $I_{r/2} = [r/2 - \alpha\sqrt{r/4}, r/2 + \alpha\sqrt{r/4}]$, here comes our main observation: for arbitrary probability p, there exists a α such that whenever a_r appears in keystream z, the following equation holds:

$$P(\sum_{i=0}^{r-1} s_i \in I_{r/2}) = p. \tag{4}$$

As in [5], we formally define two kinds of intuitive notion of imbalance.

Definition 1. *W.l.o.g, we assume the interval $I_{r/2}$ includes odd number of integers. Let $S_0 = \{z_i | i \in I_{r/2}, z_i = 0\}$, $S_1 = \{z_i | i \in I_{r/2}, z_i = 1\}$, the first kind of imbalance of the interval $I_{r/2}$, $Imb_1(I_{r/2})$, is defined as $|S_1| - |S_0|$, where $|\cdot|$ is the cardinality of a set. If $Imb_1(I_{r/2}) \neq 0$, this interval is said to be imbalanced. See Figure 2.*

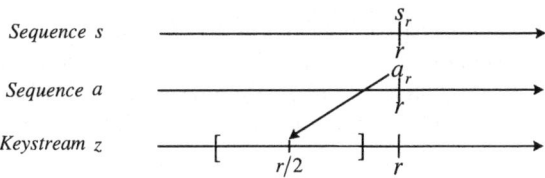

Fig. 2. The interval that a_r probably lies in.

Definition 2. *The notations are the same as those in Definition 1. Let $P_0^{(r)} = \sum_{z_i \in S_0} P(a_r = z_i)$, $P_1^{(r)} = \sum_{z_i \in S_1} P(a_r = z_i)$, the second kind of imbalance of the interval $I_{r/2}$, $Imb_2(I_{r/2})$, is defined as $P_1^{(r)} - P_0^{(r)}$. If $Imb_2(I_{r/2}) \neq 0$, this interval is also said to be imbalanced. See Figure 2.*

Now there are two kinds of construction methods of sequence $\hat{a}$ corresponding to these two kinds of imbalance. The first one is a straightforward majority poll according to Definition 1. The second one is a similar but more reasonable poll according to Definition 2.

Method 1. Following Definition 1, if $Imb_1(I_{r/2}) > 0$, let $\hat{a}_r = 1$. Otherwise, let $\hat{a}_r = 0$.

Method 2. Following Definition 2, if $Imb_2(I_{r/2}) \geq 0$, let $\hat{a}_r = 1$. Otherwise, let $\hat{a}_r = 0$.

Both theoretical analysis and experimental results show that sequence $\hat{a}$ constructed above satisfying $P(\hat{a}_i = a_i) = \frac{1}{2} + \varepsilon$ with $\varepsilon > 0$ as expected. We will give the theoretical analysis in next section and the experimental results in Section 4.

Next, we will present a brief description of the fast correlation attack [3] involved in our attack. This attack is a one-pass correlation attack consisting of two stages: pre-processing stage aiming at the construction of parity-check equations of weight k and processing stage in which a majority poll is conducted for D ($D > L - B$) considered bits other than the first B bits ($x_0, x_1, \cdots, x_{B-1}$) of the initial state ($x_0, x_1, \cdots, x_{L-1}$). In general, there are three new ideas proposed in [3]. First, a match-and-sort algorithm is proposed to construct parity-check equations of the following form with respect to a given considered bit x_i

$$x_i = x_{m_1} \oplus \ldots \oplus x_{m_{k-1}} \oplus \sum_{j=0}^{B-1} c_j x_j \tag{5}$$

where m_j ($1 \leq j \leq k - 1$) denote the indices of the keystream bits and the last sum represents a partial exhaustive search over ($x_0, \cdots, x_{B-1}$) of the initial state ($x_0, \cdots, x_{L-1}$). (5) offers plenty of suitable parity-check equations needed for high performance decoding, meanwhile avoids the low weight restriction of the feedback polynomial of the LFSR. Second, after regrouping the parity-check equations that contain the same pattern of $B - B_1$ initial bits, an application of Walsh transform is suggested to evaluate the parity-check equations in processing stage for a given z_i, i.e. when $\omega = [x_{B_1}, x_{B_1+1}, \cdots, x_{B-1}]$, $F_i(\omega) = \sum(-1)^{t_i^1 \oplus t_i^2}$ is just the difference between the number of predicted 0 and the number of predicted 1, where $t_i^1 = z_{m_1} \oplus \cdots \oplus z_{m_{k-1}} \oplus \sum_{j=0}^{B_1-1} c_j x_j$ and $t_i^2 = \sum_{j=B_1}^{B-1} c_j x_j$. Then for each of the D considered bits, if $F_i(\omega) > \theta$, let $x_i = 0$. If $F_i(\omega) < -\theta$, let $x_i = 1$, where θ is the decision threshold. Third, in order to have at least $L - B$ correctly recovered bits among the D considered bits, a check procedure is used which requires an exhaustive search on all subsets of size $L - B$ among the $L - B + \delta$ bits. The total complexity of the processing stage is:

$$O(2^B D \log_2 \Omega + (1 + p_{err}(2^B - 1)) \binom{L - B + \delta}{\delta} \frac{1}{\varepsilon^2}) \qquad (6)$$

where p_{err} is the probability that a wrong guess results in at least $L - B + \delta$ predicted bits and Ω is the expected number of parity-check equations of weight k for each considered bit. For the details of these formulae and the notations, please see the Appendix A and [3].

A summary of our attack is as follows:

1. Input: the feedback polynomial, $f(x)$, of LFSR A, a segment of keystream $z_0, z_1, \cdots, z_{N-1}$, $N' < N$, N' is determined by $N' \approx N - \alpha\sqrt{N'}/2$.
2. Construct sequence $\hat{a} = \hat{a}_0, \cdots, \hat{a}_{N'-1}$ according to Method 1 or Method 2 from keystream $z_0, z_1, \cdots, z_{N-1}$.
3. For each guess of $(a_0, \cdots, a_{B-1})$ and each bit position i, $(i = B+1, B+2, \ldots, D)$, evaluate the parity-check equations using the Walsh transform technique. Select those bits passing the majority poll to recover the initial state of LFSR A using the above check procedure.

After having recovered the initial state of LFSR A, we should also restore the initial state of LFSR S. With the knowledge of known sequence of LFSR A and keystream z, the remaining problem is much simplified compared to the original one. One way to do so is to use the method proposed in [6]. Here we do not focus on this problem.

3 Analysis of Our Attack

In this section, we will analyze our attack deeply, mainly on the two correlation analysis methods. We give two theorems on the coincidence probabilities $P(\hat{a}_r = a_r)$ under the above two methods, respectively. We will show that a special case of our method 2 is equivalent to the method proposed by Golić in [6].

3.1 The Coincidence Probability Under Method 1

Keep the assumption that both sequences generated by LFSR A and LFSR S are purely random. Theorem 1 yields the probability that sequence $\hat{a}$ equals sequence a under method 1.

Theorem 1. *Under method 1, the probability that the constructed sequence $\hat{a}$ equals sequence a is given by*

$$P(\hat{a}_r = a_r) = \frac{1}{2} + \frac{1}{2^{2E}}\binom{2E}{E}\frac{p}{4} = \frac{1}{2} + \varepsilon_r. \qquad (7)$$

where $2E+1$ satisfying $E = \lfloor(\alpha\sqrt{r}-1)/2\rfloor$, is the closest odd integer to $\alpha\sqrt{r}$ and $p = \frac{1}{\sqrt{2\pi}}\int_{-\alpha}^{\alpha} e^{-x^2/2}dx$ is the probability in (4).

Proof. According to method 1, we have

$$P(\hat{a}_r = a_r) = P(s_r = 1)P(\hat{a}_r = a_r|s_r = 1) + P(s_r = 0)P(\hat{a}_r = a_r|s_r = 0)$$

$$= \frac{1}{2}P(\hat{a}_r = a_r|s_r = 1) + \frac{1}{4}$$

$$= \frac{1}{2}P(\hat{a}_r = a_r|\sum_{i=0}^{r-1}s_i \in I_{r/2}, s_r = 1)P(\sum_{i=0}^{r-1}s_i \in I_{r/2}|s_r = 1)$$

$$+ \frac{1}{2}P(\sum_{i=0}^{r-1}s_i \bar{\in} I_{r/2}|s_r = 1)P(\hat{a}_r = a_r|\sum_{i=0}^{r-1}s_i \bar{\in} I_{r/2}, s_r = 1) + \frac{1}{4}$$

$$= \frac{1}{4} + \frac{1}{4}(1-p) + \frac{p}{2}P(\hat{a}_r = a_r|\sum_{i=0}^{r-1}s_i \in I_{r/2}, s_r = 1)$$

$$= \frac{1}{2} - \frac{p}{4} + \frac{p}{2}P^*$$

where $P^* = P(\hat{a}_r = a_r|\sum_{i=0}^{r-1}s_i \in I_{r/2}, s_r = 1)$ can be derived by the following equations.

$$P^* = P(\hat{a}_r = a_r = 0|\sum_{i=0}^{r-1}s_i \in I_{r/2}, \cdot) + P(\hat{a}_r = a_r = 1|\sum_{i=0}^{r-1}s_i \in I_{r/2}, \cdot)$$

$$= P(a_r = 0)P(\hat{a}_r = 0|a_r = 0, \sum_{i=0}^{r-1}s_i \in I_{r/2}, s_r = 1)$$

$$+ P(a_r = 1)P(\hat{a}_r = 1|a_r = 1, \sum_{i=0}^{r-1}s_i \in I_{r/2}, s_r = 1)$$

$$= \frac{1}{2}\sum_{i=E}^{2E}\binom{2E}{i}\frac{1}{2^{2E}} + \frac{1}{2}\sum_{i=E}^{2E}\binom{2E}{i}\frac{1}{2^{2E}}. \qquad (8)$$

(8) comes from the observation that if $a_r = j$ ($j = 0, 1$), then there must be at least E elements other than a_r itself in $I_{r/2}$ to be j for $\hat{a}_r = a_r = j$ holds. According to $\sum_{i=E}^{2E}\binom{2E}{i} = \sum_{i=0}^{E}\binom{2E}{i}$, we get

$$P^* = \frac{1}{2} + \frac{1}{2^{2E+1}}\binom{2E}{E}.$$

This completes the proof.

Corollary 1. *The coincidence probability $P(\hat{a}_r = a_r)$ is a function of r satisfying*

$$\frac{1}{2} < P(\hat{a}_r = a_r) \leq \frac{3}{4} \tag{9}$$

where the upper bound is achieved when $r = 0$.

Theorem 1 implies that the smaller r, the larger $P(\hat{a}_r = a_r)$ is. Note that our aim is to have a sequence $\hat{a}$ with a large enough correlation to a, which means that we should make the probability $P(\hat{a}_r = a_r)$ as large as possible. The larger ε_r is, the larger number of bits in sequence $\hat{a}$ satisfy $\hat{a}_r = a_r$. However, the above theorem shows that the probability function has an irregular form such that the classical methods for finding global maximum value of regular functions can not be used to obtain its global maximum. Instead, we try to find out the optimum numerical values of $P(\hat{a}_r = a_r)$ for each r. From Theorem 1, we can see that the bias

$$\varepsilon_r = \frac{1}{2^{2E}} \binom{2E}{E} \frac{p}{4} \tag{10}$$

is dependent on the product of p and $\binom{2E}{E}/2^{2E}$. Therefore, the optimum value of ε_r is

$$\varepsilon_{\max}^{(r)} = max_{0 \leq p \leq 1}\{\frac{1}{2^{2E}} \binom{2E}{E} \frac{p}{4}\}. \tag{11}$$

Note that $2E + 1$ is a measure of the length of $I_{r/2}$ which is determined by the probability p chosen in advance. In intuitive point of view, we should always choose p (by choosing α) rather large so that we can guarantee the interval $I_{r/2}$ always includes the indices of the elements that lie in keystream z. One easy way to do so is to choose p equals to one fixed value such as $0.90, 0.95, \cdots$, even $p = 0.99$. However, both theoretical and experimental results show that the bias ε_r drops so rapidly in this way that the average coincidence probability found is not good enough for an efficient fast correlation attack. Instead, we programmed in Mathematica to find each α that results in $\varepsilon_{\max}^{(r)}$. Figure 3 (In Figure 3, the horizontal axes represents α) shows for each r, where the optimum of α is located in the range $(0, 5)$.

Note that our construction method of sequence $\hat{a}$ is independent of the concrete LFSR structure under the purely random assumption, which means the pre-computation of the optimum values of α would be applied to arbitrary LFSR. Figure 3 shows that the optimum values of α satisfy $1 \leq \alpha \leq 2$ for $r \geq 244$. Noting the instruction Findminimum in Mathematica can only find the local minimum, we use the following two instructions to find the optimum value of α (a represents α):

Findminimum$[-\frac{\binom{2E}{E}}{2^{2E}} \frac{\int_{-a}^{a} e^{-x^2/2} dx}{4\sqrt{2\pi}}, \{a, 0, 5\}], 0 \leq r \leq 243$

or

Findminimum$[-\frac{\binom{2E}{E}}{2^{2E}} \frac{\int_{-a}^{a} e^{-x^2/2} dx}{4\sqrt{2\pi}}, \{a, 1, 5\}], r \geq 244.$

Figure 4 (In Figure 4 and 5, the horizontal axis represent keystream length N) shows the locations of the optimum values of α. With the knowledge of the

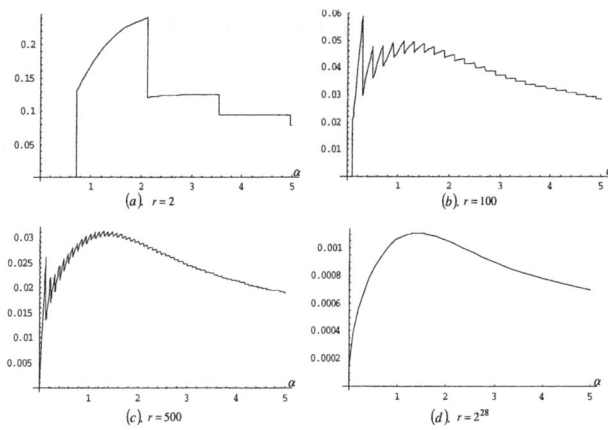

Fig. 3. The optimum position of α.

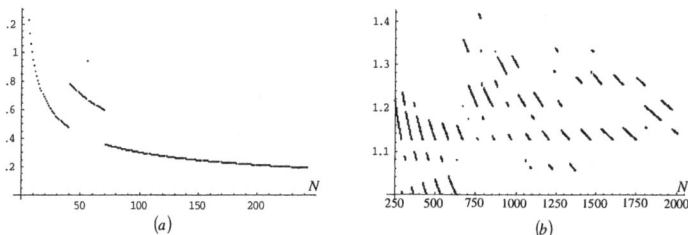

Fig. 4. The optimum value α that results in $\varepsilon_{\max}^{(r)}$. (a)-small scall, (b)-larger scale.

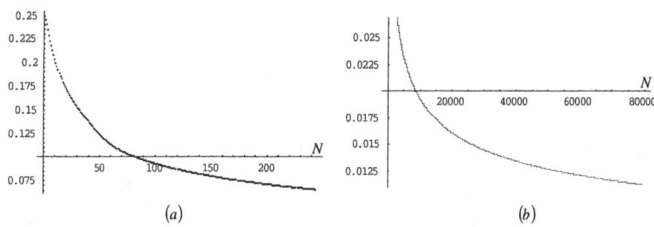

Fig. 5. The values of $\varepsilon_{\max}^{(r)}$. (a)-small scale, (b)-large scale.

optimum values of α, the biases we found are plotted in Figure 5. Let $H = \{\hat{a}_i | i \in \{0, 1, \cdots, N-1\}, \hat{a}_i = a_i\}$, the correlation found in this way is defined as $|H|/N$. We can see that the correlations is good enough for an efficient fast correlation attack against LFSR of moderate length. For example, for N=243, it amounts to 0.56555. For $N = 3000$, the correlation is 0.52748 and for $N = 8000$, it is 0.52075. See Section 4.

3.2 The Coincidence Probability Under Method 2

Next, we consider the probability $P(\hat{a}_r = a_r)$ under the construction of method 2. We will show that a special case of method 2 is equivalent to the method proposed by Golić in [6] in a sense that the numerical biases found under both methods (a special case of our method 2 and the method in [6]) are almost the same.

First note that from Definition 2 and (1), we have

$$P_0^{(r)} = \sum_{z_i \in S_0} P(a_r = z_i) = \sum_{z_i \in I_{r/2}} \binom{r}{i}(1-z_i)(\frac{1}{2})^{r+1} \qquad (12)$$

$$P_1^{(r)} = \sum_{z_i \in S_1} P(a_r = z_i) = \sum_{z_i \in I_{r/2}} \binom{r}{i} z_i (\frac{1}{2})^{r+1}. \qquad (13)$$

(12) and (13) imply that

$$E(P_0^{(r)}) = E(P_1^{(r)}) = \frac{1}{2}\sum_{z_i \in I_{r/2}} \binom{r}{i}(\frac{1}{2})^{r+1} = \frac{1}{2}(P_1^{(r)} + P_0^{(r)}), \qquad (14)$$

where $E(\cdot)$ is the mathematical expected value of the random variable. Note that method 2 actually takes into account the weight (the probability $P(a_r = z_k)$ associated with the point) of each point in $I_{r/2}$ upon making a majority poll, while in method 1, we regard each point in $I_{r/2}$ as the same, i.e. no one is more important than any other one. Therefore,

$$P(\hat{a}_r = a_r) = P(\hat{a}_r = a_r, \sum_{i=0}^{r-1} s_i \in I_{r/2}) + P(\hat{a}_r = a_r, \sum_{i=0}^{r-1} s_i \bar{\in} I_{r/2})$$

$$= \frac{1}{2} + \{\max(P_1^{(r)}, P_0^{(r)}) - \frac{1}{2}(P_1^{(r)} + P_0^{(r)})\}$$

$$= \frac{1}{2} + \{\max(P_1^{(r)}, P_0^{(r)}) - E(\max(P_1^{(r)}, P_0^{(r)}))\} = \frac{1}{2} + \varepsilon_r. \qquad (15)$$

Now we consider an important case of method 2. Let $I_{r/2} = \{0, 1, \cdots, r\}$ such that $P_1^{(r)} + P_0^{(r)} = \frac{1}{2}$, i.e. the probability that a_r lies in the interval $I_{r/2}$ is 0.5, instead of 1, due to the nature difference between method 1 and method 2. In this case, $E(P_0^{(r)}) = E(P_1^{(r)}) = \frac{1}{4}$. It follows from (14) and (15) that

$$E(\varepsilon_r) = E(\max(P_1^{(r)}, P_0^{(r)})) - \frac{1}{4}$$

$$= E((P_1^{(r)} + P_0^{(r)})/2 + |P_1^{(r)} - P_0^{(r)}|/2) - \frac{1}{4}$$

$$= E(|P_1^{(r)} - \frac{1}{4}|). \qquad (16)$$

Since $I_{r/2} = \{0, 1, \cdots, r\}$, we regard $P_1^{(r)} = \sum_{i=0}^{r} \binom{r}{i} z_i (\frac{1}{2})^{r+1}$ as the sum of $r+1$ independent random variables $\xi_0, \xi_1, \cdots, \xi_r$ satisfying $P(\xi_i = 0) = P(\xi_i =$

$\binom{r}{i}(\frac{1}{2})^{r+1}) = 0.5$. When $r \to \infty$, $P_1^{(r)}$ follows the Normal Distribution, i.e. $P_1^{(r)} \hookrightarrow N(\frac{1}{4}, \sigma^2)$, where the variance $\sigma^2 = \sum_{i=0}^{r} \binom{r}{i}^2 (\frac{1}{2})^{2r+2}\frac{1}{4} = \binom{2r}{r}(\frac{1}{2})^{2r+2}\frac{1}{4}$. Hence, we get

$$E(\varepsilon_r) = \frac{2\sigma}{\sqrt{2\pi}} = \frac{\sqrt{(\frac{1}{2})^{2r}\binom{2r}{r}}}{2\sqrt{2\pi}} \approx \frac{1}{2\sqrt{2\pi}\sqrt[4]{\pi}} \cdot \frac{1}{\sqrt[4]{r}} \approx 0.149828\frac{1}{\sqrt[4]{r}}. \quad (17)$$

Note that the corresponding bias found in [6] is $0.1515\frac{1}{\sqrt[4]{r}}$ based on approximating a binomial distribution by a uniform distribution. Both estimations are almost the same. From above, we get the following theorem.

Theorem 2. *Under method 2 and let $I_{r/2} = \{0, 1, \cdots, r\}$, the probability that the constructed sequence $\hat{a}$ equals sequence a is given approximately by*

$$P(\hat{a}_r = a_r) \approx \frac{1}{2} + 0.149828\frac{1}{\sqrt[4]{r}} \quad (18)$$

where $I_{r/2}$ is the same notation as that defined in Section 2.

Note that we obtain Theorem 2 under a special case of method 2. As in Theorem 1, we also want to maximize the probability $P(\hat{a}_r = a_r)$ under the general case of method 2. In nature, the maximization problem is to determine how long the interval $I_{r/2}$ should be chosen (by choosing α) such that the second kind of imbalance, $Imb_2(I_{r/2})$, can be maximized. The detailed analysis appears to be difficult, for the Normal Distribution may not be used in this case. We just leave this problem open. In the following, we will show that the coincidence probability obtained under Theorem 1 is approximately comparable to those got in Theorem 2 and in [6]. See Table 1. Note that the biases listed in Table 1 are not the average values, which are listed in Section 4. We can see that the bias values got from two methods are very close. Actually, such close values have almost the same inflect on the complexity of the whole fast correlation attack. Hence, any one of them can be used in practice. If all the binomial coefficients $\binom{i}{k}$ $0 \le i \le r$ are pre-computed as suggested in [6] using the recursion $\binom{i}{k} = \binom{i-1}{k-1} + \binom{i-1}{k}$ in $O(i^2)$ time and stored in $O(r^2)$ space, then method 2 will give a slightly higher coincidence. If the optimum values of α have been pre-computed in advance, method 1 is OK.

In addition, from Theorem 2 we can see that with the increase of r, the coincidence probability $P(\hat{a}_r = a_r)$ tends to 0.5 slowly. This fact can be interpreted as the reasonable result of basic design criterion of stream ciphers that the keystream z should satisfy $P(z = 0) = P(z = 1) = 0.5$ and the fact that a binomial distribution approximates a uniform distribution when $r \to \infty$.

Table 1. The one-point bias values of two methods.

r	1000	4000	8000	20000
Th. 1	0.0258843	0.018021	0.0150915	0.0119576
Th. 2	0.0266436	0.0188399	0.0158424	0.012599

4 Experimental Results

In this section we present some simulation results of our attack together with some comparisons with other known attacks on the shrinking generator. The experiments were done on a Pentium 4 PC processor.

First, we list the optimum values of α that give $\varepsilon_{\max}^{(r)}$ in Table 2. We use Mathematica to pre-compute these values in about four hours. It can be easily seen that most of the optimum values of α lie in the interval $(1.3, 1.5)$. The average value $\bar{\alpha} = 1.376395$ corresponds to the average probability $\bar{p} = 83.13\%$. It is worth noting that the optimum values of α are applicable to arbitrary LFSRs due to our purely random assumption in Section 2. Table 3 shows the average biases obtained by two theoretical methods and computer simulations. It is obvious that Theorem 1 is preferable when r is small, while Theorem 2 coincides with simulations better and offers a little better correlation when r grows large. The actual values of ε in Table 3 are found based on a shrinking generator with the following two primitive polynomials as the feedback polynomials of LFSR A and LFSR S, respectively: $f_A(x) = 1 + x + x^3 + x^5 + x^9 + x^{11} + x^{12} + x^{17} + x^{19} + x^{21} + x^{25} + x^{27} + x^{29} + x^{32} + x^{33} + x^{38} + x^{40}$ [3, 15, 16, 10] and $f_S(x) = 1 + x + x^2 + x^3 + x^4 + x^5 + x^{42}$ by method 1. The experimental results are in accordance with the theoretical expectations very well.

In order to compare our attack with other known ones, we consider another example of the shrinking generator with the generating LFSR A of length 61, as suggested in [9]. For practical considerations, we assume the length of LFSR S ≈ 61. Following the fast correlation attack in Section 2 and Appendix A,

Table 2. The optimum values of α (N=120000).

Domain	Number of α	Percent
$1.0 \sim 1.1$	248	0.2%
$1.1 \sim 1.2$	3139	2.5%
$1.2 \sim 1.3$	4308	3.6%
$1.3 \sim 1.4$	63480	53.0%
$1.4 \sim 1.5$	48221	40.2%
$1.5 \sim 1.6$	365	0.3%
others	239	0.2%
$\bar{\alpha}$ average	1.376395	100%

Table 3. The average biases ε of two methods and simulations.

N	ε(**Th. 1**)	ε(**Th. 2**)	ε(found)
240	0.0667726	0.0512096	0.054167
3000	0.02748	0.0270324	0.02100
8000	0.02075	0.0211382	0.02037
40000	0.0135484	0.014129	0.015650
80000	0.0113329	0.01188	0.012275
140000	0.00982376	0.0103285	0.008700

we choose the attack parameters as follows: $D = 36, \delta = 3, B = 46, k = 5$ for $L = 61$, the keystream length is $N = 140000 \approx 2^{17.1}$ and the coincidence probability is 0.50982376. We use the parity-check equations of weight 5, which can be obtained in $O(2^{43})$ pre-processing time and can be reused in later as many times as desirable. The expected number of parity-check equations for a given bit is $\Omega = 4.88464 \times 10^{14}$ and the probability that one parity-check equation gives the correct prediction is $q = \frac{1}{2}(1 + 0.01964752^4)$. From Appendix A, in order to have $P_1 \geq (L - B + \delta)/D = 0.5$, we choose $t = 2.4423196361 \times 10^{14}$ such that $P_1 \approx 0.500156$ and $P_v \approx 0.999999$. This gives the success probability

$$P_{succ} = \sum_{j=0}^{3} \binom{18}{3} P_v^{18-j}(1-P_v)^j \approx 99.9\%.$$

The probability of false alarm is negligible in this case. In fact, the probability P_{err} is limited to $P_{err} \approx 7.6 \times 10^{-45}$. Hence, the total processing complexity is

$$2^{46} \cdot 36 \cdot \log_2 \Omega + (1 + p_{err}(2^{46} - 1))\binom{18}{3}\frac{1}{\varepsilon^2} \approx 2^{56.7786}.$$

Table 4 shows the comparisons of different known attacks on the above example shrinking generator.

Table 4. Comparisons of different attacks on the example shrinking generator.

	[13]	[8]	A.[12]	B.[12]	C.[12]	Our attack
Length of z	few	$2^{10.23}$	few	2^{30}	$2^{30} - 2^{40}$	$2^{17.1}$
Complexity	2^{80}	2^{77}	2^{71}	2^{56}	$2^{50} - 2^{40}$	2^{56}
p_{succ}	100%	100%	66%	66%	66%	99.9%

For the detailed discussion of the concrete values in Table 4, see Appendix B. From Table 4, we can see that the attacks in [13], [8] and the attack A in [12] are all with the complexity higher than an exhaustive search. The attacks B and C in [12] are faster than an exhaustive search. But if a very high probability of success is required, we have to repeat the whole attack at least 4 times, which, for the best complexity result in [12], results in a 2^{42} keystream length and 2^{42} complexity. The required keystream length is too long for a 61-stage LFSR. In contract, the keystream length required in our attack is rather small, $2^{17.1}$, and the complexity is comparable to those in [12]. Hence, our attack offers a better trade-off between these parameters. In addition, our attack is better than the recent proposed attack on irregularly clocked generators in [17]. In that paper, a malformed shrinking generator with a LFSR S of length 26 and LFSR A of length 60 is cracked using an exhaustive search over the initial states of LFSR S with $1000000 \approx 2^{20}$ keystream bits. Besides, several fast correlation attack ideas on the (SG) have been proposed in [6]. However, few concrete results are available in that paper, making it difficult to make a comparison with it.

Some Remarks. An important fact about our attack is that the coincidence probability between a and $\hat{a}$ decreases, though rather slowly, with the increasing length of keystream. Hence, we propose two recommendations on attacking the shrinking generator.

1. It is of great importance to improve the fast correlation attack techniques by reducing the number of keystream bits required and deriving more efficient algorithm to construct parity-check equations with a little more weight. A new fast correlation attack is proposed in [18] without the detailed processing procedures, whose main advantage is the small amount of keystream necessary for a success attack with respect to a certain noise level compared to other attacks. From our experiments, the bias corresponding to $N = 3000$ keystream is 0.0274845, we think it is a promising way to apply this kind of attack to the shrinking generator.
2. Another direction is to consider the sequence $\hat{a}$ satisfying $P(\hat{a}_i = a_i) = p_i$ with different p_i, which is more closer to the truth of the construction method. Actually, such a method is used in [14] whose main disadvantage is the weight restriction of the feedback polynomials. Therefore, it is important to develop new fast correlation attacks applicable to the different p_i case, while maintaining the property that it is independent of the feedback polynomial's weight.

5 Conclusions

In this paper, we demonstrate a fast correlation attack on the shrinking generator with fixed connections. Our attack confirms that Golić's conjecture is correct. In addition, comparisons with other known attacks reveal that our attack offers a better trade-off between the required keystream length, success probability and the complexity.

Acknowledgements

We would like to thank the anonymous reviewers for very helpful comments.

References

1. A. Biryukov, "Block Ciphers and Stream Ciphers: The State of the Art", http://eprint.iacr.org/2004/094.pdf.
2. A. Canteaut, M. Trabbia, "Improved Fast Correlation Attacks Using Parity-Check Equations of Weight 4 and 5", *Advances in Cryptology-EUROCRYPT'2000*, LNCS vol. 1807, Springer-Verlag, (2000), pp. 573-588.
3. P. Chose, A. Joux, M. Mitton, "Fast Correlation Attacks: An Algorithmic Point of View", *Advances in Cryptology-EUROCRYPT'2002*, LNCS vol. 2332, Springer-Verlag, (2002), pp. 209-221.
4. D. Coppersmith, H. Krawczyk, Y. Mansour, "The Shrinking Generator", *Advances in Cryptology-Crypto'93*, LNCS vol. 773, Springer-Verlag, (1994), pp.22-39.
5. P. Ekdahl, T.Johansson, "Predicting the Shrinking Generator with Fixed Connections", *Advances in Cryptology-EUROCRYPT'2003*, LNCS vol. 2656, Springer-Verlag, (2003), pp. 330-344.

6. J. Dj. Golić, "Correlation analysis of the shrinking Generator", *Advances in Cryptology-Crypto'2001*, LNCS vol. 2139 Springer-Verlag, (2001), pp. 440-457.
7. J. Dj. Golić, "Computation of Low-weight parity-check ploynomials", *Electronic Letters*, Vol. 32, No. 21, pp. 1981-1982, October 1996.
8. J. Dj. Golić, "Embedding and probabilistic correlation attacks on clock-controlled shift registers", *Advances in Cryptology-EUROCRYPT'94*, LNCS vol. 950, Springer-Verlag, (1994), pp. 230-243.
9. H. Krawczyk, "The shrinking generator: Some practical considerations", *Fast Software Encryption-FSE'94*, LNCS vol. 809, Springer-Verlag, (1994), pp. 45-46.
10. T. Johansson, F. Jonnson, "Improved fast correlation attack on stream ciphers via convolutional codes", *Advances in Cryptology-EUROCRYPT'1999*, LNCS vol. 1592, Springer-Verlag, (1999), pp. 347-362.
11. T. Johansson, F. Jönsson, "Fast correlation attacks through reconstruction of linear polynomials", *Advances in Cryptology-Crypto'2000*, LNCS vol. 1880, Springer-Verlag, (2000), pp. 300-315.
12. T. Johansson, "Reduced complexity correlation attacks on two clock-controlled generators", *Advances in Cryptology-ASIACRYPT'98*, LNCS vol. 1514, Springer-Verlag, (1998), pp. 342-357.
13. A. Menezes, P. van Oorschot, S. Vanstone, *Handbook of Applied Cryptography*, CRC Press,1997.
14. W. Meier, O. Staffelbach, "Fast correlation attacks on certain stream ciphers", *Journal of Cryptology*, (1989) 1 pp. 159-176.
15. M. Mihaljević, P.C. Fossorier, H.Imai, "Fast correlation attack algorithm with list decoding and an application", *Fast Software Encryption-FSE'2001*, LNCS vol. 2355, Springer-Verlag, (2002), pp. 196-210.
16. M. Mihaljević, P.C. Fossorier, H.Imai, "A Low-complexity and high-performance algorithm for fast correlation attack", *Fast Software Encryption-FSE'2000*, LNCS vol. 1978, Springer-Verlag, (2001), pp. 196-212.
17. H. Molland, "Improved Linear Consistency Attack on Irregular Clocked Keystream Generators", *Fast Software Encryption-FSE'2004*, LNCS vol. 3017, Springer-Verlag, (2004), pp. 109-126.
18. M. Noorkami, F. Fekri, "A Fast Correlation Attack via Unequal Error Correcting LDPC Codes", *CT-RSA'2004*, LNCS vol. 2964, Springer-Verlag, (2004), pp. 54-66.
19. L. Simpson, J. Dj. Golić, "A probabilistic correlation attack on the shrinking generator", *ACISP'98*, LNCS vol. 1438, Springer-Verlag, (1998), pp. 147-158.
20. D. Wagner, "A Generalized Birthday Problem", *Advances in Cryptology-Crypto'2002*, LNCS vol. 2442, Springer-Verlag, (2002), pp. 288-303.
21. D. F. Zhang, W. D. Chen, "Information Leak analysing on the Shrinking Generator and the Self-Shrinking Generator", *Journal of China Institute of Communications*, Vol. 17, No. 4, pp. 15-20, July 1996.

A Notations and Formulae of a One-Pass Fast Correlation Attack

1. $P(z_i = x_i) = \frac{1}{2}(1 + \varepsilon)$.
2. N is the length of the keystream.
3. L is the length of the LFSR.
4. B is the number of bits partially exhausitive searched.

5. D is the number of bits under consideration.
6. k is the weight of the parity-check equations.
7. $q = \frac{1}{2}(1 + \varepsilon^{k-1})$ is the probability that one parity-check equation yielding the correct prediction.
8. Ω is the expected number of weight k parity-check equations for each considered bit.
9. δ is the number of bits that predicted other than the $n - B$ bits.
10. $P_1 = \sum_{j=\Omega-t}^{\Omega}(1-q)^{\Omega-j}q^j\binom{\Omega}{j}$ is the probability that at least $\Omega - t$ parity-check equations give the correct result, where t is the smallest integer satisfying $D \cdot P_1 \geq L - B + \delta$.
11. θ is the threshold such that $\theta = \Omega - 2t$.
12. $P_2 = \sum_{j=\Omega-t}^{\Omega}(1-q)^j q^{\Omega-j}\binom{\Omega}{j}$ is the probability that at least $\Omega - t$ parity-check equations give the wrong result.
13. $P_v = P_1/(P_1 + P_2)$ is the probability that a bit is correctly predicted with at least $\Omega - t$ parity-check equations give the same prediction.
14. $P_{succ} = \sum_{j=0}^{\delta}\binom{L-B+\delta}{j}P_v^{L-B+\delta-j}(1-P_v)^j$ is the probability that at most δ bits are wrong among the $n - B + \delta$ predicted bits.
15. $E = \frac{1}{2^{\Omega-1}}\sum_{j=\Omega-t}^{\Omega}\binom{\Omega}{j}$ is the probability that a wrong guess yields at least $\Omega - t$ identical predictions for a given bit.
16. $P_{err} = \sum_{j=L-B+\delta}^{D}\binom{D}{j}E^j(1-E)^{D-j}$ is the probability that false alarm occurs.
17. When $k = 4$, the time complexity of the pre-processing stage is $O(N^2 \log N)$. When $k = 5$, the time complexity is $O(DN^2 \log N)$. In both cases, the memory complexities are $O(N)$.

B Remarks on the Concrete Values in Table 4

The attack in [13] is a divide-and-conquer attack on LFSR S requiring $O(2^{L_S}L_A^3)$ operations. For $L_S \approx L_A = 61$, it amounts to 2^{80}. The probabilistic attack proposed in [8] is also an exhaustive attack with complexity around $2^{L_A}(4L_A)^2$. As in [12], here we choose $4L_A$ for unique decoding. For $L_A = 61$, the complexity is 2^{77}. There are three attacks proposed in [12]. Attack A is an exhaustive search using the decoding algorithm given in that paper. Both attack B and C are based on searching for specific weak subsequences in the keystream z. The difference between B and C is that several weak subsequences are required in attack C, which results in the very long length of the required keystream, i.e. 2^{40}. Though the complexity of C is the lowest, 2^{40}, the required keystream length, 2^{40}, is absolutely unrealistic for a LFSR A of length 61. Besides, the decoding algorithm in [12] has a failure probability 0.34, when its complexity is assumed to be 2^{10}.

Improved Efficiency for CCA-Secure Cryptosystems Built Using Identity-Based Encryption

Dan Boneh[1,*] and Jonathan Katz[2,**]

[1] Computer Science Department, Stanford University, Stanford CA 94305
dabo@cs.stanford.edu
[2] Dept. of Computer Science, Univ. of Maryland
jkatz@cs.umd.edu

Abstract. Recently, Canetti, Halevi, and Katz showed a general method for constructing CCA-secure encryption schemes from identity-based encryption schemes in the standard model. We improve the efficiency of their construction, and show two specific instantiations of our resulting scheme which offer the most efficient encryption (and, in one case, key generation) of any CCA-secure encryption scheme to date.

Keywords: Chosen-ciphertext security, Identity-based encryption, Public-key encryption.

1 Introduction

Security against adaptive chosen-ciphertext attacks (i.e., "CCA-security") [29, 17, 1] has become the *de facto* level of security for public-key encryption schemes. The reasons for this are many: CCA security helps protect against subtle attacks that have been demonstrated against schemes *not* meeting this notion of security [3, 24, 23]; is helpful in defending against "active" attackers who may modify messages in transit (see [32]); and, finally, allows encryption schemes to be developed and then securely "plugged in" to higher-level protocols which may then be executed in arbitrary environments (see, e.g., [8, Sec. 8.2.2]).

Nevertheless, only a relatively small number of encryption schemes have been rigorously proven secure against adaptive chosen-ciphertext attacks *in the standard model*[1] (i.e., without resorting to the use of random oracles [2]). Schemes based on general assumptions are known [17, 30, 27], but these rely on generic non-interactive zero-knowledge proofs [4, 18] and do not currently lead to practical solutions. More interesting from a practical point of view are efficient schemes based on specific number-theoretic assumptions; two general methodologies for constructing such schemes are known. The first methodology is based on the "smooth hash proof systems" of Cramer and Shoup [14], and has led to a variety

---

[*] Supported by NSF and the Packard Foundation.
[**] This research was supported by NSF Trusted Computing Grant #0310751.
[1] From now on, we use "CCA security" to refer by default to security which is proven in the standard model.

of constructions [13, 14, 19, 15, 25]. The second, and more recent, method [11] constructs a CCA-secure encryption scheme from any semantically-secure (or, "CPA-secure") identity-based encryption (IBE) scheme [7, 12] (which can in turn be constructed in the standard model based on specific number-theoretic assumptions [10, 5, 6, 34]). Overall, the most efficient CCA-secure encryption scheme currently known is a hybrid encryption system due to Kurosawa and Desmedt [25] which builds on the original proposal of Cramer and Shoup [13] and relies on the decisional Diffie-Hellman assumption.

In this paper, we suggest a new method which allows for the construction of very efficient CCA-secure encryption schemes. Our technique modifies the approach of Canetti, Halevi, and Katz [11], who (as noted above) show a transformation from any semantically-secure "weak" IBE scheme to a CCA-secure public-key encryption scheme. Briefly and somewhat informally, their transformation from an IBE scheme[2] (Setup, Der, Enc, Dec) to a CCA-secure scheme proceeds as follows: key generation is performed by running Setup and letting the public (resp. secret) key be the master public key PK (resp., master secret key msk) output by this algorithm. To encrypt a message m using public key PK, a sender generates a random key-pair (vk, sk) for a one-time signature scheme and sends the ciphertext $\langle vk, \mathsf{Enc}_{PK}(vk, m), \sigma \rangle$, where $\mathsf{Enc}_{PK}(vk, m)$ represents an encryption of message m for the "identity" vk using master public parameters PK, and σ represents a signature on the second component of this ciphertext using sk. To decrypt ciphertext $\langle vk, C, \sigma \rangle$, the receiver first verifies whether $\mathsf{Vrfy}_{vk}(C, \sigma) \stackrel{?}{=} 1$. If so, the receiver then decrypts C with respect to the "identity" vk (it can do this since it has the master secret key msk).

Though conceptually simple, this transformation does add noticeable overhead to the underlying IBE scheme: encryption requires the sender to generate keys for a one-time signature scheme [26] and also to compute a signature using the keys just generated; decryption requires the receiver to verify a signature with respect to the verification key included as part of the ciphertext. Although one-time signatures are "easy" to construct in theory, and are more efficient than "full-blown" signatures (i.e., those which are existentially unforgeable under an adaptive chosen-message attack [20]), they still have their price. In particular:

- One-time signatures based on cryptographic hash functions such as SHA-1 can be designed to allow very efficient *signing*; key generation, on the other hand, typically requires hundreds of hash function evaluations and is relatively expensive (though not as expensive as key generation in schemes based on number-theoretic assumptions). More problematic, perhaps, is that such schemes have very long public keys and signatures, which would result in very long ciphertexts in the scheme of [11].
- One-time signatures based on number-theoretic assumptions (say, by adapting "full-blown" signature schemes) yield schemes whose computational cost – both for key generation and signing – is more expensive, but which have the advantage of short(er) public keys and signatures.

---
[2] Definitions of IBE schemes and their security, as well as definitions of CCA-secure encryption, are reviewed in Section 2.

Either way, the transformation of Canetti, Halevi, and Katz results in a CCA-secure encryption scheme which is less efficient than the underlying IBE system.

1.1 Our Contribution

We describe a transformation from any CPA-secure "weak" IBE system to a CCA-secure encryption scheme which adds essentially no overhead. The efficiency advantage of our approach arises from our observation that the one-time signature in the construction of Canetti, et al. (as described earlier) can be replaced by a message-authentication code (MAC) along with an appropriate "encapsulation" of a MAC key (for the purposes of this informal description, one can think of an encapsulation as a commitment). Using the notation introduced earlier, encryption using our approach is now performed (informally) by first "encapsulating" a key r which results in an encapsulation com along with a de-commitment string dec. The final ciphertext is $\langle \mathsf{com}, \mathsf{Enc}_{PK}(\mathsf{com}, m \circ \mathsf{dec}), \mathsf{tag}\rangle$, where tag is now a message authentication code computed on the second component of the ciphertext using key r. Decryption of ciphertext $\langle \mathsf{com}, C, \mathsf{tag}\rangle$ is done in the natural way, but note that here the receiver must first decrypt C (with respect to "identity" com) and only then can the receiver verify the correctness of tag. Indeed, this feature of our scheme complicates the security proof somewhat (and in particular we must be careful to avoid circular arguments).

Adapting [16, 21], we show how encapsulation of the MAC key can be done both efficiently and securely using, e.g., SHA-1: encapsulation requires only a single hash function evaluation, and is secure under the assumption that SHA-1 is second-preimage resistant (the scheme can be easily modified so as to be secure under the weaker assumption of the existence of UOWHFs [28]). This encapsulation scheme may have other applications, and thus the scheme – as well as the relatively simple proof of security we provide for this encapsulation scheme here (cf. Theorem 2) – may be of independent interest. Furthermore, our technique of replacing a one-time signature by a MAC seems applicable to other constructions (e.g., those of [17, 30] as well as the various extensions mentioned in [11]), giving efficiency improvements in those cases as well.

In addition to the general method discussed above, we also show two specific instantiations of our approach based on two IBE schemes recently introduced by Boneh and Boyen [5]. Our resulting schemes are quite efficient: in particular, the times required for key generation and encryption are as fast as (or faster than) the most efficient previous CCA-secure schemes to date.

1.2 Hybrid Encryption

In practice, public-key encryption is almost never used to encrypt actual data. Instead, *hybrid encryption* is typically used, whereby a public-key scheme is used to encrypt a random key, and the data is then encrypted using some symmetric-key encryption scheme and this key. In fact, "encryption" of the symmetric key is not required; "encapsulation" (cf. [33]) – which may be more efficient – is enough. It is well known that if both the public-key encapsulation scheme and the underlying symmetric-key encryption scheme are CCA-secure, then the resulting hybrid scheme is CCA-secure as well.

Interestingly, Kurosawa and Desmedt have recently shown [25] that the public-key encapsulation scheme does not necessarily need to be CCA-secure in order for the resulting hybrid scheme to be CCA-secure. In particular, they show a hybrid encryption scheme which is based on, but more efficient than, the Cramer-Shoup scheme [13] *when used for hybrid encryption*. The specific hybrid schemes proposed here are as efficient as the Kurosawa-Desmedt scheme in terms of encryption (and, in one case, key generation), but somewhat less efficient in other measures; we provide detailed comparisons in Section 4. It is somewhat surprising that constructions based on completely different approaches end up having such similar performance for both encryption and key generation.

1.3 Outline

In Section 3, we present and prove secure a generic construction of a CCA-secure encryption scheme based on a variety of primitives (IBE, MACs, and encapsulation) formally defined in Section 2. Section 4 describes in more detail two specific instantiations of the various primitives; the efficiency of the resulting schemes are then compared with previous work.

2 Basic Definitions

We review the standard definitions of public-key encryption schemes and their security against adaptive chosen-ciphertext attacks. This is followed by definitions of identity-based encryption, message authentication, and "encapsulation" as needed for our construction.

Definition 1. (Public-Key Encryption) *A public-key encryption scheme* PKE *is a triple of* PPT *algorithms* (Gen, Enc, Dec) *such that:*

- *The randomized key generation algorithm* Gen *takes as input a security parameter* 1^k *and outputs a public key* PK *and a secret key* SK. *We write* $(PK, SK) \leftarrow \mathsf{Gen}(1^k)$.
- *The randomized encryption algorithm* Enc *takes as input a public key* PK *and a message* $m \in \{0,1\}^*$, *and outputs a ciphertext* C. *We write* $C \leftarrow \mathsf{Enc}_{PK}(m)$.
- *The decryption algorithm* Dec *takes as input a ciphertext* C *and a secret key* SK. *It returns a message* $m \in \{0,1\}^*$ *or the distinguished symbol* $\bot$. *We write* $m \leftarrow \mathsf{Dec}_{SK}(C)$.

We require that for all (PK, SK) *output by* Gen, *all* $m \in \{0,1\}^*$, *and all* C *output by* $\mathsf{Enc}_{PK}(m)$ *we have* $\mathsf{Dec}_{SK}(C) = m$.

Definition 2. (CCA Security) *A public-key encryption scheme* PKE *is secure against adaptive chosen-ciphertext attacks (i.e., is "CCA-secure") if the advantage of any* PPT *adversary A in the following game is negligible in the security parameter* k:

1. Gen(1^k) outputs (PK, SK). Adversary A is given 1^k and PK.
2. The adversary may make polynomially-many queries to a decryption oracle $\text{Dec}_{SK}(\cdot)$.
3. At some point, A outputs two messages m_0, m_1 with $|m_0| = |m_1|$. A bit b is randomly chosen and the adversary is given a "challenge ciphertext" $C^* \leftarrow \text{Enc}_{PK}(m_b)$.
4. A may continue to query its decryption oracle $\text{Dec}_{SK}(\cdot)$ except that it may not request the decryption of C^*.
5. Finally, A outputs a guess b'.

We say that A succeeds if $b' = b$, and denote the probability of this event by $\Pr_{A,\mathsf{PKE}}[\mathsf{Succ}]$. The adversary's advantage is defined as $|\Pr_{A,\mathsf{PKE}}[\mathsf{Succ}] - 1/2|$.

2.1 Identity-Based Encryption

Informally, an IBE scheme is a public-key encryption scheme in which any string (i.e., identity) can serve as a public key. In more detail, a setup algorithm is first run to generate "master" public and secret keys. Given the master secret key and any string $ID \in \{0,1\}^*$ (which can be viewed as an identity), it is possible to derive a "personal secret key" SK_{ID}. Any sender can encrypt a message for "identity" ID using only the master public key and the string ID. The resulting ciphertext can be decrypted using the derived secret key SK_{ID}, but the message remains hidden from an adversary who does not know SK_{ID} even if that adversary is given $SK_{ID'}$ for multiple identities $ID' \neq ID$. The concept of identity-based encryption was introduced by Shamir [31], and provably-secure IBE schemes in the random oracle model were demonstrated by Boneh and Franklin [7] and Cocks [12]. More recently, provably-secure IBE schemes in the standard model have been developed [10, 5, 6, 34]; see further discussion below.

In the original definition of security for IBE proposed and achieved by Boneh and Franklin [7], the adversary may choose the "target identity" (ID in the above discussion) in an adaptive manner, based on the master public key and any keys $SK_{ID'}$ the adversary has obtained thus far. A weaker notion of security, proposed and achieved by Canetti, Halevi, and Katz [10], requires the adversary to specify the target identity *before* the public-key is published; we will refer to this notion of security as "weak" IBE. As in [11], our construction only requires weak IBE schemes secure against chosen-plaintext attacks. We therefore only recall this definition of security.

Definition 3. (IBE) *An identity-based encryption scheme* IBE *is a 4-tuple of* PPT *algorithms* (Setup, Der, Enc, Dec) *such that:*

- *The randomized setup algorithm* Setup *takes as input a security parameter* 1^k *and a value ℓ for the identity length. It outputs some system-wide parameters PK along with a master secret key* msk. *(We assume that k and ℓ are implicit in PK.)*
- *The (possibly randomized) key derivation algorithm* Der *takes as input the master key* msk *and an identity $ID \in \{0,1\}^\ell$. It returns the corresponding decryption key SK_{ID}. We write $SK_{ID} \leftarrow \text{Der}_{\mathsf{msk}}(ID)$.*

- The randomized encryption algorithm Enc takes as input the system-wide public key PK, an identity $ID \in \{0,1\}^\ell$, and a message $m \in \{0,1\}^*$; it outputs a ciphertext C. We write $C \leftarrow \mathsf{Enc}_{PK}(ID, m)$.
- The decryption algorithm Dec takes as input an identity ID, its associated decryption key SK_{ID}, and a ciphertext C. It outputs a message $m \in \{0,1\}^*$ or the distinguished symbol $\bot$. We write $m \leftarrow \mathsf{Dec}_{SK_{ID}}(ID, C)$.

We require that for all (PK, msk) output by Setup, all $ID \in \{0,1\}^\ell$, all SK_{ID} output by $\mathsf{Der}_{\mathsf{msk}}(ID)$, all $m \in \{0,1\}^*$, and all C output by $\mathsf{Enc}_{PK}(ID, m)$ we have $\mathsf{Dec}_{SK_{ID}}(ID, C) = m$.

As mentioned earlier, we provide a definition of security only for the case of "weak" IBE, as considered in [10, 5]. (Of course, a scheme satisfying the stronger definition of [7, 6] is trivially a weak IBE scheme as well.)

Definition 4. (Selective-ID IBE) *An identity-based scheme* IBE *is secure against selective-identity, chosen-plaintext attacks if for all polynomially-bounded functions $\ell(\cdot)$ the advantage of any* PPT *adversary A in the following game is negligible in the security parameter k:*

1. $A(1^k, \ell(k))$ outputs a target identity $ID^* \in \{0,1\}^{\ell(k)}$.
2. $\mathsf{Setup}(1^k, \ell(k))$ outputs (PK, msk). The adversary is given PK.
3. The adversary A may make polynomially-many queries to an oracle $\mathsf{Der}_{\mathsf{msk}}(\cdot)$, except that it may not request the secret key corresponding to the target identity ID^*.
4. At some point, A outputs two messages m_0, m_1 with $|m_0| = |m_1|$. A bit b is randomly chosen and the adversary is given a "challenge ciphertext" $C^* \leftarrow \mathsf{Enc}_{PK}(ID^*, m_b)$.
5. A may continue to query its oracle $\mathsf{Der}_{\mathsf{msk}}(\cdot)$, but still may not request the secret key corresponding to the identity ID^*.
6. Finally, A outputs a guess b'.

We say that A succeeds if $b' = b$, and denote the probability of this event by $\Pr_{A,\mathsf{IBE}}[\mathsf{Succ}]$. The adversary's advantage is defined as $|\Pr_{A,\mathsf{IBE}}[\mathsf{Succ}] - 1/2|$.

For completeness, we remark that a slightly weaker definition – in which $\ell = \Omega(\log k)$ is *a priori* bounded, rather than being given as a parameter to Setup – suffices for our construction.

2.2 Message Authentication

We view a *message authentication code* as a pair of PPT algorithms (Mac, Vrfy). The authentication algorithm Mac takes as input a key sk and a message M, and outputs a string tag. The verification algorithm Vrfy takes as input a key sk, a message M, and a string tag; it outputs either 0 ("reject") or 1 ("accept"). We require that for all sk and M we have $\mathsf{Vrfy}_{sk}(M, \mathsf{Mac}_{sk}(M)) = 1$. For simplicity, we assume that Mac and Vrfy are deterministic.

We give a definition of security tailored to the requirements of our construction; in particular, we require only "one-time" security for our message authentication code. We remark that efficient schemes satisfying this definition can be constructed without any computational assumptions using, e.g., almost strongly universal hash families [35].

Definition 5. (Message Authentication) *A message authentication code (Mac, Vrfy) is secure against a one-time chosen-message attack if the success probability of any* PPT *adversary A in the following game is negligible in the security parameter k:*

1. *A random key $sk \in \{0,1\}^k$ is chosen.*
2. *$A(1^k)$ outputs a message M and is given in return $\mathsf{tag} = \mathsf{Mac}_{sk}(M)$.*
3. *A outputs a pair (M', tag').*

We say that A succeeds if $(M, \mathsf{tag}) \neq (M', \mathsf{tag}')$ and $\mathsf{Vrfy}_{sk}(M', \mathsf{tag}') = 1$.

In the above, the adversary succeeds even if $M = M'$ but $\mathsf{tag} \neq \mathsf{tag}'$. Thus, the definition corresponds to what has been termed "strong" security in the context of signature schemes.

2.3 Encapsulation

We define a notion of "encapsulation" which may be viewed as a weak variant of commitment. (Note that our definition is unrelated to that of *key encapsulation* which was discussed in Section 1.2.) In terms of functionality, an encapsulation scheme commits the sender to a *random string* as opposed to a chosen message as in the case of commitment. In terms of security, our construction only requires binding to hold for *honestly-generated encapsulations*; this is analogous to assuming an honest sender during the first phase of a commitment scheme.

Definition 6. (Encapsulation) *An encapsulation scheme is a triple of* PPT *algorithms* (Setup, $\mathcal{S}, \mathcal{R}$) *such that:*

- Setup *takes as input the security parameter 1^k and outputs a string* pub.
- $\mathcal{S}$ *takes as input 1^k and* pub, *and outputs $(r, \mathsf{com}, \mathsf{dec})$ with $r \in \{0,1\}^k$. We refer to* com *as the public commitment string and* dec *as the de-commitment string.*
- $\mathcal{R}$ *takes as input* (pub, com, dec) *and outputs an $r \in \{0,1\}^k \cup \{\bot\}$.*

We require that for all pub *output by* Setup *and for all $(r, \mathsf{com}, \mathsf{dec})$ output by $\mathcal{S}(1^k, \mathsf{pub})$, we have $\mathcal{R}(\mathsf{pub}, \mathsf{com}, \mathsf{dec}) = r$. We also assume for simplicity that* com *and* dec *have fixed lengths for any given value of the security parameter.*

As in the case of commitment, an encapsulation scheme satisfies notions of both binding and hiding. Informally, "hiding" requires that com should leak no information about r; more formally, the string r should be indistinguishable from random even when given com (and pub). "Binding" requires that an honestly-generated com can be "opened" to only a single (legal) value of r; see below.

Definition 7. (Secure Encapsulation) *An encapsulation scheme* $(\mathsf{Setup}, \mathcal{S}, \mathcal{R})$ *is secure if it satisfies both hiding and binding as follows:*

Hiding: *The following is negligible for all* PPT A:

$$\left| \Pr \left[\begin{array}{l} \mathsf{pub} \leftarrow \mathsf{Setup}(1^k); r_0 \leftarrow \{0,1\}^k; \\ (r_1, \mathsf{com}, \mathsf{dec}) \leftarrow \mathcal{S}(1^k, \mathsf{pub}); b \leftarrow \{0,1\} \end{array} : A(1^k, \mathsf{pub}, \mathsf{com}, r_b) = b \right] - \frac{1}{2} \right|.$$

Binding: *The following is negligible for all* PPT A:

$$\Pr \left[\begin{array}{l} \mathsf{pub} \leftarrow \mathsf{Setup}(1^k); \\ (r, \mathsf{com}, \mathsf{dec}) \leftarrow \mathcal{S}(1^k, \mathsf{pub}); \\ \mathsf{dec}' \leftarrow A(1^k, \mathsf{pub}, r, \mathsf{com}, \mathsf{dec}) \end{array} : \mathcal{R}(\mathsf{pub}, \mathsf{com}, \mathsf{dec}') \notin \{\bot, r\} \right].$$

In the above, both hiding and binding are required to hold only computationally. In Section 4 we show a novel encapsulation scheme which is both simple and efficient, and which achieves *statistical* hiding (and computational binding).

3 A Generic Construction

We now describe our construction of a CCA-secure encryption scheme from the primitives introduced in the previous section. Let $(\mathsf{Setup}', \mathsf{Der}', \mathsf{Enc}', \mathsf{Dec}')$ be an IBE scheme, $(\mathsf{Setup}, \mathcal{S}, \mathcal{R})$ be an encapsulation scheme, and $(\mathsf{Mac}, \mathsf{Vrfy})$ be a message authentication code. Our scheme is constructed as follows:

Key Generation: Keys for our scheme are generated by running $\mathsf{Setup}'(1^k)$ to generate (PK, msk) and $\mathsf{Setup}(1^k)$ to generate pub. The public key is (PK, pub), and the secret key is msk.

Encryption: To encrypt a message m using public key (PK, pub), a sender first encapsulates a random value by running $\mathcal{S}(1^k, \mathsf{pub})$ to obtain $(r, \mathsf{com}, \mathsf{dec})$. The sender then encrypts the "message" $m \circ \mathsf{dec}$ with respect to the "identity" com; that is, the sender computes $C \leftarrow \mathsf{Enc}'_{PK}(\mathsf{com}, m \circ \mathsf{dec})$. The resulting ciphertext C is then authenticated by using r as a key for a message authentication code; i.e., the sender computes $\mathsf{tag} = \mathsf{Mac}_r(C)$. The final ciphertext is $\langle \mathsf{com}, C, \mathsf{tag} \rangle$.

Decryption: To decrypt a ciphertext $\langle \mathsf{com}, C, \mathsf{tag} \rangle$, the receiver derives the secret key SK_com corresponding to the "identity" com, and uses this key to decrypt the ciphertext C as per the underlying IBE scheme; this yields a "message" $m \circ \mathsf{dec}$ (if decryption fails, the receiver outputs $\bot$). Next, the receiver runs $\mathcal{R}(\mathsf{pub}, \mathsf{com}, \mathsf{dec})$ to obtain a string r; if $r \neq \bot$ and $\mathsf{Vrfy}_r(C, \mathsf{tag}) = 1$, the receiver outputs m. Otherwise, the receiver outputs $\bot$.

Intuition for the security of the above encryption scheme against chosen-ciphertext attacks is similar to [11]. Let $\langle \mathsf{com}^*, C^*, \mathsf{tag}^* \rangle$ be the challenge ciphertext (cf. Definition 2). In the absence of any decryption queries, it is clear that the value of the bit b remains hidden from the adversary due to the security of the underlying IBE scheme. Decryption queries of the form $\langle \mathsf{com}, C, \mathsf{tag} \rangle$ with com $\neq$ com* do not further help the adversary since the adversary would be unable to determine b *even if it had the secret key* SK_{com} *corresponding to* com (this follows again from the security of the underlying IBE scheme). Thus, it is left to examine decryption queries of the form $\langle \mathsf{com}^*, C, \mathsf{tag}\rangle$. The crux of our proof is to show that all queries of this form are rejected (i.e., the decryption oracle returns $\bot$ in response to all queries of this form) with all but negligible probability. A formal proof of this statement is somewhat involved, as it requires avoiding the apparent "circularity" arising from the IBE scheme, the message authentication code, and the encapsulation scheme; the details are given in the proof below.

Theorem 1. *Assuming the IBE scheme, message authentication code, and encapsulation scheme used above satisfy Definitions 4, 5, and 7, respectively, the above construction is a* PKE *scheme which is secure against adaptive chosen-ciphertext attacks.*

Proof. Given any PPT adversary $\mathcal{A}$ attacking the above encryption scheme in an adaptive chosen-ciphertext attack, we construct a PPT adversary $\mathcal{A}'$ attacking the underlying IBE scheme in a selective-identity, chosen-plaintext attack. Relating the success probabilities of these adversaries gives the desired result.

Let $\ell(k)$ denote the length of strings com output by $\mathcal{S}$. Define adversary $\mathcal{A}'$ as follows:

1. $\mathcal{A}'(1^k, \ell(k))$ runs $\mathsf{Setup}(1^k)$ to generate pub, and runs $\mathcal{S}(1^k, \mathsf{pub})$ to obtain $(r^*, \mathsf{com}^*, \mathsf{dec}^*)$. The adversary $\mathcal{A}'$ then outputs the "target identity" com*.
2. $\mathcal{A}'$ is then given IBE parameters PK. Adversary A', in turn, runs $\mathcal{A}$ on inputs 1^k and (PK, pub).
3. When $\mathcal{A}$ submits the ciphertext $\langle \mathsf{com}, C, \mathsf{tag}\rangle$ to its decryption oracle, $\mathcal{A}'$ proceeds as follows:
   – If com $=$ com*, then $\mathcal{A}'$ returns $\bot$.
   – If com $\neq$ com*, then $\mathcal{A}'$ makes the oracle query $\mathsf{Der}'_{\mathsf{msk}}(\mathsf{com})$ to obtain SK_{com}. It then computes $m \circ \mathsf{dec} = \mathsf{Dec}'_{SK_{\mathsf{com}}}(\mathsf{com}, C)$, followed by $r = \mathcal{R}(\mathsf{pub}, \mathsf{com}, \mathsf{dec})$. If $r \neq \bot$ and $\mathsf{Vrfy}_r(C, \mathsf{tag}) = 1$, it returns m to $\mathcal{A}$. Otherwise, it returns $\bot$.
4. At some point, $\mathcal{A}$ outputs two messages m_0, m_1. Adversary $\mathcal{A}'$ outputs the messages $m_0 \circ \mathsf{dec}^*$ and $m_1 \circ \mathsf{dec}^*$, and receives in return a ciphertext C^*. It computes $\mathsf{tag}^* = \mathsf{Mac}_{r^*}(C^*)$ and returns $\langle \mathsf{com}^*, C^*, \mathsf{tag}^*\rangle$ to $\mathcal{A}$.
5. $\mathcal{A}$ may continue to make decryption oracle queries, and these are answered as before. (Recall, $\mathcal{A}$ may not query the decryption oracle on the challenge ciphertext itself.)
6. Finally, $\mathcal{A}$ outputs a guess b'; this same guess is output by $\mathcal{A}'$.

Note that $\mathcal{A}'$ represents a legal strategy for attacking the underlying IBE scheme in a selective-identity, chosen-plaintext attack; in particular, $\mathcal{A}'$ never requests the secret key corresponding to "target identity" com*.

Before analyzing the success probability of A', we prove a claim bounding the probability of a certain event. Say a ciphertext $\langle \mathsf{com}, C, \mathsf{tag}\rangle$ is *valid* if decryption of this ciphertext would not result in $\bot$. Let Valid denote the event that A ever submits a ciphertext $\langle \text{com}^*, C, \text{tag} \rangle$ to its decryption oracle which is valid. (We always implicitly assume that $\langle \text{com}^*, C, \text{tag} \rangle \neq \langle \text{com}^*, C^*, \text{tag}^* \rangle$ since this event is disallowed after A is given the challenge ciphertext, and occurs with only negligible probability before A is given the challenge ciphertext.)

Claim. Pr[Valid] is negligible.

Proof. Let Game 0 denote the original experiment in which A interacts with a real decryption oracle (and not the simulated decryption oracle provided by A'); we are interested in bounding $\text{Pr}_0[\text{Valid}]$. Let Equiv be the event that the adversary ever submits a ciphertext $\langle \text{com}^*, C, \text{tag} \rangle$ for which (1) C decrypts to some arbitrary $m \circ \text{dec}$ (using the secret key SK_{com^*}) and (2) $\mathcal{R}(\text{pub}, \text{com}^*, \text{dec}) = r$ with $r \notin \{r^*, \bot\}$. Let Forge be the event that Equiv does *not* occur, and A at some point submits a ciphertext $\langle \text{com}^*, C, \text{tag} \rangle$ such that $\text{Vrfy}_{r^*}(C, \text{tag}) = 1$. Clearly, we have $\text{Pr}_0[\text{Valid}] \leq \text{Pr}_0[\text{Equiv}] + \text{Pr}_0[\text{Forge}]$.

We first show that $\text{Pr}_0[\text{Equiv}]$ is negligible, by the binding property of the encapsulation scheme. Consider an adversary B acting as follows: given input $(1^k, \text{pub}, r^*, \text{com}^*, \text{dec}^*)$, adversary B generates (PK, msk) for the IBE scheme and runs A on inputs 1^k and (PK, pub). Whenever A makes a decryption oracle query, B can legitimately answer this query since B knows msk. When A submits its two messages m_0, m_1, adversary B simply chooses $b \in \{0, 1\}$ at random and encrypts m_b in the expected way to generate a completely valid ciphertext $\langle \text{com}^*, C^*, \text{tag}^* \rangle$ (B can easily do this since it has both r^* and dec^*). Now, if Equiv ever occurs then B learns dec such that $\mathcal{R}(\text{pub}, \text{com}^*, \text{dec}) \notin \{\bot, r^*\}$. But this exactly violates the binding property of $(\text{Setup}, \mathcal{S}, \mathcal{R})$.

We next show that $\text{Pr}_0[\text{Forge}]$ is negligible. Let $q(k)$ be a polynomial upper bound on the number of decryption queries made by A, and let Forge_i denote the event that Forge occurs for the first time on the ith decryption query of A. Let Forge'_i denote the event that the ith decryption query is of the form $\langle \text{com}^*, C, \text{tag} \rangle$ and $\text{Vrfy}_{r^*}(C, \text{tag}) = 1$ *when all previous decryption queries of the form $\langle \text{com}^*, C', \text{tag}' \rangle$ are answered with $\bot$* (without checking whether they are valid or not). We refer to this latter "game" (which formally depends on the i under consideration) as Game $0'$.

Note that $\text{Pr}_0[\text{Forge}] = \sum_{i=1}^{q(k)} \text{Pr}_0[\text{Forge}_i]$. Furthermore, for all i we have $\text{Pr}_{0'}[\text{Forge}'_i] \geq \text{Pr}_0[\text{Forge}_i]$. Letting $\text{Forge}' \stackrel{\text{def}}{=} \cup_i \text{Forge}'_i$, we obtain $\text{Pr}_0[\text{Forge}] \leq \text{Pr}_{0'}[\text{Forge}']$.

Define Game 1 which proceeds exactly as Game $0'$, except that A is now given a random encryption of $m_b \circ 0^{n(k)}$ instead of a random encryption of $m_b \circ \text{dec}^*$ (here, $n(k) \stackrel{\text{def}}{=} |\text{dec}^*|$; recall that Definition 6 requires the length of dec^* to be fixed for a given value of k). We claim that $|\text{Pr}_{0'}[\text{Forge}'] - \text{Pr}_1[\text{Forge}']|$ is negligible. Indeed, if this is not the case then we can easily construct an algorithm B attacking the security of the underlying IBE scheme:

- Given input 1^k, algorithm B runs $\text{Setup}(1^k)$ to generate pub and then runs $\mathcal{S}(1^k, \text{pub})$ to obtain $(r^*, \text{com}^*, \text{dec}^*)$. It outputs com^* as the target identity

and is then given the IBE parameters PK. Finally, it runs A on inputs 1^k and (PK, pub).
- Decryption queries of A are answered as follows:
  - Queries of the form $\langle \mathsf{com}, C, \mathsf{tag} \rangle$ with $\mathsf{com} \neq \mathsf{com}^*$ are answered by first querying $\mathsf{Der}'_{\mathsf{msk}}(\mathsf{com})$ to obtain SK_{com}, and then decrypting in the usual way.
  - Upon receiving a query of the form $\langle \mathsf{com}^*, C, \mathsf{tag} \rangle$, first check whether $\mathsf{Vrfy}_{r^*}(C, \mathsf{tag}) = 1$. If so, abort the experiment and output 1. Otherwise, return $\bot$ to A.
- Eventually, A sends a pair of messages m_0, m_1 to its encryption oracle. B selects a bit b at random, and sends $m_b \circ \mathsf{dec}^*$ and $m_b \circ 0^{n(k)}$ to its encryption oracle. It receives in return a challenge ciphertext C^*, and uses this to generate a ciphertext $\langle \mathsf{com}^*, C^*, \mathsf{tag}^* \rangle$ in the natural way.
- Further decryption queries of A are answered as above.
- If A halts and B has not previously aborted the experiment, then B outputs a random bit.

The probability that B outputs 1 when given an encryption of $m_b \circ \mathsf{dec}^*$ is $\frac{1}{2} + \frac{1}{2} \cdot \Pr_{0'}[\mathsf{Forge}']$. On the other hand, the probability that B outputs 1 when given an encryption of $m_b \circ 0^{n(k)}$ is $\frac{1}{2} + \frac{1}{2} \cdot \Pr_1[\mathsf{Forge}']$. Since the difference between these two probabilities must be negligible if the underlying IBE scheme is secure, this proves the current claim.

Define Game 2 which proceeds exactly as Game 1, except that the challenge ciphertext given to A is now constructed as follows: $\mathcal{S}(1^k, \mathsf{pub})$ is run to give $(r, \mathsf{com}^*, \mathsf{dec}^*)$ but an independent random key $r^* \in \{0,1\}^k$ is chosen as well. Compute $C^* \leftarrow \mathsf{Enc}_{PK}(\mathsf{com}^*, m \circ 0^{n(k)})$, followed by $\mathsf{tag}^* = \mathsf{Mac}_{r^*}(C^*)$. The challenge ciphertext, as usual, is $\langle \mathsf{com}^*, C^*, \mathsf{tag}^* \rangle$. We claim that the difference $|\Pr_1[\mathsf{Forge}'] - \Pr_2[\mathsf{Forge}']|$ is negligible. To see this, consider the following algorithm B breaking the hiding property of the encapsulation scheme:

- B is given input 1^k and $(\mathsf{pub}, \mathsf{com}^*, \tilde{r})$. It then runs $\mathsf{Setup}'(1^k)$ to generate (PK, msk), and runs A on input 1^k and (PK, pub).
- Decryption queries of A are answered as follows:
  - Queries of the form $\langle \mathsf{com}, C, \mathsf{tag} \rangle$ with $\mathsf{com} \neq \mathsf{com}^*$ are answered by running $\mathsf{Der}'_{\mathsf{msk}}(\mathsf{com})$ to obtain SK_{com}, and then decrypting in the usual way.
  - Upon receiving a query of the form $\langle \mathsf{com}^*, C, \mathsf{tag} \rangle$, first check whether $\mathsf{Vrfy}_{\tilde{r}}(C, \mathsf{tag}) = 1$. If so, abort the experiment and output 1. Otherwise, return $\bot$ to A.
- Eventually, A sends a pair of messages m_0, m_1 to its encryption oracle. B selects a bit b at random and proceeds as follows: it computes $C^* \leftarrow \mathsf{Enc}_{PK}(\mathsf{com}^*, m_b \circ 0^{n(k)})$, computes $\mathsf{tag}^* = \mathsf{Mac}_{\tilde{r}}(C^*)$, and returns the challenge ciphertext $\langle \mathsf{com}^*, C^*, \mathsf{tag}^* \rangle$ to A.
- Further decryption queries of A are answered as above.
- If A halts and B has not previously aborted the experiment, then B outputs a random bit.

Now, if $\tilde{r}$ is such that $(\tilde{r}, \mathsf{com}^*, \mathsf{dec}^*)$ was output by $\mathcal{S}(1^k, \mathsf{pub})$ then the view of A is exactly as in Game 1 and so the probability that B outputs 1 in this case is $\frac{1}{2}(1 + \Pr_1[\mathsf{Forge}'])$. On the other hand, if $\tilde{r}$ is chosen independently of com^* then the view of A is exactly as in Game 2 and so the probability that B outputs 1 in this case is $\frac{1}{2}(1 + \Pr_2[\mathsf{Forge}'])$. Since the difference between these two probabilities must be negligible by the hiding property of the encapsulation scheme, this proves the current claim.

Finally, we claim that $\Pr_2[\mathsf{Forge}']$ is negligible. This follows quite easily from the security of the message authentication code, and we omit the details here. This completes the proof of the claim. □

Given the preceding claim, we see that the simulation which A' provides for A is statistically close to a real execution of A: in particular, the only difference occurs when Valid occurs. We therefore conclude that the advantage of A' is negligibly close to the advantage of A. Since the advantage of A' is negligible under the assumed security of the underlying IBE, the advantage of A must be negligible as well. This completes the proof of Theorem 1. □

4 Efficient Instantiations

Here, we describe two particular instantiations of our scheme by describing specific instantiations of the various primitives.

IBE Schemes. Boneh and Boyen [5] recently proposed two efficient IBE schemes suitable for our purposes. We refer to [5] for the full details and content ourselves with giving only a high-level description of their first scheme here. Let $\mathbb{G}$ and $\mathbb{G}_1$ be two (multiplicative) cyclic groups of prime order q for which there exists an efficiently-computable map $\hat{e} : \mathbb{G} \times \mathbb{G} \to \mathbb{G}_1$ which is *bilinear* and *non-degenerate*. Namely, (1) for all $\mu, \nu \in \mathbb{G}$ and $a, b \in \mathbb{Z}_q$ we have $\hat{e}(\mu^a, \nu^b) = e(\mu, \nu)^{ab}$, and (2) $\hat{e}(g, g) \neq 1$ for some generator g of $\mathbb{G}$. The IBE scheme is defined as follows:

Setup: Pick random generators g, g_1, g_2 of $\mathbb{G}$ and a random $x \in \mathbb{Z}_q$. Set $g_3 = g^x$ and $Z = \hat{e}(g_1, g_3)$. The master public key is $PK = (g, g_1, g_2, g_3, Z)$ and the master secret key is $\mathsf{msk} = x$.
Derive: To derive the secret key for "identity" $ID \in \mathbb{Z}_q$ using $\mathsf{msk} = x$, choose a random $r \in \mathbb{Z}_q$ and return the key $SK_{ID} = (g_1^x g_2^r g_3^{r \cdot ID}, g^r)$.
Encrypt: To encrypt a message $M \in \mathbb{G}_1$ with respect to "identity" $ID \in \mathbb{Z}_q$, choose a random $s \in \mathbb{Z}_q$ and output the ciphertext $(g^s, g_2^s g_3^{s \cdot ID}, M \cdot Z^s)$.
Decrypt: To decrypt ciphertext (A, B, C) using private key (K_1, K_2), output $C \cdot \hat{e}(B, K_2)/\hat{e}(A, K_1)$.

Correctness can be easily verified. Security of the above scheme is based on the decisional bilinear Diffie-Hellman (decision-BDH) problem. For efficiency, we assume that the master secret key msk contains the discrete logarithms of g_1, g_2, and g_3 with respect to base g, in which case generating SK_{ID} requires only two exponentiations.

The second IBE scheme of Boneh and Boyen [5] is more efficient than the above in terms of both key-generation and decryption time (the time required for encryption is essentially the same), but is based on a cryptographic assumption which is less standard.

Note that when the above scheme is used for key encapsulation (in the sense of Section 1.2), the sender need only send $(g^s, g_2^s g_3^{s \cdot ID})$ and compute the key $H_\alpha(Z^s)$ where H is a keyed hash function (see below); the receiver, given ciphertext $\langle A, B \rangle$, computes the matching key $H_\alpha(\hat{e}(A, K_1)/\hat{e}(B, K_2))$, where K_1, K_2 are as before. In this description, H represents a keyed hash function where the key α is included as part of the receiver's public key. Under the decisional-BDH assumption, it suffices for H to be chosen from a pairwise-independent hash family in order for the scheme to be secure. We remark, however, that this encapsulation scheme is also secure under a potentially weaker "hash BDH" assumption as well (and a similar remark holds also for the second IBE scheme of [5]). See further discussion at the end of this section.

Message Authentication Codes. A number of efficient message authentication codes are known, and we do not suggest any particular one. We stress that we only require "one-time" security (cf. Definition 5) and so efficient schemes which do not rely on any computational assumptions (e.g., [35]) may be used. Furthermore, messages to be authenticated have a (known) fixed length; this enables slight optimizations and/or simplifications of known schemes.

Encapsulation Schemes. We suggest an encapsulation scheme based on a fixed cryptographic hash function $H: \{0,1\}^{448} \to \{0,1\}^{128}$ (constructed, e.g., by suitably modifying the output length of SHA-1), and for a particular choice of security parameters; it is easy to adapt the scheme for the more general case. Our scheme works as follows:

- Setup chooses a hash function h from a family of pairwise independent hash functions mapping 448-bit strings to 128-bit strings, and outputs pub = h.
- The encapsulation algorithm $\mathcal{S}$ takes pub as input, chooses a random $x \in \{0,1\}^{448}$, and then outputs $(r = h(x), \text{com} = H(x), \text{dec} = x)$.
- The recovery algorithm $\mathcal{R}$ takes as input (pub = h, com, dec) and outputs $h(\text{dec})$ if $H(\text{dec}) = \text{com}$, and $\perp$ otherwise.

Note that binding holds as long as it is infeasible to find a dec' $\neq$ dec such that $H(\text{dec}') = H(\text{dec})$, *where* dec *is chosen uniformly at random* (cf. Definition 7). Thus, binding holds as long as H is second-preimage resistant (the construction can be easily modified so as to be based on UOWHFs by simply having Setup choose a key h' for a UOWHF and including h' in pub); collision-resistance is not necessary[3]. Furthermore, the above scheme satisfies statistical hiding. More specifically:

---

[3] This also explains why an output length of 128 bits for H should provide a sufficient level of security.

Theorem 2. *For the encapsulation scheme described above, the statistical difference between the following distributions is at most* 2^{-63}:

(1) $\{\mathsf{pub} \leftarrow \mathsf{Setup}; (r, \mathsf{com}, \mathsf{dec}) \leftarrow \mathcal{S}(\mathsf{pub}) : (\mathsf{pub}, \mathsf{com}, r)\}$
(2) $\{\mathsf{pub} \leftarrow \mathsf{Setup}; (r, \mathsf{com}, \mathsf{dec}) \leftarrow \mathcal{S}(\mathsf{pub}); r' \leftarrow \{0,1\}^{128} : (\mathsf{pub}, \mathsf{com}, r')\}$.

Proof (Sketch). The idea is loosely based on [16,21], but our proof is much simpler. For any $x \in \{0,1\}^{448}$, let $N_x \stackrel{\text{def}}{=} \{x' \mid H(x') = H(x)\}$ (this is simply the set of elements hashing to $H(x)$). Call x *good* if $|N_x| \geq 2^{255}$, and *bad* otherwise. Since the output length of H is 128 bits, there are at most $2^{255} \cdot 2^{128} = 2^{383}$ bad x's; thus, the probability that an x chosen uniformly at random from $\{0,1\}^{448}$ is bad is at most 2^{-65}.

Assuming x is good, the min-entropy of x – given pub and com – is at least 255 bits since every $\tilde{x} \in N_x$ is equally likely. Viewing h as a strong extractor (or, equivalently, applying the leftover-hash lemma [22]) we see that $\{h, H(x), h(x)\}$ has statistical difference at most 2^{-64} from $\{h, H(x), U_{128}\}$, where U_{128} represents the uniform distribution over $\{0,1\}^{128}$. The theorem follows easily. □

A Concrete Scheme. Given the primitives above, we may construct a CCA-secure encryption scheme as described in the previous section. However, as discussed in Section 1.2, improved efficiency can be obtained by directly constructing a hybrid encryption scheme; we do so here.

Key Generation requires running the key-generation algorithm for the underlying IBE scheme and then choosing a hash function h from a family of pairwise independent hash functions.

Encryption of a message M involves (1) running the encapsulation scheme to obtain $(k = h(x), ID = H(x), x)$; (2) using the underlying IBE as a key encapsulation scheme, with identity ID, to generate a ciphertext C_1 encapsulating a key k'; (3) using k' to encrypt $M \circ x$ by, for example, computing $C_2 = G(k') \oplus (M \circ x)$, where G is a PRG; (4) computing a MAC on C_1, C_2 using key k.

The ciphertext consists of ID, C_1, C_2, and the tag output by the MAC.

Decryption of ciphertext $(ID, C_1, C_2, \mathsf{tag})$ is done in the obvious way: recover k' from C_1 (using identity ID), recover $M \circ x$ from C_2, and compute $k = h(x)$. If $H(x) = ID$ and $\mathsf{Vrfy}_k((C_1, C_2), \mathsf{tag}) = 1$, then output M; otherwise, output $\bot$.

We tabulate the efficiency of our schemes, and compare them to the scheme of Kurosawa-Desmedt [25], in Table 1. Scheme 1 is instantiated using the first IBE from [5], as described above; scheme 2 is instantiated using the second IBE from [5]. During encryption all bases of exponentiation are fixed which potentially enables further speed-up by pre-computation. In Scheme 1 we assume that g_1, g_3 are generated by raising the fixed generator g to a random power. Hence, computing $\hat{e}(g_1, g_3)$ requires only a single exponentiation assuming $\hat{e}(g, g)$ is pre-computed.

In addition to comparing the efficiency of these various schemes, it is interesting also to compare the cryptographic assumptions on which they are based.

Table 1. Efficiency comparison for CCA-secure hybrid encryption schemes. When tabulating computational efficiency, "private-key" operations (hash function/block cipher evaluations) are ignored, and one multi-exponentiation is counted as 1.5 exponentiations. Ciphertext overhead represents the difference (in bits) between the ciphertext length and the message length, and $|p|$ is the length (in bits) of a group element. "p-exp" refers to an exponentiation relative to a fixed base.

	Encryption	Decryption	Key generation	Ciphertext overhead
Scheme 1	3.5 p-exps.	2 p-exps. + 2 pairings	3 exps.	$2|p| + 704$
Scheme 2	3.5 p-exps.	1.5 exps. + 1 pairing	2 exps.	$2|p| + 704$
KD [25]	3.5 p-exps.	1.5 exps.	3 exps.	$2|p| + 128$

Security of the Kurosawa-Desmedt scheme (as in the case of the Cramer-Shoup scheme [13] on which it is based) *inherently* relies on the decisional Diffie-Hellman assumption, and it does not seem possible to obtain provable security using a weaker variant of this assumption. In contrast, as noted earlier, our schemes may be proven secure under "hash BDH"-type assumptions which are potentially weaker than the decisional-BDH assumption[4].

5 Conclusions

We present an efficient methodology for constructing CCA-secure public-key cryptosystems from weak identity-based encryption schemes. Our construction adds only a MAC and a weak "commitment" to the original IBE system. Consequently, performance of the resulting public-key system is very close to the performance of the underlying IBE scheme. This improves on a previous transformation of Canetti, et al. which relies on the use of one-time signature schemes.

Applying our construction to recent IBE systems of Boneh and Boyen we obtain an efficient CCA-secure public-key cryptosystem without random oracles. Encryption (and, in one case, key generation) in the resulting systems are more efficient than in the Cramer-Shoup scheme, and on par with the recent proposal of Kurosawa and Desmedt. Decryption time and ciphertext size are comparable, though a bit worse. Our schemes are also somewhat more flexible than the Kurosawa-Desmedt scheme in terms of the cryptographic assumptions needed to obtain a proof of security. Our results show that building CCA-secure systems from IBE can produce very efficient schemes. The resulting schemes, as well as the proofs of security, are very different from those based on the work of Cramer and Shoup.

---

[4] In fact, we may base security of our constructions on purely *computational* – rather than *decisional* – assumptions; e.g., the computational-BDH assumption (using hard-core bits to encrypt one bit at a time). Although this no longer yields a practical scheme, it achieves CCA-secure encryption based on a computational assumption while avoiding the extreme inefficiency of NIZK proofs.

References

1. M. Bellare, A. Desai, D. Pointcheval, and P. Rogaway. Relations Among Notions of Security for Public-Key Encryption Schemes. *Adv. in Cryptology – Crypto 1998*, LNCS vol. 1462, Springer-Verlag, pp. 26–45, 1998.
2. M. Bellare and P. Rogaway. Random Oracles are Practical: a Paradigm for Designing Efficient Protocols. *First ACM Conf. on Computer and Comm. Security*, ACM, pp. 62–73, 1993.
3. D. Bleichenbacher. Chosen Ciphertext Attacks Against Protocols Based on the RSA Encryption Standard PKCS#1. *Adv. in Cryptology – Crypto 1998*, LNCS vol. 1462, Springer-Verlag, pp. 1–12, 1998.
4. M. Blum, P. Feldman, and S. Micali. Non-Interactive Zero-Knowledge and its Applications. *20th ACM Symposium on Theory of Computing (STOC)*, ACM, pp. 103–112, 1988.
5. D. Boneh and X. Boyen. Efficient Selective-ID Secure Identity Based Encryption Without Random Oracles. *Adv. in Cryptology – Eurocrypt 2004*, LNCS vol. 3027, Springer-Verlag, pp. 223–238, 2004. Full version available from http://eprint.iacr.org/2004/172
6. D. Boneh and X. Boyen. Secure Identity Based Encryption Without Random Oracles. *Adv. in Cryptology – Crypto 2004*, LNCS vol. 3152, Springer-Verlag, pp. 443–459, 2004.
7. D. Boneh and M. Franklin. Identity-Based Encryption from the Weil Pairing. *Adv. in Cryptology – Crypto 2001*, LNCS vol. 2139, Springer-Verlag, pp. 213–229, 2001. Full version in *SIAM J. Computing* 32(3): 586–615, 2003 and available from http://crypto.stanford.edu/~dabo/pubs.html
8. R. Canetti. Universally Composable Security: A New Paradigm for Cryptographic Protocols. *42nd IEEE Symp. on Foundations of Computer Science (FOCS)*, IEEE, pp. 136–145, 2001. Full version available at http://eprint.iacr.org/2000/067/
9. R. Canetti, O. Goldreich, and S. Halevi. The Random Oracle Methodology, Revisited. *30th ACM Symp. on Theory of Computing (STOC)*, ACM, pp. 209–218, 1998.
10. R. Canetti, S. Halevi, and J. Katz. A Forward-Secure Public-Key Encryption Scheme. *Adv. in Cryptology – Eurocrypt 2003*, LNCS vol. 2656, Springer-Verlag, pp. 255–271, 2003. Full version available at http://eprint.iacr.org/2003/083
11. R. Canetti, S. Halevi, and J. Katz. Chosen-Ciphertext Security from Identity-Based Encryption. *Adv. in Cryptology – Eurocrypt 2004*, LNCS vol. 3027, Springer-Verlag, pp. 207–222, 2004.
12. C. Cocks. An Identity-Based Encryption Scheme Based on Quadratic Residues. *Cryptography and Coding*, LNCS vol. 2260, Springer-Verlag, pp. 360–363, 2001.
13. R. Cramer and V. Shoup. A Practical Public Key Cryptosystem Provably Secure Against Chosen Ciphertext Attack. *Adv. in Cryptology – Crypto 1998*, LNCS vol. 1462, Springer-Verlag, pp. 13–25, 1998.
14. R. Cramer and V. Shoup. Universal Hash Proofs and a Paradigm for Adaptive Chosen Ciphertext Secure Public-Key Encryption. *Adv. in Cryptology – Eurocrypt 2002*, LNCS vol. 2332, Springer-Verlag, pp. 45–64, 2002.
15. J. Camenisch and V. Shoup. Practical Verifiable Encryption and Decryption of Discrete Logarithms. *Adv. in Cryptology – Crypto 2003*, LNCS vol. 2729, Springer-Verlag, pp. 126–144, 2003.
16. I. Damgård, T.P. Pedersen, and B. Pfitzmann. On the Existence of Statistically-Hiding Bit Commitment Schemes and Fail-Stop Signatures. *Adv. in Cryptology – Crypto 1993*, LNCS vol. 773, Springer-Verlag, pp. 250–265, 1993.

17. D. Dolev, C. Dwork, and M. Naor. Non-Malleable Cryptography. *SIAM J. Computing* 30(2): 391–437, 2000.
18. U. Feige, D. Lapidot, and A. Shamir. Multiple Non-Interactive Zero-Knowledge Proofs Under General Assumptions. *SIAM J. Computing* 29(1): 1–28, 1999.
19. R. Gennaro and Y. Lindell. A Framework for Password-Based Authenticated Key Exchange. *Adv. in Cryptology – Eurocrypt 2003*, LNCS vol. 2656, Springer-Verlag, pp. 524–543, 2003.
20. S. Goldwasser, S. Micali, and R. Rivest. A Digital Signature Scheme Secure against Adaptive Chosen-Message Attacks. *SIAM J. Computing* 17(2): 281–308, 1988.
21. S. Halevi and S. Micali. Practical and Provably-Secure Commitment Schemes from Collision-Free Hashing. *Adv. in Cryptology – Crypto 1996*, LNCS vol. 1109, Springer-Verlag, pp. 201–215, 1996.
22. J. Håstad, R. Impagliazzo, L. Levin, and M. Luby. Construction of a Pseudorandom Generator from any One-Way Function. *SIAM J. Comp.* 28(4): 1364–1396, 1999.
23. N. Howgrave-Graham, P. Q. Nguyen, D. Pointcheval, J. Proos, J. H. Silverman, A. Singer, and W. Whyte. The Impact of Decryption Failures on the Security of NTRU Encryption. *Adv. in Cryptology – Crypto 2003*, LNCS vol. 2729, Springer-Verlag, pp. 226–246, 2003.
24. M. Joye, J.-J. Quisquater, and M. Yung. On the Power of Misbehaving Adversaries and Security Analysis of the Original EPOC. *Cryptographers' Track – RSA 2001*, LNCS vol. 2020, Springer-Verlag, pp. 208–222, 2001.
25. K. Kurosawa and Y. Desmedt. A New Paradigm of Hybrid Encryption Scheme. *Adv. in Cryptology – Crypto 2004*, LNCS vol. 3152, Springer-Verlag, pp. 426–442, 2004.
26. L. Lamport. Constructing Digital Signatures from a One-Way Function. Technical Report CSL-98, SRI International, 1978.
27. Y. Lindell. A Simpler Construction of CCA-Secure Public-Key Encryption Under General Assumptions. *Adv. in Cryptology – Eurocrypt 2003*, LNCS vol. 2656, Springer-Verlag, pp. 241–254, 2003.
28. M. Naor and M. Yung. Universal One-Way Hash Functions and Their Cryptographic Applications. *21st ACM Symposium on Theory of Computing (STOC)*, ACM, pp. 33–43, 1989.
29. C. Rackoff and D. Simon. Non-Interactive Zero-Knowledge Proof of Knowledge and Chosen Ciphertext Attack. *Adv. in Cryptology – Crypto 1991*, LNCS vol. 576, Springer-Verlag, pp. 433–444, 1992.
30. A. Sahai. Non-Malleable Non-Interactive Zero Knowledge and Adaptive Chosen-Ciphertext Security. *40th IEEE Symposium on Foundations of Computer Science (FOCS)*, IEEE, pp. 543–553, 1999.
31. A. Shamir. Identity-Based Cryptosystems and Signature Schemes. *Adv. in Cryptology – Crypto 1984*, LNCS vol. 196, Springer-Verlag, pp. 47–53, 1985.
32. V. Shoup. Why Chosen Ciphertext Security Matters. IBM Research Report RZ 3076, November, 1998. Available at http://www.shoup.net/papers.
33. V. Shoup. Using Hash Functions as a Hedge Against Chosen Ciphertext Attack. *Adv. in Cryptology – Eurocrypt 2000*, LNCS vol. 275–288, Springer-Verlag, pp. 1807, 2000.
34. B. Waters. Efficient Identity-Based Encryption Without Random Oracles. Available at http://eprint.iacr.org/2004/180
35. M.N. Wegman and J.L. Carter. New Hash Functions and Their Use in Authentication and Set Equality. *J. Computer System Sciences* 22(3): 265–279, 1981.

A Generic Conversion with Optimal Redundancy

Yang Cui[1], Kazukuni Kobara[2], and Hideki Imai[2]

[1] Dept. of Information & Communication Engineering, University of Tokyo
cuiyang@imailab.iis.u-tokyo.ac.jp
[2] Institute of Industrial Science, University of Tokyo,
Komaba 4-6-1, Meguro-Ku, Tokyo, 153-8505, Japan
{kobara,imai}@iis.u-tokyo.ac.jp
http://imailab-www.iis.u-tokyo.ac.jp/imailab.html

Abstract. In this paper, we present a generic asymmetric encryption conversion ROC, namely *Redundancy Optimal Conversion*, which has the *optimal* message redundancy for one-way trapdoor function in the random oracle model. To our best knowledge, it is the first generic conversion to achieve such an optimal redundancy result for both one-way trapdoor permutation and not length-preserving function.

To obtain IND-CCA security, the conversion only needs the weaker requirement of the one-wayness, than the partial-domain one-wayness, which succeeds to greatly extend the application area of the generic conversion. Further, plaintext awareness property of the encryption is not required any more, which also contributes to reduce the message redundancy and hence removes the re-encryption step of the decryption process, considerably reducing the computational burden. Finally, it has simple construction of two cryptographic hash functions and two bitwise XORs, as same as the widely used OAEP conversion, but more generally useful.

Keywords: Optimal redundancy, IND-CCA, conversion, plaintext awareness.

1 Introduction

The requirement of the strong confidentiality in active attack scenarios motivates the cryptographic encryption scheme to be IND-CCA (*i.e. polynomial-time indistinguishability against adaptive chosen ciphertext attacks*) [15, 24] secure. Actually, the underlying security notion has been respected as the standard by both academics and industrials. Rather than thwarting the inversion of the one-way trapdoor function wholly in polynomial-time, it is definitely harder preventing any single bit of the ciphertext from being distinguished, against the adversary without the knowledge of the trapdoor. Cryptographic researchers pay more randomness or redundancy to cope with the hardness of achieving this top level security. While at the same time, too much randomness will aggravate the load of the cryptosystem, especially in communication cost, which is extremely disliked in the bandwidth limited network.

From a view point of redundancy, requirement of desired security varies much. There have been some general methodologies to achieve IND-CCA [10, 3, 8, 5] for the asymmetric (public key) encryption. The first general methodology employs the technique of Naor and Yung's paradigm [18], where the plaintext is encrypted twice by distinct keys, and a non-interactive zero-knowledge (NIZK) proof is built to prove that the two ciphertexts are indeed encryptions of the same plaintext, which intuitively makes the decryption oracle useless. Although their paradigm only achieves *indifferent* security, the NIZK technique has become a powerful tool to build a CCA secure scheme, as the first CCA secure scheme by Dolev, Dwork and Naor [10]. However, schemes by this technique are not practical at all, in that the NIZK proof is involved, thus this methodology is only considered meaningful in theoretical aspect, unless some efficient NIZK proof is invented.

Another methodology, Cramer-Shoup's paradigm [8, 9], yields the first practical and provably CCA secure encryption scheme in the standard model (i.e. without using the random oracle), and has been extended to more general case by universal hash proofs. The proof is technically based on specific assumptions, i.e. *Decision Diffie Hellman* problem or *Deciding Quadratic Residuosity problem*. Note that the above scheme is practical, but it seems not able to beat the performance of random oracle based schemes in speed and message expansion efficiency.

The random oracle model, by Bellare and Rogaway [2], is the most widely used methodology, powerful to design provably secure and very efficient schemes. Since original asymmetric encryption primitive has rarely achieved CCA security, some famous generic conversion like *Optimal Asymmetric Encryption Padding* (OAEP) [3] is constructed in this model, and used to preprocess the message and additional randomness, together with some asymmetric encryption primitive, for instance, RSA or Rabin, yielding IND-CCA security.

Obviously, among the above main methodologies[1], generic conversions in the random oracle model are entitled the most practical and the best efficient. In the present paper, we study the least message redundancy of generic conversions to achieve IND-CCA security in the random oracle model, not only for specific one-way trapdoor permutation (deterministic), but also for not length-preserving one-way trapdoor function (probabilistic), and provide such a conversion that needs nearly the same elements as original OAEP (i.e. easy to be modified from OAEP), but bases on a distinct assumption and gets a better performance.

1.1 Related Work

OAEP conversion has several variants, such as SAEP [4], OAEP+ [25], OAEP++ [17] etc., which are all equipped with plaintext awareness (PA), by means of consequent zero bits or one more hash function, which will definitely raise the redundancy. And more significantly these variants are not capable to deal with not

---

[1] Recently, Canetti et al. [5] presents a generic CCA conversion from any weak secure Identity Based Encryption (IBE), which is in another context, and we do not need the IBE property in particular.

length-preserving function, like ElGamal. And thanks to Fujisaki and Okamoto's generic conversions [12, 13], the not length-preserving function can be enhanced to CCA security. Their first conversion is only applicable to IND-CPA (*i.e. polynomial-time indistinguishability against chosen plaintext attacks*) primitive, and soon improved to the second one, which obtains CCA security from a weak requirement OW-CPA, applicable to numerous primitives. Later, Pointcheval [23] gets a similar conversion.

More recently, conversions dependent on Gap problem [19], such as, REACT [20], GEM [6] and two schemes by [7] are introduced for the merits of "on the fly" speed or size efficiency. Except [7], all of these conversions can deal with both not length-preserving function and permutation, however, unfortunately they get a redundant construction for validity check, or the property PA. The notion of PA is presented in [3] and improved in [1]. Roughly speaking, it means that if an adversary has successfully created a valid ciphertext, she must know that corresponding plaintext. It probably derives from the intuition to force the adversary not able to get the help from the decryption oracle, but appears artificial. The recent work [11] has presented another notion that may be sufficient and necessary to generate IND-CCA security in the random oracle model. And our motivation of this work focuses on removing the redundancy from plaintext awareness.

In Asiacrypt'03, Phan and Pointcheval [21] have put forward two CCA secure schemes, one is in the so-called random permutation model, stronger required than random oracle; the other is a 3-round OAEP without PA, which they claim to be "without redundancy". Both of them can only be applied to one-way trapdoor permutation, and the practical 3-round OAEP still suffers from the same restriction as the original OAEP that only partial-domain one-wayness secure primitive is applicable [14].

Note that their "without redundancy" means the expanded ciphertext size is zero beyond both the plaintext size and necessary randomness size for CCA security. We hereby define the redundancy just as the expanded size of ciphertext over the plaintext. We say optimal redundancy if the ciphertext size over plaintext can not be reduced given that the IND-CCA security is guaranteed corresponding to the security parameter κ. Indeed, we find out that our conversion also achieves their property of "without redundancy" in the occasion of length-preserving permutation, and more generally implicate the analogous scenario for not length-preserving trapdoor function.

1.2 Our Contribution

In this paper, we build the first generic conversion scheme ROC, which is able to generate IND-CCA security from both one-way trapdoor permutation and not length-preserving trapdoor function primitives with optimal redundancy, given that the underlying primitives are OW-PCA (*one-wayness against plaintext check attack*) secure in the random oracle model.

Further, plaintext awareness property of the encryption is not required any more, which also contributes to reduce the message redundancy and hence removes the re-encryption step of the decryption process, considerably reducing the computational burden. Finally, it has simple construction of two cryptographic hash functions and two bitwise XORs, as same as the widely used OAEP conversion, easily obtained from original OAEP conversion, but makes it fit for almost all asymmetric primitives.

The paper will first start with some basic security notions in the next section, then introduce the new construction of ROC in the section 3, and prove the security of the conversion in the random oracle model in section 4, at last explain the optimal result by a comparison of the generic conversions in section 5 and draw a conclusion.

2 Security Notions

Here, in this section, we recall some basic security definitions so that we can conclude the security requirements of our schemes. At first, we give some notations and conventions. We define that $x \xleftarrow{R} X$ as sampling randomly. The Ω and $\mathcal{M}$ are the random coin space and message space, respectively. We define $|\cdot|$ as the length. Further, if we say that $\epsilon(k)$ is negligible, we mean that for any constant c, there exists $k_0 \in \mathbb{N}$, such that $\epsilon < (1/k)^c$ for any $k > k_0$.

Definition 1 (Public Key Encryption). *Let $\Pi = (\mathcal{K}, \mathcal{E}, \mathcal{D})$ be a triple of algorithms:*

- *the key generation algorithm $\mathcal{K}$: on a secret input 1^k ($k \in \mathbb{N}$), in polynomial time in k, it probabilistically produces a pair of keys (pk, sk), publicly and secretly known respectively.*
- *the encryption algorithm $\mathcal{E}$: on input of message $m \in \{0,1\}^n$ and public key pk, the algorithm $\mathcal{E}(m; r)$ produces the ciphertext c of m, where $c \in \{0,1\}^*$, $r \in \Omega$ (for one-way permutation algorithm, r is omitted)*
- *the decryption algorithm $\mathcal{D}$: By using a ciphertext c and the secret key sk, $\mathcal{D}$ returns the plaintext m, or outputs $\bot$, when it is an invalid ciphertext. This algorithm is deterministic.*

Definition 2 (One Wayness). *In the notion of one wayness, there exists no such an adversary $\mathcal{A}$, with the public data only, that can get the whole preimage of the ciphertext in a polynomial time bound t, and with an inverting probability $\mathsf{Succ}^{\mathsf{ow}}$ more than negligible ϵ:*

$$\mathsf{Succ}^{\mathsf{ow}} = \Pr_{\substack{m \leftarrow \{0,1\}^n \\ r \xleftarrow{R} \Omega}} \left[(\mathsf{pk}, \mathsf{sk}) \leftarrow \mathcal{K}(1^k) : \mathcal{A}(\mathcal{E}_{\mathsf{pk}}(m; r)) = m \right]$$

Remark. It is known that the one-wayness notion is the least requirement for a encryption algorithm, but another notion called partial-domain one-wayness is obviously stronger requirement that need even part of ciphertext impossibly to be computed out by the adversary. Fujisaki et al. [14] have pointed that OAEP indeed need a partial-domain one-wayness assumption, but RSA-OAEP is still secure precisely because the random self-reducibility of RSA. Fortunately, in this paper, we do not need the strong requirement of partial-domain one-wayness even applying our conversion to RSA.

Definition 3 (IND-CCA). *A public key encryption scheme Π is IND-CCA secure, if there exists no polynomial time adversary $\mathcal{A} = (\mathcal{A}_1, \mathcal{A}_2)$ who, under the help of the decryption oracle, can distinguish the encryption c_b of two equal-length, distinct plaintexts m_0, m_1, $b \xleftarrow{R} \{0,1\}$, with the probability significantly greater than 1/2 (the only restriction is that the target ciphertext cannot be sent to the decryption oracle directly). More formally, the scheme is IND-CCA secure, if the advantage of adversary is less than negligible ϵ, (s is the last state information of $\mathcal{A}_1$, and $\hat{b} \in \{0,1\}$):*

$$\mathsf{Adv}_\Pi^{\mathsf{ind-cca}} = 2 \times \Pr_{\substack{b \xleftarrow{R} \{0,1\} \\ r \xleftarrow{R} \Omega}} \begin{bmatrix} (\mathsf{pk}, \mathsf{sk}) \leftarrow \mathcal{K}(1^k), \\ (m_0, m_1, s) \leftarrow \mathcal{A}_1(\mathsf{pk}), c \leftarrow \mathcal{E}_{\mathsf{pk}}(m_b; r), \\ \hat{b} \leftarrow \mathcal{A}_2(c, m_0, m_1, s) : \hat{b} = b \end{bmatrix} - 1$$

The intuition behind IND-CCA security is that any polynomial-time adversary can not gain even one bit useful information through the strongest attack. However, when it is derived from the random oracle model, a controversial notion called plaintext awareness has been built.

Definition 4 (Plaintext Awareness). *[1] For a public key encryption scheme $\Pi = (\mathcal{K}, \mathcal{E}, \mathcal{D})$, oracles family $\mathcal{H}$, B and K are algorithms called adversary and knowledge extractor, respectively. Let $\mathcal{Q}_H$ be all the queries between B and randomly chosen H, $\mathcal{C}$ as the set of queries answer by $\mathcal{E}_{\mathsf{pk}}^H$, and y be the output of B. For any $k \in \mathbb{N}$,*

$$\mathsf{Succ}_{\Pi,B,K}^{\mathsf{PA}} = \Pr \begin{bmatrix} H \xleftarrow{R} \mathcal{H}, (\mathsf{pk}, \mathsf{sk}) \leftarrow \mathcal{K}(1^k), \\ (\mathcal{Q}_H, \mathcal{C}, y) \leftarrow \mathsf{run} B^{H, \mathcal{E}_{\mathsf{pk}}^H}(\mathsf{pk}) : \\ K(\mathcal{Q}_H, \mathcal{C}, y, \mathsf{pk}) = \mathcal{D}_{\mathsf{sk}}^H(y) \end{bmatrix}$$

where $y \notin \mathcal{C}$, and K is a $\epsilon(k)$-extractor if K has run in polynomial time in the input length and, for any adversary B, $\mathsf{Succ}_{\Pi,B,K}^{\mathsf{PA}} \geq 1 - \epsilon(k)$. Then Π is secure in PA, if Π is IND-CPA secure and there exists a $\epsilon(k)$-extractor K with negligible $\epsilon(k)$.

Remark. The powerful PA directly denies the decryption oracle to help the adversary with useful information. An IND-CPA scheme together with PA obtains IND-CCA consequently. However, it is unnecessarily strong, and we will prove that our scheme is IND-CCA secure but not in PA.(see Sect. 4)

3 The New Construction

3.1 Intractability of Gap Problem

In this section, we present the new construction of generic conversion, which removes the artificial plaintext awareness assumption, and thus obtains an optimal encryption size overhead, where we solely need a weak security condition for both the permutation and not length-preserving primitives, called one-wayness against plaintext checking attack, as follows:

Definition 5 (OW-PCA). *[20] A public key encryption scheme Π is said to be OW-PCA, if there exists no such an adversary $\mathcal{A}$, with the public data and the help of the corresponding plaintext checking oracle $\mathcal{O}_{\sf pca}$, can get the whole preimage of the ciphertext with at most q queries to the $\mathcal{O}_{\sf pca}$ oracle, in a polynomial time bound t and a winning probability more than negligible ϵ, where the plaintext check oracle $\mathcal{O}_{\sf pca}$ takes as input a plaintext m and a ciphertext c, output 1 or 0 for checking whether c is the encryption of m or not:*

$$\text{Succ}_{\Pi,\mathcal{A}}^{\sf ow-pca} = \Pr_{\substack{m,m' \leftarrow \{0,1\}^n \\ r \stackrel{R}{\leftarrow} \Omega}} \left[\begin{array}{l} ({\sf pk}, {\sf sk}) \leftarrow \mathcal{K}(1^k), c \leftarrow \mathcal{E}_{\sf pk}(m;r), \\ \mathcal{O}_{\sf pca}(m';c) \stackrel{?}{=} 1 : m \leftarrow (t,q) \mathcal{A}^{\sf ow}(c) \end{array} \right]$$

Remark. For any deterministic public key encryption primitive, for example RSA, the OW-PCA provides an exact OW-CPA security immediately, because the adversary can encrypt the testing message by herself. And for the probabilistic ones, they should be reduced to a kind of Gap problem [19, 20], which is generally based upon the "gap" of one problem between in computation case and in decision case. Although it seems a little easier to solve than the computation problem, there is not yet any attack known. For instance, ElGamal just can employ our conversion based on the following GDH problem in a multiplicative group $\mathcal{G}$ of prime order p with a generator g:

1. *The Computational Diffie-Hellman Problem* (CDH): given a triple of $\mathcal{G}$ elements, (g, g^a, g^b), to find $g^c = g^{ab}$.
2. *The Decision Diffie-Hellman Problem* (DDH): given a quadruple of (g, g^a, g^b, g^c), to decide whether $c = ab \mod p$.
3. *The Gap Diffie-Hellman Problem* (GDH): given a oracle which can solve DDH problem, to find a solution of CDH problem.

Indeed, if taking the Discrete Logarithm problem as intractable, the above gap problem is yet considered not to be a polynomially solvable one. And since such kind of decision problem and computational problem are both well studied, we can use the intractability of them in appropriate group.

Certainly, there are still other gap problems, and due to no significant weakness is known for the gap problems, many encryption primitives can be based on them and proven to be OW-PCA secure. Thus, it is convenient to implement both deterministic primitive like RSA or Rabin, and probabilistic primitive like ElGamal with our conversion, even NTRU [16].

3.2 Generic Conversion

We build a generic conversion, ROC, which security relies on the OW-PCA assumption, and has an optimal redundancy.

Setup. k, k_1, n is equivalent to $|r|, |r'|, |m|$ respectively, s.t. $r, r' \stackrel{R}{\leftarrow} \Omega$, and $r \in \{0,1\}^k, r' \in \{0,1\}^{k_1}$; $m \in \mathcal{M} = \{0,1\}^n$. The primitive is a public key encryption algorithm Π satisfying the OW-PCA security, mapping the message to the domain $\mathcal{Z}$, $\ell = |\Pi_{output}|$, $\Pi : \mathcal{M} \to \mathcal{Z}$, $\mathcal{Z} \subseteq \{0,1\}^\ell$.

Assume the random oracle family $\mathcal{H}$, and $G, H \xleftarrow{R} \mathcal{H}$,
$$G : \{0,1\}^k \to \{0,1\}^n, H : \{0,1\}^{\ell+n} \to \{0,1\}^k.$$

Construction. The conversion ROC is defined as the following:
- it runs Π's key generation algorithm $\mathcal{K}$: output a pair (pk, sk).
- $\mathcal{E}_{\text{roc}}$: on input of m and r, first compute $x := m \oplus G(r)$. And by optional randomness r', the algorithm $\mathcal{E}_{\text{pk}}(x; r')$ produces the output z, the input and output of Π will be hashed by random oracle H, $t := H(z||x) \oplus r$. Finally, get the ciphertext $c := z||t$.
- $\mathcal{D}_{\text{roc}}$: On the input of c, parse cipher and use $\mathcal{D}_{\text{sk}}(z)$ to obtain message m, note that there is no re-encryption for checking.

Remark. This underlying scheme does not require the plaintext awareness, because anyone can make a valid ciphertext by himself, and any bit strings in $\mathcal{M}$ has a valid corresponding ciphertext according to our definition. This greatly reduces the redundancy cost of the encryption, and for our conversion scheme, one more merit lies in the deletion of re-encryption process, speeding up the scheme much, while some conversion has to run the burdened public key encryption once more to check the validity.

The conversion needs as the exactly OAEP does, and both the structure are similar, which also shows the merit of two-round Feistel network. But ROC can do with the not length-preserving functions as well, and assumes the one-wayness instead of partial-domain one-wayness. When Π is one-way trapdoor permutation, the conversion can be simplified further, in that the x and z have a one-to-one relation, thus the hash of (x) is sufficient this time. When the Π is not length-preserving function, which is the original OAEP or 3-round OAEP [21] can not cope with, our conversion can still base the security on the OW-PCA assumption.

And in both occasions, the message redundancy is optimal, let mLen and cLen be the input and output of the trapdoor function, respectively. Redundancy RE is as follows:

- for not length-preserving function, $\text{RE}_{\text{roc}} = |c| - |m| = \text{cLen} - \text{mLen} + k = \text{RE}_\Pi + k$.
- for permutation, $\text{RE}_{\text{roc}} = |c| - |m| = k$.

$\mathcal{E}nc$:
$$\begin{aligned} x &:= m \oplus G(r) \\ z &:= \mathcal{E}_{\text{pk}}(x; r') \\ t &:= H(z||x) \oplus r \\ c &:= z||t \end{aligned}$$

$\mathcal{D}ec$:
$$\begin{aligned} z||t &:= c \\ x &:= \mathcal{D}_{\text{sk}}(z) \\ r &:= H(z||x) \oplus t \\ m &:= x \oplus G(r) \end{aligned}$$

Fig. 1. Generic conversion ROC.

Our claimed optimal result derives from the observation that RE_Π is the self redundancy of the primitive, which obviously can not be reduced if the underlying encryption primitive is fixed. And k is the random number length, capable to be adapted to different security reduction cost. We can, in practice, choose $k = c(\kappa)$, (i.e. κ is the security parameter), where $c(\kappa)$ is the lower bound which makes the conversion IND-CCA secure corresponding to the required security level. Due to the quadratic reduction cost (see 4), ROC has to pay about 2κ-bit randomness for CCA security, which is the same as [21].

4 Security Analysis

The underlying generic conversion ROC is IND-CCA secure, but not in the sense of PA according to the following theorems.

4.1 Chosen Ciphertext Security

Theorem 1. *Let $\mathcal{A}$ be a CCA adversary who breaks the indistinguishability of the conversion ROC, with non-negligible advantage $Adv_{roc,\mathcal{A}}^{ind-cca}$ and polynomial-bounded running time τ, making at most q_G, q_H and q_D queries to the hash functions G, H and decryption oracle respectively. Then there exists an algorithm $\mathcal{B}$, breaking the OW-PCA security of the asymmetric encryption scheme Π with successful probability $Succ_{\Pi,\mathcal{B}}^{ow-pca}$ and running time τ', where*

$$Adv_{roc,\mathcal{A}}^{ind-cca} \leq Succ_{\Pi,\mathcal{B}}^{ow-pca} + q_G \cdot (\frac{2q_D+1}{2^k} + \frac{q_D}{2^n})$$
$$\tau' \leq (q_G + q_H) \cdot T_{\mathcal{O}_{pca}} + \tau$$

$T_{\mathcal{O}_{pca}}$ *is the execution time of the plaintext checking oracle $\mathcal{O}_{pca}$.*

Proof. Refer to the appendix A for the proof of reduction cost.

4.2 No Plaintext Awareness Property

For the sake of self-containess of the paper, although intuitively the proposed conversion is not satisfied with PA, precisely due to the observation that any ciphertext could be decoded to some valid plaintext, we still give a short explanation that is rigorously proved.

Theorem 2. *The proposed conversion scheme ROC does not meet the requirement of the plaintext awareness PA. Formally, there exists a polynomial time adversary B, for which the knowledge extractor K cannot output the plaintext with the following winning probability,*

$$Succ_{roc,B,K}^{PA} > 2^{-n}$$

Proof. If there exists a knowledge extractor K for the underlying conversion ROC, then depending upon the output of adversary B, (pk, z, t), K output a plaintext m'. Obviously, because K is disallowed to query the oracle G, H, the success probability is not overwhelmingly large with security parameter k, but only $Succ_{roc,B,K}^{PA} \leq 2^{-n}$.

Table 1. Comparison of the Data Redundancy.

	PA	IND-CCA	Re-enc	Redundancy*
FO	Yes	OW-CPA	Yes	cLen
Pointcheval	Yes	PDOW-CPA	Yes	cLen+R★
REACT	Yes	OW-PCA	No	cLen+H★
GEM	Yes	OW-PCA	No	cLen
ROC	No	OW-PCA	No	$RE_\Pi + c(\kappa)$

∗: Data Redundancy=Ciphertext Size − Plaintext Size.
★: R and H are randomness and hash output size respectively.

5 Comparison of Generic Conversions

In this section, we will have a rough comparison of the generic conversion in the literature.

First, our proposal is the first optimally redundant generic conversion for not length-preserving function[2], thus the advantage is obvious. (see Table 1). Recall that the cLen is the output size of the primitive, which must be larger than the function self redundancy cLen-mLen, we can conclude the winning on the message redundancy. And the absence of re-encryption speeds up the conversion as well.

Second, our scheme is also applicable to one-way permutations, like RSA. Although due to the security reduction, we can also get exactly the same $|m| + 2\kappa$ optimal bandwidth as [21] did, we should clarify that in this occasion our conversion is advantageous merely for encrypting an arbitrary long message, say 1024-bit long, which the 3-round OAEP can not transfer at one time, but ours does. And one more advantage is that security depends on the one-wayness rather than partial-domain one-wayness in that the OW-PCA is equivalent to OW-CPA in the deterministic encryption scenario.

6 Conclusion

We build the first generic conversion ROC with optimal redundancy in the random oracle model, which achieves IND-CCA security for any one-way trapdoor function given that the underlying primitives are OW-PCA secure solely. The optimally redundant conversion has many useful applications, and the security bases on the one-wayness rather than partial-domain one-wayness.

Acknowledgement

The authors would like to thank Goichiro Hanaoka, and anonymous referees for helpful comments.

---

[2] Recently, Fujisaki introduced another solution independently [11]; and in Asiacrypt'04 [22], the security of 3-round OAEP for probabilistic primitive is improved, but not IND-CCA secure.

References

1. M. Bellare, A. Desai, D. Pointcheval, P. Rogaway. Relations Among Notions of Security for Public-Key Encryption Schemes. In *Crypto'98*, LNCS 1462, pages 26-45, Springer-Verlag, 1998.
2. M. Bellare and P. Rogaway. Random Oracles Are Practical: A paradigm for designing efficient protocols, in Proc. First Annual Conference on Computer and Communications Security, ACM, 1993.
3. M. Bellare and P. Rogaway. Optimal Asymmetric Encryption - How to Encrypt with RSA. In *Eurocrypt'94*, LNCS 950, pages 92-111. Springer-Verlag, 1995.
4. D. Boneh. Simplified OAEP for the RSA and Rabin Functions. In *Crypto'01*, LNCS 2139, pages 275-291. Springer-Verlag, 2001.
5. R. Canetti, S. Halevi and J. Katz. Chosen-Ciphertext Security from Identity-Based Encryption. In *Eurocrypt'04*, LNCS 3027, pages 207-222. Springer-Verlag, 2004.
6. J. Coron, H. Handschuh, M. Joye, P. Paillier, D. Pointcheval, C. Tymen. GEM: A Generic Chosen-Ciphertext Secure Encryption Method. In *CT-RSA'02*, LNCS 2271, pages 263-276. Springer-Verlag, 2002.
7. Y. Cui, K. Kobara, H. Imai. Compact Conversion Schemes for the Probabilistic OW-PCA Primitives. In *ICICS 2003*, LNCS 2836, pages 269-279. Springer-Verlag, 2003.
8. R.Cramer, V.Shoup. A Practical Public Key Cryptosystem Provably Secure Against Adaptive Chosen Ciphertext Attack. In *Crypto'98*, LNCS 1462, pages 13-25, Springer-Verlag, 1998.
9. R. Cramer, V. Shoup. Universal Hash Proofs and a Paradigm for Adaptive Chosen Ciphertext Secure Public-Key Encryption. In *Eurocrypt'02*, LNCS 2332, pages 45-64, Springer-Verlag, 2002.
10. D. Dolev, C. Dwork, M. Naor: Nonmalleable Cryptography.(Extended Abstract). STOC 1991: 542-552.
11. E. Fujisaki. Plaintext-Simulatability. Cryptology ePrint Archive: Report 2004/218.
12. E. Fujisaki, T. Okamoto. How to Enhance the Security of Public-Key Encryption at Minimum Cost. In *PKC'99*, LNCS 1560, pages 53-68. Springer-Verlag, 1999.
13. E. Fujisaki, T. Okamoto. Secure Integration of Asymmetric and Symmetric Encryption Schemes, In *Crypto'99*, Springer-Verlag, LNCS 1666, pp.537-554, 1999.
14. E. Fujisaki, T. Okamoto, D.Pointcheval and J.Stern. RSA-OAEP Is Secure under the RSA Assumption. In *Crypto'01*, LNCS 2139, pages 260-274, Springer-Verlag, 2001.
15. S. Goldwasser, S. Micali. Probabilistic encryption. Journal of Computer Security, 28:270–299, 1984.
16. J. Hoffstein, J. Pipher, and J.H. Silverman. NTRU: a Ring based Public Key Cryptosystem. In Proc. of ANTS III, LNCS 1423, pages 267-288. Springer-Verlag, 1998. First presented at the rump session of Crypto '96.
17. K. Kobara, H. Imai. OAEP++ : A Very Simple Way to Apply OAEP to Deterministic OW-CPA Primitives. Cryptology ePrint Archive: Report 2002/130
18. M. Naor and M. Yung. Public-Key Cryptosystems Provably-Secure against Chosen-Ciphertext Attacks. 22nd ACM Symposium on Theory of Computing, pp. 427-437, 1990.
19. T. Okamoto, D. Pointcheval. The Gap-Problems: A New Class of Problems for the Security of Cryptographic Schemes. In *PKC'01*, LNCS 1992, pages 104-118, Springer-Verlag, 2001.

20. T. Okamoto, D. Pointcheval. REACT: Rapid Enhanced-Security Asymmetric Cryptosystem Transform. In CT-RSA 2001, LNCS 2020, pages 159-175, Springer-Verlag, 2001.
21. D.H. Phan, D. Pointcheval. Chosen-Ciphertext Security without Redundancy. In *Asiacrypt'03*, LNCS 2894, pages 1-18, Springer-Verlag, 2003.
22. D.H. Phan, D. Pointcheval. OAEP 3-Round: A Generic and Secure Asymmetric Encryption Padding. In *Asiacrypt'04*, LNCS, Springer-Verlag.
23. D.Pointcheval. Chosen-Ciphertext Security for Any One-Way Cryptosystem. In *PKC'00*, LNCS 1751, pages 129-146, Springer-Verlag, 2000.
24. C.Rackoff and D.Simon. Non-interactive Zero-knowledge Proof of Knowledge and Chosen Ciphertext Attack. In *Crypto'91*, LNCS 576, pages 433-444, Springer-Verlag, 1992.
25. V.Shoup. OAEP Reconsidered. In *Crypto'01*, LNCS 2139, pages 239-259, Springer-Verlag, 2001.

Appendix A: Proof of Theorem 1

The proof will be shown by a series of games as follows.

Initialization. We will define several games $\mathbf{G_i}$, $0 \leq i \leq 5$, which incrementally modifies the simulation process, at last leads to the challenge bit b completely independent of the view of the adversary. Note that the probability process is always on the same space, and the changes made to the game rules is step by step. In Game G_i, S_i is considered to be the event that $\hat{b} = b$, for $0 \leq i \leq 5$. And we make S_G as the set at which the r has been queried by adversary, and answer of G is g, (r, g) is kept in the S_G; similarly, S_H as the set of queries of H, in which $(z||x, h)$ is kept. Obviously, pairs appeared in the S_G, S_H include just part of the queries, and we want to show our simulation is still doable when the new query comes.

Game $\mathbf{G_0}$. This is the original IND-CCA game (definition 3),

$$(\mathsf{pk}, \mathsf{sk}) \leftarrow \mathcal{K}(1^\kappa), G, H \stackrel{R}{\leftarrow} \mathcal{H},$$
$$(m_0, m_1, s) \leftarrow \mathcal{A}_1^{G,H,\mathcal{D}_{\mathsf{sk}}}(\mathsf{pk}), m_0, m_1 \in \{0,1\}^n, m_0 \neq m_1,$$
$$b \stackrel{R}{\leftarrow} \{0,1\}, r^* \leftarrow \{0,1\}^k,$$
$$g^* \leftarrow G(r^*), m^* \leftarrow m_b,$$
$$x^* \leftarrow m^* \oplus g^*, z^* \leftarrow \mathcal{E}_{\mathsf{pk}}(x^*),$$
$$h^* \leftarrow H(z^*||x^*), t^* \leftarrow r^* \oplus h^*,$$
$$c^* \leftarrow (z^*||t^*),$$
$$\hat{b} \leftarrow \mathcal{A}_2^{G,H,\mathcal{D}_{\mathsf{sk}}}(c^*, m_0, m_1, s)$$

Note that $\Pr[S_0] = \Pr[\hat{b} = b]$, and the adversary has the access to decryption oracle $\mathcal{D}_{\mathsf{sk}}$, with the restriction that the challenge ciphertext c^* cannot be queried to it. The random oracles G, H are available for the adversary in the game.

Game $\mathbf{G_1}$. This game slightly modifies the original IND-CCA game, simulating the game as the challenger's view. The underlying oracles are all available. Holding the querying list of random oracles G, H, we answer the occurred queries just as what the lists tell us, and new query in the next way.

If a ciphertext $c = z||t$ ($c \neq c^*$), is queried, answer m as long as (c,m) has been asked to the decryption oracle $\mathcal{D}_{sk}$; otherwise, return a $m \in \{0,1\}^n$ randomly. More precisely, to check whether $(z||\sigma)$ in the S_H, $\sigma \in \{0,1\}^n$. If the (z, σ, h) pair is in the S_H, then let

$$r = t \oplus h$$

look up the list S_G, if $r \in S_G$, make the following setting $\mathsf{Ans}^{c \neq c^*}$:

$$h = H(z||x),\ r = t \oplus h,\ g = G(r),\ m = x \oplus g$$

If $r \notin S_G$, the same setting as the $\mathsf{Ans}^{c \neq c^*}$ is also able to return a random value hold the relation which cannot be distinguished by the adversary. Similarly, the underlying setting can be applied to deal with the event that $(z||\sigma)$ is not in the S_H, which finishes the simulation of the game $\mathbf{G_1}$. Thus, the successful probability of $\mathbf{G_1}$ is equivalent to that of $\mathbf{G_0}$.

$$\Pr[S_1] = \Pr[S_0]$$

Game $\mathbf{G_2}$. Now this game $\mathbf{G_2}$ generates the ciphertext c^* beforehand, and makes the view of corresponding plaintext m^* independent.

$$c^\sharp \xleftarrow{R} \{0,1\}^{k+z},\ c^* = c^\sharp$$

And we use such a lemma showing that $\Pr[S_1]$ is bounded.

Lemma 1. $\Pr[S_1]$ *is bounded by the following, where* $\Pr[AskH_{1.2}]$ *denotes the probability of event that* $z^\sharp || \sigma^\sharp$ *has been asked to the* H *oracle.*

$$\Pr[S_1] \leq 1/2 + q_G/2^k + \Pr[AskH_{1.2}]$$

Proof. When we simulate the game $\mathbf{G2}$, there are some events that error is output, but we show that this can be bounded by another event's probability, and so on. More precisely, the random chosen values beforehand mean other values fixed implicitly, thus the bound is computed by the following sub-games.

Game 1.1. First randomly choose the $r^\sharp \xleftarrow{R} \{0,1\}^k$, and $g^\sharp \xleftarrow{R} \{0,1\}^n$, then

$$r^* = r^\sharp, g^* = g^\sharp,$$
$$x^* = m^* \oplus g^\sharp,\ z^* = \mathcal{E}_{\mathsf{pk}}(x^*),\ h^* = H(z^*||x^*),\ t^* = r^\sharp \oplus h^*$$

$AskG_{1.1}$ denotes the event that r^* has been asked to the G, $r^* \in S_G$. Because the $\Pr[S_{1.1}] = \Pr[S_1]$ unless the $AskG_{1.1}$ occurred, there is

$$\Pr[S_{1.1} \wedge \neg AskG_{1.1}] = \Pr[S_1 \wedge \neg AskG_{1.1}]$$

therefore,

$$|\Pr[S_{1.1}] - \Pr[S_1]|$$
$$= |\Pr[S_{1.1} \wedge AskG_{1.1}] + \Pr[S_{1.1} \wedge \neg AskG_{1.1}] -$$
$$\Pr[S_1 \wedge AskG_{1.1}] - \Pr[S_1 \wedge \neg AskG_{1.1}]|$$
$$= |\Pr[S_{1.1} \wedge AskG_{1.1}] - \Pr[S_1 \wedge AskG_{1.1}]|$$
$$\leq \Pr[AskG_{1.1}]$$

Game 1.2. Randomly choose $r, x, h, r^\sharp \xleftarrow{R} \{0,1\}^k$, $x^\sharp \xleftarrow{R} \{0,1\}^n$, $h^\sharp \xleftarrow{R} \{0,1\}^k$. then implicitly there is,

$$r^* = r^\sharp, x^* = x^\sharp, h^* = h^\sharp,$$
$$g^* = m^* \oplus x^\sharp, t^* = r^\sharp \oplus h^\sharp$$

$AskH_{1.2}$ denotes that x^* has been asked to H, then there is,

$$|\Pr[AskG_{1.2}] - \Pr[AskG_{1.1}]| \le \Pr[AskH_{1.2}]$$

where, $\Pr[AskG_{1.2}]$ is at most $q_G/2^k$.

Game 1.3. Since randomly choose $t^\sharp \xleftarrow{R} \{0,1\}^k$, $x^\sharp \xleftarrow{R} \{0,1\}^n$, then

$$c^* = \mathcal{E}_{\mathsf{pk}}(x^\sharp) \| t^\sharp$$

c^* is totally random, thus

$$\Pr[AskH_{1.3}] = \Pr[AskH_{1.2}]$$

Noticing the probability of game 1.1 is just $1/2$, all above sub-games of $\mathbf{G_2}$ lead to an inequation that finishes the proof of Lemma 1.

$$\Pr[S_1] = 1/2 + q_G/2^k + \Pr[AskH_{1.2}]$$

Game $\mathbf{G_3}$. In this game, we try to simulate even though the elements of ciphertext appearing in the queries list in different order. The event $AskH_{2.2}$(same as $AskH_3$) denotes that certain $c = z\|t$ has been queried to the decryption oracle, but $z\|x \notin S_H$. Then the next lemma holds.

Lemma 2. *The following probability of the event is bounded by,*

$$|\Pr[AskH_{2.2}] - \Pr[AskH_{1.2}]| \le q_D \cdot q_G/2^k + q_D \cdot q_G/2^n$$

Proof. Similarly we build some sub-games to prove the above lemma.

Game 2.1. To assume that this game rules the as the previous $\mathbf{G_2}$ except that the event r has been asked to G before the $z\|t$ is asked to H, because the latter h is uniformly distributed, thus,

$$|\Pr[AskH_{2.1}] - \Pr[AskH_{1.2}]| \le q_D \cdot q_G/2^k$$

Game 2.2. Compared to *Game 2.1*, x has been asked to the H oracle, but there is no record in the S_G.

$$|\Pr[AskH_{2.2}] - \Pr[AskH_{2.1}]| \le q_D \cdot q_G/2^n$$

Summarizing the *Game 2.1, 2.2*, the proof of Lemma 2. has been proven.

Game G_4. After simulating the decryption oracle with different queries order to G, H oracles, we further try to deal with the adversary without the underlying oracles, but just using the record lists S_G, S_H. Note that the simulation is perfect except that the only bad event, due to the absence of random oracles, is the query to G before that query to H is made. Let us denote $F_4 = [AskG \wedge \neg AskH]$. Since until the $z\|t$ is asked to the H oracle, h is not fixed and uniformly distributed, thus

$$r = h \oplus t$$

and the probability of r has been asked to G is bounded by $q_G/2^k$, then the total probability of occurrence is bounded by $q_D \cdot q_G/2^k$. We get the following inequation:

$$|\Pr[AskH_4] - \Pr[AskH_3]| \leq \Pr[AskG \wedge \neg AskH] \leq q_D \cdot q_G/2^k$$

Game G_5. At last, we simulate the decryption oracle without the sk, which can show us the query-answer pair correctly, if it has been recorded in the S_H. More precisely, the probability of $AskH_4$ in $\mathbf{G_4}$ is the same as $AskH_5$ in $\mathbf{G_5}$.

$$\Pr[AskH_5] = \Pr[AskH_4]$$

Finally, the probability of $AskH$ in $\mathbf{G_5}$ is bounded by the successful probability of one-wayness adversary $\mathsf{Succ}_{\Pi,\mathcal{B}}^{\mathsf{ow-pca}}$, and the running time of adversary is bounded by the $(q_G + q_H)$ queires to the plaintext check oracle,

$$\Pr[AskH_5] \leq \mathsf{Succ}_{\Pi,\mathcal{B}}^{\mathsf{ow-pca}}$$

which has proven the theorem 1.

Choosing Parameter Sets for **NTRUEncrypt** with **NAEP** and **SVES-3**

Nick Howgrave-Graham, Joseph H. Silverman, and William Whyte

NTRU Cryptosystems,
5 Burlington Woods, MA 01803

Abstract. We present, for the first time, an algorithm to choose parameter sets for NTRUEncrypt that give a desired level of security.

1 Introduction

Different descriptions of NTRUEncrypt, and different proposed parameter sets, have been in circulation since 1996 [9–11, 5]. However, the method for choosing parameter sets has always been something of a black art. No single paper has ever described a machine which takes as input a desired security level k and outputs a parameter set that gives k bits of security.

It is the aim of this paper to provide such a machine. This paper presents a fixed algorithm to generate all required parameters for NTRUEncrypt with the SVES-3 encryption scheme, starting from a single parameter, k. Additionally, we present a more flexible framework generalizing the fixed algorithm in order to allow for architecture and efficiency tradeoffs, while still maintaining security against all known attacks. The fixed algorithm presented earlier always produces parameters consistent with this framework. We arrive at the parameter bounds specified in this framework by reviewing the effectiveness of each known attack.

2 A Specific Algorithm

This section gives a specific instantiation of the parameter generation algorithm for NTRUEncrypt-SVES-3 with binary underlying polynomials and $p = 2$. The input to this algorithm is the security parameter k. We denote this algorithm $\mathcal{P}^1_{\text{ntru}}$, and denote by $\mathcal{P}^1_{\text{ntru}}(k)$ the parameter set produced by this algorithm with input k. Table 2 gives $\mathcal{P}^1_{\text{ntru}}(k)$ for various common values of k. We present a more general parameter generation framework later.

1. Set N to be the first prime greater than $3k + 8$.
2. Set d to be the smallest integer such that

$$\frac{1}{\sqrt{N}} \binom{N/2}{d/2} > 2^k \ .$$

Set $d_F = d_r = d$. Set $d_g = \lfloor N/2 \rfloor$.

3. Set d_{m0} to be the largest integer such that

$$2^{N-1} \sum_{i=0}^{d_{m0}} \binom{N}{i} < 2^{-40} .$$

If $\frac{1}{\sqrt{N}} \binom{N/2}{d_{m0}} < 2^k$, increase N to the next largest prime and return to step 2.
4. Set q to be the first prime greater than $4d + 1$.
5. Verify that the order of $q \pmod{N}$ is $N-1$ or $(N-1)/2$. If it is not, increase q to the next prime value until a q with a sufficiently high order is found.
6. Calculate $c = \sqrt{4\pi e \sqrt{d(N-d)/N} \sqrt{d_{m0}(N-d_{m0})/N}/q}$.
If $c < 2.6$ or

$$0.42N - 3.4 - \max_r \left(\log_2 \left(1 - \left(1 - \prod_{i=0}^{d-1} \left(1 - \frac{r}{N-i} \right) \right)^N \right) + 0.21r \right) < k ,$$

increase N to the next largest prime and return to step 2. Otherwise, output $\{N, q, p = 2, d_F, d_r, d_g, d_{m0}\}$ and stop.

The rest of this paper derives and justifies this algorithm.

3 Bit Strength

We quantify security in terms of bit strength k, evaluating how much effort an attacker has to put in to break a scheme. All the attacks we consider here have variable running times, so we describe the strength of a parameter set using the notion of *cost*. For an algorithm $\mathcal{A}$ with running time t and probability of success ε, the cost is defined as

$$C_\mathcal{A} = t/\varepsilon .$$

This definition of cost is not the only one that could be used. For example, consider *indistinguishability against adaptive chosen-ciphertext attack*. In this attack, an attacker with access to encryption and decryption oracles chooses two messages M_0 and M_1, and is given the encryption of one of them. The attacker's output is a single bit $i \in \{0,1\}$. She wins if M_i was in fact the encrypted message. Here, the relevant measure of the attacker's power is the advantage over a random guess, defined as

$$\mathrm{adv}(\mathcal{A}(\mathsf{ind})) = 2.(\Pr[\mathsf{Succ}[\mathcal{A}]] - 1/2) .$$

We will use either measure as appropriate.

Our notion of cost is derived from [19] and related work. An alternate notion of cost, which is the definition above multiplied by the amount of memory used, is proposed in [28]. The use of this measure would allow significantly more efficient parameter sets, as the meet-in-the-middle attack described in Section 5.1 is essentially a time-memory tradeoff that keeps the product of time and memory constant. However, current practice is to use the measure of cost above.

We also acknowledge that the notion of comparing public-key security levels with symmetric security levels, or of reducing security to a single headline measure, is inherently problematic – see an attempt to do so in [24], and useful comments on this in [17]. In particular, extrapolation of breaking times is an inexact science, the behavior of breaking algorithms at high security levels is by definition untested, and one can never disprove the existence of an algorithm that attacks NTRUEncrypt (or any other system) more efficiently than the best currently known method. However, estimates of security have to start somewhere, and we consider this paper to provide a useful starting point for NTRUEncrypt.

4 The NTRUEncrypt One-Way Function

An implementation of the NTRUEncrypt encryption primitive is specified by the following parameters:

- N *Degree Parameter.* A positive integer. The associated NTRU lattice has dimension $2N$.
- q *Large Modulus.* A positive integer. The associated NTRU lattice is a convolution modular lattice of modulus q.
- p *Small Modulus.* An integer or a polynomial.
- $\mathcal{D}_f, \mathcal{D}_g$ *Private Key Spaces.* Sets of polynomials from which the private keys are selected.
- $\mathcal{D}_m$ *Plaintext Space.* Set of polynomials that represent encryptable messages. It is the responsibility of the encryption scheme to provide a method for encoding the message that one wishes to encrypt into a polynomial in this space.
- $\mathcal{D}_r$ *Blinding Value Space.* Set of polynomials from which the temporary blinding value used during encryption is selected.
- center *Centering Method.* A means of performing mod q reduction on decryption.

Definition 1. *The* Ring of Convolution Polynomials *is*
$$\mathcal{R} = \frac{\mathbb{Z}[X]}{(X^N - 1)}.$$

Multiplication of polynomials in this ring corresponds to the convolution product of their associated vectors. We also use the notation $\mathcal{R}_q = \frac{(\mathbb{Z}/q\mathbb{Z})[X]}{(X^N-1)}$.

Definition 2. *A polynomial* $a(X) = a_0 + a_1 X + \cdots + a_{N-1} X^{N-1}$ *is identified with its vector of coefficients* $\mathbf{a} = [a_0, a_1, \ldots, a_{N-1}]$. *The* centered norm $\|\mathbf{a}\|$ *of a polynomial or vector is defined by*

$$\|\mathbf{a}\|^2 = \sum_{i=0}^{N-1} a_i^2 - \frac{1}{N}\left(\sum_{i=0}^{N-1} a_i\right)^2. \tag{1}$$

Definition 3. *The* width Width(a) *of a polynomial or vector is defined by*
$$\text{Width}(\mathbf{a}) = \text{Max}(a_0, \ldots, a_{N-1}) - \text{Min}(a_0, \ldots, a_{N-1}).$$

4.1 Key Generation

NTRUEncrypt key generation consists of the following operations:

1. Randomly generate "small" polynomials f and g in $\mathcal{D}_f$, $\mathcal{D}_g$ respectively.
2. Invert f in $\mathcal{R}_q$ to obtain f_q, invert f in $\mathcal{R}_p$ to obtain f_p, and check that g is invertible in $\mathcal{R}_q$ [12].
3. The public key $h = p * g * f_q \pmod{q}$. The private key is the pair (f, f_p).

4.2 Encryption

NTRUEncrypt encryption consists of the following operations:

1. Randomly select a "small" polynomial $r \in \mathcal{D}_r$.
2. Calculate the ciphertext e as $e \equiv r * h + m \pmod{q}$.

4.3 Decryption

NTRUEncrypt decryption consists of the following operations:

1. Calculate $a \equiv \texttt{center}(f * e)$, where the centering operation `center` reduces its input into the interval $[A, A + q - 1]$.
2. Recover m by calculating $m \equiv f_p * a \pmod{p}$.

To see why decryption works, use $h \equiv p * g * f_q$ and $e \equiv r * h + m$ to obtain

$$a \equiv p * r * g + f * m \pmod{q} . \tag{2}$$

For appropriate choices of parameters and `center`, this is an equality over $\mathbb{Z}$, rather than just over $\mathbb{Z}_q$. Therefore step 2 recovers m: the $p * r * g$ term vanishes, and $f_p * f * m = m \pmod{p}$.

4.4 The NTRU Hard Problem and One-Way Function

The one-way function underlying NTRU is:

$$F : \mathcal{D}_m \times \mathcal{D}_r \to \mathcal{R}_q$$
$$F(m, r) = m + r * h,$$

where $q, N \in \mathbb{Z}$, $p \in \mathbb{Z}[X]$, $h \in \mathcal{R}_q$ are given by the output of key generation.

Definition 4. *(The $\mathcal{P}$-NTRU problem) For a parameter set $\mathcal{P}$, we denote by $\text{Succ}^{\text{ow}}_{\text{ntru}}(\mathcal{A}, \mathcal{P})$ the success probability of any adversary $\mathcal{A}$ for finding a preimage of F,*

$$\text{Succ}^{\text{ow}}_{\text{ntru}}(\mathcal{A}, \mathcal{P}) = \Pr \left[\begin{array}{l} (m', r') \leftarrow \mathcal{A}(e, h) \\ \text{s.t. } F(m', r') = e \end{array} \middle| \begin{array}{l} (pk = h, sk) \leftarrow \mathcal{K}, m \xleftarrow{R} \widetilde{R} \\ r \leftarrow \texttt{genr}(\rho), \rho \xleftarrow{R} \mathcal{R}_r, e = F(m, r) \end{array} \right].$$

Assumption 1 *(The $\mathcal{P}^1_{\text{ntru}}$-NTRU assumption) For every probabalistic polynomial (in k) time algorithm $\mathcal{A}$ there exists a negligible function $\nu_\mathcal{A}$ such that for sufficiently large k, we have*

$$\text{Succ}^{\text{ow}}_{\text{ntru}}(\mathcal{A}, \mathcal{P}^1_{\text{ntru}}(k)) \leq \nu_\mathcal{A}(k).$$

5 Security of the NTRU One-Way Function

Most public key cryptosystems, such as RSA [26] or ECC [18, 22], are based on a one-way function for which there is one best-known method of attack: factoring in the case of RSA, Pollard-rho in the case of ECC. In the case of NTRU, there are *two* primary methods of approaching the one-way function, both of which must be considered when selecting a parameter set.

5.1 Combinatorial Security

Polynomials are drawn from a known space $\mathcal{S}$. This space can best be searched by using a combinatorial technique originally due to Odlyzko [16], which can be used to recover f or g from h or r and m from e. We denote the combinatorial security of polynomials drawn from $\mathcal{S}$ by Comb[$\mathcal{S}$], and the set of binary polynomials of degree $N-1$ with exactly d coefficients equal to 1 by $\mathcal{B}_N(d)$. Then

$$\mathrm{Comb}[\mathcal{B}_N(d)] \geq \frac{\binom{N/2}{d/2}}{\sqrt{N}}.$$

5.2 Lattice Security

The *NTRU Lattice* $\mathcal{L}_h$ associated to a polynomial $h \in \mathcal{R}$ is the lattice

$$\mathcal{L}_h = \{(u,v) \in \mathcal{R}^2 : v \equiv h * u/p \ (\mathrm{mod}\ q)\}, \quad \text{satisfying}$$
$$\dim(\mathcal{L}_h) = 2N \quad \text{and} \quad \mathrm{Disc}(\mathcal{L}_h) = q^N.$$

Lattice-based attacks may be mounted against a ciphertext e to recover the plaintext, or against a public key h to recover the private key. This section treats lattice-based attacks on the public key; the analysis for attacks on a ciphertext is almost identical. More details can be found in [13].

An NTRUEncrypt public key h describes a $2N$-dimensional NTRU lattice containing the private key (f, g). When f is of the form f = 1 + pF. the best lattice attack on the private key involves solving a Close Vector Problem (CVP)[1]. Experimentally, it has been found that an NTRU lattice of this form can usefully be characterized by two quantities

$$a = N/q, \quad c = \sqrt{4\pi e \|F\| \|g\|/q}.$$

This is to say that for constant (a, c), the experimentally observed running times for lattice reduction behave roughly as

$$\log(T) = AN + B,$$

for some experimentally-determined constants A and B.

---

[1] Coppersmith and Shamir [6] propose related approaches which turn out not to materially affect security.

Table 1 summarizes results for breaking times from [9, 13], and more recent experiments, giving breaking times for inhomogenous NTRU lattices with different (a, c) values. We represent the breaking time in terms of bit security, which may be converted to time in MIPS-years using the equality 80 bits $\sim$ 10^{12} MIPS-years.

Table 1. Extrapolated bit security depending on (c, a).

c	a	Bit Security
1.73	0.53	$0.3563N - 2.263$
2.6	0.8	$0.4245N - 3.440$
3.7	2.7	$0.4512N + 0.218$
5.3	1.4	$0.6492N - 5.436$

For constant (a, c), increasing N increases the breaking time exponentially. For constant (a, N), increasing c increases the breaking time. For constant (c, N), increasing a decreases the breaking time, although the effect is slight. More details on this table are given in [13]. We write

$$\text{Lattice Bit Security } b_{\text{latt}} \equiv \alpha N + \beta .$$

The technique known as *zero-forcing* [13, 20] can be used to reduce the dimension of an NTRU lattice problem. The precise amount of the expected performance gain is heavily dependent on the details of the parameter set; we refer the reader to [13, 20] for more details. In this paper we use the formula[2]

$$\text{Gain} \sim \left(1 - \left(1 - \prod_{i=0}^{d-1} \left(1 - \frac{r}{N-i}\right)\right)^N\right) 2^{\alpha r/2} \qquad (3)$$

to determine the expected gain due to picking a pattern of r zeroes, if f has d non-zero entries, and the lattice breaking bit security goes as $\alpha N + \beta$. This will typically overestimate the gain, but we use this formula for reasons of prudence.

5.3 Decryption Failure Security

NTRU decryption can fail on validly encrypted messages if the center method returns the wrong value of A, or if the coefficients of prg + fm do not lie in an interval of width q. Decryption failures leak information about the decrypter's private key [14, 25], so a center method must make the chance of a decryption failure vanishingly small.

The parameter sets recommended in [5] allow a decryption failure probability of about 2^{-104} for 80-bit security. In this paper, we will pick parameter sets such that there will be no decryption failure, by selecting q to be greater than the maximum possible value of prg + fm. Centering then becomes simply a matter of reducing into the interval $[0, q-1]$.

---
[2] Note that this formula, used in [13], corrects the equivalent formula given in [20].

5.4 Other Security Considerations

The following parameter selection criteria must also be taken into account, although encryption and decryption will work even if they are violated.

Choosing N – The degree parameter N must be prime. (See [7].)

N, q and p – The small and large moduli p and q must be relatively prime in the ring $\mathcal{R}$. Equivalently, the three quantities

$$p, \quad q, \quad X^N - 1$$

must generate the unit ideal in the ring $\mathbb{Z}[X]$. (As an example of why this is necessary, in the extreme case that p divides q, the plaintext is equal to the ciphertext reduced modulo p.)

Factorization of $X^N - 1 \pmod{q}$ – If $\mathsf{F}(X)$ is a factor of $X^N - 1 \pmod{q}$, and if $\mathsf{h}(X)$ is a multiple of $\mathsf{F}(X)$, i.e., if $\mathsf{h}(X)$ is zero in the field $K = (\mathbb{Z}/q\mathbb{Z})[X]/\mathsf{F}(X)$, then an attacker can recover the value of $\mathsf{m}(X)$ in the field K.

If q has order $t \pmod{N}$, then

$$X^N - 1 \equiv (X-1)\mathsf{F}_1(X)\mathsf{F}_2(X)\cdots\mathsf{F}_{(N-1)/t}(X) \quad \text{in } (\mathbb{Z}/q\mathbb{Z})[X],$$

where each $\mathsf{F}_i(X)$ has degree t and is irreducible mod q. If $\mathsf{F}_i(X)$ has degree t, the probability that $\mathsf{h}(X)$ or $\mathsf{r}(X)$ is divisible by $\mathsf{F}_i(X)$ is presumably $1/q^t$. To avoid attacks based on the factorization of h or r, we will require that for each prime divisor P of q, the order of $P \pmod{N}$ must be $N-1$ or $(N-1)/2$. This requirement has the useful side-effect of increasing the probability that randomly chosen f will be invertible in $\mathcal{R}_q$ [27].

Information Leakage from Encrypted Messages – The transformation $\mathsf{a} \to \mathsf{a}(1)$ is a ring homomorphism, and so the ciphertext e has the property that

$$\mathsf{e}(1) = \mathsf{r}(1)\mathsf{h}(1) + \mathsf{m}(1).$$

An attacker will know $\mathsf{h}(1)$, and for many choices of parameter set $\mathsf{r}(1)$ will also be known. Therefore, the attacker can calculate $\mathsf{m}(1)$. The larger $|\mathsf{m}(1) - N/2|$ is, the easier it is to mount a combinatorial or lattice attack to recover the msssage, so the sender should always ensure that $\|\mathsf{m}\|$ is sufficiently large. This will double the encryption time, but does not appear to lead to any attacks. One of our inputs into the parameter generation algorithm will be a lower bound for the probability that a randomly generated m will be too small.

6 Encryption Schemes: NAEP

In order to protect against adaptive chosen ciphertext attacks, we must use an appropriately defined *encryption scheme*. The scheme described in [15] gives provable security in the random oracle model [2,3]. We briefly outline it here.

NAEP uses two hash functions:

$$G : \{0,1\}^{N-l} \times \{0,1\}^l \to \mathcal{D}_r \quad H : \{0,1\}^N \to \{0,1\}^N$$

In terms of the security parameter, we wish $l = \Theta(k)$, and also $N - l = \Theta(k)$. To encrypt a message $M \in \{0,1\}^{N-l}$ using NAEP one uses the functions

$$\text{compress}(x) = (x \pmod q) \pmod 2,$$

$$\text{B2P} : \{0,1\}^N \to \mathcal{D}_m \cup \text{``error''}, \quad \text{P2B} : \mathcal{D}_m \to \{0,1\}^N$$

The function compress puts the coefficients of the modular quantity $x \pmod q$ in to the interval $[0, q)$, and then this quantity is reduced modulo 2. The role of compress is simply to reduce the size of the input to the hash function H for gains in practical efficiency. The function B2P converts a bit string into a binary polynomial, or returns "error" if the bit string does not fulfil the appropriate criteria – for example, if it does not have the appropriate level of combinatorial security. The function P2B converts a binary polynomial to a bit string.

The encryption algorithm is then specified by:

1. Pick $b \xleftarrow{R} \{0,1\}^l$.
2. Let $\mathsf{r} = G(M, b)$, $\mathsf{m} = \text{B2P}(\ (M||b) \oplus H(\text{compress}(\mathsf{r} * \mathsf{h}))\ )$.
3. If B2P returns "error", go to step 1.
4. Let $\mathsf{e} = \mathsf{r} * \mathsf{h} + \mathsf{m} \in \mathcal{R}_q$.

Step 3 ensures that only messages of the appropriate form will be encrypted. To decrypt a message $\mathsf{e} \in \mathcal{R}_q$ one does the following:

1. Let $\mathsf{a} = \text{center}(\mathsf{f} * \mathsf{e} \pmod q)$.
2. Let $\mathsf{m} = \mathsf{f}_p^{-1} * \mathsf{a} \pmod p$.
3. Let $\mathsf{s} = \mathsf{e} - \mathsf{m}$.
4. Let $M||b = \text{P2B}(\mathsf{m}) \oplus H(\text{compress}(\text{P2B}(\mathsf{s})))$.
5. Let $\mathsf{r} = G(M, b)$.
6. If $\mathsf{r} * \mathsf{h} = \mathsf{s} \pmod q$, and $\mathsf{m} \in \mathcal{D}_m$, then return the message M, else return the string "invalid ciphertext".

The use of the scheme NAEP introduces a single additional parameter:

> l *Random Padding Length.* The length of the random padding b concatenated with M in step 1.

The ind game requires an attacker to identify the message encrypted in a single, specific ciphertext. Therefore, the random padding does not require collision resistance, but it does require preimage resistance. We therefore set $l = k$ to ensure that attacks based on guessing the random padding have a k-bit cost (where cost is defined relative to the attacker's advantage).

6.1 Instantiating NAEP: SVES-3

The EESS#1 v2 standard [5] specifies an instantiation of NAEP known as SVES-3. In SVES-3, the following specific design choices are made:

- To allow variable-length messages, a one-byte encoding of the message length in bytes is prepended to the message. The message is padded with zeroes to fill out the message block.
- The hash function G which is used to produce r takes as input M; b; an OID identifying the encryption scheme and parameter set; and a string h_{trunc} derived by truncating the public key to length l_h bits.

SVES-3 includes h_{trunc} in G so that r depends on the specific public key. Even if an attacker were to find an (M, b) that gave an r with an increased chance of a decryption failure, that (M, b) would apply only to a single public key and could not be used to attack other public keys. In the case of the parameter sets proposed in this document, there are no decryption failures and so no need to input h_{trunc} to G. We will therefore use SVES-3 but set $l_h = 0$.

7 Selecting Parameter Sets for SVES-3: Framework

Having completed our review of security considerations for NTRU parameter sets, we can now specify an algorithm that generates a parameter set for NTRUEncrypt-NAEP with a desired bit security level k. First, we specify our overall framework. Then we apply it to specific sets of constraints on the parameters.

1. Determine μ, the number of bits that must be transported in m. Pick an initial candidate N, a prime number that allows μ bits to be transported.
2. For this value of N, find values of d_F, d_g, d_r that give the required level of combinatorial security.
3. Using the bound P_{reject} given in Constraint 6, calculate the minimum integer d_{m0} and the maximum integer d_{m1} such that $N/2 - d_{m0} = d_{m1} - N/2$ and the probability that a randomly chosen binary vector will have between d_{m0} and d_{m1} 1s is greater than $1 - P_{\text{reject}}$. If d_{m0} does not give sufficient combinatorial security, increase N to the next prime and repeat this step.
4. Calculate the maximum possible width of prg + m + pFm. Set q to be the first prime greater than this number.
5. Verify that the order of $q \pmod{N}$ is $N-1$ or $(N-1)/2$. If it is not, increase q to the next prime value until a q with a sufficiently high order is found.
6. Verify whether the lattice strength is greater than 2^k for the selected N, q, $\mathcal{D}_f, \mathcal{D}_g, \mathcal{D}_r, \mathcal{D}_m$. (In the case of $\mathcal{D}_m$, the check is performed for the m with $d_m = d_{m0}$, or in other words the weakest m that will occur). If the strength is greater than 2^k, terminate. Otherwise, increase N to the next highest prime number and return to step 2.

The analysis below will explain why this process is likely to terminate after a very small number of iterations.

7.1 Binary Polynomials

We illustrate the method using binary polynomials. In this case, we use the following constraints.

1. Take p = 2. Require q to be prime.
2. f will be of the form $1 + \mathsf{p}F$.
3. The polynomials F, g, r, m will be binary. Product form polynomials will not be used.
4. F, g, r will have d_F, d_g, d_r 1s respectively.
5. The system must be capable of transporting $2k$ bits of message.
6. The chance that a message representative m will be rejected due to having insufficient security, P_{reject}, will be less than 2^{-40}.
7. Subject to the constraints above, minimize bandwidth.
8. Subject to the constraints above, maximize lattice security.

Select N – For k-bit security, we require $l \geq k$, as stated in the discussion of the security of NAEP. We also want to transport $2k$ bits of message, as stated in constraint 5, and to use 8 bits to encode the length of the transported message. The total number of bits to be transported in m is therefore $3k + 8$. We set N to be the first prime greater than $3k + 8$.

Select Polynomial Spaces – We select values for d_F, d_g, d_r so that $\text{Comb}[\mathcal{B}_N(d_F)]$, $\text{Comb}[\mathcal{B}_N(d_g)]$, $\text{Comb}[\mathcal{B}_N(d_r)] > 2^k$. The smaller d_F, d_r are, the faster operations will be. We select d_F, d_r such that $d_F = d_r = d$, d the smallest value for which $\text{Comb}[\mathcal{B}_N(d)] \geq 2^k$. The results are shown in table 2. For all values of N in the given range, $d \sim 0.19N$; in other words d increases (slightly slower than) linearly with N. Therefore, NTRUEncrypt encryption and decryption times scale roughly as N^2 for our parameter sets.

There is no particular advantage, in performance or bandwidth, to taking g to be small, so long as it is binary. Following constraint 8, we therefore take $d_g = \lfloor N/2 \rfloor$. This is a change from practice in previous parameter sets, where d_g has typically been taken to be the same as d_f.

Select $\mathcal{D}_m$ – Table 2 gives the value of d_{m0} (and $d_{m1} = N - d_{m0}$) for each N that gives a chance of 2^{-40} of having to re-encrypt. In all cases, d_{m0} is comfortably above d_f, and so m will have sufficient combinatorial security. If d_{m0} had been below d_f, increasing N will both (a) reduce the value of d_f that gives combinatorial security and (b) increase the d_{m0} that gives the desired probability of having to re-encrypt. The process of increasing N in this step will therefore eventually terminate.

Select q – We select q subject to the requirements

$$q > \text{Max}(\text{Width}(\mathsf{prg} + \mathsf{fm})),$$
$$\text{Order}(q \ (\text{mod } N)) \geq (N-1)/2.$$

We now consider how to calculate the width of prg + fm.

Each term in the polynomial obtained by multiplying a polynomial a by a binary polynomial b with d_b 1s can be thought of as the result of selecting d_b terms from a and summing them. If b is also binary, with d_b 1s, clearly the minimum possible value of any term in a $*$ b is $\max(0, d_a + d_b - N)$, and the maximum is $\min(d_a, d_b)$ [3].

In this case $d_F = d_g = d_r = d$. The number of 1s in m, d_m, is variable, but if it is less than d_F then $\max(\text{Width}(F * m)) < d_F$, and if it is greater than d_F then $\max(\text{Width}(F * m)) = d_F$. So

$$\max(\text{Width}(\text{prg} + m + \text{pfm})) = 1 + 2pd = 1 + 4d$$
$$\Rightarrow q > 1 + 4d \sim 0.76N.$$

Taking q to be the first prime greater than $1 + 4d$ gives a q with a large enough order (mod N) for almost all the (d, N) pairs under consideration. The exception is $k = 256$, $N = 787$, $d = 140$, for which the first q implied, 563, has order only 131 (mod N), and the lowest q that satisfies the order requirement turns out to be 587. The values of q obtained are given in Table 2.

Check Lattice Strength – Having calculated d and q, we can now calculate the lattice characteristics (a, c). For a binary polynomial b with d 1s, the centered norm is given by $|b| = \sqrt{d(N - d)/N}$. and ranges from $\sqrt{d}$, when d is small, to $\sqrt{d/2}$, when $d = N/2$. The thicker f, g, r, m are, the harder the lattice problem is. We therefore calculate c for lattice attacks on (r, m) when the number of 1s in m is d_{m0} to give a lower bound on the lattice security. All the parameter sets under consideration give $c \geq 2.77$, so we can use the $c = 2.6, a = 0.8$ experimental lattice strengths in Table 1 to extrapolate the strength of (r, m) and (F, g).

For interest, we briefly consider extreme cases. If $d = 0.001N$, then we have

$$q \sim 1 + 0.004N, \quad c \sim 11.6, \quad a \sim 250 . \qquad (4)$$

If $d = N/2$, then we have

$$q \sim 2N, \quad c = 2.066, \quad a = 0.5 . \qquad (5)$$

This shows that as d/N increases, c will decrease to a minimum of 2.066.

As table 2 below shows, the suggested parameters clearly give a sufficient level of lattice security, even taking zero-forcing into account.

Increase N if Necessary – If the strength against lattice attacks is insufficient, we increase N. This will decrease (or at worst not increase) the value of d necessary to give combinatorial security, reducing d/N. As noted in equations 4 and 5 above, as d/N decreases, c will increase. Even if c were to stay constant, increasing N would increase the lattice strength; since we increase both c and N, lattice strength will certainly increase, eventually reaching the desire strength. The process of increasing N will therefore eventually terminate.

---

[3] The maximum width of the product of two binary polynomials is therefore $N/2$.

Summary – This has rederived and justified the algorithm $\mathcal{P}^1_{ntru}$ presented at the start of this paper. Table 2 summarizes the results. For different values of k, we give the corresponding N, d, and q values. These, along with $\mathsf{p} = 2$, the definition of center as reduction into $[0, q-1]$, and the specification $l = k$, fully parameterize the system. For interoperability, other design decisions must be made, such as the exact instantiations of the random oracles; we do not address that question in this paper.

We also present the lattice bit security b_{latt}, the number of zeroes an adversary should guess when zero-forcing r, the lattice bit security including zero-forcing $b^{\text{zf}}_{\text{latt}}$, the number of additions required for a convolution by f or r, the public key size $N\lceil \log_2 q \rceil$ and, for comparison, the sizes of RSA and ECC keys that give a similar level of security. We also include the number of Adds With Carry required for an ECC point operation at the same security level: details of how this figure was calculated, and discussion of the appropriate figures to compare, can be found in Appendix A. The bandwidth given is the minimum bandwidth. In the case where q is a nine-bit quantity, for example, an implementation may decide to encode each coefficient in 16 bits rather than 9.

7.2 Product-Form Polynomials

Next we use the method above to generate parameter sets that make use of product form polynomials for efficiency advantages. We take $\mathsf{F} = \mathsf{f}_1 * \mathsf{f}_2 + \mathsf{f}_3$, with $\mathsf{f}_1, \mathsf{f}_2, \mathsf{f}_3$ all random binary with d 1s, r to have the same form, and g to be binary with $d_g = \lfloor N/2 \rfloor$. Full details of the process are given in Appendix B. Table 3 summarizes the results, including the speedup relative to the parameters for binary polynomials investigated above, the public key size $N\lceil \log_2 q \rceil$ and the RSA and ECC figures as above.

8 Conclusions

We presented a framework for generation of NTRUEncrypt parameter sets and used it to generate parameter sets for different levels of bits security. The framework is robust and adaptable: if future developments in lattice analysis significantly affect breaking times, it will be possible to calculate new parameter sets that give an appropriate level of security. With different inputs to the framework, different parameter sets would be possible. For example, one might take $\mathsf{p} = 2 + X$ and q a power of 2 for efficiency in performing reductions; one might require $q < 256$, increasing N as necessary, for use on 8-bit processors; one might consider an alternate encryption scheme that transported fewer bits to save bandwidth. We have also demonstrated that NTRUEncrypt remains more efficient than other well-studied cryptosystems, and shown that for increasing security levels the bandwidth required for NTRUEncrypt is less than for RSA.

This paper is merely a contribution to the systematic study of how to generate NTRUEncrypt parameter sets, but we hope a useful one.

Table 2. Final Parameter Sets for different values of k using binary polynomials.

k	N	d	d_{m0}	q	$c(f,g)$	$c(r,m)$	b_{latt}	r	b_{latt}^{zf}	adds	size	RSA	ECC	ECC AWC
80	251	48	70	197	2.93	2.77	103.1	29	97.98	12048	2008	1024	163	112210
112	347	66	108	269	2.94	2.83	143.9	31	138.26	22902	3033	~ 2048	224	170356
128	397	74	128	307	2.93	2.84	165.1	33	159.17	29378	3501	3072	256	277280
160	491	91	167	367	2.98	2.90	205.0	35	198.75	44681	4383	4096	320	–
192	587	108	208	439	2.97	2.91	245.7	37	239.21	63396	5193	7680	384	936618
256	787	140	294	587	2.95	2.91	330.6	41	323.45	110180	7690	15360	512	1595434

Table 3. Final Parameter Sets for different values of k using product form polynomials.

k	N	d	q	adds	speedup	size	RSA	ECC	ECC AWC
80	251	8	293	6024	2.00	2259	1024	163	112210
112	347	11	541	11451	2.00	3370	~ 2048	224	170356
128	397	12	659	14292	2.06	3890	3072	256	277280
160	491	15	967	22095	2.02	4870	4096	320	–
192	587	17	1229	29937	2.12	6347	7680	384	936618
256	787	22	2027	51942	2.12	8459	15360	512	1595434

Acknowledgements

We would like to thank the anonymous referees for their comments, and Philip Hirschhorn for his help with lattice reduction experiments.

References

1. ANSI X9.62, Public Key Cryptography for the Financial Services Industry: The Elliptic Curve Digital Signature Algorithm (ECDSA), 1999.
2. M. Bellare and P. Rogaway. Optimal asymmetric encryption. In Proc. of Eurocrypt '94, volume 950 of LNCS, pages 92–111. IACR, Springer-Verlag, 1995.
3. D. Boneh, Simplified OAEP for the RSA and Rabin functions, In proceedings of Crypto '2001, Lecture Notes in Computer Science, Vol. 2139, Springer-Verlag, pp. 275-291, 2001
4. M. Brown, D. Hankerson, J. López, and A. Menezes, Software Implementation of the NIST Elliptic Curves Over Prime Fields, *CT-RSA 2001*, D. Naccache (Ed.), LNCS 2020, 250–265, Springer-Verlag, 2001.
5. Consortium for Efficient Embedded Security, *Efficient Embedded Security Standard #1 version 2*, available from http://www.ceesstandards.org.
6. D. Coppersmith and A. Shamir, *Lattice Attack on NTRU*, Advances in Cryptology - Eurocrypt'97, Springer-Verlag
7. C. Gentry, Key recovery and message attacks on NTRU-composite, *Advances in Cryptology – Eurocrypt '01*, LNCS 2045. Springer-Verlag, 2001
8. D. Hankerson, J. Hernandez, A. Menezes, *Software implementation of elliptic curve cryptography over binary fields*, Proceedings of CHES 2000, Lecture Notes in Computer Science, 1965 (2000), 1-24
9. J. Hoffstein, J. Pipher, J.H. Silverman, *NTRU: A new high speed public key cryptosystem*, in Algorithmic Number Theory (ANTS III), Portland, OR, June 1998, Lecture Notes in Computer Science 1423 (J.P. Buhler, ed.), Springer-Verlag, Berlin, 1998, 267–288. See also http://www.ntru.com.

10. J. Hoffstein and J. H. Silverman. Optimizations for NTRU. In Public-key Cryptography and Computational Number Theory. DeGruyter, 2000. Available at [4].
11. J. Hoffstein and J. H. Silverman, Random Small Hamming Weight Products With Applications To Cryptography, Discrete Applied Mathematics, to appear, Available from http://www.ntru.com.
12. J. Hoffstein and J. H. Silverman. Invertibility in truncated polynomial rings. Technical report, NTRU Cryptosystems, October 1998. Report #009, version 1, available at http://www.ntru.com.
13. J. Hoffstein, J. H. Silverman, W. Whyte, Estimated Breaking Times for NTRU Lattices, Technical report, NTRU Cryptosystems, June 2003 Report #012, version 2, available at http://www.ntru.com.
14. N. A. Howgrave-Graham, P. Nguyen, D. Pointcheval, J. Proos, J. H. Silverman, A. Singer, W. Whyte, *The Impact of Decryption Failures on the Security of NTRU Encryption*, Advances in Cryptology – Crypto 2003, Lecture Notes in Compputer Science 2729, Springer-Verlag, 2003, 226-246.
15. N. Howgrave-Graham, J. H. Silverman, A. Singer and W. Whyte. NAEP: Provable Security in the Presence of Decryption Failures IACR ePrint Archive, Report 2003-172, http://eprint.iacr.org/2003/172/
16. N. A. Howgrave-Graham, J. H. Silverman, W. Whyte, A Meet-in-the-Middle Attack on an NTRU Private key, Technical report, NTRU Cryptosystems, June 2003. Report #004, version 2, available at http://www.ntru.com.
17. B. Kaliski, Comments on SP 800-57, Recommendation for Key Management, Part 1: General Guidelines. Available from
   http://csrc.nist.gov/CryptoToolkit/kms/CommentsSP800-57Part1.pdf.
18. N. Koblitz. *Elliptic curve cryptosystems*. Mathematics of Computation, 48, pages 203–209, 1987.
19. A. K. Lenstra, E. R. Verheul, *Selecting cryptographic key sizes*, Journal of Cryptology vol. 14, no. 4, 2001, 255-293. Available from http://www.cryptosavvy.com.
20. A. May, J.H. Silverman, *Dimension reduction methods for convolution modular lattices*, in Cryptography and Lattices Conference (CaLC 2001), J.H. Silverman (ed.), Lecture Notes in Computer Science 2146, Springer-Verlag, 2001
21. T. Meskanen and A. Renvall. Wrap Error Attack Against NTRUEncrypt. Proc. of WCC '03.
22. V. Miller. *Uses of elliptic curves in cryptography*. In Advances in Cryptology: Crypto '85, pages 417–426, 1985.
23. NIST, Digital Signature Standard, FIPS Publication 186-2, February 2000.
24. NIST Special Publication 800-57, Recommendation for Key Management, Part 1: General Guideline, January 2003. Available from
   http://csrc.nist.gov/CryptoToolkit/kms/guideline-1-Jan03.pdf.
25. J. Proos *Imperfect Decryption and an Attack on the NTRU Encryption Scheme*, IACR ePrint Archive, report 02/2003. http://eprint.iacr.org/2003/002/.
26. R. Rivest, A. Shamir, L. M. Adleman, *A method for obtaining digital signatures and public-key cryptosystems*, Communications of the ACM, 21 (1978), 120-126.
27. J. H. Silverman, Invertibility in Truncated Polynomial Rings, Technical report, NTRU Cryptosystems, October 1998 Report #009, version 1, available at http://www.ntru.com.
28. Robert D. Silverman, A Cost-Based Security Analysis of Symmetric and Asymmetric Key Lengths. *RSA Labs Bulletin 13*, April 2000. available from http://www.rsasecurity.com/rsalabs.

A Comparing ECC Times to NTRUEncrypt Times

In this section we give a comparison, in terms of basic operations, of elliptic curve point multiplication and NTRUEncrypt polynomial operations. The speed of elliptic curve point multiplications (in software) for the prime fields given in [1, 23] is analyzed in [4]. The analysis in [4] is most complete for the smallest NIST prime field $F_{p_{192}}$, where $p_{192} = 2^{192} - 2^{64} - 1$. The prime p_{192} is slightly less than a power of 2^{32} (a typical machine word size) and has a sparse bit representation, yielding added efficiencies for various operations such as modular reduction.

The first three columns of Table 4 appear in [4, Table 10] and give the number of basic operations required for an average point multiplication on the NIST p_{192} elliptic curve. The authors of [4] note that "95.4% of the total execution time was spent on these basic operations."

Operations in the first three columns of Table 4 are modulo p. We denote by AWC. the amount of time it takes to perform a single 32 bit addition-with-carry, and estimate the time for an elliptic curve point multiplication in terms of AWC. For example, a single addition modulo p_{192} takes 6 AWC[4].

To estimate AWC numbers for other security levels, we follow [4, Table 9], which gives running times for point multiplications for all the recommended

Table 4. Average number of function calls and percentage of time spent on the basic field operations in executions of [4, Algorithm 10] for elliptic curve point multiplication for the p_{192} curve. Data in first three columns is from [4, Table 10]. The algorithm numbers in Table 4 refer to the algorithms described in [4].

Field operation	# of calls	Percentage of total time	AWC per call	Number of AWC
Addition (Alg 1)	1137	5.8%	6.00	6822.00
Subtraction (Alg 2)	1385	7.4%	6.28	8703.93
Integer multiplication (Alg 3)	1213	38.3%	37.14	45048.72
Integer squaring (Alg 4)	934	28.20%	35.5	33169.03
Fast reduction (Alg 7)	2147	14.8%	8.11	17407.86
Modular inversion (Alg 8)	1	0.9%	1058.59	1058.59
Total		95.4%		112210.14

Table 5. Estimated number of AWC for each of the NIST recommended finite fields, derived from [4, Table 9].

Field	Time	AWC
p_{192}	2144	112210
p_{224}	3255	170356
p_{256}	5298	277280
p_{384}	17896	936618
p_{521}	30484	1595433

---

[4] Plus potentially one subtraction of p, which we consider free due to the form of p_{192}.

fields. The estimated number of AWC for field p_i is simply the number of AWC for p_{192} times the ratio of the running times for p_i and p_{192}. The results are shown in Table 5. Our figures are based on the figures of [4] that do not use precomputation, as ECC encryption and decryption typically involve one point multiplication on an arbitrary base point for which precomputation cannot have been performed[5].

The NTRUEncrypt performance figures given are the time for a single convolution, calculated as

$$\text{Time} = dN \text{ (Binary polynomials)},$$
$$\text{Time} = 3dN \text{ (Product-form polynomials)}.$$

This slightly underestimates the time for a convolution, in which each of the N coefficients is produced by a combination of d additions (without carry) and 1 to d reductions mod q. However, the overhead due to the reductions is not great; on a 32-bit machine, for example, each reduction can be accomplished in $\log_2(d)$ subtractions. We therefore consider the figures presented to be a good first-order approximation to the actual running times.

In summary, NTRUEncrypt convolution operations with binary polynomials are $7.5 - 15$ times faster, and NTRUEncrypt convolutions with product-form polynomials are $15 - 30$ times faster than ECC point multiplications, at the same security level. This figure leaves out the time required for any hash function operations. For encryption, ECC requires an additional point multiplication to a known base point, which increases encryption times by a factor of $1.3 - 2$. For decryption, NTRUEncrypt-NAEP requires an additional encryption operation for the consistency, increasing decryption times by a factor of 2.

B Details of Product Form Calculations

B.1 Combinatorial Security of Product-Form Polynomials

Product-form polynomials [10, 11] are polynomials of the form $a_1 * a_2$ or $a_1 * a_2 + a_3$. The advantage of polynomials of this form is that they can be specified more compactly, and multiplied by more quickly, than binary polynomials with the same level of combinatorial security, though at the cost of requiring more RAM.

In this paper we will only consider the combinatorial security of polynomials of the form $a = a_1 * a_2 + a_3$, where a_1, a_2, a_3 are all binary with $d_{a_1}, d_{a_2}, d_{a_3}$ 1s respectively, $d_{a_1} = d_{a_2} = d_{a_3} = d_a$, and there are no further constraints on a. If $\mathcal{P}_N(d)$ is the set of all polynomials of this form, then $\text{Comb}[\mathcal{P}_N(d)] \geq \min\Big($

$\binom{N-\lceil N/d \rceil}{d-1}^2, \max\left(\binom{N-\lceil \frac{N}{d} \rceil}{d-1}\binom{N-\lceil \frac{N}{d} \rceil}{d-2}, \binom{N}{2d}\right), \max\left(\binom{N}{d}\binom{N}{d-1}, \binom{N-\lceil \frac{N}{2d} \rceil}{2d-1}\right)\Big)$.

Previous parameter sets [5] have suggested using product-form polynomials $a = a_1 * a_2 + a_3$, where the product polynomial a is constrained to be binary. However, this increases the time to generate those polynomials, and more so

---

[5] Signing is more likely to use precomputation, increasing speeds about 3.5-fold.

as the security parameter k increases. For reasons of efficiency the parameter generation algorithm in this paper does not require binary output polynomials.

B.2 Parameter Set Generation

We work from the same constraints as in section 7.1, except that:

1. F will be of the form $f_1 * f_2 + f_3$, with f_1, f_2, f_3 all random binary.
2. r will be of the form $r_1 * r_2 + r_3$, with r_1, r_2, r_3 all random binary.
3. $f_1, f_2, f_3, r_1, r_2, r_3$ will have d 1s; g will have d_g 1s.

As before, we set N to be the first prime greater than $3k + 8$. The parameter set generation then proceeds as follows.

Select Polynomial Spaces – Select the smallest d that gives the desired level of combinatorial security, and take $d_g = N/2$ to give the greatest possible lattice security. Table 6 shows the resulting values for d and the corresponding Hamming weight of F, Hw(F)(= Hw(r)). For all values of N, $d_F = d_r \sim 0.03N$. As in the previous section, d increases slightly slower than linearly with N, so NTRU-Encrypt encryption and decryption times scale approximately as N^2. The value obtained for d_{m0} only depends on N and does not change.

Select q – Select the smallest prime q such that $\mathrm{Order}(q \pmod N) \geq (N-1)/2$ and $q > \mathrm{Max}(\mathrm{Width}(\mathrm{prg} + \mathrm{fm}))$. For both $f * m$ and $r * g$, one of the operands is binary but the other is product-form, so the width of the product $a * b$ is no longer bounded by $\min(\mathrm{Hw}(a, b))$. However, since one of the operands is binary, the width is certainly bounded by $\mathrm{Hw}(a)$, where a is the non-binary input polynomial. Therefore,

$$\max(\mathrm{Width}(\mathrm{prg} + \mathrm{m} + \mathrm{pfm})) = 1 + 2pd(d+1) = 1 + 4d(d+1)$$
$$\Rightarrow q > 1 + 4d(d+1) \Rightarrow q \sim cN^2.$$

Applying the requirement that the order of $q \pmod N$ be large, we increase $q(N = 397)$ and $q(N = 491)$. The other values of q are unaffected.

Check Lattice Strength – We now calculate the lattice characteristics (a, c). For g we use the standard centered norm. For f, the situation is more complicated. Roughly speaking, centered norms obey the *pseudo-multiplicative* and *pseudo-additive rules*

$$|a * b| \sim |a| * |b|, \quad |a + b| \sim \sqrt{|a|^2 + |b|^2}.$$

The centered norm $|F|$ will in general be $|F| \sim \sqrt{d^2(N-d)^2/N^2 + d(N-d)/N}$. However, in the case where F is binary, $|F|$ will take the considerably lower value $|F| = \sqrt{D(N-D)/N}$, $D = d^2 + d$. Although it will be extremely rare for randomly generated product-form F to be binary, we will use this lower value in calculating c. For all parameter sets under consideration we obtain the result $c > 1.73$, so we can use the extrapolation line obtained at $c = 1.73$, presented in table 1 above, to estimate lattice strength. Estimating the effects of zero-forcing is also harder in this case, because the number of zeroes in f, r is now variable.

We will assume that the product-form polynomials can be approximated by dropping $d(d+1)$ balls into N boxes. The expected number of empty boxes is

$$\mathsf{E}(\text{zeroes}) = N(1 - 1/N)^{d(d+1)} .$$

We use this expected number of zeroes in our zero-forcing calculations[6,7].

We can also estimate $\|\mathsf{F}\|$ by estimating the expected number of 0s, 1s, 2s, and so on, and calculating the centered norm using Equation 1. This third estimate of $\|\mathsf{F}\|$ gives a higher c value than the other methods described above. We denote it by $c_{0,1,2,3}$ in Table 6.

Table 6 summarizes the results for lattice strength for product form polynomials. The value r is the number of zeroes an adversary should guess when zero-forcing. We also give c_{f_1,f_2,f_3}, the expected value of c as calculated from the norms of f_1, f_2, f_3, and $c_{0,1,2,3}$, the expected value of c as calculated from the expected numbers of 0s, 1s, 2s and 3s in F. Both of these measures give a higher value for c than the one we use, demonstrating that in general the lattice security will be considerably above the extrapolation line based on $c = 1.73$. The final parameter sets are given in Table 3.

Table 6. Lattice constant c for different values of k.

k	N	d	d/N	Hw(F)	q	$c(f,g)$	$c(r,m)$	b_{latt}	r	$b_{\text{latt}}^{\text{zf}}$	c_{f_1,f_2,f_3}	E_0	E_1	E_2	E_3	E_4	$c_{0,1,2,3}$
80	251	8	0.032	72	293	2.57	2.43	87.2	20	80.1	2.76	188	55	7	1	0	2.79
112	347	11	0.032	132	541	2.21	2.13	117.8	16	118.7	2.59	237	90	18	2	0	2.56
128	397	12	0.030	156	659	2.24	2.17	136.3	17	136.6	2.50	268	105	21	4	0	2.53
160	491	15	0.031	210	967	2.08	2.02	171.3	16	170.1	2.42	301	146	38	4	0	2.52
192	587	17	0.029	306	1229	2.02	1.97	203.3	14	204.6	2.39	348	180	51	8	0	2.41
256	787	22	0.028	462	2027	1.78	1.75	278.1	14	276.7	2.27	414	261	93	17	1	2.28

---

[6] If a given polynomial has more than the expected number of zeroes, this will help the attacker by improving their chances of guessing a pattern, but also harm them because the fewer entries a polynomial has the greater its norm, and the harder the associated lattice problem, will in general be.

[7] An attacker could also attempt zero-forcing by inverting h and looking for patterns of zeroes in g. This approach would be worthwhile if there were fewer zeroes in F than in g, but for the parameter sets under consideration, this is not the case and zero-forcing on F will always be more effective.

Foundations of Group Signatures: The Case of Dynamic Groups

Mihir Bellare, Haixia Shi, and Chong Zhang

Dept. of Computer Science & Engineering, University of California, San Diego,
9500 Gilman Drive, La Jolla, CA 92093, USA
{mihir,hashi,c2zhang}@cs.ucsd.edu
http://www-cse.ucsd.edu/users/{mihir,hashi,c2zhang}

Abstract. Recently, a first step toward establishing foundations for group signatures was taken [5], with a treatment of the case where the group is static. However the bulk of existing practical schemes and applications are for dynamic groups, and these involve important new elements and security issues. This paper treats this case, providing foundations for dynamic group signatures, in the form of a model, strong formal definitions of security, and a construction proven secure under general assumptions. We believe this is an important and useful step because it helps bridge the gap between [5] and the previous practical work, and delivers a basis on which existing practical schemes may in future be evaluated or proven secure.

1 Introduction

The purpose of foundational work is to provide strong, formal definitions of security for cryptographic primitives, thereby enabling one to unambiguously assess and prove the security of constructs and their use in applications, and then prove the existence of schemes meeting the given definitions. As evidenced by the development of the foundations of encryption [20, 24, 19, 25, 27, 17], however, this program can require several steps and considerable effort.

This paper takes the next step in the foundational effort in group signatures begun by [5]. Below we provide some background and then discuss our contributions.

1.1 Background and Motivation

GROUP SIGNATURES. The setting, introduced by Chaum and Van Heyst [15], is of a group of entities, each having its own private signing key, using which it can produce signatures on behalf of the group, meaning verifiable under a single public verification key associated to the group as a whole. The basic security requirements are that the identity of the group member producing a particular signature not be discernible from this signature (anonymity), except to an authority possessing a special "opening" key (traceability).

With time, more security requirements were added, including unlinkability, unforgeability, collusion resistance [4], exculpability [4], and framing resistance [16]. Many practical schemes were presented, some with claims of proven security in the random oracle model [1]. However, it is often unclear what the schemes or claimed proofs in these works actually deliver in terms of security guarantees, due largely to the fact that the requirements are informal and sometimes ambiguous, not precisely specifying adversary capabilities and goals. It would be beneficial in this context to have proper foundations, meaning strong formal definitions and rigorously proven-secure schemes.

FOUNDATIONS FOR STATIC GROUPS. The first step toward this end was taken by [5], who consider the case where the group is *static*. In their setting, the number of group members and their identities are fixed and frozen in the setup phase, where a trusted entity chooses not only the group public key and an opening key for the opening authority, but also, for each group member, chooses a signing key and hands it to the member in question. Within this framework, they formalize two (strong) security requirements that they call full-anonymity and full-traceability, and show that these imply all the informal existing requirements in the previous literature. They then present a static group signature scheme shown to meet these requirements, assuming the existence of trapdoor permutations.

DYNAMIC GROUPS. However, static groups limit applications of group signatures, since they do not allow one to add members to the group with time. They also require an uncomfortably high degree of trust in the party performing setup, since the latter knows the signing keys of all members and can thus frame any group member. These limitations were in fact recognized early in the development of the area, and the practical literature has from the start focused on the case where the group is *dynamic*. In this setting, neither the number nor the identities of group members are fixed or known in the setup phase, which now consists of the trusted entity choosing only a group public key and a key for the authority. An entity can join the group, and obtain a private signing key at any time, by engaging in an appropriate join protocol with the authority.

CLOSING THE GAP. We thus have the following gap: foundations have been provided for the static case [5], but the bulk of applications and existing practical schemes are for the dynamic case [15, 16, 11, 14, 26, 13, 4, 3, 1]. Since the ultimate goal is clearly to have proven secure schemes in settings suitable for applications, it is important to bridge the above-mentioned gap by providing foundations for dynamic group signatures.

However, an extension of the existing treatment of static groups [5] to the dynamic case does not seem to be immediate. Dynamic groups are more complex, bringing in new elements, security requirements and issues. A dedicated and detailed treatment is required to resolve the numerous existing issues and ambiguities. This paper provides such a treatment.

1.2 Model and Definitions for the Dynamic Group Setting

The first contribution of this paper is to provide a model and strong, formal definitions of a small number of key security requirements for dynamic group signatures that, in keeping with [5], are then shown to imply the large number of existing informal requirements.

SELECTED FEATURES. We highlight a few important features of the model and definitions:

- Two authorities. As suggested in some previous works, we separate the authority into two, an opener (who can open signatures) and an issuer (who interacts with a user to issue the latter a signing key). Each has its own secret key. This provides more security (compared to having a single authority) in the face of the possibility that authorities can be dishonest.
- Trust levels. We consider different levels of trust in each authority, namely that it may be uncorrupt (trusted), partially corrupt (its secret key is available to the adversary but it does not deviate from its prescribed program) or fully corrupt (the adversary controls it entirely, so that it may not follow its program). In order to protect group members against dishonest authorities to the maximum extent possible, we formulate security requirements to require the lowest possible level of trust in each authority.
- Three key requirements. We formulate three key requirements, namely anonymity, traceability and non-frameability. The levels of trust for each authority for each requirement are summarized in Figure 1, and, as we explain in Section 4, are the minimum possible in each case. (In the static setting, the single full-traceability requirement covered both traceability and non-frameability [5]. We separate them here because we can ask for and achieve non-frameability with lower levels of trust in the authorities than traceability.)
- PKI. We assume that each group member or potential group member has a personal public key, established and certified, for example by a PKI, independently of any group authority, so that it has a means to sign information, using a matching personal private key that it retains. This is *necessary* in order for group members to protect themselves from being framed by a partially or fully corrupt issuer, and makes explicit what were called "long-term credentials" in [1].
- Publicly verifiable proofs of opening. In order to be protected against a fully corrupt opener, the opener is required to accompany any claim that a particular identity produced a particular signature with a publicly verifiable proof to this effect (cf. [13]).
- Concurrent join protocols. In an Internet-based system, we would expect that many entities may concurrently engage in the join protocol with the issuer. Our model captures this by allowing the adversary to schedule all message delivery in any number of concurrent join sessions.

Requirement	Opener	Issuer
Anonymity	uncorrupt	fully corrupt
Traceability	partially corrupt	uncorrupt
Non-frameability	fully corrupt	fully corrupt

Fig. 1. Levels of trust in authorities for each of our three security requirements. In each case, these are the lowest levels of trust achievable.

DEFINITIONAL APPROACH. In order to provide clear, succinct yet formal definitions, and also allow for easy additions of more definitions, we take a modular approach that follows the paradigm of [7]. We first specify a model that consists of defining various oracles that provide the adversary with various attack capabilities. Each of the formal definitions then provides the adversary with some appropriate subset of these oracles, depending on the type of attack capabilities the definition wishes to give the adversary.

As research in this area has shown, requirements for group signatures tend to grow and evolve with time (cf. [4, 16, 23]). The benefit of the modular definitional approach we employ here is that it is easy to add new requirements, first by introducing new oracles to capture new attack capabilities if necessary, and then by formulating new definitions in terms of adversaries that call on the old and new oracles.

1.3 A Construction of a Secure Dynamic Group Signature Scheme

Given the stringency of our security requirements, the first and most basic question that should be considered is whether a secure dynamic group signature scheme even exists, and, if so, under what assumptions its existence can be proved. Although the setting and requirements for dynamic groups are more complex and demanding than for static groups, we can prove the existence of a secure dynamic group signature scheme under the same assumptions as used to prove the existence of a secure static group signature scheme [5], namely the existence of trapdoor permutations. As is not uncommon with foundational schemes, ours is polynomial-time but not efficient, and should be taken as a proof of concept only.

The construction uses as building blocks the following: trapdoor permutation based public-key encryption schemes secure against chosen-ciphertext attack [17], trapdoor permutation based (ordinary) digital signature schemes secure against chosen-message attack [6], and trapdoor permutation based simulation-sound adaptive non-interactive zero-knowledge (NIZK) proofs for NP [28]. We provide a way to define a group public key, keys for the two authorities, and a join protocol so that the private signing key of any group member, as well as the signature created, have essentially the same format as in the scheme of [5], thereby enabling us to build on the latter. We then augment the opening algorithm to also produce NIZK proofs of its claims, and define a judge algorithm to check such proofs. To provide traceability and non-frameability, the join protocol requires, on the one hand, that the group member provide the issuer with a signature (relative to the personal public key that the group member has via the PKI) of some information related to the private signing key it is issued. (This signature is stored by the issuer in the registration table and can later be accessed by the opener.) However, it also ensures that the issuer does not know the private signing key of the group member. We note that in our scheme, the length of signatures and the size of keys do not depend on the number of members in the group. (The registration table has size proportional to the number of users but is not considered part of the keys.)

We remark that the join protocol is simple and uses no zero-knowledge (ZK) proofs. This is important because it facilitates showing security under arbitrary concurrent executions. But it may be surprising because the join protocols in practical schemes such as that of [1] use ZK proofs even though the security requirements there are milder than in our case.

1.4 Discussion and Related Work

We do not consider revocation of group members[1]. Different solutions tend to require or depend on different model elements [10, 2, 29] and we believe it is restrictive to pin down features geared toward some solution as part of what is supposed to be a general model. However, as noted above, our model has an extensible format, and can be extended in different ways to accommodate different revocation approaches and requirements.

In specifying our model and definitions we have built on numerous elements of previous works, including informal discussions in [5] about extensions to the dynamic setting. We remark however that we were not always able to follow the suggestions of the latter. For example they suggested that a proof of opening could consist of the coins underlying a certain ciphertext in the signature. But the decryption algorithms of existing trapdoor permutation based, chosen-ciphertext secure encryption schemes [17, 28] do not recover the coins, and, even if one had a scheme that did, one would need to know whether it was secure against a stronger type of chosen-ciphertext attacks in which the decryption oracle returns not just the message but also the coins underlying a given ciphertext. Instead, we use NIZK proofs.

Our model assumes that the issuer and opener are provided their keys by a trusted initialization process that chooses these keys along with the group public key. Naturally, if so desired, such a process may be implemented by a secure distributed computation protocol in which the authorities jointly compute their keys and the group public key. This would enable one to dispense with the trusted initialization.

There may be schemes or setting in which there is a single authority that plays the roles of both issuer and opener, rather than there being two separate authorities as in our model. This case is simpler than the one we consider, and

---

[1] Our terminology may thus be misleading. In some previous works, what we are considering are called partially dynamic groups rather than dynamic groups. The term monotonically growing groups has also been suggested.

our definitions and scheme can easily be "dropped down" to handle it. Of course, the security achieved will be weaker.

Our model captures the functionality of current efficient proposals for group signature schemes, in particular that of [1]. Although we do not know whether their scheme can be proven secure in our model, providing the model at least enables one to address this question rigorously in the future.

Camenisch and Lysyanskaya [12] present simulation-based definitions for identity-escrow schemes with appointed verifiers, which are related to group signature schemes. We believe however that models like that of [5] and ours are easier to use.

In concurrent and independent work, Kiayis, Tsiounis and Yung [23] introduce an extension of group signatures called traceable signatures. However, in the dynamic group signature setting, their model is different from ours. In particular, they consider a single authority rather than separate issuing and opening authorities. This means that they cannot consider authority behavior that is as adversarial as the ones we consider, namely fully corrupt, and, in some cases, even partially corrupt authorities. Not only does this mean their requirements are weaker than ours, but also this is where most of the novel issues, as compared with [5], arise. We also note that their model does not include a PKI, yet some such structure would appear to be required to realize certain assumptions they make. (Namely that the authority is not given the power to modify transcripts of the join protocol in non-frameability).

Finally we note that the traceable signature scheme of [23] is in the random-oracle model, being derived as the Fiat-Shamir transform [18] of a traceable identification scheme. Note that it may be *impossible* to "implement" the random oracle of the Fiat-Shamir transform with a "real" function in a way that results in a secure real-world scheme (Goldwasser and Tauman [22]). In contrast our scheme is in the standard model.

2 Notation

We let $\mathbb{N} = \{1, 2, 3, \ldots\}$ be the set of *positive* integers. If x is a string, then $|x|$ denotes its length, while if S is a set then $|S|$ denotes its size. The empty string is denoted by ε. If $k \in \mathbb{N}$ then 1^k denotes the string of k ones. If n is an integer then $[n] = \{1, \ldots, n\}$. If S is a set then $s \xleftarrow{\$} S$ denotes the operation of picking an element s of S uniformly at random.

Unless otherwise indicated, algorithms are randomized. We write $A(x, y, \ldots)$ to indicate that A is an algorithm with inputs $x, y, \ldots$,, and by $z \xleftarrow{\$} A(x, y, \ldots)$ we denote the operation of running A with inputs $x, y, \ldots$ and letting z be the output. We write $A(x, y, \ldots : \mathcal{O}_1, \mathcal{O}_2, \ldots)$ to indicate that A is an algorithm with inputs $x, y, \ldots$ and access to oracles $\mathcal{O}_1, \mathcal{O}_2, \ldots$, and by $z \xleftarrow{\$} A(x, y, \ldots : \mathcal{O}_1, \mathcal{O}_2, \ldots)$ we denote the operation of running A with inputs $x, y, \ldots$ and access to oracles $\mathcal{O}_1, \mathcal{O}_2, \ldots$, and letting z be the output.

3 A Model for Dynamic Group Signature Schemes

Here we provide a model in which definitions can later be formulated. We begin with a discussion of the syntax, namely the algorithms that constitute a dynamic group signature scheme.

ALGORITHMS AND THEIR USAGE. Involved in a group signature scheme are a trusted party for initial key generation, an authority called the issuer, an authority called the opener, and a body of users, each with a unique identity $i \in \mathbb{N}$, that may become group members. The scheme is specified as a tuple $\mathcal{GS} = $ (GKg, UKg, Join, Iss, GSig, GVf, Open, Judge) of polynomial-time algorithms whose intended usage and functionality are as follows. Throughout, $k \in \mathbb{N}$ denotes the security parameter.

GKg – In a setup phase, the trusted party runs the *group-key generation* algorithm GKg on input 1^k to obtain a triple (gpk, ik, ok). The *issuer key ik* is provided to the issuer, and the *opening key ok* is provided to the opener. The *group public key gpk*, whose possession enables signature verification, is made public.

UKg – A user that wants to be a group member should begin by running the *user-key generation* algorithm UKg on input 1^k to obtain a *personal public and private key pair* (**upk**$[i]$, **usk**$[i]$). We assume that the table **upk** is public. (Meaning, anyone can obtain an authentic copy of the personal public key of any user. This might be implemented via a PKI.)

Join, Iss – Once a user has its personal key pair, it can join the group by engaging in a *group-joining protocol* with the issuer. The *interactive algorithms* Join, Iss implement, respectively, the user's and issuer's sides of this interaction. Each takes input an incoming message (this is ε if the party is initiating the interaction) and a current state, and returns an outgoing message, an updated state, and a decision which is one of accept, reject, cont. The communication is assumed to take place over secure (i.e. private and authenticated) channels, and we assume the user sends the first message. If the issuer accepts, it makes an entry for i, denoted **reg**$[i]$, in its *registration table* **reg**, the contents of this entry being the final state output by Iss. If i accepts, the final state output by Join is its private signing key, denoted **gsk**$[i]$.

GSig – A group member i, in possession of its signing key **gsk**$[i]$, can apply the *group signing* algorithm GSig to **gsk**$[i]$ and a message $m \in \{0,1\}^*$ to obtain a quantity called a signature on m.

GVf – Anyone in possession of the group public key gpk can run the deterministic *group signature verification* algorithm GVf on inputs gpk, a message m, and a candidate signature σ for m, to obtain a bit. We say that σ is a *valid* signature of m with respect to gpk if this bit is one.

Open – The opener, who has read-access to the registration table **reg** being populated by the issuer, can apply the deterministic *opening* algorithm Open to its opening key ok, the registration table **reg**, a message m, and a valid signature

σ of m under gpk. The algorithm returns a pair (i, τ), where $i \geq 0$ is an integer. In case $i \geq 1$, the algorithm is claiming that the group member with identity i produced σ, and in case $i = 0$, it is claiming that no group member produced σ. In the former case, τ is a proof of this claim that can be verified via the Judge algorithm.

Judge – The deterministic *judge* algorithm Judge takes inputs the group public key gpk, an integer $j \geq 1$, the public key **upk**$[j]$ of the entity with identity j (this is ε if this entity has no public key), a message m, a valid signature σ of m, and a proof-string τ. It aims to check that τ is a proof that j produced σ. We note that the judge will base its verification on the public key of j.

THE ORACLES. The correctness and security definitions will be formulated via experiments in which an adversary's attack capabilities are modeled by providing it access to certain oracles. We now introduce the oracles that we will need. (Different experiments will provide the adversary with different subsets of this set of oracles.)

The oracles are specified in Figure 2 and explained below. It is assumed that the overlying experiment has run GKg on input 1^k to obtain keys gpk, ik, ok that are used by the oracles. It is also assumed that this experiment maintains the following global variables which are manipulated by the oracles: a set HU of honest users; a set CU of corrupted users; a set GSet of message-signature pairs; a table **upk** such that **upk**$[i]$ contains the public key of $i \in \mathbb{N}$; a table **reg** such that **reg**$[i]$ contains the registration information of group member i. The sets HU, CU, GSet are assumed initially empty, and all entries of the tables **upk**, **reg** are assumed initially to be ε. Randomized oracles or algorithms use fresh coins upon each invocation unless otherwise indicated.

AddU($\cdot$) – By calling this *add user* oracle with argument an identity $i \in \mathbb{N}$, the adversary can add i to the group as an honest user. The oracle adds i to the set HU of honest users, and picks a personal public and private key pair (**upk**$[i]$, **usk**$[i]$) for i. It then executes the group-joining protocol by running Join (on behalf of i, initialized with gpk, **upk**$[i]$, **usk**$[i]$) and Iss (on behalf of the issuer, initialized with gpk, ik, i, **upk**$[i]$). When Iss accepts, its final state is recorded as entry **reg**$[i]$ in the registration table. When Join accepts, its final state is recorded as the private signing key **gsk**$[i]$ of i. The calling adversary is returned upk$[i]$.

CrptU($\cdot, \cdot$) – By calling this *corrupt user* oracle with arguments an identity $i \in \mathbb{N}$ and a string upk, the adversary can corrupt user i and set its personal public key **upk**$[i]$ to the value upk chosen by the adversary. The oracle initializes the issuer's state in anticipation of a group-joining protocol with i.

SndToI($\cdot, \cdot$) – Having corrupted user i, the adversary can use this *send to issuer* oracle to engage in a group-joining protocol with the honest, Iss-executing issuer, itself playing the role of i and not necessarily executing the interactive algorithm Join prescribed for an honest user. The adversary provides the oracle with i and a message M_{in} to be sent to the issuer. The oracle, which maintains the issuer's

AddU(i)
   If $i \in$ CU or $i \in$ HU then return ε
   HU $\leftarrow$ HU $\cup \{i\}$
   $dec^i \leftarrow$ cont; $\mathbf{gsk}[i] \leftarrow \varepsilon$
   $(\mathbf{upk}[i], \mathbf{usk}[i]) \stackrel{\$}{\leftarrow} \mathsf{UKg}(1^k)$
   $\mathrm{St}_{jn}^i \leftarrow (gpk, \mathbf{upk}[i], \mathbf{usk}[i])$
   $\mathrm{St}_{iss}^i \leftarrow (gpk, ik, i, \mathbf{upk}[i])$; $M_{jn} \leftarrow \varepsilon$
   $(\mathrm{St}_{jn}^i, M_{iss}, dec^i) \leftarrow \mathsf{Join}(\mathrm{St}_{jn}^i, M_{jn})$
   While $dec^i =$ cont do
      $(\mathrm{St}_{iss}^i, M_{jn}, dec^i) \leftarrow \mathsf{Iss}(\mathrm{St}_{iss}^i, M_{iss}, dec^i)$
      If $dec^i =$ accept then $\mathbf{reg}[i] \leftarrow \mathrm{St}_{iss}^i$
      $(\mathrm{St}_{jn}^i, M_{iss}, dec^i) \leftarrow \mathsf{Join}(\mathrm{St}_{jn}^i, M_{jn})$
   Endwhile
   $\mathbf{gsk}[i] \leftarrow \mathrm{St}_{jn}^i$
   Return $\mathbf{upk}[i]$

SndToI(i, M_{in})
   If $i \notin$ CU then return ε
   $(\mathrm{St}_{iss}^i, M_{out}, dec^i) \leftarrow \mathsf{Iss}(\mathrm{St}_{iss}^i, M_{in}, dec^i)$
   If $dec^i =$ accept then $\mathbf{reg}[i] \leftarrow \mathrm{St}_{iss}^i$
   Return M_{out}

SndToU(i, M_{in})
   If $i \notin$ HU then
      HU $\leftarrow$ HU $\cup \{i\}$
      $(\mathbf{upk}[i], \mathbf{usk}[i]) \stackrel{\$}{\leftarrow} \mathsf{UKg}(1^k)$
      $\mathbf{gsk}[i] \leftarrow \varepsilon$; $M_{in} \leftarrow \varepsilon$
      $\mathrm{St}_{jn}^i \leftarrow (gpk, \mathbf{upk}[i], \mathbf{usk}[i])$
   $(\mathrm{St}_{jn}^i, M_{out}, dec) \leftarrow \mathsf{Join}(\mathrm{St}_{jn}^i, M_{in})$;
   If $dec =$ accept then $\mathbf{gsk}[i] \leftarrow \mathrm{St}_{jn}^i$
   Return (M_{out}, dec)

CrptU(i, upk)
   If $i \in$ HU $\cup$ CU then return ε
   CU $\leftarrow$ CU $\cup \{i\}$
   $\mathbf{upk}[i] \leftarrow upk$
   $dec^i \leftarrow$ cont
   $\mathrm{St}_{iss}^i \leftarrow (gpk, ik, i, \mathbf{upk}[i])$
   Return 1

USK(i)
   Return $(\mathbf{gsk}[i], \mathbf{usk}[i])$

RReg(i)
   Return $\mathbf{reg}[i]$

WReg(i, ρ)
   $\mathbf{reg}[i] \leftarrow \rho$

Open(m, σ)
   If $(m, \sigma) \in$ GSet then return $\bot$
   Return $\mathsf{Open}(gpk, ok, \mathbf{reg}, m, \sigma)$

GSig(i, m)
   If $i \notin$ HU then return $\bot$
   If $\mathbf{gsk}[i] = \varepsilon$ then return $\bot$
   Else return $\mathsf{GSig}(gpk, \mathbf{gsk}[i], m)$

Ch$_b$(i_0, i_1, m)
   If $i_0 \notin$ HU or $i_1 \notin$ HU then
      return $\bot$
   If $\mathbf{gsk}[i_0] = \varepsilon$ or $\mathbf{gsk}[i_1] = \varepsilon$ then
      return $\bot$
   $\sigma \leftarrow \mathsf{GSig}(gpk, \mathbf{gsk}[i_b], m)$
   GSet $\leftarrow$ GSet $\cup \{(m, \sigma)\}$
   Return σ

Fig. 2. Oracles provided to adversaries in the experiments of Figure 3.

state (the latter having been initialized by an earlier call to CrptU($i, \cdot$)), computes a response as per Iss, returns the outgoing message to the adversary, and sets entry $\mathbf{reg}[i]$ of the registration table to Iss's final state if the latter accepts.

SndToU($\cdot, \cdot$) – In some definitions we will want to consider an adversary that has corrupted the issuer. The *send to user* oracle SndToU($\cdot, \cdot$) can be used by such an adversary to engage in a group-joining protocol with an honest, Join-executing user, itself playing the role of the issuer and not necessarily executing the interactive algorithm Iss prescribed for the honest issuer. The adversary provides the oracle with i and a message M_{in} to be sent to i. The oracle maintains the state of user i, initializing this the first time it is called by choosing a personal public and private key pair for i, computes a response as per Join, returns the outgoing message to the adversary, and sets the private signing of i to Join's final state if the latter accepts.

USK($\cdot$) – The adversary can call this *user secret keys* oracle with argument the identity $i \in \mathbb{N}$ of a user to expose both the private signing key $\mathbf{gsk}[i]$ and the personal private key $\mathbf{usk}[i]$ of this user.

RReg($\cdot$) – The adversary can read the contents of entry i of the registration table $\mathbf{reg}$ by calling this *read registration table* oracle with argument $i \in \mathbb{N}$.

WReg($\cdot, \cdot$) – In some definitions we will allow the adversary to write/modify the contents of entry i of the registration table $\mathbf{reg}$ by calling this *write registration table* oracle with argument $i \in \mathbb{N}$.

GSig($\cdot, \cdot$) – A *signing* oracle, enabling the adversary to specify the identity i of a user and a message m, and obtain the signature of m under the private signing key $\mathbf{gsk}[i]$ of i, as long as i is an honest user whose private signing key is defined.

Ch($b, \cdot, \cdot, \cdot$) – A *challenge* oracle provided to an adversary attacking anonymity, and depending on a challenge bit b set by the overlying experiment. The adversary provides a pair i_0, i_1 of identities and a message m, and obtains the signature of m under the private signing key of i_b, as long as both i_0, i_1 are honest users with defined private signing keys. The oracle records the message-signature pair in GSet to ensure that the adversary does not later call the opening oracle on it.

Open($\cdot, \cdot$) – The adversary can call this *opening* oracle with arguments a message m and signature σ to obtain the output of the opening algorithm on m, σ, computed under the opener's key ok, as long as σ was not previously returned in response to a query to Ch($b, \cdot, \cdot, \cdot$).

REMARKS. We are assuming the existence of a secure (private and authentic) channel between any prospective group member and the issuer, as in [1]. The privacy assumption is reflected in the fact that the adversary is not provided the transcript of an interaction generated by the AddU($\cdot$) oracle. The authenticity assumption is reflected in the fact that a party is initialized with the correct identity and personal public key of its partner if relevant. (When the issuer is fully corrupted, reflected by the adversary having a SndToU($\cdot, \cdot$) oracle, the adversary does get the transcript of the communication, via its oracle queries and answers.) We note however that the secure channels assumption is made more for simplicity than anything else, and protocols are easily modified to avoid it.

4 Notions of Correctness and Security

Here we provide the definitions of correctness and security of a dynamic group signature scheme, based on the model of an adversary with oracles introduced above. We begin with correctness and then define three security requirements: anonymity, traceability and non-frameability.

Experiment $\mathbf{Exp}^{\text{corr}}_{\mathcal{GS},A}(k)$

$(gpk, ik, ok) \xleftarrow{\$} \mathsf{GKg}(1^k)$; $\mathrm{CU} \leftarrow \emptyset$; $\mathrm{HU} \leftarrow \emptyset$; $(i, m) \xleftarrow{\$} A(gpk : \mathsf{AddU}(\cdot), \mathsf{RReg}(\cdot))$
If $i \notin \mathrm{HU}$ then return 0 ; If $\boldsymbol{gsk}[i] = \varepsilon$ then return 0
$\sigma \leftarrow \mathsf{GSig}(gpk, \boldsymbol{gsk}[i], m)$; If $\mathsf{GVf}(gpk, m, \sigma) = 0$ then return 1
$(j, \tau) \leftarrow \mathsf{Open}(gpk, ok, \boldsymbol{reg}, m, \sigma)$; If $i \neq j$ then return 1
If $\mathsf{Judge}(gpk, i, \boldsymbol{upk}[i], m, \sigma, \tau) = 0$ then return 1 else return 0

Experiment $\mathbf{Exp}^{\text{anon-}b}_{\mathcal{GS},A}(k)$ // $b \in \{0, 1\}$

$(gpk, ik, ok) \xleftarrow{\$} \mathsf{GKg}(1^k)$; $\mathrm{CU} \leftarrow \emptyset$; $\mathrm{HU} \leftarrow \emptyset$; $\mathrm{GSet} \leftarrow \emptyset$
$d \xleftarrow{\$} A(gpk, ik : \mathsf{Ch}(b, \cdot, \cdot, \cdot), \mathsf{Open}(\cdot, \cdot), \mathsf{SndToU}(\cdot, \cdot), \mathsf{WReg}(\cdot, \cdot), \mathsf{USK}(\cdot), \mathsf{CrptU}(\cdot, \cdot))$
Return d

Experiment $\mathbf{Exp}^{\text{trace}}_{\mathcal{GS},A}(k)$

$(gpk, ik, ok) \xleftarrow{\$} \mathsf{GKg}(1^k)$; $\mathrm{CU} \leftarrow \emptyset$; $\mathrm{HU} \leftarrow \emptyset$
$(m, \sigma) \xleftarrow{\$} A(gpk, ok : \mathsf{SndToI}(\cdot, \cdot), \mathsf{AddU}(\cdot), \mathsf{RReg}(\cdot), \mathsf{USK}(\cdot), \mathsf{CrptU}(\cdot, \cdot))$
If $\mathsf{GVf}(gpk, m, \sigma) = 0$ then return 0 ; $(i, \tau) \leftarrow \mathsf{Open}(gpk, ok, \boldsymbol{reg}, m, \sigma)$
If $i = 0$ or $\mathsf{Judge}(gpk, i, \boldsymbol{upk}[i], m, \sigma, \tau) = 0$ then return 1 else return 0

Experiment $\mathbf{Exp}^{\text{nf}}_{\mathcal{GS},A}(k)$

$(gpk, ik, ok) \xleftarrow{\$} \mathsf{GKg}(1^k)$; $\mathrm{CU} \leftarrow \emptyset$; $\mathrm{HU} \leftarrow \emptyset$
$(m, \sigma, i, \tau) \xleftarrow{\$} A(gpk, ok, ik : \mathsf{SndToU}(\cdot, \cdot), \mathsf{WReg}(\cdot, \cdot), \mathsf{GSig}(\cdot, \cdot), \mathsf{USK}(\cdot), \mathsf{CrptU}(\cdot, \cdot))$
If $\mathsf{GVf}(gpk, m, \sigma) = 0$ then return 0
If the following are all true then return 1 else return 0:
- $i \in \mathrm{HU}$ and $\boldsymbol{gsk}[i] \neq \varepsilon$ and $\mathsf{Judge}(gpk, i, \boldsymbol{upk}[i], m, \sigma, \tau) = 1$
- A did not query $\mathsf{USK}(i)$ or $\mathsf{GSig}(i, m)$

Fig. 3. Experiments used to define correctness, anonymity, traceability and non-frameability of a dynamic group signature scheme $\mathcal{GS} = (\mathsf{GKg}, \mathsf{UKg}, \mathsf{Join}, \mathsf{Iss}, \mathsf{GSig}, \mathsf{GVf}, \mathsf{Open}, \mathsf{Judge})$.

4.1 Correctness

The correctness condition pertains to signatures generated by honest group members, and asks the following: the signature should be valid; the opening algorithm, given the message and signature, should correctly identify the signer; the proof returned by the opening algorithm should be accepted by the judge. Formalizing these conditions in the dynamic group setting is more involved than formalizing them in a static setting in that these conditions must hold for all honest users under any "schedule" under which these users join the group. Accordingly, we formalize correctness via an experiment involving an adversary. To dynamic group signature scheme $\mathcal{GS}$, any adversary A and any $k \in \mathbb{N}$ we associate the experiment $\mathbf{Exp}^{\text{corr}}_{\mathcal{GS},A}(k)$ depicted in Figure 3. We let

$$\mathbf{Adv}^{\text{corr}}_{\mathcal{GS},A}(k) = \Pr\left[\mathbf{Exp}^{\text{corr}}_{\mathcal{GS},A}(k) = 1\right].$$

We say that dynamic group signature scheme $\mathcal{GS}$ is *correct* if $\mathbf{Adv}^{\text{corr}}_{\mathcal{GS},A}(k) = 0$ for any adversary A and any $k \in \mathbb{N}$. Note that the adversary is not computationally restricted.

4.2 Anonymity

FORMAL DEFINITION. To dynamic group signature scheme $\mathcal{GS}$, any adversary A, a bit $b \in \{0,1\}$ and any $k \in \mathbb{N}$ we associate the experiment $\mathbf{Exp}_{\mathcal{GS},A}^{\text{anon-}b}(k)$ depicted in Figure 3. We let

$$\mathbf{Adv}_{\mathcal{GS},A}^{\text{anon}}(k) \;=\; \Pr\left[\,\mathbf{Exp}_{\mathcal{GS},A}^{\text{anon-1}}(k) = 1\,\right] - \Pr\left[\,\mathbf{Exp}_{\mathcal{GS},A}^{\text{anon-0}}(k) = 1\,\right].$$

We say that dynamic group signature scheme $\mathcal{GS}$ is *anonymous* if the function $\mathbf{Adv}_{\mathcal{GS},A}^{\text{anon}}(\cdot)$ is negligible for any polynomial-time adversary A.

DISCUSSION. The definition is liberal with regard to what it means for the adversary to win. It need not recover the identity of a signer from a signature, but, following [5], need only distinguish which of two signers of its choice signed a target message of its choice. Formally, this means it wins if it guesses the value of the bit b in the $\mathsf{Ch}(b,\cdot,\cdot,\cdot)$ oracle. In the process, the adversary is provided with extremely strong attack capabilities, including the ability to fully corrupt the issuer. (The adversary is not only given the issuer key ik, but is provided access to the $\mathsf{SndToI}(\cdot,\cdot)$ oracle, which enables it to play the role of issuer in interacting with users in the join protocol.) The adversary is additionally allowed to obtain both the personal private key and the private signing key of any user (via the USK oracle); read, write or modify the content of the registration table (via the $\mathsf{RReg}, \mathsf{WReg}$ oracles); corrupt users and interact with the issuer on their behalf (via the $\mathsf{CrptU}, \mathsf{SndToU}$ oracles); and obtain the identity of the signer of any signature except the challenge one (via the Open oracle).

We do not provide the adversary access to the GSig and AddU oracles because they are redundant given the capabilities already provided to the adversary. Naturally, the adversary is also denied the opener's key ok, since the latter would enable it to run the Open algorithm. (Meaning the opener must be assumed uncorrupt.)

4.3 Traceability

FORMAL DEFINITION. To dynamic group signature scheme $\mathcal{GS}$, any adversary A and any $k \in \mathbb{N}$ we associate the experiment $\mathbf{Exp}_{\mathcal{GS},A}^{\text{trace}}(k)$ depicted in Figure 3. We let

$$\mathbf{Adv}_{\mathcal{GS},A}^{\text{trace}}(k) \;=\; \Pr\left[\,\mathbf{Exp}_{\mathcal{GS},A}^{\text{trace}}(k) = 1\,\right].$$

We say that dynamic group signature scheme $\mathcal{GS}$ is *traceable* if the function $\mathbf{Adv}_{\mathcal{GS},A}^{\text{trace}}(\cdot)$ is negligible for any polynomial-time adversary A.

DISCUSSION. Traceability asks that the adversary be unable to produce a signature such that either the honest opener declares itself unable to identify the origin of the signature (meaning the Open algorithm returns (i,τ) with $i = 0$), or, the honest opener believes it has identified the origin but is unable to produce a correct proof of its claim (meaning the Open algorithm returns (i,τ) with $i > 0$ but the proof τ is rejected by the judge). In the process, the adversary is allowed to create honest group members (via the AddU oracle); obtain both the personal private key and the private signing key of any user (via the USK oracle); read the content of the registration table (via the RReg oracles); and corrupt users and interact with the issuer on their behalf (via the CrptU, SndToU oracles).

Note that traceability cannot be achieved in the presence of even a partially corrupt issuer, for such an issuer can create dummy users with valid signing keys and thus create untraceable signatures. (That is, the assumption that the issuer is uncorrupt is minimal). Accordingly, in the definition, the adversary is not given *ik* as input and not given a SndToU oracle. Also it is not allowed to write to the registration table (meaning it is not given a WReg oracle) since it could otherwise remove the information enabling a group member to be traced. Also, the assumption that the opener is partially but not fully corrupt is minimal, for a fully corrupt opener could simply refuse to trace.

4.4 Non-frameability

FORMAL DEFINITION. To dynamic group signature scheme $\mathcal{GS}$, any adversary A and any $k \in \mathbb{N}$ we associate the experiment $\mathbf{Exp}^{\mathrm{nf}}_{\mathcal{GS},A}(k)$ depicted in Figure 3. We let

$$\mathbf{Adv}^{\mathrm{nf}}_{\mathcal{GS},A}(k) = \Pr\left[\mathbf{Exp}^{\mathrm{nf}}_{\mathcal{GS},A}(k) = 1\right].$$

We say that dynamic group signature scheme $\mathcal{GS}$ is *non-frameable* if the function $\mathbf{Adv}^{\mathrm{nf}}_{\mathcal{GS},A}(\cdot)$ is negligible for any polynomial-time adversary A.

DISCUSSION. Non-frameability asks that the adversary be unable to create a judge-accepted proof that an honest user produced a certain valid signature unless this user really did produce this signature. (This implies the more usual formulation, namely that it cannot produce a signature that an honest opener would attribute to a user unless the latter really did produce it, because, if it could produce such a signature, it could also produce the judge-accepted proof. The latter is true because we give it the secret key of the opener). The adversary outputs a message m, a signature σ, an identity i and a proof τ. It wins if σ is a valid signature of m, i is an honest user, and the judge accepts τ as a proof that i produced σ, yet the adversary did not query the signing oracle GSig with i, m and did not obtain i's signing key $gsk[i]$ via the USK oracle. Barring these restrictions, the adversary is extremely powerful, and in particular much stronger than for traceability (which is why, unlike [5], we separate the two). In particular it may fully corrupt both the opener and the issuer. (Reflected in its getting input *ok, ik* and having access to the SndToU oracle.) Additionally, it may create a colluding subset of users by using its USK oracle to obtain signing keys of all users except the target one it outputs, and also corrupt users via CrptU.

4.5 Remarks

Recall that in [5] the issuing process was static and trusted, and their single authority played the role of opener. Their full-traceability requirement, which covered both traceability and non-frameability, allowed the opener to be partially but not fully corrupt. We are asking for traceabiltiy under the same conditions, which, as we have argued above, are minimal in the dynamic setting. But we ask for non-frameability under much more adverse conditions, namely when both authorities may be fully corrupt. (In achieving this, the PKI is crucial). This is the motivation for separating their single requirement into two.

We note that a reader might find that what is intuitively regarded as traceability is covered by the combination of traceability and non-frameability rather than by the formal traceability alone.

In [8] we point out that, as in the static case [5], the key requirements that we define (anonymity, traceability and non-frameability) are strong enough to capture and imply all existing informal security requirements in the literature.

5 Our Construction

We begin by describing the primitives we use, and then overview our construction. A full description of the construction, together with definitions of security for the primitives, can be found in [8].

PRIMITIVES. We use a digital signature scheme $\mathcal{DS} = (\mathsf{K_s}, \mathsf{Sig}, \mathsf{Vf})$ specified, as usual, by algorithms for key generation, signing and verifying. It should satisfy the standard notion of unforgeability under chosen message attack [21], the definition of which is recalled in [8].

We use a public-key encryption scheme $\mathcal{AE} = (\mathsf{K_e}, \mathsf{Enc}, \mathsf{Dec})$ specified, as usual, by algorithms for key generation, encryption and decryption. It should satisfy the standard notion of indistinguishability under adaptive chosen-ciphertext attack (IND-CCA) [27], the definition of which is recalled in [8].

The last building block we need are simulation-sound NIZK proofs of membership in NP languages. We use the following terminology. An NP-*relation over domain* $\mathrm{Dom} \subseteq \{0,1\}^*$ is a subset ρ of $\{0,1\}^* \times \{0,1\}^*$ such that membership of $(x,w) \in \rho$ is decidable in time polynomial in the length of the first argument for all x in domain Dom. The language associated to ρ is the set of all $x \in \{0,1\}^*$ such that there exists a w for which $(x,w) \in \rho$. Often we will just use the term NP-relation, the domain being implicit. If $(x,w) \in \rho$ we will say that x is a *theorem* and w is a *proof* of x.

Fix an NP relation ρ over domain Dom. Consider a pair of polynomial time algorithms (P,V), where P is randomized and V is deterministic. They have access to a *common reference string*, R. In [8] we recall the definition of (P,V) being a simulation-sound, non-interactive zero-knowledge proof system for ρ over domain Dom.

OVERVIEW OF OUR CONSTRUCTION. We fix a digital signature scheme $\mathcal{DS} = (\mathsf{K_s}, \mathsf{Sig}, \mathsf{Vf})$ and a public-key encryption scheme $\mathcal{AE} = (\mathsf{K_e}, \mathsf{Enc}, \mathsf{Dec})$ as above. We now show that the building blocks above can be used to construct a group signature scheme $\mathcal{GS} = (\mathsf{GKg}, \mathsf{UKg}, \mathsf{GSig}, \mathsf{GVf}, \mathsf{Join}, \mathsf{Iss}, \mathsf{Open}, \mathsf{Judge})$ that is anonymous, traceable and non-frameable. We now present an overview of our construction.

The group public key gpk consists of the security parameter k, a public encryption key pk_e, a verification key pk_s for digital signatures which we call the *certificate verification* key, and two reference strings R_1 and R_2 for NIZK proofs. We denote by sk_s the signing key corresponding to pk_s, and call it the *certificate creation* key. The issuer secret key ik is the certificate creation key sk_s. The opener secret key ok is the decryption key sk_e corresponding to pk_e, together with the random coins r_e used to generate (sk_e, pk_e). The certificate creation key sk_s is however denied to the group opener. (This prevents the latter from issuing certificates for keys it generates itself, and is important to attain traceability.)

In the group-joining protocol, user i generates a verification key pk_i and the corresponding signing key sk_i. It uses its personal private key $\mathbf{usk}[i]$ to produce a signature sig_i on pk_i. The signature sig_i prevents the user from being framed by a corrupt issuer. (The personal public and private key pair $(\mathbf{upk}[i], \mathbf{usk}[i])$ were obtained by running the user-key generation algorithm prior to the group-joining protocol. This is handled by the oracles.) The users sends pk_i, sig_i to the issuer, who issues membership to i by signing pk_i using the certificate creation key sk_s. The issuer then stores (pk_i, sig_i) in the registration table. Later, sig_i can be used by the opener to produce proofs for its claims.

A group member i can produce a signature for a message m under pk_i by using its secret signing key sk_i. To make this verifiable without losing anonymity, it encrypts the verification key pk_i under pk_e and then proves in zero-knowledge that verification succeeds with respect to pk_i. However, to prevent someone from simply creating their own key pair sk_i, pk_i and doing this, it also encrypts i and its certificate $cert_i$, and proves in zero-knowledge that this certificate is a signature of $\langle i, pk_i \rangle$ under the certificate verification key pk_s present in the group public key. Group signature verification comes down to verification of the NIZK proofs.

Opening is possible because the group opener has the decryption key sk_e. It obtains the user identity i by decrypting the ciphertext in the signature. When i is indeed an existing user, the opener proves its claim by supplying evidence that it decrypts the ciphertext correctly, and the user public key it obtained from decryption is authentic (i.e. signed by user i using $\mathbf{usk}[i]$). The former is accomplished by a zero-knowledge proof. The judge algorithm simply checks if these proofs are correct.

SPECIFICATION OF OUR CONSTRUCTION. We now specify witness relations ρ_1 and ρ_2 underlying the zero-knowledge proofs. Relation ρ_1 is defined as follows: $((pk_e, pk_s, m, C), (i, pk', cert, s, r)) \in \rho_1$ iff

$\mathsf{Vf}(pk_s, \langle i, pk' \rangle, cert) = 1$, $\mathsf{Vf}(pk', m, s) = 1$ and $\mathsf{Enc}(pk_e, \langle i, pk', cert, s \rangle; r) = C$.

Here m is a k-bit message, C a ciphertext and s a signature. We are writing $\mathsf{Enc}(pk_e, m; r)$ for the encryption of message m under key pk_e using coins r, and assume that $|r| = k$. The domain Dom_1 corresponding to ρ_1 is the set of all (pk_e, pk_s, m, C) such that pk_e (resp. pk_s) is a public key having non-zero probability of being produced by $\mathsf{K_e}$ (resp. $\mathsf{K_s}$) on input k, and m is a k-bit string. It is immediate that ρ_1 is an NP relation over Dom_1. Relation ρ_2 is defined as follows: $((pk_e, C, i, pk, cert, s), (sk_e, r_e)) \in \rho_2$ iff

$$\mathsf{K_e}(1^k; r_e) = (pk_e, sk_e) \text{ and } \mathsf{Dec}(sk_e, C) = \langle i, pk, \text{cert}, s \rangle$$

Here C is a ciphertext, i an identity and s a signature. The domain Dom_2 corresponding to ρ_2 is the set of all $(pk_e, C, i, pk, \text{cert}, s)$ such that pk_e is a public key having non-zero probability of being produced by $\mathsf{K_e}$ on input k. It is immediate that ρ_2 is an NP relation over Dom_2.

We fix a proof system (P_1, V_1) for ρ_1 and (P_2, V_2) for ρ_2. A detailed specification of the algorithms GKg, UKg, GSig, GVf, Join, Iss, Open, Judge that comprise our dynamic group signature scheme $\mathcal{GS}$, based on the above, can be found in [8].

SECURITY RESULTS. Fix digital signature scheme $\mathcal{DS} = (\mathsf{K_s}, \mathsf{Sig}, \mathsf{Vf})$, public-key encryption scheme $\mathcal{AE} = (\mathsf{K_e}, \mathsf{Enc}, \mathsf{Dec})$, NP-relations ρ_1 over domain Dom_1, ρ_2 over domain Dom_2, and their non-interactive proof systems (P_1, V_1) and (P_2, V_2) as above, and let $\mathcal{GS} = (\mathsf{GKg}, \mathsf{UKg}, \mathsf{GSig}, \mathsf{GVf}, \mathsf{Join}, \mathsf{Iss}, \mathsf{Open}, \mathsf{Judge})$ denote the dynamic group signature scheme associated to them as per our construction. We derive our main result (Theorem 1) via the following three lemmas proved in [8].

Lemma 1. *If $\mathcal{AE}$ is an IND-CCA secure encryption scheme, (P_1, V_1) is a simulation sound, computational zero-knowledge proof system for ρ_1 over Dom_1 and (P_2, V_2) is a computational zero-knowledge proof system for ρ_2 over Dom_2, then group signature scheme $\mathcal{GS}$ is anonymous.* □

Lemma 2. *If digital signature scheme $\mathcal{DS}$ is secure against forgery under chosen-message attack, (P_1, V_1) is a sound non-interactive proof system for ρ_1 over Dom_1 and (P_2, V_2) is a sound non-interactive proof system for ρ_2 over Dom_2, then group signature scheme $\mathcal{GS}$ is traceable.* □

Lemma 3. *If digital signature scheme $\mathcal{DS}$ is secure against forgery under chosen-message attack, (P_1, V_1) is a sound non-interactive proof system for ρ_1 over Dom_1 and (P_2, V_2) is a sound non-interactive proof system for ρ_2 over Dom_2, then group signature scheme $\mathcal{GS}$ is non-frameable.* □

We know that if trapdoor permutations exist then so do secure digital signature schemes [6], IND-CCA secure encryption schemes [17, 28] and simulation sound NIZK proofs for NP [28]. As a consequence we have:

Theorem 1. *If there exists a family of trapdoor permutations, then there exists a dynamic group signature scheme that is anonymous, traceable and non-frameable.*

Acknowledgments

We thank Bogdan Warinschi for comments on a previous draft. The authors are supported in part by NSF grants CCR-0098123, ANR-0129617 and CCR-0208842, and an IBM Faculty Partnership Development Award.

References

1. G. Ateniese, J. Camenisch, M. Joye, and G. Tsudik. A practical and provably secure coalition-resistant group signature scheme. *Advances in Cryptology – CRYPTO '00*, Lecture Notes in Computer Science Vol. 1880, M. Bellare ed., Springer-Verlag, 2000.
2. G. Ateniese and G. Tsudik. Quasi-efficient revocation in group signature schemes. *Financial Cryptography '02*, Lecture Notes in Computer Science Vol. 2357, M. Blaze ed., Springer-Verlag, 2002.
3. G. Ateniese and G. Tsudik. Group signatures à la carte. *Proceedings of the 10th Annual Symposium on Discrete Algorithms*, ACM-SIAM, 1999.
4. G. Ateniese and G. Tsudik. Some open issues and directions in group signature. *Financial Cryptography '99*, Lecture Notes in Computer Science Vol. 1648, M. Franklin ed., Springer-Verlag, 1999.
5. M. Bellare, D. Micciancio and B. Warinschi. Foundations of group signatures: Formal definitions, simplified requirements, and a construction based on general assumptions. *Advances in Cryptology – EUROCRYPT '03*, Lecture Notes in Computer Science Vol. 2656, E. Biham ed., Springer-Verlag, 2003.
6. M. Bellare and S. Micali. How to sign given any trapdoor permutation. *JACM*, 39(1):214–233, 1992.
7. M. Bellare and P. Rogaway. Entity authentication and key distribution. *Advances in Cryptology – CRYPTO '93*, Lecture Notes in Computer Science Vol. 773, D. Stinson ed., Springer-Verlag, 1993.
8. M. Bellare, H. Shi, and C. Zhang. Foundations of group signatures: the case of dynamic groups. Full version of this abstract. http://www-cse.ucsd.edu/users/mihir.
9. M. Blum, A. DeSantis, S. Micali, and G. Persiano. Non-interactive zero-knowledge proof systems. *SIAM J. on Computing*, 20(6):1084–1118, 1991.
10. E. Bresson and J. Stern. Efficient revocation in group signatures. *Public-Key Cryptography '01*, Lecture Notes in Computer Science Vol. 1992, K. Kim ed., Springer-Verlag, 2001.
11. J. Camenisch. Efficient and generalized group signature. *Advances in Cryptology – EUROCRYPT '97*, Lecture Notes in Computer Science Vol. 1233, W. Fumy ed., Springer-Verlag, 1997.
12. J. Camenisch and A. Lysyanskaya. An identity-escrow scheme with appointed verifiers. *Advances in Cryptology – CRYPTO '01*, Lecture Notes in Computer Science Vol. 2139, J. Kilian ed., Springer-Verlag, 2001.
13. J. Camenisch and M. Michels. A group signature scheme with improved efficiency. *Advances in Cryptology – ASIACRYPT '98*, Lecture Notes in Computer Science Vol. 1514, D. Pei ed., Springer-Verlag, 1998.
14. J. Camenisch and M. Stadler. Efficient group signatures schemes for large groups. *Advances in Cryptology – CRYPTO '97*, Lecture Notes in Computer Science Vol. 1294, B. Kaliski ed., Springer-Verlag, 1997.
15. D. Chaum and E. van Heyst. Group signatures. *Advances in Cryptology – EUROCRYPT '91*, Lecture Notes in Computer Science Vol. 547, D. Davies ed., Springer-Verlag, 1991.
16. L. Chen and T. P. Pedersen. New group signature schemes. *Advances in Cryptology – EUROCRYPT '94*, Lecture Notes in Computer Science Vol. 950, A. De Santis ed., Springer-Verlag, 1994.

17. D. Dolev, C. Dwork, and M. Naor. Nonmalleable cryptography. *SIAM J. on Computing*, 30(2):391–437, 2000.
18. A. Fiat and A. Shamir. How to prove yourself: Practical solutions to identification and signature problems. *Advances in Cryptology – CRYPTO '86*, Lecture Notes in Computer Science Vol. 263, A. Odlyzko ed., Springer-Verlag, 1986.
19. O. Goldreich. A uniform-complexity treatment of encryption and zero-knowledge. *J. of Cryptology*, 6(1):21–53, 1993.
20. S. Goldwasser and S. Micali. Probabilistic encryption. *JCSS*, 28:270–299, 1984.
21. S. Goldwasser, S. Micali, and R. Rivest. A digital signature scheme secure against adaptive chosen-message attacks. *SIAM J. on Computing*, 17(2):281–308, 1988.
22. S. Goldwasser and Y. Tauman. On the (In)security of the Fiat-Shamir paradigm. *Proceedings of the 44th Symposium on Foundations of Computer Science*, IEEE, 2003.
23. A. Kiayias, Y. Tsiounis and M. Yung. Traceable signatures. *Advances in Cryptology – EUROCRYPT '04*, Lecture Notes in Computer Science Vol. 3027, C. Cachin and J. Camenisch ed., Springer-Verlag, 2004.
24. S. Micali, C. Rackoff, and B. Sloan. The notion of security for probabilistic cryptosystems. *SIAM J. on Computing*, 17(2):412–426, 1988.
25. M. Naor and M. Yung. Public-key cryptosystems provably secure against chosen ciphertext attacks. *Proceedings of the 22nd Annual Symposium on the Theory of Computing*, ACM, 1990.
26. H. Petersen. How to convert any digital signature scheme into a group signature scheme. *Proceedings of Security Protocols Workshop '97*.
27. C. Rackoff and D. Simon. Non-interactive zero-knowledge proof of knowledge and chosen ciphertext attack. *Advances in Cryptology – CRYPTO '91*, Lecture Notes in Computer Science Vol. 576, J. Feigenbaum ed., Springer-Verlag, 1991.
28. A. Sahai. Non-malleable non-interactive zero knowledge and adaptive chosen-ciphertext security. *Proceedings of the 40th Symposium on Foundations of Computer Science*, IEEE, 1999.
29. D. Song. Practical forward-secure group signature schemes. *Proceedings of the 8th Annual Conference on Computer and Communications Security*, ACM, 2001.

Time-Selective Convertible Undeniable Signatures

Fabien Laguillaumie[1,2] and Damien Vergnaud[2]

[1] France Telecom Research and Development,
42, rue des Coutures, B.P. 6243, 14066 Caen Cedex 4, France
[2] Laboratoire de Mathématiques Nicolas Oresme,
Université de Caen, Campus II, B.P. 5186,
14032 Caen Cedex, France
{laguillaumie,vergnaud}@math.unicaen.fr

Abstract. Undeniable signatures were introduced in 1989 by Chaum and van Antwerpen to limit the self-authenticating property of digital signatures. An extended concept – the convertible undeniable signatures – proposed by Boyar, Chaum, Damgård and Pedersen in 1991, allows the signer to convert undeniable signatures to ordinary digital signatures. We present a new efficient convertible undeniable signature scheme based on bilinear maps. Its unforgeability is tightly related, in the random oracle model, to the computational Diffie-Hellman problem and its anonymity to a non-standard decisional assumption. The advantages of our scheme are the short length of the signatures, the low computational cost of the signature and the receipt generation. Moreover, a variant of our scheme permits the signer to universally convert signatures pertaining only to a specific time period. We formalize this notion as the *time-selective conversion*.

Keywords: Convertible undeniable signatures, bilinear maps, anonymity, exact security, time-selective conversion.

1 Introduction

Digital signatures aim at recover *in silico* the usual properties of the traditional *in vivo* signatures, namely authentication, integrity, non-repudiation of the signed document and universal verifiability of the signatures. However, unlike handwritten signatures, digital signatures can be copy-cloned and therefore authenticated confidential documents (*e.g.* software certificates, contracts, dishonorable bills) can easily be disseminated.

For privacy reasons, it is preferable, in many applications, that the verification of signatures be controlled or (at least) limited by the signer. This remark justifies the concept of undeniable signatures, introduced at the very end of the eighties by Chaum and van Antwerpen [14]. In this setting, the verification (and the denial) of a signature requires the cooperation of the signer. The security of their protocol relies on the discrete logarithm problem, but suffers from the fact that the interactive protocols were not zero-knowledge. One year later, Chaum improved significantly the initial proposal in [12] by providing a zero-knowledge version. Moreover there exist documents whose authentication must be limited at first, but which will require ordinary digital signatures after some period of time. In 1991, the concept has been refined by giving the possibility to transform an undeniable signature into a *self-authenticating* signature. These *convertible undeniable signatures*, proposed in [6] by Boyar, Chaum, Damgård and Pedersen, provide individual and universal conversions of the signatures. Unfortunately, this El Gamal-like scheme has been broken in 1996 by Michels, Petersen, and Horster [25] who proposed a repaired version with heuristic security. Since then, many schemes have then been proposed, based upon classical signatures, such as Schnorr [26], El Gamal [17] and RSA [18–20]. Very recently, Monnerat and Vaudenay [28] proposed short undeniable signatures based on the computation of characters which do not provide the conversion property. Convertible undeniable signatures have given rise to many applications in cryptography [6, 7, 16].

In all these protocols, the universal conversion consists in revealing a part of the signer's secret key. This conversion makes all signatures, *past as well as future*, be universally verifiable. This property may be undesirable in some context and furthermore the corresponding keys cannot be used to generate undeniable signatures any more. As in a classical public key infrastructure the public key needs to be certified by an authority (as well as in any asymmetric cryptographic protocol), this approach leads to the registration of a large number of public/secret key pairs for the signer, and the need for the verifiers to check the validity of these certificates. Besides, in the identity-based paradigm[1], the problem is even more serious since there is a unique secret key by identity.

Our Contributions. In this article, we propose a new convertible undeniable signature scheme, in the spirit of both the original paper of Chaum and van Antwerpen [6] and the short signatures from bilinear maps proposed by Boneh, Lynn and Shacham [5]. The amusing fact is that the idea underlying these two papers is actually the same. In both cases, a signature consists in an exponentiation of (a hash-value of) the message by the signer's secret key : $h(m)^s$. In Chaum and van Antwerpen's scheme, the anonymity of signatures [18] comes from the difficulty of the decisional Diffie-Hellman problem in a prime order subgroup of the multiplicative group of a finite field, whereas the efficiency of Boneh *et al.* signatures comes from the ease of this problem on certain elliptic curves.

We combine the best of the two worlds and, introducing Zhang, Safavi-Naini and Susilo's technique from [33,34], we obtain a convertible undeniable signature protocol which is one of the most efficient. As in [26], the signer not only can selectively convert valid signatures into ordinary digital signatures, but he can also convert any invalid signature into an universally verifiable statement about this fact. Moreover, to overcome the difficulty mentionned above, we introduce and formalize the *time-selective convertible undeniable signatures* which supports

---

[1] An identity-based undeniable signature scheme has been proposed by Libert and Quisquater in 2004 [24], and, like our scheme, it is built with bilinear maps. However it does not provide the universal conversion property.

signers in gradually converting the undeniable signatures in a controlled fashion. A slight variant of our new scheme permits the signer to universally convert signatures pertaining only to a specific time period.

The new convertible undeniable signature scheme is designed for devices with constrained computation capabilities or with low bandwidth. It can be embedded in smart cards for example, as the main computation for a signature is a scalar multiplication on an elliptic curve, and the signature is essentially one point (with some additional random salt). The unforgeability of our scheme is tightly related, in the random oracle model, to the computational Diffie-Hellman problem and its anonymity to a non-standard decisional assumption.

The article is organized as follows: first we formally define the concept of time-selective convertible undeniable signature scheme and its security model. Then, we review the cryptographic properties of bilinear maps and describe the problems upon which depend our schemes. In the following section, we describe our new scheme and its time-selective convertible variant, and finally we prove its security in the random oracle model.

2 Formal Definition and Security Model

2.1 Definition

In this subsection, we formalize the concept of time-selective convertible undeniable signatures.

Definition 1 (Time-Selective Convertible Undeniable Signature).
Given an integer k, a time-selective convertible undeniable signature scheme TSCUS with security parameter k is defined by the following:

- a **common parameter generation algorithm** *TSCUS.Setup: it is a probabilistic algorithm which takes as input k. The outputs are the public parameters;*
- a **key generation algorithm** *TSCUS.KeyGen: it is a probabilistic algorithm which takes as inputs the public parameters and outputs a pair of keys (pk, sk) and a public number of time periods $T \in \mathbb{N}$;*
- a **signing algorithm** *TSCUS.Sign: it is a probabilistic algorithm which takes as inputs T, a message m, a secret key sk, an integer $t \in [\![1,T]\!]$, and the public parameters. The output σ is a time-selective convertible undeniable signature on m for the time period t;*
- **confirming/denying protocols** *TSCUS.{Confirm, Deny}: they are protocols which take as inputs T, a message m, an integer $t \in [\![1,T]\!]$, a bit string σ, a pair of keys (pk, sk) and the public parameters. The output is a (possibly non-interactive) non-transferable proof that σ is actually a valid/invalid time-selective convertible undeniable signature on m for the time period t, with respect to the key pk;*
- an **individual receipt generation algorithm** *TSCUS.IReceipt: it is an algorithm which takes as inputs T, a message m, an integer $t \in [\![1,T]\!]$, a bit string σ, a secret key sk and the public parameters. It outputs an individual receipt $\tilde{\sigma}$ which makes it possible to universally verify whether σ is valid or not.*

- a **verifying algorithm for individually converted signature** TSCUS.IVerify: it is a deterministic algorithm which takes as inputs T, a message m, an integer $t \in [\![1, T]\!]$, a bit string σ, a bit string $\tilde{\sigma}$, the signer's public key pk, and the public parameters. It tests whether $\tilde{\sigma}$ is a valid individual receipt with respect to σ and the public key pk. If it does, the algorithm states whether σ is a valid time-selective convertible undeniable signature on m for the time period t with respect to the key pk or not, else it outputs Error;
- a **universal receipt generation algorithm** TSCUS.UReceipt: it is a deterministic algorithm which takes as inputs T, a secret key sk, an integer $t \in [\![1, T]\!]$, and the public parameters and outputs a universal receipt $\mathcal{I}_t$ which makes it possible to universally verify all time-selective convertible undeniable signature σ on m for any time period $t' \leq t$;
- a **verifying algorithm for universally converted signature** TSCUS.UVerify: it is a deterministic algorithm which takes as inputs T, a message m, two integers $(t, t') \in [\![1, T]\!]^2$ such that $t' \leq t$, a bit string σ, a public key pk, a bit string $\mathcal{I}_t$ and the public parameters. It tests whether $\mathcal{I}_t$ is a valid universal receipt for the time period t with respect to the key pk. If it does, it states whether σ is a valid time-selective convertible undeniable signature on m for the time period t' with respect to the key pk or not, else it outputs Error;

and must satisfy the following properties (formally discussed in the next section):

1. *completeness and soundness:* the confirming and denying protocols and the verifying algorithms are complete and sound, where completeness means that valid (invalid) signatures can always be proved valid (invalid), and soundness means that no valid (invalid) signature can be proved invalid (valid);
2. *Unforgeability:* given a public key pk, it is computationally infeasible, without the knowledge of the corresponding secret key to produce a time-selective convertible undeniable signature for any time period $t \in [\![1, T]\!]$ which is accepted by the verification algorithms or by the confirming protocols;
3. *Anonymity:* given a message m and a time-selective convertible undeniable signature σ on m for a time period $t' \in [\![1, T]\!]$, it is computationally infeasible without the knowledge of the signing secret key or of some universal receipt $\mathcal{I}_t$ for some $t \geq t'$, to determine which secret key sk was used to generate σ;
4. *Non-transferability:* a verifier participating in an execution of the confirming/denying protocols does not obtain information that could be used to convince a third party about the validity/invalidity of a signature.

Definition 2 (Convertible Undeniable Signature). *Given an integer k, a convertible undeniable signature scheme with security parameter k is a time-selective convertible undeniable signature scheme with security parameter k, and whose key generation algorithm always outputs 1 as the number of time periods.*

2.2 Security Model

In this subsection, we define the quantitative notions of unforgeability and anonymity of a time-selective convertible undeniable signature scheme. The proofs of security are carried in the random oracle model proposed by Bellare and Rogaway [2]. In this model, hash functions are idealized as oracles which output a random value for each new query.

Security Against Existential Forgery Under Chosen Message Attack.
The *de facto* standard notion of security for digital signatures was defined by Goldwasser, Micali and Rivest [21] as *existential forgery against adaptive chosen message attacks* (EF-CMA). For time-selective convertible undeniable signatures, the unforgeability security is defined along the same lines, with the notable difference that, while mounting a chosen-message attack, we suppose that the adversary has access to the universal receipts $\mathcal{I}_t$ for all $t \in [\![1, T]\!]$. Moreover, he is allowed to query a receipt generating oracle Υ and a confirming/denying oracle Ξ on any couple message/signature of his choice, in addition to the classical access to the signing oracle Σ and to the random oracle $\mathcal{H}$. As usual, in the adversary answer, there is the natural restriction that in the returned triple message/signature/time period $(m^\star, \sigma^\star, t^\star)$, the signature $\sigma^\star$ on $m^\star$ has not been obtained from Σ for the time period $t^\star$. However, $\mathcal{A}$ is allowed to query the signing oracle on the returned message $m^\star$ in any time period $t \in [\![1, T]\!]$ (especially for $t = t^\star$).

Definition 3 (Unforgeability). *Let TSCUS be a time-selective convertible undeniable signature scheme and let $\mathcal{A}$ be an EF-CMA-adversary against TSCUS. We consider the following random experiment, where k is a security parameter:*

Experiment $\mathbf{Exp}^{\text{ef-cma}}_{TSCUS, \mathcal{A}}(k)$

params $\xleftarrow{R}$ *TSCUS.Setup*(k)
$(pk, sk, T) \xleftarrow{R}$ *TSCUS.KeyGen*(params)
For j from 1 to T do $\mathcal{I}_j \leftarrow$ *TSCUS.UConvert*(params, pk, sk, j)
$(m^\star, \sigma^\star, t^\star) \leftarrow \mathcal{A}^{\mathcal{H}, \Sigma, \Upsilon, \Xi}$(params, $pk, \{\mathcal{I}_j\}_{j \in [\![1, T]\!]}$)
Return *TSCUS.UVerify*(params, $pk, m^\star, \sigma^\star, t^\star, T, \mathcal{I}_T$)

We define the success of the adversary $\mathcal{A}$, via

$$\mathbf{Succ}^{\text{ef-cma}}_{TSCUS, \mathcal{A}}(k) = \Pr\left[\mathbf{Exp}^{\text{ef-cma}}_{TSCUS, \mathcal{A}}(k) = valid\right].$$

Given $(k, \tau) \in \mathbb{N}^2$ and $\varepsilon \in [0, 1]$, the scheme TSCUS is said to be (k, τ, ε)-EF-CMA secure, if no EF-CMA-adversary $\mathcal{A}$ running in time τ has a success $\mathbf{Succ}^{\text{ef-cma}}_{TSCUS, \mathcal{A}}(k) \geq \varepsilon$.

Anonymity. In this section, we state the precise definition of *anonymity* under a chosen message attack (Ano-CMA) which captures the notion that an attacker cannot determine under which key a signature was performed. Our formalization follows the notion, introduced by Galbraith and Mao in [18] for undeniable signatures. We consider an Ano-CMA-adversary $\mathcal{A}$ that runs in two stages. In the **find** stage, it takes as input two public keys pk_0 and pk_1 and outputs a message $m^\star$, a time period $t^\star$ together with some state information s. In the **guess** stage it gets a challenge time-selective convertible undeniable signature $\sigma^\star$ formed by signing the message $m^\star$ at random under one of the two keys for the time period $t^\star$, and must say which key was chosen. In both stages, the adversary has access to the signing oracles Σ_0, Σ_1, to the receipt generating oracles Υ_0 and Υ_1

to the confirming/denying oracles Ξ_0 and Ξ_1, and to the random oracle(s) $\mathcal{H}$. The attacker is also given the universal receipts $\mathcal{I}_1, \ldots, \mathcal{I}_{t^*-1}$ of both potential signers. The only restriction of the attacker is that he cannot query the couple (m^*, σ^*) on the converting and confirming/denying oracles.

Definition 4 (Anonymity). *Let TSCUS be a time-selective convertible undeniable signature scheme and let $\mathcal{A}$ be an Ano-CMA-adversary against TSCUS. We consider the following random experiments, for $r \in \{0,1\}$, where k is a security parameter:*

$Experiment\ \mathbf{Exp}_{TSCUS, \mathcal{A}}^{ano-cma-r}(k)$

---

params $\xleftarrow{R}$ TSCUS.Setup(k)
$(pk_0, sk_0, T_0) \xleftarrow{R}$ TSCUS.KeyGen(params)
$(pk_1, sk_1, T_1) \xleftarrow{R}$ TSCUS.KeyGen(params)
$(m^*, t^*, s) \leftarrow \mathcal{A}^{\mathcal{H}, \Sigma_0, \Sigma_1, \Upsilon_0, \Upsilon_1, \Xi_0, \Xi_1}(find, \text{params}, pk_0, pk_1)$ with $t^* \leq \min(T_0, T_1)$
$\sigma^* \leftarrow$ TSCUS.Sign(params, m, sk_r, t^*)
For j from 1 to $t^* - 1$ do
$\quad \mathcal{I}_j^0 \leftarrow$ TSCUS.UConvert(params, pk_0, sk_0, j)
$\quad \mathcal{I}_j^1 \leftarrow$ TSCUS.UConvert(params, pk_1, sk_1, j)
$d \leftarrow \mathcal{A}^{\mathcal{H}, \Sigma_0, \Sigma_1, \Upsilon_0, \Upsilon_1, \Xi_0, \Xi_1}(guess, \text{params}, pk_0, pk_1, m^*, \sigma^*, t^*, s, \{\mathcal{I}_j^0, \mathcal{I}_j^1\}_{j \in [\![1, t^*-1]\!]})$
Return d

We define the advantage of the adversary $\mathcal{A}$, via

$$\mathbf{Adv}_{TSCUS, \mathcal{A}}^{ano-cma}(k) = \left| \Pr\left[\mathbf{Exp}_{TSCUS, \mathcal{A}}^{ano-cma-1}(k) = 1\right] - \Pr\left[\mathbf{Exp}_{TSCUS, \mathcal{A}}^{ano-cma-0}(k) = 1\right] \right|.$$

Given $(k, \tau) \in \mathbb{N}^2$ and $\varepsilon \in [0,1]$, the scheme TSCUS is said to be (k, τ, ε)-Ano-CMA secure, if no Ano-CMA-adversary $\mathcal{A}$ running in time τ has an advantage $\mathbf{Adv}_{TSCUS, \mathcal{A}}^{ano-cma}(k) \geq \varepsilon$.

Remark 1. To obtain the security results, the executions of the confirming/denying protocols must be simulated in the random oracle model. In our scheme, these protocols are achieved by interactive zero-knowledge proofs of equality/inequality of (root of) discrete logarithms, which simulation is impossible in the random oracle model [18]. Usually [18, 29], this problem is overcome with the use of designated-verifier proofs [22]. By definition of these proofs, the adversary gains no information other than the validity/invalidity of the signature from its interaction with the signer. As the adversary can obtain this conviction by querying the receipt generating oracle, there is no loss of generality to suppose that it does not have access to the confirming/denying oracles.

3 Background

3.1 Proof of Equality or Inequality of Two Discrete Logarithms

Let $(\mathbb{G}, +)$ and $(\mathbb{H}, \cdot)$ be two groups of the same prime order q and let P and g be generators of $\mathbb{G}$ and $\mathbb{H}$ (respectively). We denote, as in [11], the zero-knowledge proof of equality of the discrete logarithm of $Y \in \mathbb{G}$ in base P and the one of $y \in \mathbb{H}$ in base g by $PK(a : y = g^a \wedge Y = aP)$. We use the notation $PK(a : y \neq g^a \wedge Y = aP)$ for the proof of inequality of the discrete logarithms. We refer the reader to Chaum and Pedersen's paper [16] for the proof of equality, and these of Camenisch and Shoup [10] for the proof of inequality.

In the design of the new scheme, we also need zero-knowledge proofs of equality/inequality of a root of a discrete logarithm: given $t \in \mathbb{N}$, $PK(a : y = g^{a^t} \wedge Y = aP)$ and $PK(a : y \neq g^{a^t} \wedge Y = aP)$. Efficient protocols can be found in [8, 11].

3.2 Bilinear Maps

Recently admissible bilinear maps have allowed the opening up of new territories in cryptography, making possible the realisation of protocols that were previously unknown or impractical.

Definition 5 (Admissible Bilinear Map [4]). *Let $(\mathbb{G}, +)$ and $(\mathbb{H}, \times)$ be two groups of the same prime order q and let us denote by P a generator of $\mathbb{G}$. An admissible bilinear map is a map $e : \mathbb{G} \times \mathbb{G} \longrightarrow \mathbb{H}$ satisfying the following properties:*

- *bilinear: $e(aQ, bR) = e(Q, R)^{ab}$ for all $(Q, R) \in \mathbb{G}^2$ and all $(a, b) \in \mathbb{Z}^2$;*
- *non-degenerate: $e(P, P) \neq 1$.*
- *computable: there exists a polynomial time algorithm to compute e.*

Definition 6 (Prime-Order-BDH-Parameter-Generator [4]). *A prime-order-BDH-parameter-generator is a probabilistic algorithm which takes as input a security parameter k and outputs a 5-tuple $(q, P, \mathbb{G}, \mathbb{H}, e)$ where q is a prime with $2^k < q < 2^{k+1}$, $\mathbb{G}$ and $\mathbb{H}$ are groups of order q, P generates $\mathbb{G}$, and $e : \mathbb{G}^2 \longrightarrow \mathbb{H}$ is an admissible bilinear map.*

Usually $\mathbb{G}$ can be considered as a subgroup of points on a (hyper)elliptic curve over a finite field, $\mathbb{H}$ as a subgroup of the multiplicative group of a related finite field and e as the Weil or Tate pairing [4].

3.3 The xyz-Decisional Diffie-Hellman Problem

The unforgeability of our scheme is related to the classical Diffie-Hellman problem:

Computational Diffie-Hellman (CDH): Let $(\mathbb{G}, +)$ be a group of prime order q, and let a, b be in $[\![1, q-1]\!]$. Given $(P, aP, bP) \in \mathbb{G}^3$, compute abP.

The design of the new scheme is connected to the following decisional problem:
xyz-Decisional Diffie-Hellman Problem (xyz-DDH): Let $(\mathbb{G}, +)$ be a group of prime order q, and let x, y and z be in $[\![1, q-1]\!]$. Given $(P, xP, yP, zP, Q) \in \mathbb{G}^5$, decide whether $Q = xyzP$.

At first glance, this may seem very similar to the classical decisional Diffie-Hellman (DDH) problem (in fact, the associated computational problem is equivalent to the CDH problem). The DDH assumption, underlying the security of many cryptographic protocols does not hold in the bilinear setting. Even if it is easier than the DBDH problem [4], the xyz-DDH problem seems intractable. Considering the xyz-DDH problem and assuming its difficulty (combined with the ease of the DDH problem), we are able to design cryptographic protocols achieving a trade-off between authenticity and privacy. The time-selective convertible undeniable signature scheme is based on the following observations:

- Assuming the difficulty of the xyz-DDH problem, given $(P, xP, yP, zP, Q) \in \mathbb{G}^5$, no one can efficiently decide whether $Q = xyzP$.
- Everyone can be convinced by someone knowing either x, y or z that $Q = xyzP$. This can be done thanks to the proofs of equality of two discrete logarithms mentionned above, and the equalities $e(Q, P) = e(yP, zP)^x = e(xP, zP)^y = e(xP, yP)^z$.
- After the publication of xyP, everyone can decide whether $Q = xyzP$.

Our new protocol is designed according to this idea but relies on a stronger assumption.

3.4 A New Decisional Problem

Recently, Boneh and Boyen [3] proposed an efficient digital signature scheme whose security (in the standard security model) relies on the so-called ℓ-Strong Diffie-Hellman problem (ℓ-SDH). This computational problem is slightly weaker than the ℓ-CAA problem[2] introduced by Mitsunari, Sakai, Kasahara in 2002 in relation with the security of a traitor tracing scheme [27]. Using the tricky polynomial construction from [27], it is easy to prove that the ℓ-CAA problem is polynomial time equivalent[3] to the:

(ℓ, T)-**Computational CAA Problem** $((\ell, T)$-**CCAA**): Let $(\mathbb{G}, +)$ be a group of prime order q, and let x be in $[\![1, q-1]\!]$. Given two integers ℓ and T and

$$\left[(x^i P)_{i \in [\![0, 2T-1]\!]}, \left(\frac{x^T}{x + h_j} P, h_j \right)_{j \in [\![1, \ell]\!]} \right] \in \mathbb{G}^{2T} \times (\mathbb{G} \times [\![1, q-1]\!])^\ell,$$

compute a pair $(Q, h) \in \mathbb{G} \times ([\![1, q-1]\!] \setminus \{h_1, \ldots, h_\ell\})$ verifying $(x+h)Q = x^T P$.

Like the DDH problem, thanks to the bilinear map, the decisional problem associated to (ℓ, T)-CCAA is easy. Therefore, we introduce a decisional variant of (ℓ, T)-CCAA, similar to the xyz-DDH problem which runs in 3 stages:

(ℓ, T)-xyz **Decisional CAA Problem** $((\ell, T)$-xyz-**DCAA**): Let $(\mathbb{G}, +)$ be a group of prime order q, let x, y and z be in $[\![1, q-1]\!]$ and ℓ and T be in $\mathbb{N}$.

---
[2] collusion attack algorithm with ℓ traitors
[3] In fact, the ℓ-CAA problem from [27] is exactly the $(\ell, 1)$-CCAA problem

Input: $\left[(x^i P)_{i\in[\![0,2T-1]\!]}, yP, zP, h, \left(\dfrac{x^T y}{x+h_k}P, h_k\right)_{k\in[\![1,\ell]\!]}\right]$
in $\mathbb{G}^{2T+2} \times [\![1, q-1]\!] \times (\mathbb{G} \times [\![1, q-1]\!])^\ell$, with $h \notin \{h_1, \ldots, h_\ell\}$
Oracle: for a request $t^\star \in [\![1,T]\!]$, the oracle answers $\left[(x^i yP)_{i\in[\![1,t^\star-1]\!]}, Q\right] \in \mathbb{G}^{t^\star}$
Output: decide whether $(x+h)Q = x^{t^\star} yzP$

The anonymity of our convertible undeniable signature scheme (*i.e.* with one time period) is related to the $(\ell, 1)$-*xyz*-DCAA problem. The link between this problem and the ℓ-CAA problem from [27] is analogous to the one between the *xyz*-DDH problem and the CDH problem. We want to stress that, though non-standard, these problems are random-self reducible and achieve generic security (as defined by Shoup [31]). The proof is similar to the one of the generic security of ℓ-SDH [3].

To conclude this section, we quantify the new algorithmic assumption.

Definition 7 (((ℓ, T)-*xyz*-DCAA Assumption**). *Let $\mathcal{G}en$ be a prime-order-BDH-parameter-generator and let ℓ and T be two integers. Let D be an adversary that takes on input $(q, P, \mathbb{G}, \mathbb{H}, e)$ a 5-tuple generated by $\mathcal{G}en$ and $[(X_i)_{i\in[\![1,2T-1]\!]}, Y, Z, h, (R_j, h_j)_{j\in[\![1,\ell]\!]}]$ in $\mathbb{G}^{2T+1} \times [\![1, q-1]\!] \times (\mathbb{G} \times [\![1, q-1]\!])^\ell$ and returns a bit. We consider the following random experiments, where k is a security parameter, for $r \in \{0, 1\}$:*

$\boxed{Experiment\ \mathbf{Exp}_{\mathcal{G}en,D}^{(\ell,T)-\text{xyz-dcaa-}r}(k)}$

$\mathtt{setup} = (q, P, \mathbb{G}, \mathbb{H}, e) \xleftarrow{R} \mathcal{G}en(k)$
$x \xleftarrow{R} [\![1, q-1]\!]$. For i from 1 to $2T-1$ do $X_i \leftarrow x^i P$
$(y, z) \xleftarrow{R} [\![1, q-1]\!]^2$, $Y \leftarrow yP$, $Z \leftarrow zP$
$h_1, \ldots, h_\ell, h \xleftarrow{R} [\![1, q-1]\!]$
For j from 1 to ℓ do $R_j = x^T y(x+h_j)^{-1} P$
$t^\star \leftarrow D(\mathtt{setup}, (X_i)_{i\in[\![1,2T-1]\!]}, Y, Z, h, (R_j, h_j)_{j\in[\![1,\ell]\!]})$
For k from 1 to $t^\star - 1$ do $Y_k \leftarrow x^k yP$
If $r=0$ then $Q \leftarrow x^T yz(x+h)^{-1}P$ else $Q \xleftarrow{R} \mathbb{G}$
$d \leftarrow D(\mathtt{setup}, t^\star, (X_i)_{i\in[\![1,2T-1]\!]}, Y, Z, h, (R_j, h_j)_{j\in[\![1,\ell]\!]}, (Y_k)_{k\in[\![1,t^\star-1]\!]}, Q)$
Return d

We define the corresponding advantage of D in solving the $(\ell, T)-xyz$-DCAA problem via:

$$\mathbf{Adv}_{\mathcal{G}en,D}^{(\ell,T)\text{-xyz-dcaa}}(k) = \left|\Pr\left[\mathbf{Exp}_{\mathcal{G}en,D}^{(\ell,T)\text{-xyz-dcaa-0}}(k) = 1\right]\right.$$
$$\left. - \Pr\left[\mathbf{Exp}_{\mathcal{G}en,D}^{(\ell,T)\text{-xyz-dcaa-1}}(k) = 1\right]\right|.$$

Given $(k, \tau) \in \mathbb{N}^2$ and $\varepsilon \in [0, 1]$, $\mathcal{G}en$ is said to be (k, τ, ε)-(ℓ, T)-*xyz*-DCAA-secure if no adversary D running in time τ has advantage $\mathbf{Adv}_{\mathcal{G}en,D}^{(\ell,T)-\text{xyz-dcaa}}(k) \geq \varepsilon$.

4 A New Convertible Time-Selective Undeniable Signature Scheme

4.1 The New Convertible Undeniable Signature Scheme: CUSBM

In this section, we describe the new convertible undeniable signature scheme CUSBM, based on bilinear map. It is designed as follows:

Setup and Key Generation

Setup: Let k be a security parameter, $\mathcal{G}en$ be a prime-order-BDH-parameter-generator and $(q, P, \mathbb{G}, \mathbb{H}, e)$ some output of $\mathcal{G}en(k)$. Let $f_r : \mathbb{N} \to \mathbb{N}$ be a function. We denote $n_r = f_r(k)$. Let $[\{0,1\}^* \times \{0,1\}^{n_r} \longrightarrow \mathbb{G}]$ be a hash function family, and H be a random member of this family. Let $[\{0,1\}^* \times \{0,1\}^{n_r} \longrightarrow [\![1, q-1]\!]]$ be a hash function family, and h be a random member of this family. The public parameters are $[(q, P, \mathbb{G}, \mathbb{H}, e), H, h]$

KeyGen: Alice picks randomly two integers $a_1, a_2 \in [\![1, q-1]\!]$ and computes the points $P_1 = a_1 P$ and $P_2 = a_2 P$. Alice's public key is the pair (P_1, P_2) and her secret key is (a_1, a_2).

Signing Algorithm

Sign: Given a message $m \in \{0,1\}^*$, Alice picks at random $r \in \{0,1\}^{n_r}$ and computes the point
$$\sigma = \frac{a_1 a_2}{(a_2 + h(m||r))} H(m||r).$$

The convertible undeniable signature of the message m is (σ, r).

Confirmation/Denial Protocols

Confirm: Given a message m and a signature (σ, r), Alice can confirm (σ, r) with the following interactive proof of knowledge: $PK(a_2 : e(\sigma, P_2+h(m||r)P) = e(H(m||r), P_1)^{a_2} \wedge P_2 = a_2 P)$.

Deny: Given a message m and an invalid signature (σ, r), Alice can deny (σ, r) with the following interactive proof of knowledge: $PK(a_2 : e(\sigma, P_2+h(m||r)P) \neq e(H(m||r), P_1)^{a_2} \wedge P_2 = a_2 P)$.

Receipt Generation and Verification

IReceipt: Given a message $m \in \{0,1\}^*$ and a putative signature (σ, r) on m, Alice computes the point $\tilde{\sigma} = a_2 H(m||r) \in \mathbb{G}$. The individual receipt with respect to σ is $\tilde{\sigma}$.

IVerify: Given a message $m \in \{0,1\}^*$, a putative signature (σ, r) on m and a putative individual receipt $\tilde{\sigma}$ on (σ, r), the validity of the receipt is decided by checking whether $e(\tilde{\sigma}, P) = e(P_2, H(m||r))$ or not. If $\tilde{\sigma}$ is valid, then the validity of (σ, r) is decided by checking whether $e(\sigma, P_2 + h(m||r)P) = e(\tilde{\sigma}, P_1)$ or not.

UReceipt: Alice publishes the point $I = a_1 a_2 P$.

UVerify: The validity of the universal receipt I is decided by verifying that $e(P_1, P_2) = e(I, P)$. If it is valid, given a signature (σ, r) on a message $m \in \{0,1\}^*$ and I, everyone checks the validity of this signature by verifying that $e(\sigma, P_2 + h(m||r)P) = e(H(m||r), I)$.

Efficiency Considerations. Comparing with previous convertible undeniable signature schemes, CUSBM has a number of advantages. The signature only consists in an element of $\mathbb{G}$ and some additional random salt. In practice, the size of an element of $\mathbb{G}$ can be reduced by a factor 2 with compression techniques and the random salt has size $n_r = 112$. Therefore, the size of the signature is only 272 bits. Furthermore, a receipt (individual and universal) is also an element of $\mathbb{G}$, and therefore has bit size 160. From an efficiency point of view, the signature generation and the individual and universal receipts generation algorithms require only one exponentiation as the most expensive operation. Unfortunately, it turns out that the signature verification is slightly more time consuming, as it requires 2 pairing evaluations.

4.2 The First Time-Selective Convertible Undeniable Signature Scheme: TSCUSBM

TSCUSBM is a time-selective convertible undeniable signature scheme which is a variant of CUSBM.

Setup and Key Generation

TSCUSBM.Setup = CUSBM.Setup

KeyGen: Alice picks randomly two integers $a_1, a_2 \in [\![1, q-1]\!]$, and computes the points $P_1 = a_1 P$ and $P_2 = a_2 P$. She chooses a number of time periods $T \in \mathbb{N}$. Alice's public key is the pair (P_1, P_2, T) and her secret key is (a_1, a_2).

Signing Algorithm

Sign: Given a message $m \in \{0,1\}^*$ and a time period t, Alice picks at random $r \in \{0,1\}^{n_r}$ and computes the point $\sigma = a_1 a_2^t (a_2 + h(m||r))^{-1} H(m||r)$. The signature of the message m is (σ, r, t).

Confirmation/Denial Protocols

Confirm: Given a message m and a signature (σ, r, t), Alice can confirm (σ, r, t) with the following interactive proof of knowledge: $PK(a_2 : e(\sigma, P_2 + h(m||r)P) = e(H(m||r), P_1)^{a_2^t} \wedge P_2 = a_2 P)$.

Deny: Given a message m and an invalid signature (σ, r, t), Alice can deny (σ, r, t) with the following interactive proof of knowledge:

$$PK(a_2 : e(\sigma, P_2 + h(m||r)P) \neq e(H(m||r), P_1)^{a_2^t} \wedge P_2 = a_2 P).$$

Receipt Generation and Verification

IReceipt: Given an integer $t \in [\![1, T]\!]$, a message $m \in \{0,1\}^*$ and a signature (σ, r) on m for the time period t, Alice computes the t-tuple

$$\tilde{\sigma} = (a_2 H(m||r), a_2^2 H(m||r), \ldots, a_2^t H(m||r)).$$

The individual receipt with respect to σ is $\tilde{\sigma}$.

IVerify: Given an integer $t \in [\![1,T]\!]$, a message $m \in \{0,1\}^*$, a putative signature (σ, r) on m for the time period t and a putative individual receipt $\tilde{\sigma} = (\tilde{\sigma}^{(1)}, \ldots, \tilde{\sigma}^{(t)})$ on (σ, r), the validity of the receipt is decided by checking whether $e(\tilde{\sigma}^{(1)}, P) = e(P_2, H(m\|r))$ and $e(\tilde{\sigma}^{(i)}, P) = e(P_2, \tilde{\sigma}^{(i-1)})$ for $i \in [\![2, t]\!]$ or not. If $\tilde{\sigma}$ is valid, then the validity of (σ, r) is decided by checking whether $e(\sigma, P_2 + h(m\|r)P) = e(\tilde{\sigma}^{(t)}, P_1)$ or not.

UReceipt: Given an integer $t \in [\![1,T]\!]$, this protocol consists for Alice in publishing the t-tuple $I_t = (a_1 a_2 P, a_1 a_2^2 P, \ldots, a_1 a_2^t P) \in \mathbb{G}^t$.

UVerify: Given an integer $t \in [\![1,T]\!]$, the validity of the universal receipt $I_t = (I^{(1)}, \ldots, I^{(t)})$ is decided by verifying that $e(P_1, P_2) = e(I^{(1)}, P)$ and $e(I^{(i-1)}, P_2) = e(I^{(i)}, P)$ for $i \in [\![2, t]\!]$. If it is valid, given a signature (σ, r) on a message $m \in \{0,1\}^*$ for a time period $t' \leq t$ and I_t, everyone checks the validity of this signature by verifying that $e(\sigma, P_2 + h(m\|r)P) = e(H(m\|r), I^{(t')})$.

Remark 2. In a sequential use of these signatures (*i.e.* signatures for the time period t are individually converted only after the publication of I_{t-1}), the verification processes can be considerably improved (as in the CUSBM scheme). Otherwise, the signer's operations are still efficient, but the verifier has more computations to perform. The scheme remains nevertheless reasonable.

5 Security Results

Unforgeability. The theorem below states that TSCUSBM is EF-CMA secure assuming the intractibility of the CDH problem in the random oracle model (we replace the hash function H by random oracle $\mathcal{H}$; there is no need to do ideal assumptions on h).

Theorem 1 (Unforgeability of TSCUSBM). *Let $\mathcal{G}en$ be a prime-order-BDH-parameter-generator, let $f_r : \mathbb{N} \to \mathbb{N}$ and let $\mathcal{A}$ be an EF-CMA-adversary against CUSBM in the random oracle model, that produces an existential forgery with probability $\varepsilon = \mathbf{Succ}_{CUSBM,\mathcal{A}}^{ef-cma}$, within time τ, making $q_\mathcal{H}$, q_Σ and q_Υ queries to the random oracle, to the signing oracle and to the receipt generating oracle. Then there exist $\varepsilon' \in [0,1]$ and $\tau' \in \mathbb{N}$ verifying*

$$\begin{cases} \varepsilon' \geq \varepsilon - \dfrac{q_\Sigma(q_\mathcal{H} + q_\Sigma)}{2^{n_r}} \\ \tau' \leq \tau + (q_\mathcal{H} + q_\Sigma + O(1))T_{Exp-\mathbb{G}} + (q_\Sigma + O(1))T_{Exp-q} \end{cases}$$

such that CDH can be solved with probability ε', within time τ', and where $T_{Exp-\mathbb{G}}$ (resp. T_{Exp-q}) denotes the time for an exponentiation in $\mathbb{G}$ (resp. modulo q) and $n_r = f_r(k)$.

Proof. The proof is very similar to Boneh, Lynn and Shacham's unforgeability proof in [5]. For the reader's convenience, it is supplied in appendix A.

Corollary 1 (Unforgeability of CUSBM). *Under the CDH assumption, the scheme CUSBM is EF-CMA secure in the random oracle model.*

Remark 3. In order to shorten the public keys, we can modify the key generation of CUSBM as follows : the key generation algorithm produces a public/private pair of keys (aP, a). This variant of CUSBM (setting $a_1 = a_2 = a$) is unforgeable under the ℓ-CAA [27] assumption.

Anonymity. The next theorem claims TSCUSBM's anonymity against a chosen message attack under the (ℓ, T)-xyz-DCAA assumption in the random oracle model (the hash functions h and H are replaced by random oracles $\hbar$ and $\mathcal{H}$):

Theorem 2 (Anonymity of TSCUSBM). *Let $\mathcal{G}en$ be a prime-order-BDH-parameter-generator, let $f_r : \mathbb{N} \to \mathbb{N}$ and let $\mathcal{A}$ be an Ano-CMA-adversary against TSCUSBM in the random oracle model, that breaks anonymity with advantage* $\mathbf{Adv}^{ano-cma}_{CUS,\mathcal{A}}$, *within time τ, making $q_\mathcal{H}$, $q_\hbar$ q_Σ and q_Υ queries to the random oracles, to the signing oracles and to the receipt generating oracles. Then there exist $T \in \mathbb{N}$, $\varepsilon' \in [0,1]$ and $\tau' \in \mathbb{N}$ verifying*

$$\begin{cases} \varepsilon' \geq \dfrac{\varepsilon}{2} - \dfrac{(q_\Sigma + 1)(q_\mathcal{H} + q_\hbar)}{2^{n_r - 1}} - \dfrac{1}{2^k} \\ \tau' \leq \tau + (q_\mathcal{H} + 2q_\Sigma + Tq_\Upsilon + O(1))T_{Exp-\mathbb{G}} + (q_\Sigma + Tq_\Upsilon + O(1))T_{Exp-q} \end{cases}$$

such that $(q_\mathcal{H}, T)$-xyz-DCAA can be solved with probability ε', within time τ', and where $T_{Exp-\mathbb{G}}$ (resp. T_{Exp-q}) denotes the time for an exponentiation in $\mathbb{G}$ (resp. modulo q) and $n_r = f_r(k)$.

Proof. Our method of proof is inspired by Shoup [32]: we define a sequence of game Game$_0$, ..., Game$_5$ starting from the actual Ano-CMA adversary $\mathcal{A}$ and modify it step by step, until we reach a final game whose success probability has an upper bound related to solving the $(q_\mathcal{H}, T)$-xyz-DCAA problem. All the games operate on the same underlying probability space: the public and private keys of the signature scheme, the coin tosses of $\mathcal{A}$ and the random oracles.

Let k be a security parameter, T and ℓ be two integers and let $(q, P, \mathbb{G}, \mathbb{H}, e)$ be a 5-tuple generated by $\mathcal{G}en(k)$ and $[(X_i)_{i \in [\![1, 2T-1]\!]}, Y, Z, h), (R_j, h_j)_{j \in [\![1, \ell]\!]}]$ in $\mathbb{G}^{2T+1} \times [\![1, q-1]\!] \times (\mathbb{G} \times [\![1, q-1]\!])^\ell$ be a random instance of the $(q_\mathcal{H}, T)$-xyz-DCAA problem. We denote $X_0 = P$. We construct a simulation which solves this instance.

Game$_0$. We consider an Ano-CMA-adversary $\mathcal{A}$ with advantage $\mathbf{Adv}^{ano-cma}_{TSCUSBM, \mathcal{A}}(k)$, within time τ. The key generation algorithm is run twice and produces the following pairs of keys (sk_0, pk_0, T_0), and (sk_1, pk_1, T_1). The adversary $\mathcal{A}$ is fed with pk_0, pk_1, T_0 and T_1 and, querying the random oracles $\mathcal{H}$ and $\hbar$, the signing oracles Σ_0 and Σ_1, and the receipt generating oracles Υ_0 and Υ_1, outputs a message m^* and a time period $t^* \in [\![1, \min(T_0, T_1)]\!]$. A challenge signature is then produced by flipping a coin $b \in \{0, 1\}$ and signing the message under the key sk_b. The adversary is given this challenge signature (σ^*, r^*, t^*), and outputs a bit b^* at the end of the **guess** stage.

We denote by $q_\mathcal{H}$, $q_\hbar$, q_Σ, and q_Υ the number of queries from the random oracles, from the signing oracles, and from the receipt generating oracles and we assume that $T \geq \min(T_0, T_1)$ and $q_\Sigma \leq \ell$. The only requirement is that the challenge signature $(\sigma^\star, r^\star, t^\star)$ cannot be queried to a receipt generating oracle.

In any game Game_i, we denote by Guess_i the event $b^\star = b$. By definition,

$$|2\Pr[\mathsf{Guess}_0] - 1| = \mathbf{Adv}^{\mathsf{psi-cma}}_{\mathsf{TSCUSBM},\mathcal{A}}(k).$$

Game_1. First, we pick $(\alpha, \beta) \in [\![1, q-1]\!]^2$ at random and modify the simulation by replacing pk_0 by (X_1, Y_1) and pk_1 by $(\alpha X_1, \beta Y_1)$. The distributions of pk_0 and pk_1 are unchanged since we consider a random instance of the $(q_\mathcal{H}, T)$-xyz-DCAA problem. Therefore we have

$$\Pr[\mathsf{Guess}_1] = \Pr[\mathsf{Guess}_0].$$

Game_2. In this game, we simulate the random oracles $\mathcal{H}$ and $\hbar$ and maintain appropriate lists, which we denote by H-List and h-List. For any fresh query $(m, r) \in \{0,1\}^\star \times \{0,1\}^{n_r}$

- to the oracle $\mathcal{H}$, we pick at random $s \in [\![1, q-1]\!]$, compute $sP \in \mathbb{G}$, store (m, r, s, sP, T) in the H-List and return sP as the answer to the oracle call;
- to the oracle $\hbar$, we pick at random $u \in [\![1, q-1]\!]$, store (m, r, u) in the h-List and returns u as the answer to the oracle call.

In the random oracle model, this game is clearly identical to the previous one. Hence, we obtain

$$\Pr[\mathsf{Guess}_2] = \Pr[\mathsf{Guess}_1].$$

Game_3. Now we simulate the signing oracles. We initialize a counter to $i = 1$, and for each new request $m \in \{0,1\}^\star$, $t \in [\![1, T]\!]$ we pick $r \in \{0,1\}^{n_r}$ at random. If there exists a triple $(m, r, ?)$ in the h-List or a 5-tuple $(m, r, ?, ?, ?)$ in the H-List, we abort the simulation. We pick at random $s \in [\![1, q-1]\!]$, compute $sX_{T-t} \in \mathbb{G}$ and store (m, r, s, sX_{T-t}, t) in the H-List.
If the query is to Σ_0, the signature is (sR_i, r, t), we refresh the h-List with (m, r, h_i), then we increment the counter. If the query is to Σ_1 the signature is $(s\alpha^{t-1}\beta R_i, r, t)$, we refresh the h-List with $(m, r, \alpha h_i)$, then we incremente the counter. This game perfectly simulates the signing oracle if we do not abort. As we abort with probability at most $(q_\mathcal{H} + q_\hbar)2^{-n_r}$, we have

$$|\Pr[\mathsf{Guess}_3] - \Pr[\mathsf{Guess}_2]| \leq \frac{q_\Sigma(q_\mathcal{H} + q_\hbar)}{2^{n_r}}$$

Game_4. We simulate the receipt generating oracles. When the adversary requests a couple message/putative signature $(m, (\sigma, r, t))$ on m to Υ_0 with $t \in [\![1, T]\!]$, we look in the H-List for a 5-tuple $(m, r, s, sX_{T-t'}, t')$, then the answer is the t-tuple $(sX_{T-t'+1}, \ldots, sX_{T-t'+t})$. If this request is on Υ_1, we answer with $(s\alpha^{T-t'+1}X_{T-t'+1}, \ldots, s\alpha^{T-t'+t}X_{T-t'+t})$ This simulation is perfect, therefore we have

$$\Pr[\mathsf{Guess}_4] = \Pr[\mathsf{Guess}_3].$$

Game_5. Finally, in this game, in the challenge generation: once the adversary $\mathcal{A}$ asks for a signature on a message $m^\star$ for a time period $t^\star$, we query the problem challenger with $t^\star$. It outputs $(Y_j)_{j \in [\![1, t^\star - 1]\!]} \in \mathbb{G}^{t^\star - 1}$ and $Q \in \mathbb{G}$.

We pick a bit $b \in \{0,1\}$ and $r^* \in \{0,1\}^{n_r}$ at random. If there exists a 5-tuple $(m^*, r^*, ?, ?, ?)$ in the H-List or a triple $(m^*, r^*, ?)$ in the h-List, we abort the simulation, else we update the H-List by storing $(m^*, r^*, \bot, Z, \bot)$ and the h-List by storing $(m, r, \alpha^b h)$. We output $((\alpha^{t^*-1}\beta)^b Q, r^*, t^*)$ as the challenge signature and $(Y_j)_{j \in [\![1, t^*-1]\!]}$ as the universal receipts for the challenge..

If $Q = Q_{\text{real}} = x^T yz(x+h)^{-1}P$, this game perfectly simulates the challenge generation if we do not abort (which happens with probability at most $(q_\mathcal{H} + q_h)2^{-n_r}$). Therefore

$$|\Pr[\text{Guess}_5 | Q = Q_{\text{real}}] - \Pr[\text{Guess}_4]| \leq \frac{q_\mathcal{H} + q_h}{2^{n_r}}$$

If $Q = Q_{\text{random}}$ is a random element from $\mathbb{G}$, the adversary gains no information on b, in an information theoretic sense, therefore

$$\Pr[\text{Guess}_5 | Q = Q_{\text{random}}] \leq \frac{1}{2} + \frac{q_\mathcal{H} + q_h}{2^{n_r}} + \frac{1}{2^k}.$$

The last term accounts for the probability that $Q_{\text{random}} = Q_{\text{real}}$. By definition, the advantage in the Game$_5$ simulation in solving the $(q_\mathcal{H}, T) - xyz$-DCAA problem is: $\mathbf{Adv}^{(q_\mathcal{H},T)-\text{xyz}-\text{dcaa}}_{\text{Gen, Game}_5}(k) = |\Pr[\text{Guess}_5 | Q = Q_{\text{real}}] - \Pr[\text{Guess}_5 | Q = Q_{\text{random}}]|$. A simple computation gives the claimed bounds for ε' and τ'. □

Remark 4. Using Boneh-Boyen's technique [3], it is possible to avoid the ideal hypothesis of perfect randomness of the hash function h.

Corollary 2 (Anonymity of CUSBM). *Under the $(\ell, 1)$-xyz-DCAA assumption, the scheme CUSBM is Ano-CMA secure in the random oracle model.*

Final Remarks and Conclusion

Time-selective convertible signatures are introduced to eliminate the burden of registration of new public keys after the universal receipt publication. We formalized the security notions of a time-selective convertible undeniable signature scheme and thanks to a new technique to design cryptographic protocols achieving a tradeoff between authenticity and privacy, we proposed the first scheme meeting this definition.

The new schemes offer the advantage of issuing short signatures. Moreover, the computational costs for the signer in the signature generation, the confirmation/denial protocols and the receipt generation algorithms, are the lowest of all known convertible undeniable signature schemes.

The *xyz-technique* has other applications. For example, we are also able to design very efficient universally convertible directed signatures – such a construction had remained open since 1993. Full details will appear elsewhere. Other applications will certainly appear: the use of this trick to design distributed contract signing and verifiable signature sharing protocols seems to be an interesting topic for further research.

Ackowledgements

We express our gratitude to Pascal Paillier for his helpful comments.

References

1. P. S. L. M. Barreto, H. Y. Kim: Fast hashing onto elliptic curves over fields of characteristic 3, Cryptology ePrint Archive, Report 2001/098 (2001)
2. M. Bellare, P. Rogaway: Random Oracles are Practical: a Paradigm for Designing Efficient Protocols. Proc. of 1st ACM Conference on Computer and Communications Security. 62–73 (1993)
3. D. Boneh, X. Boyen: Short Signatures Without Random Oracles. Proc. of Eurocrypt'04, Springer LNCS Vol. 3027, 56–73 (2004)
4. D. Boneh, M. Franklin: Identity-based Encryption from the Weil Pairing. SIAM J. Computing, 32(3), 586–615 (2003)
5. D. Boneh, B. Lynn, H. Shacham: Short signatures from the Weil pairing. Proc. of Asiacrypt'01, Springer LNCS Vol. 2248, 514–532 (2001)
6. J. Boyar, D. Chaum, I. B. Damgård, T.P. Pedersen: Convertible undeniable signatures. Proc. of Crypto'90, Springer LNCS Vol. 537, 189–205 (1991)
7. C. Boyd, E. Foo: Off-line Fair Payment Protocols using Convertible Signatures. Proc. of Asiacrypt'98, Springer LNCS Vol. 1514, 271–285 (1998)
8. E. Bresson and J. Stern: Proofs of Knowledge for Non-Monotone Discrete-Log Formulae and Applications. Proc. of ISC 2002, Springer LNCS Vol. 2433, 272–288 (2002)
9. J. Camenisch, M. Michels: Confirmer Signature Schemes Secure against Adaptative Adversaries. Proc. of Eurocrypt'00, Springer LNCS Vol. 1807, 243–258 (2000)
10. J. Camenisch, V. Shoup: Practical Verifiable Encryption and Decryption of Discrete Logarithms. Proc. of Crypto'03, Springer LNCS Vol. 2729, 126–144 (2003)
11. J. Camenisch, M. Stadler: Efficient Group Signature Schemes for Large Groups. Proc. of Crypto'97, Springer LNCS Vol. 1296, 410–424 (1997)
12. D. Chaum: Zero-Knowledge undeniable signatures. Proc. of Eurocrypt'90, Springer LNCS Vol. 473, 458–464 (1991)
13. D. Chaum: Designated Confirmer Signatures. Proc. of Eurocrypt'94, Springer LNCS Vol. 950, 86–91 (1995)
14. D. Chaum, H. van Antwerpen: Undeniable Signatures. Proc. of Crypto'89, Springer LNCS Vol. 435, 212–216 (1989)
15. D. Chaum, E. van Heijst, and B. Pfitzmann: Cryptographically strong undeniable signatures, unconditionally secure for the signer. Proc. of Crypto'91, Springer LNCS Vol. 576, 470–484 (1992)
16. D. Chaum, T.P. Pedersen: Wallet Databases with Observers. Proc. of Crypto'92, Springer LNCS Vol. 740, 89–105 (1993)
17. I. Damgard, T.P. Pedersen: New convertible undeniable signature schemes. Proc. of Eurocrypt'96, Springer LNCS Vol. 1070, 372–386 (1996)
18. S. Galbraith, W. Mao: Invisibility and anonymity of undeniable and confirmer signatures. Proc. of CT-RSA 2003, Springer LNCS Vol. 2612 80–97 (2003)
19. S. Galbraith, W. Mao, K.G. Paterson: RSA-based undeniable signatures for general moduli. Proc. of CT-RSA 2002, Springer LNCS Vol. 2271, 200–217 (2002)
20. R. Gennaro, H. Krawczyk, T. Rabin: RSA-based undeniable signatures. Proc. of Crypto'97, Springer LNCS Vol. 1294, 132–149 (1997)

21. S. Goldwasser, S. Micali, R. Rivest: A Digital Signature Scheme Secure against Adaptive Chosen-Message Attacks. SIAM J. Computing, 17 (2), 281–308 (1988)
22. M. Jakobsson, K. Sako, R. Impagliazzo: Designated Verifier Proofs and their Applications. Proc.of Eurocrypt'96, Springer LNCS Vol. 1070, 142–154 (1996)
23. H. Krawczyk, T. Rabin: Chameleon Signatures. Proc. of NDSS 2000, 143–154 (2000)
24. B. Libert, J.-J. Quisquater: Identity Based Undeniable Signatures. Proc. of CT-RSA 2004, Springer LNCS Vol. 2964, 112–125 (2004)
25. M. Michels, H. Petersen, P. Horster: Breaking and repairing a convertible undeniable signature scheme. Proc. of ACM Conference on Computer and Communications Security 1996, 148–152, ACM Press (1996)
26. M. Michels, M. Stadler: Efficient Convertible Undeniable Signature Schemes. Proc. of SAC'97, 231–244 (1997)
27. S. Mitsunary, R. Sakai, M. Kasahara: A New Traitor Tracing. IEICE Trans. Fundamentals, Vol. E85-A (2), 481–484 (2002)
28. J. Monnerat, S. Vaudenay: Undeniable Signatures Based on Characters: How to Sign with One Bit. Proc. of PKC 2004, Springer LNCS Vol. 2947, 69–85 (2004)
29. D. Pointcheval: Self-Scrambling Anonymizers. Proc. of Financial Cryptography 2000, Springer LNCS Vol. 1962, 259–275 (2000)
30. D. Pointcheval, J. Stern: Security Arguments for Digital Signatures and Blind Signatures. J. Cryptology, Vol. 13 (3), 361–396 (2000)
31. V. Shoup: Lower bounds for discrete logarithms and related problems. Proc. of Eurocrypt'97, Springer LNCS Vol. 1233, 256–266 (1997)
32. V. Shoup: OAEP reconsidered. Manuscript, November 16, 2000. Revised September 18, 2001. Full length version of the extended abstract in Proc. Crypto'01 (2001)
33. F. Zhang, R. Safavi-Naini, W. Susilo: Efficient Verifiably Encrypted Signature and Partially Blind Signature from Bilinear Pairings. Proc. of Indocrypt 2003, Springer LNCS Vol. 2904, 191–204 (2003) Revised version available from the authors.
34. F. Zhang, R. Safavi-Naini, W. Susilo: An efficient Signature Scheme from Bilinear Pairings and its Applications. Proc. of PKC 2004, Springer LNCS Vol. 2947, 277–290 (2004)

A Proof of Unforgeability

We define a sequence of games $\mathsf{Game}_0, \ldots, \mathsf{Game}_3$ starting from the actual EF-CMA adversary $\mathcal{A}$ and modify it step by step, until we reach a final game whose success probability has an upper bound related to solving CDH. All the games operate on the same underlying probability space: the public and private keys of the signature scheme, the coin tosses of the adversary $\mathcal{A}$ and the random oracle $\mathcal{H}$. Let k be a security parameter, let $(q, P, \mathbb{G}, \mathbb{H}, e)$ be a 5-tuple generated by $\mathcal{G}en(k)$ and let (X, Y) be a random instance of the CDH problem.

Game_0. We consider an EF-CMA-adversary $\mathcal{A}$ with success $\mathsf{Succ}^{ef-cma}_{\mathsf{TSCUSBM}, \mathcal{A}}$, within time τ. The key generation algorithm is run and produces a pair of keys (pk, sk, T). The universal receipt generation algorithm produces the information $(\mathcal{I}_1, \ldots, \mathcal{I}_T)$. The adversary $\mathcal{A}$ is fed with pk and $(\mathcal{I}_1, \ldots, \mathcal{I}_T)$ and, querying the random oracle $\mathcal{H}$, the signing oracle Σ, and the receipt generating oracle Υ, outputs a couple $(m^\star, (\sigma^\star, r^\star, t^\star))$.

We denote by $q_\mathcal{H}$, q_Σ and q_Υ the number of queries from the random oracle $\mathcal{H}$, from the signing oracle Σ and from the receipt generating oracle Υ. The only requirement is that the output signature $(\sigma^\star, r^\star, t^\star)$ has not been obtained from the signing oracle. For a signing query on a message m, we first ask an hash value of m with some additional random salt r and when the adversary outputs its forgery $(m^\star, (\sigma^\star, r^\star, t^\star))$, we ask a hash value of $(m^\star, r^\star)$. Therefore at most $q_\mathcal{H} + q_\Sigma + 1$ queries are asked to the hash oracle during this game.

In any Game$_j$, we denote by Forge$_j$ the event

$$\mathsf{TSCUSBM.UVerify}(m^\star, (\sigma^\star, r^\star, t^\star), pk, \mathcal{I}_T) = 1.$$

By definition, we have $\Pr[\mathsf{Forge}_0] = \mathbf{Succ}_{\mathsf{TSCUSBM}, \mathcal{A}}^{\mathsf{ef-cma}}$.

Game$_1$. In this game, we pick an element $a_2 \in [\![1, q-1]\!]$ at random, and we modify the simulation by replacing pk by $(X, a_2 P)$ and the T-tuple $(\mathcal{I}_1, \ldots, \mathcal{I}_T)$ by $(a_2 X, a_2^2 X, \ldots, a_2^T X)$. The distribution of pk is unchanged since (X, Y) is a random instance of the CDH problem and a_2 is picked at random. From now on, thanks to the knowledge of a_2, we can simulate the receipt generating oracle Υ. We have $\Pr[\mathsf{Forge}_1] = \Pr[\mathsf{Forge}_0]$.

Game$_2$. In this game, we simulate the random oracle $\mathcal{H}$. For any fresh query $(m, r) \in \{0,1\}^\star \times \{0,1\}^{n_r}$ to the oracle $\mathcal{H}$, we pick at random $u \in [\![1, q-1]\!]$ and compute $Q = (a_2 + h(m||r))uY$. We store (m, r, u, Q) in the H-List and returns Q as the answer to the oracle call. In the random oracle model, this game is clearly identical to the previous one. Hence, we get $\Pr[\mathsf{Forge}_2] = \Pr[\mathsf{Forge}_1]$.

Game$_3$. In this game, we simulate the signing oracle Σ: for any message m, whose signature is queried for the time period t, we pick at random two elements $r \in \{0,1\}^{n_r}$, $v \in [\![1, q-1]\!]$, and compute $\sigma = a_2^t v(a_2 + h(m||r))^{-1} X$. If the H-List includes a quadruple $(m, r, ?, ?)$ we abort the simulation, else we store (m, r, v, vP) in the H-List and outputs σ as the signature. As we abort with probability at most $(q_\mathcal{H} + q_\Sigma)2^{-n_r}$, summing up the inequalities for all signature queries, we have $|\Pr[\mathsf{Forge}_3] - \Pr[\mathsf{Forge}_2]| \leq q_\Sigma(q_\mathcal{H} + q_\Sigma)2^{-n_r}$

When the game Game$_3$ terminates, outputting a valid pair message/signature $(m^\star, (\sigma^\star, r^\star, t^\star))$, by definition of existential forgery, the H-List includes a quadruple $(m^\star, r^\star, u^\star, Q^\star)$ with $Q^\star = (a_2 + h(m||r))u^\star Y$ By the simulation, we have $e(\sigma^\star, P_2 + h(m||r)P) = e(Q^\star, a_2^{t^\star} P_1)$, and the point $R = (u^\star a_2^{t^\star})^{-1}\sigma^\star$ is a solution to our instance of the Computational Diffie-Hellman Problem. This concludes the proof. □

On Tolerant Cryptographic Constructions

Amir Herzberg

Computer Science Department, Bar Ilan University, Ramat Gan, Israel
herzbea@cs.biu.ac.il
http://AmirHerzberg.com

Abstract. Cryptographic schemes are often constructed using multiple component cryptographic modules. A construction is *tolerant* for a (security) specification if it meets the specification, provided a majority (or other threshold) of the components meet their specifications. We define tolerant constructions, and investigate 'folklore', practical cascade and parallel constructions. In particular, we show that cascading encryption schemes provides tolerance under chosen plaintext attack, non-adaptive chosen ciphertext attack (CCA1) and a weak form of adaptive chosne ciphertext attack (weak CCA2), but not under the 'standard' CCA2 attack. Similarly, certain parallel constructions ensure tolerance for unforgeability of Signature/MAC schemes, OWF, ERF, AONT and certain collision-resistant hash functions. We present (new) tolerant constructions for (several variants of) commitment schemes, by composing simple constructions, and general method of composing tolerant constructions. Our constructions are simple, efficient and practical. To ensure practicality, we use concrete security analysis (in addition to the simpler asymptotic analysis).

1 Introduction

Most cryptographic schemes do not have an unconditional proof of security. The classical method to establish security is by cryptanalysis, i.e. accumulated evidence of failure of experts to find weaknesses in the function. However, cryptanalysis is an expensive, time-consuming and fallible process. In particular, since a seemingly-minor change in a cryptographic function may allow an attack which was previously impossible, cryptanalysis allows only validation of specific functions and development of engineering principles and attack methodologies and tools, but does not provide a solid theory for designing cryptographic functions. Indeed, it is impossible to precisely predict the rate of future cryptanaltical successes. Prudent designers are usually able to ensure security by using sufficient margins and conservative to allow for unexpected breakthroughs, e.g. [LV01]; however, there is often resistance to replace widely deployed standards which were not broken yet, 'just' since the safety margins are eroded.

Hence, it is desirable to design cryptographic schemes to be tolerant of cryptanalysis, failure of assumptions and other vulnerabilities (including known trapdoors). A *tolerant cryptographic scheme* remains secure following successful cryptanalysis of some subset of its cryptographic 'components', or successful counterexample to one of the assumptions underlying its security (e.g. factoring). Tolerance does not imply unconditional-security; however, it would hopefully provide sufficient advanced-warning time to replace broken cryptographic components.

Many cryptographic systems and constructions use redundant components in the hope of achieving tolerance. The most familiar such construction is cascade. Cascading of cryptosystems is very natural; novices and experts alike believe that the cascade $\mathcal{E} \circ \mathcal{E}'$ of two encryption schemes $\mathcal{E}$, $\mathcal{E}'$ is at least as secure as the more secure of the two, hopefully even more secure than both. Indeed, cascading of cryptosystems has been a common practice in cryptography for hundreds of years.

However, so far, there are few publications analyzing tolerant cryptographic constructions. In [1], Asmuth and Blakely present a simple 'parallel' construction of a randomized cryptosystem from two component ciphers, with the hope of achieving tolerance; proof of security was given only in [23]. A similar 'parallel' construction for block ciphers appears in [28]. More attention was given to *cascading* of block ciphers . Even and Goldreich showed that keyed cascade ensures tolerance against message recovery attacks on block ciphers [17, Theorem 5], and conjectured that the result holds for other specifications of ciphers. Damgard and Knudsen [13] proved that it holds for security against key-recovery under chosen-plaintext attacks. Maurer and Massey [32] claimed that the proof in [17] holds only under the uninterestingly restrictive assumption that the enemy cannot exploit information about the plaintext statistics, but we disagree. We extend the proof of [17] and show that, as expected intuitively and in [17], keyed cascading provides tolerance to many confidentiality specifications, not only of block ciphers but also of other schemes such as public key and shared key cryptosystems.

Our proof holds for three definitions of security under indistinguishability test: the well-known notions of plaintext only attack and non-adaptive chosen ciphertext attack (CCA1), as well as a new, weak version of adaptive chosen ciphertext attack (wCCA2), which may be of independant interest. We note that cascading does *not* provide tolerance for the well-known notion of adaptive chosen ciphertext attack (CCA2). Also, cascading does not provide tolerance if the length of the output may differ for two messages of the same length. These observations demonstrate the importance of backing intuition with analysis and proof.

Tolerance is relevant to any cryptographic scheme, not just for confidentiality. In particular, it is widely accepted that the parallel construction $g(x) \| f(x)$, using the same input x to both functions, ensures tolerance for several integrity properties, such as (several variants of) collision-resistant hashing as well as Message Authentication Codes (MAC) and digital signatures. We prove that the parallel construction indeed provide tolerance for such integrity specifications. The parallel construction is used, for tolerance, in practical designs and standards, e.g. in the W3C XML-DSIG specifications and in the TLS protocol [35].

We present few simple tolerant constructions for some basic cryptographic goals (specifications); further research is required to find tolerant constructions for other goals.

Efficiency is critical for practical tolerant constructions; implementors will rarely be willing to tolerate non-negligible performance loss, 'just' in order to tolerate potential vulnerabilities in a cryptographic function. In fact, ignoring efficiency, we can ensure tolerance by using provable constructions of cryptographic mechanisms from few 'basic' cryptographic mechanisms, which have simple tolerant designs. For example, many cryptographic mechanisms can be constructed from one-way functions; and it is sufficient that one of $\{g, f\}$ is a one-way function, to ensure that $g(x) \| f(x)$ is also a one-way function. Provably-secure constructions based on one-way functions exist for many cryptographic mechanisms, e.g. pseudo-random generators [19, 26] and signature schemes [34]. Therefore, by using a tolerant construction of one way function (from multiple candidate one-way functions) as the basis of some cryptographic scheme, the scheme retains the proven security properties even if one of the candidate one-way functions is not secure. However, such constructions are often inefficient, and involve unacceptable degradation in security parameters (e.g., require absurd key and/or block sizes) [27]. To quantify loss in security and efficiency due to the constructions, we use concrete security measures [7, 8].

Our contributions. We identify and define cryptographic tolerance as a criteria for cryptographic specifications, and goal for constructions. Some additional contributions include:

- *Precise analysis of the security of several 'folklore' constructions.* In particular, we show that cascade encryption indeed ensures tolerance as long as each component encryption has fixed output length (for fixed length input), and for several variants of indistinguishability including a weak form of adaptive chosen ciphertext attack (weak CCA2), but not for the 'regular' CCA2 specification. We note that the 'multiple encryption' construction of [14] seems to ensure tolerant encryption for CCA2, but at significant overhead (ciphertext length more than doubles), which may be unacceptable for many applications.
- *Efficient, practical constructions for commitment schemes.* To our knowledge, these are the first provably-secure tolerant constructions of general cryptographic functions, beyond the folklore constructions, and few additional cryptographic constructions proven secure based on validity of either of two (specific) 'hardness' assumptions, e.g. [39].
- *Compositions of constructions.* Cryptographic constructions are usually studied in isolation; however, sometimes one construction is good for ensuring some properties, e.g. confidentiality, while another is good for other properties, e.g. integrity. We define compositions of constructions, to combine the benefits of different constructions (when possible), and present a generic composition based on a simple combinatorial object (composition structure). Finally, we use the generic composition and two simple composition objects to compose the cascade and parallel constructions, creating two new efficient composite constructions for commitment schemes.

2 Notations and Definitions

We first fix some notations and recall the (standard) cryptographic definitions, for schemes to which we present tolerant constructions. We then define the concept of tolerant constructions.

2.1 Cryptographic Functions and Schemes

We find it convenient to define tolerant constructions for cryptographic schemes which are finite sets of functions, taken from some general set F of (cryptographic) functions[1]. If a function is 'randomized', we write the randomization bits as explicit input bits, for clarity and to facilitate concrete security analysis. We denote schemes by calligraphical font e.g. $\mathcal{S}$, and refer to particular function π in the scheme using *dot notation* i.e. $\mathcal{S}.\pi$. We next define some cryptographic functions and schemes, beginning with encryption; more definitions will appear in the final version.

Encryption. An encryption scheme $\mathcal{E}$ consists of three functions $<KG, E, D>$ (for key generation, encryption and decryption, respectively). The key generation function $\mathcal{E}.KG$ accepts as input a random string, and its output is a pair of keys: e, d for encryption and decryption, respectively. (For symmetric encryption, $e = d$.) We again use dot notation to refer to particular key, e.g. $\mathcal{E}.KG.e(r)$ is the encryption key generated by $\mathcal{E}.KG$ on input r. We use subscripts to denote keys, and the random input to the encryption function. Encryption of message m, where $m \in M$ for some message space M, using key e and randomness r, is $\mathcal{E}.E_{e,r}(m) \in C$, where C is the ciphertext space. The decryption function $\mathcal{E}.D$ accepts as input ciphertext $c \in C$, and private decryption key d, and returns a message $m' \in M$ or a failure indicator $\bot$. The correctness requirement is $\mathcal{E}.D_d(\mathcal{E}.E_{e,r}(m)) = m$, for any $m \in M, r \in \{0,1\}^*$ and $r_{KG} \in \{0,1\}^*$ with $e = E.KG.e(r_{KG}), d = E.KG.d(r_{KG})$.

Security of Encryption. To define security for encryption schemes, we use the quantitative 'indistinguishability experiment' approach of [7], but extend their definitions as follows:

- Our definitions and results apply to both shared-key and public-key encryption. We use a flag $ISPUB$ to signal when the encryption key is public.
- In our quantitative security analysis, we bound, in addition to the 'standard' capabilities of the adversary, also the length of ciphertext (l) and plaintext ($l - \Delta$) in queries. We use this to bound the length of the inputs to the encryption schemes used in our reductions, in order to limit the computation time.
- Cascade encryption is insecure for adaptive chosen ciphertext attack (CCA2). Recently, [14] presented constructions for 'multiple encryption' schemes that appear to be tolerant for CCA2, but have significant overhead; we believe that in many applied scenarios, this overhead would not be acceptable.

---
[1] Some definitions allow cryptographic schemes to be stateful machines.

We found that cascade encryption is secure under a weak-CCA2 attack model, where the attacker can chose ciphertext adaptively, but if the decrypted plaintext is one of the two chosen 'test' plaintexts, then the oracle returns a special 'bingo' signal, but does not identify the plaintext; we call this the 'weak decrypt' oracle. Namely: $\mathcal{E}.wD_{d,p[0],p[1]}(c) = \{m := \mathcal{E}.D_d(c);$ if $m = p[0]$ or $m = p[1]$ return 'bingo'; else return $m;\}$. Weak-CCA2 follows the criticism of [3] on 'regular' CCA2, but is even weaker than their gCCA2 notion; still, it may be sufficient in practice, in particular it allows the practical 'feedback only CCA' attacks of [6, 29]. The Weak-CCA2 notion may have additional applications; for example, it may be possible to design a universal re-encryption scheme [24] secure under Weak-CCA2 attack, while current definitions and constructions allow only CCA1 attacks (since re-encryption makes it impossible to ensure security under CCA2).

Definition 1 (Indistinguishability Experiment). *Let $\mathcal{E}$ be an encryption scheme and let $k, l, t, \rho, \Delta, \rho_A \in \mathbb{N}, ISPUB \in \{T/F\}$ and $q : \{select, find\} \times \{E, D, wD\} \to \mathbb{N}$. Let $A \in PPT$ be an (adversarial) machine with access to oracle O. Let $IndExp_{A,\mathcal{E},ISPUB}(k, q, l, \rho, t, \Delta, \rho_A)$ be the following experiment:*

1. $r_{ed} \in_R \{0,1\}^k$; $e = \mathcal{E}.KG.e(r_{ed})$; $d = \mathcal{E}.KG.d(r_{ed})$
2. *Let O be an oracle to the functions:* $\{\mathcal{E}.E_{e,r}\ ,\ \mathcal{E}.D_d\ ,\ \mathcal{E}.wD_{d,p[0],p[1]}\}$
3. *If $ISPUB = T$ then $e' = e$ else $e' = \lambda$*
4. $(p[0], p[1], state) \leftarrow A^O(``select", e', 1^k)$; /* select phase */
5. $b \in_R \{0, 1\}$;
6. $r \in_R \{0,1\}^\rho$; $c = \mathcal{E}.E_{e,r}(p[b])$;
7. $\beta = A^O(``find", c, state)$; /* find phase */
8. *Return win iff the following conditions hold:*
   (a) $\beta = b$
   (b) $|p[1]| = |p[0]| \leq l - \Delta$
   (c) *total running time of A is less than t*
   (d) *A makes at most $q[\phi, f]$ calls to oracle $\mathcal{E}.f$ at phase $\phi \in \{select, find\}$*
   (e) *A uses at most ρ_A random bits*
   (f) *In its oracle queries, A uses m, c s.t. $|m| \leq l - \Delta$ and $|c| \leq l$.*

The confidentiality specifications depend on the maximal *advantage* a for the adversary.

Definition 2. *$\mathcal{E}$ satisfies specification $IND_{ISPUB}(a, k, q, l, \Delta, t, \rho, \rho_A)$ if for every adversary $A \in PPT$ holds $Pr[IndExp_{A,\mathcal{E},ISPUB}(k, q, l, \Delta, t, \rho, \rho_A) = win] < \frac{1}{2} + a$.*

We now also present asymptotic, polynomial-time complexities. Allowing polynomial number of each type of queries, possibly restricting queries to oracle D for the 'weaker' notions (cf. to CCA2), i.e. CPA, CCA1 and wCCA2.

Definition 3. *$\mathcal{E}$ satisfies specification $CCA2-IND_{ISPUB}$ if $\mathcal{E} \in PPT$ and for any strictly positive polynomials $l, \Delta, t, \rho, \rho_A, a$ and positive polynomials $q[\phi, f]$*

for $\phi \in \{select, find\}$ and $f \in \{E, D, wD\}$, exists some integer k_0 such that for every $k \geq k_0$, holds: $\Pr[IndExp_{A,\mathcal{E},ISPUB}(k, q(k), l(k), \Delta(k), t(k), \rho(k), \rho_A(k)) = win] < \frac{1}{2} + a(k)$. We say that $\mathcal{E}$ satisfies specifications $wCCA2 - IND_{ISPUB}$, $CCA1 - IND_{ISPUB}$, $CPA - IND_{ISPUB}$, respectively, if $\mathcal{E}$ satisfies $CCA2 - IND_{ISPUB}$ restricted to $q[find, D] = 0$, $q[find, D] = q[find, wD] = 0$, or $q[select, D] = q[select, wD] = 0$, respectively.

Commitment. A (non-interactive) commitment scheme $\mathcal{C}$ consists of four functions[2] $< KG, C, D, V >$ (for Key Generation, Commit, Decommit and Validate, respectively). The key generation function $\mathcal{C}.KG$ accepts as input a random string, and its output is a public commitment key ck. The commitment and decommit functions $\mathcal{C}.C, \mathcal{C}.D$ have both three inputs: a message $m \in M$, a public commitment key ck and randomness r, and their respective outputs are: a commitment $\mathcal{C}.C_{ck,r}(m)$ and a decommitment $\mathcal{C}.D_{ck,r}(m)$. The validate function $\mathcal{C}.V$ has the same inputs, plus the commitment and decommitment values $(c, d$ respectively), and outputs $True$ iff c, d are a correct commitment and decommitment values for m. The correctness requirement is $\mathcal{C}.V_{ck}(m, \mathcal{C}.C_{ck,r}(m), \mathcal{C}.D_{ck,r}(m)) = True$, for any $m \in M$, r and r_{KG} such that $ck = \mathcal{C}.KG(r_{KG})$.

Security of Commitment. Commitment schemes have a confidentiality property, called *hiding*, and an integrity property, called *binding*. We only sketch the asymptotic definitions; the final version of this manuscript will contain complete (and concrete) definitions.

Hiding. No PPT adversary can distinguish the commitments of any two messages of its choice.

Binding. Given (random) ck, every PPT adversary A has negligible probability of finding a collision, i.e. values c, d, d', m, m' s.t. $\mathcal{C}.V_{ck}(m, c, d) = \mathcal{C}.V_{ck}(m', c, d') = True$ (notice the commitment c is the same!).

Following [3], we also consider relaxed binding, where A has negligible probability of finding a message m s.t. when given $c = \mathcal{C}.C_{ck,r}(m)$ and $d = \mathcal{C}.D_{ck,r}(m)$, the PPT adversary A can find m', d' s.t. $\mathcal{C}.V_{ck}(m', c, d') = True$. As motivated in [3], known constructions for commitment schemes can use $UOWHF$ for relaxed binding, but require the (strictly stronger) $CRHF$ for (strict) binding. We show later a construction which is tolerant for both versions of binding ('strict' or relaxed).

Our construction is also tolerant for trapdoor commitments [40], or chameleon hash functions [30]. In these schemes, the key generation produces also a secret trapdoor key $\mathcal{C}.KG.t$. The schemes also define a $Switch$ algorithm, which uses the trapdoor key, to transform any valid commitment to any message m to an indistinguishable commitment to any other message m' (adversary may choose both m and m').

---

[2] Most existing definitions of commitment schemes are slightly different, and in particular use 'open' to recover the message, rather than 'validate'. However, as a result, the constructions use long decommitment strings, which contain the original message. By explicitly using the message as separate input, we can compare the actual overhead of schemes.

2.2 Performance Specifications

For asymptotic security analysis, it is sufficient to require all algorithms to be probabilistic polynomial time. However, to allow concrete security analysis of constructions, we need concrete bounds on the complexities of the schemes. In this version of the work, we present such bounds (and concrete analysis) only for encryption schemes, as follows.

Definition 4 (Concrete Complexity Bounds for Encryption Schemes). *Let $k, k', l, \Delta, \rho \in \mathbb{N}$, with $l > \Delta$, and let $\tau : \{KG, E, D\} \to \mathbb{N}$. For every encryption scheme $\mathcal{E} \in PPT$, let $bounds[k, k', l, \Delta, \rho, \tau](\mathcal{E}) = 1$ iff:*

1. *For inputs of length up to k, $\mathcal{E}.KG(k)$ is computable in time $\tau[KG]$ and $|\mathcal{E}.KG(k)| \geq k'$.*
2. *There is a deterministic algorithm that computes $\mathcal{E}.E_{e,r}(m)$ in time $\tau[E]$, for every $e \in \{0,1\}^{k'}$ and every m s.t. $|m| \leq l - \Delta$, and reads up to the first ρ bits of r; also, $|\mathcal{E}.E_{e,r}(m)| \leq l$.*
3. *There is a deterministic algorithm that computes $\mathcal{E}.D_d(c)$ in time $\tau[D]$, for every $d \in \{0,1\}^{k'}$ and every c s.t. $|\mathcal{E}.D_{d,r}(c)| \leq |c| \leq l$.*

2.3 Tolerant Constructions

We are interested in specifications (properties) of functions, including concrete security specifications and asymptotic security specifications. We define specifications simply as binary predicates over the set of functions F. Let $S(F) = \{s : F \to \{0,1\}\}$ be the set of all specifications (predicates) over F. We say that $f \in F$ satisfies $s \in S(F)$ iff $s(f) = 1$.

We say that a mapping c of p functions $f_1, \ldots f_p$ into a single function $c(f_1, \ldots, f_p)$, i.e. $c : F^p \to F$, is a *construction of plurality p over F*. Construction c is tolerant if $c(f_1, \ldots, f_p)$ satisfies some specification s' as long as a sufficient subset of $f_1, \ldots f_p$ satisfy specifications $s_1, \ldots s_p$, respectively (often, all specifications are identical, i.e. $s = s_1 = \cdots = s_p$ and also often $s = s'$). To complete this definition, we need to identify the sufficient subset of $f_1, \ldots f_p$; following the works on secret-sharing, we define two variants of tolerance: based on threshold t ($0 \leq t < p$), and based on general access structure $\Lambda \subseteq P(\{1, \ldots p\})$ (where Λ is a set of subsets of $\{1, \ldots p\}$). In addition, we often require that *all* of the candidate functions f_i satisfy some minimal specifications $b \in S(F)$, such as bounds on their complexities.

Definition 5. *Consider some set of functions F, integer p, predicates $s', b, s_1, \ldots, s_p \in S(F)$ and construction $c : F^p \to F$ of plurality p over F. Construction c is t-tolerant for $(s_1, \ldots s_p) \to s'$, with threshold $t < p$, if $s'(c(f_1, \ldots, f_p))$ holds provided $\sum_{i=1}^{p} s_i(f_i) \geq p - t$. Construction c is Λ-tolerant for $(s_1, \ldots s_p) \to s'$, with access structure $\Lambda \subseteq P(1, \ldots p)$, if $s'(c(f_1, \ldots, f_p))$ holds provided for some $\lambda \in \Lambda$ holds $(i \in \lambda) \to s_i(f_i) = 1$. Construction c is t – tolerant with prerequisite b if $s'(c(f_1, \ldots, f_p))$ holds provided $(\sum_{i=1}^{p} b(f_i) = p) \land (\sum_{i=1}^{p} s_i(f_i) \geq p - t)$. Similarly, c is Λ – tolerant with prerequisite b if $s'(c(f_1, \ldots, f_p))$ holds provided $(\sum_{i=1}^{p} b(f_i) = p) \land (\exists \lambda \in \Lambda) (\sum_{i \in \lambda} s_i(f_i) = |\lambda|)$. If c is 0-tolerant for $(s_1, \ldots s_p) \to s'$*

then we say that c preserves $(s_1, \ldots s_p) \to s'$ (with or without prerequisites). If c is t-tolerant for $(s_1, \ldots s_p) \to s'$, then we say that c is t-tolerant for $s \to s'$; if $t = 0$ then we say that c preserves $s \to s'$. If $s = s'$ then we say that c is t-tolerant for (or preserves) s.

3 Cascade Constructions and Their Tolerance

The most basic tolerant construction of cryptographic functions is the cascade construction $c_\circ$. We begin by discussing cascading of functions with a single input and output, such as hash functions, namely $c_\circ(f, g)[x] = f \circ g(x) = f(g(x))$. In the following subsections we discuss cascading of keyed schemes.

3.1 Simple Cascade of Keyless Cryptographic Functions

Consider any two functions $g : D_g \to R_g$, $f : D_f \to R_f$ s.t. $R_g \subseteq D_f$. The simple cascade of f and g, denoted $f \circ g$ or $c_\circ(f, g)$, is a construction of plurality 2 defined as $c \circ (f, g) = f \circ g(x) = f(g(x))$.

Unfortunately, simple cascade rarely ensures tolerance, and often does not even preserve cryptographic specifications. So far, we found simple cascade ensures tolerance only to the one-way property, and that with a prerequisite requirement $perm(f)$, which is true only if f is a permutation when restricted to input domains $\{0, 1\}^l$ for some length l.

Lemma 1. *Simple cascade of two functions is...*

1. *1-tolerant with prerequisite perm for specifications OWF.*
2. *1-tolerant with prerequisite perm for specifications:*

$$concrete - OWF^f(a, k, \rho_A, \tau_A) \wedge [Time(f, k) \leq \tau_F]$$
$$\to concrete - OWF^f(a, k, \rho_A, \tau_A + \tau_F)$$

3. *Preserves (0-tolerant), but not 1-tolerant, for specifications ERF (exposure resilient function) and AONT (all or nothing transform).*
4. *Not (even) 0-tolerant for specifications OWF and WCRHF.*

Proof: The negative claims (4 and part of 3) follow by simple examples, e.g. to prove claim 4, let h be a OWF and/or $WCRHF$. Let $g(x) = h(x)||0^{|h(x)|}$ and $f(x) = \begin{cases} 0 & \text{if } x = y0^{|x|/2} \\ h(x) & \text{otherwise} \end{cases}$. Trivially, both f and g are OWF and/or $WCRHF$, respectively, yet $f \circ g$ is neither OWF not $WCRHF$; in fact, $f \circ g(x) = 0$ for every x. Claims 1 and 2 follow from a simple reduction argument; the proof of claim 2 is in the full version (and claim 1 immediately follows from claim 2). □

3.2 Cascade Encryption Is Tolerant

The cascade encryption, i.e. cascade of two[3] encryption schemes $\mathcal{E}, \mathcal{E}'$, is denoted $c_E(\mathcal{E}, \mathcal{E}')$ or $\mathcal{E} \circ \mathcal{E}'$ and defined as follows. *Notation:* For convenience we explicitly

---
[3] Extension to arbitrary number of schemes is trivial.

write the inputs and outputs to the cascade (or any composition) as a tuple of inputs or outputs when appropriate, e.g. $<r,r'>$ to denote the pair of two random inputs (r to $\mathcal{E}$ and r' to $\mathcal{E}'$).

- $\mathcal{E} \circ \mathcal{E}'.KG.e(<r,r'>) = <\mathcal{E}.KG.e(r), \mathcal{E}'.KG.e(r')>$
- $\mathcal{E} \circ \mathcal{E}'.KG.d(<r,r'>) = <\mathcal{E}.KG.d(r), \mathcal{E}'.KG.d(r')>$
- $\mathcal{E} \circ \mathcal{E}'.E_{<e,e'>,<r,r'>}(m) = \mathcal{E}.E_{e,r}(\mathcal{E}'.E_{e',r'}(m))$
- $\mathcal{E} \circ \mathcal{E}'.D_{<d,d'>}(c) = \mathcal{E}'.D_{d'}(\mathcal{E}.D_d(c))$

Cascade encryption is a construction of plurality 2; the following lemma bounds the complexities:

Lemma 2. *Let $\mathcal{E}, \mathcal{E}'$ be a pair of encryption schemes such that for $s \in \{\mathcal{E}, \mathcal{E}'\}$ holds $bounds[k, k', l, \Delta, \rho, \tau](s) = True$ with $l > 2\Delta$. Then $\mathcal{E} \circ \mathcal{E}'$ is an encryption scheme with $bounds[2k, 2k', l, 2\Delta, 2\rho, 2\tau](\mathcal{E} \circ \mathcal{E}') = True$.* □

We now investigate the security and tolerance of cascade encryption. As noted in the introduction, cascade encryption is an ancient, widely-deployed technique, usually in the hope of improving security - e.g., providing tolerance to weaknesses of one of the two cascaded encryption schemes. Is this secure? This depends on the adversary capabilities ('attack model'). Cascade encryption is not tolerant for adaptive chosen ciphertext attack (CCA2); simply consider $\mathcal{E}'$ which ignores the least significant bit of the ciphertext, allowing adversary to decrypt the challenge ciphertext (by flipping the LSb and invoking the decryption oracle). However, as [3] argued, this 'attack' is so contrived, that it may indicate that CCA2 is overly restrictive, rather than a problem with cascade encryption. In [3], the authors present a slightly weaker definition, gCCA, but we do not think cascade is tolerant under that definition, either; on the other hand, the following lemma shows that cascade encryption is tolerant under the more relaxed *weak CCA (wCCA)* definition.

Also, note that the indistinguishability experiment restricts the adversary to select plaintexts of the same length. Obviously, the length of the ciphertext should be indistinguishable between any two plaintexts (of the same length). For simplicity, we define a predicate *FixedExtra* over encryption schemes, such that $FixedExtra(\mathcal{E}')$ holds if the length of the ciphertext depends only on the length of the plaintext and on the security parameter; this holds for all practical cryptosystems. Clearly, if the length of the output of $\mathcal{E}'$ differs for two plaintexts of the same length, then cascading it with a secure $\mathcal{E}$ may not suffice to ensure indistinguishability. We therefore require $FixedExtra(\mathcal{E}')$ to hold.

Lemma 3. *Cascade encryption is 1-tolerant with prerequisite FixedExtra, for specifications $wCCA2 - IND_{ISPUB}$, $CCA1 - IND_{ISPUB}$, $CPA - IND_{ISPUB}$. Furthermore, let $k, k', l, \Delta, \rho, t°, \rho_A° \in \mathbb{N}$ s.t. $l > 2\Delta, \tau : \{KG, E, D\} \to \mathbb{N}$, $q : \{select, find\} \times \{E, D, wD\} \to \mathbb{N}$ s.t. $q[find, D] = 0$. Then, c_E is also 1-tolerant with the additional prerequisite $bounds[k, k', l, \Delta, \rho, \tau]$, for specifications $IND_{ISPUB}(a, k, q, l, \Delta, t, \rho, \rho_A) \to IND_{ISPUB}(a, k, q, l, 2\Delta, t°, \rho, \rho_A°)$, where $\rho_A = \rho_A° + 2k + 2\rho$ and*

$$t = t° + \tau[KG] + 2\tau[E] + \sum_{j \in \{find, select\}} \sum_{f \in \{E, D, wD\}} q[j, f]\tau[f].$$ □

Cascading is a natural candidate construction for many cryptographic mechanisms; we now define and investigate tolerance of cascade of commitment and MAC/Signature schemes.

3.3 Cascade Commitment

We define cascade commitment $c_c(\mathcal{C}, \mathcal{C}')$ (or $\mathcal{C} \circ \mathcal{C}'$), i.e. cascade of two commitment schemes $\mathcal{C}, \mathcal{C}'$, as follows. (The final version will also contain the simple extension to trapdoor commitment schemes.) We again wrote inputs and outputs as tuples.

- $\mathcal{C} \circ \mathcal{C}'.KG(<r,r'>) = <\mathcal{C}.KG(r), \mathcal{C}'.KG(r')>$
- $\mathcal{C} \circ \mathcal{C}'.C_{<ck,ck'>,<r,r'>}(m) = \mathcal{C}.C_{ck,r}(\mathcal{C}'.C_{ck',r'}(m))$
- $\mathcal{C} \circ \mathcal{C}'.D_{<ck,ck'>,<r,r'>}(m)$
  $= <\mathcal{C}.D_{ck,r}(\mathcal{C}'.C_{ck,r}(m)), \mathcal{C}'.D_{ck',r'}(m), \mathcal{C}'.C_{ck',r'}(m)>$
- $\mathcal{C} \circ \mathcal{C}'.V_{<ck,ck'>}(m,c,<d,d',c'>) = \mathcal{C}'.V_{ck'}(m,c',d') \wedge \mathcal{C}.V_{ck}(c',c,d)$.

As the following lemma shows, cascade ensures the privacy (hiding) property of commitment schemes, but only preserves the integrity (binding) property.

Lemma 4. *Cascade commitment is tolerant for the hiding specification, and preserves (but is not tolerant for) the binding specification.*

We next show that cascade also preserves, but does not ensure tolerance, for other integrity properties, specifically of MAC/Signature schemes.

3.4 Cascading Preserves Unforgeability of MAC/Signature Schemes

We define cascade $\mathcal{S} \circ \mathcal{S}'$ of two MAC/Signature schemes $\mathcal{S}, \mathcal{S}'$, as follows. We again write inputs and outputs as tuples.

1. $\mathcal{S} \circ \mathcal{S}'.KG.s(<r,r'>) = <\mathcal{S}.KG.s(r), \mathcal{S}'.KG.s(r')>$
2. $\mathcal{S} \circ \mathcal{S}'.KG.v(<r,r'>) = <\mathcal{S}.KG.v(r), \mathcal{S}'.KG.v(r')>$
3. $\mathcal{S} \circ \mathcal{S}'.S_{<s,s'>,<r,r'>}(m) = <\mathcal{S}.S_{s,r}(m)(\mathcal{S}'.S_{s',r'}(m))>$
4. $\mathcal{S} \circ \mathcal{S}'.V_{<v,v'>}(\sigma, <m, \sigma'>) = \mathcal{S}.V_v(\sigma, \sigma') \wedge \mathcal{S}'.V_{v'}(\sigma', m)$

Lemma 5. *Cascade MAC/Signature is 0-tolerant for (i.e. preserves) the existential unforgeability under adaptive chosen message attack specification.*

4 Parallel Constructions and Their Tolerance

We now consider another important family of constructions, which are parallel applications of two or more cryptographic functions or schemes. Parallel constructions may use the same input to all functions, use different parts of the input to each function, or use some combination of the inputs to create the input to each function, often involving XOR or secret-sharing. Similarly, the output of some parallel constructions is simply the concatenation of the outputs of each function, while others 'merge' the outputs, by XOR or secret-sharing.

4.1 Split-Parallel-Concat Construction for OWF

Possibly the simplest parallel construction 'splits' the input among several functions, and concatenates the result. In particular, the *Split-Parallel-Concat (sc)* construction for two keyless functions f, f' is defined as $sc(f, f')[<x, x'>] = f||_{sc}f'(<x, x'>) = <f(x), f(x')>$. This trivial construction is tolerant for One-Way Functions specifications, using two or more functions.

Lemma 6. *The Split-Parallel-Concat (sc) construction is tolerant for OWF specifications.*

Proof Sketch: use argument for transforming a weak OWF into a strong OWF (see e.g. [19]). □

4.2 Copy-Parallel-Concat Construction for Integrity Specifications

The *Copy-Parallel-Concat (cc)* construction is also trivial and well-known, but it is very practical and widely deployed. Here, the input to the construction is 'copied' and used as input to each of the components; and the output is simply the concatenation of the output of all components. This simple, folklore construction provides tolerance for the integrity properties of collision-resistant hash functions, signature/MAC schemes and commitment schemes.

Let us first define the cc construction for keyless functions, e.g. (weakly collision resistant) hash functions. The Copy-Parallel-Concat (cc) parallel construction of single-input (keyless) functions f, g is denoted as $f||g$ or $c_{||}(f, g)$, and defined as $c_{||}(f, g)(x) = f||g(x) = f(x)||g(x)$. When the functions have inputs for random bits and/or keys, these are selected independently for the two functions, and the parallel construction is $f_{k,r}||g_{k',r'}(x) = f_{k,r}(x)||g_{k',r'}(x)$.

The Copy-Parallel-Concat (cc) construction of two Signature/MAC schemes $\mathcal{S}, \mathcal{S}'$, denoted $c_{||}(\mathcal{S}, \mathcal{S}') = \mathcal{S}||\mathcal{S}'$, is defined as follows. The definitions and proofs extend trivially to arbitrary number of schemes.

- $\mathcal{S}||\mathcal{S}'.KG(<r, r'>) = <\mathcal{S}.KG(r), \mathcal{S}'.KG(r')>$
- $\mathcal{S}||\mathcal{S}'.S_{<s,s'>,<r,r'>}(m) = <\mathcal{S}.S_{s,r}(m), \mathcal{S}'.S_{s',r'}(m)>$
- $\mathcal{S}||\mathcal{S}'.V_{v,v'}(m, <\sigma, \sigma'>) = \mathcal{S}.V_v(m, \sigma) \land \mathcal{S}'.V_{v'}(m, \sigma')$

Similarly, the Copy-parallel-Concat (cc) parallel construction of two commitment schemes $\mathcal{C}, \mathcal{C}'$, denoted $c_{||}(\mathcal{C}, \mathcal{C}') = \mathcal{C}||\mathcal{C}'$, is defined as follows. The definition extends trivially to trapdoor commitment.

- $\mathcal{C}||\mathcal{C}'.KG(<r, r'>) = <\mathcal{C}.KG(r), \mathcal{C}'.KG(r')>$
- $\mathcal{C}||\mathcal{C}'.C_{<ck,ck'>,<r,r'>}(m) = <\mathcal{C}.C_{ck,r}(m), \mathcal{C}'.C_{ck',r'}(m)>$
- $\mathcal{C}||\mathcal{C}'.D_{<ck,ck'>,<r,r'>}(m) = <\mathcal{C}.D_{ck,r}(m), \mathcal{C}'.D_{ck',r'}(m)>$
- $\mathcal{C}||\mathcal{C}'.V_{<ck,ck'>}(m, <c, c'>, <d, d'>) = \mathcal{C}.V_{ck}(m, c, d) \land \mathcal{C}.V_{ck}(m, c, d)$

As the following lemma shows, the parallel construction ensures tolerance for many integrity properties / specifications, but clearly is quite bad for privacy.

Lemma 7. *The Copy-Parallel-Concat (cc) construction is...*

1. *Tolerant for the 'integrity' specifications WCRHF.*
2. *Tolerant for the existential unforgeability under adaptive chosen message attack specification of Signature/MAC schemes.*

Secret sharing schemes support different *thresholds*, for tolerating exposure or corruption of shares. In particular, in our case, we are interested in the following two thresholds. First, secret sharing schemes have a privacy threshold, t_p, which determines the maximum number of shares which reveal 'no information' about the message m. Second, they have a soundness threshold t_s, which determines the minimum number of correct shares which ensures it is impossible to recover an *incorrect* message $m' \neq m$ (and $m' \neq \bot$).

For simplicity, we present the share-parallel-concat (sc) construction for ensuring tolerance from three candidate commitment schemes, $\mathcal{C}_1, \mathcal{C}_2$ and $\mathcal{C}_3$, and using an arbitrary secret sharing scheme $<Share, Reconstruct>$ with $n = 3, t_p = 1, t_s = 2$, e.g. Shamir's scheme [37]. Generalizations allowing threshold to $t_p > 1$ insecure components (by using $2t_p + 1$ components and shares) are straightforward. We use the notation $s_i = Share_{i,r}(m,n)$.

$sc(\mathcal{C}_1, \mathcal{C}_2, \mathcal{C}_3).KG(<r_1, r_2, r_3>) =$
$=<\mathcal{C}_1.KG(r_1), \mathcal{C}_2.KG(r_2), \mathcal{C}_3.KG(r_3)>$
$sc(\mathcal{C}_1, \mathcal{C}_2, \mathcal{C}_3).C_{<ck_1,ck_2,ck_3>,<r,r_1,r_2,r_3>}(m) =$
$=<\mathcal{C}_1.C_{ck_1,r_1}(s_1), \mathcal{C}_2.C_{ck_2,r_2}(s_2), \mathcal{C}_3.C_{ck_3,r_3}(s_3)>$
$sc(\mathcal{C}_1, \mathcal{C}_2, \mathcal{C}_3).D_{<ck_1,ck_2,ck_3>,<r,r_1,r_2,r_3>}(m) =$
$=<\mathcal{C}_1.D_{ck_1,r_1}(s_1), \mathcal{C}_2.D_{ck_2,r_2}(s_2), \mathcal{C}_3.D_{ck_3,r_3}(s_3), s_1, s_2, s_3>$
$sc(\mathcal{C}_1, \mathcal{C}_2, \mathcal{C}_3).V_{<ck_1,ck_2,ck_3>}(m, <c_1, c_2, c_3>, <d_1, d_2, d_3, s_1, s_2, s_3>) =$
$= \{True \text{ iff } (m = Reconstruct(s_1, s_2, s_3)) \land ((\forall_{i=1,2,3}) \mathcal{C}_i.V_{ck_i}(s_i, c_i, d_i))\}$

The tolerance of the share-parallel-concat scheme follows easily from the properties of the secret sharing scheme. Essentially, the shared-parallel-concat is a hybrid or generalization of the copy-parallel-concat and the XOR-parallel-concat constructions. The construction and lemma extend trivially to trapdoor commitment schemes.

Lemma 9. *The Share-Parallel-Concat (sc) construction of $2t+1$ commitment schemes is t-tolerant for* Binding, Relaxed-Binding *and* Hiding *specifications, for every $t \geq 1$.* □

Comment. In most practical commitment schemes, decommitment requires mainly the original message, and the additional decommitment strings d_i are quite short. However, the Share-Parallel-Concat construction uses long decommitment string; specifically, the decommitment includes $<d_1, d_2, d_3, s_1, s_2, s_3>$. Using [37], each share is as long as the message; namely the decommit string is three times as long as the message. This may be substantial overhead for some applications. The scheme we present in the next section avoids this overhead.

Comment. By using robust secret sharing and other tools, [14] achieve tolerant construction for the *CCA2-IND* specification of encryption schemes. However, their construction is very wasteful in the length of the ciphertext, which may rule it unacceptable in most applications; we expect cascade would remain the preferred tolerant construction for encryption in practice (although it is not tolerant for CCA2).

3. Tolerant for the Binding and Relaxed-Binding *specifications of commitment schemes*.
4. Preserving, but NOT tolerant, for the 'confidentiality' specifications Hiding *of commitment schemes, and* CCA1-IND, CPA-IND, CCA2-IND *and* wCCA-IND, *of encryption schemes*.
5. NOT tolerant, for (the 'confidentiality' specifications) OWF.
6. NOT even preserving for the 'confidentiality' specifications AONT.

The quantitative versions of the claims and the (simple) proofs will be included in the final version.

4.3 XOR-Parallel-Concat Construction for Encryption and AONT

Another classical tolerant construction, originally proposed in [1] for encryption schemes, takes two inputs: a message (plaintext) and a random bit string of the same length, and applies one function to the random string, and the other function to the exclusive-OR of the message with the random string. Namely, the simple *XOR-Parallel-Concat (xc)* construction for two keyless functions f, f' is defined as $xc(f, f')[< m, x >] = f||_{xc}f'(< m, x >) = < f(m \oplus x), f'(x) >$; generalization to more than two functions is trivial.

The definition for xc construction for encryption schemes $\mathcal{E}, \mathcal{E}'$, is similar:

- $\mathcal{E}||_{xc}\mathcal{E}'.KG.e(<r, r'>) = < \mathcal{E}.KG.e(r), \mathcal{E}'.KG.e(r') >$
- $\mathcal{E}||_{xc}\mathcal{E}'.KG.d(<r, r'>) = < \mathcal{E}.KG.d(r), \mathcal{E}'.KG.d(r') >$
- $\mathcal{E}||_{xc}\mathcal{E}'.E_{<e,e'>,<r,r',x>}(m) = < \mathcal{E}.E_{e,r}(x), \mathcal{E}'.E_{e',r'}(x \oplus m) >$
- $\mathcal{E}||_{xc}\mathcal{E}'.D_{<d,d'>}(<c, c'>) = \mathcal{E}'.D_{d'}(c') \oplus \mathcal{E}.D_d(c)$

Lemma 8. *The xc construction of encryption schemes is tolerant for specifications* $CCA1 - IND_{ISPUB}$ *and* $CPA - IND_{ISPUB}$, *but does not even preserve* $wCCA2 - IND_{ISPUB}$ *or* $CCA2 - IND_{ISPUB}$. *The simple xc construction is tolerant for specifications* $AONT(k, s, l) \to AONT(k, 2s, s+l)$. □

Comment. The xc construction seems unacceptably wasteful for $AONT$, as the number of bits in the secret part doubles, and the number of bits which the adversary is allowed to expose does not increase (remains $s - l$).

4.4 Share-Parallel-Concat Construction for Tolerant Commitment

In the Share-Parallel-Concat construction, the inputs to each component commitment scheme are shares of the input to the construction. A secret sharing scheme is a pair of algorithms $< Share, Reconstruct >$. The *Share* algorithm accepts a message m as input, and outputs n secret values $s_1, \ldots, s_n$ which we call shares; it is randomized, i.e. it also accepts some random input r. For convenience, let $Share_{i,r}(m, n)$ denote the ith output of *Share* on input m, number of shares n and randomness r. *Reconstruct* is a deterministic algorithm which takes n shares, $s'_1, \ldots, s'_n$, some of which may have the special value $\bot$ (for a missing share), and outputs a message m' (or $\bot$ for failure). The correctness property is that for every message m and randomness r holds $m = Reconstruct(Share_{1,r}(m, n), \ldots, Share_{n,r}(m, n))$.

5 Composing Constructions, and Tolerant Commitment

Often, we may want to combine multiple constructions, e.g. to ensure tolerance to multiple specifications. We restrict our attention to compositions of two constructions. In the first subsection we present two ways to compose the cascade construction (tolerant for *hiding*) and the Copy-Parallel-Concat (*cc*) construction (tolerant for *binding*), resulting in efficient tolerant constructions for commitment schemes (ensuring both hiding and binding specifications). In the second subsection, we generalize these results, by defining a composition as a mapping of (two) constructions, presenting a generic composition based on a combinatorial 'composition structure' variable, and showing that the compositions for commitment schemes are a special case. In particular, we use the general lemmas of the second subsection, to prove the tolerance of the constructions for commitment in the first subsection.

5.1 'Composite' Tolerant Constructions for Commitment Scheme

The Share-parallel-Concat (*sc7*) construction provides tolerant design for commitment schemes, but results in a long decommitment string (three times the original message), which may be problematic for many applications. Can we construct efficient tolerant commitment schemes, with short decommitment (and commitment) strings? In this subsection we show two such constructions, with different tradeoffs, both of which are compositions of the cascade and copy-parallel-concat (*cc*) constructions. This builds on the fact that cascade is tolerant for the hiding specification, and copy-parallel-concat (*cc*) is tolerant for the binding specifications. It therefore makes sense to combine them, e.g. use *four* candidate commitment schemes, C_{11}, C_{12}, C_{21}, and C_{22}, cascading C_{11} and C_{12} and connecting this in parallel to the cascade of C_{21} and C_{22}. We call the result the D construction, after its 'shape', and define it as follows. We use the notation $C_{ij}(m) = C_{ij}.C_{k_{ij},r_{ij}}(m)$, $\mathcal{D}_{ij}(m) = C_{ij}.\mathcal{D}_{k_{ij},r_{ij}}(m)$, $\mathcal{V}_{ij}(m) = C_{ij}.\mathcal{V}_{k_{ij}}(m, c_{ij}, d_{ij})$, $R = <r_{11}, r_{12}, r_{21}, r_{22}>$, $K = <k_{11}, k_{12}, k_{21}, k_{22}> = D.KG(R)$.

- $D.KG(R) = <C_{11}.KG(r_{11}), C_{12}.KG(r_{12}), C_{21}.KG(r_{21}), C_{22}.KG(r_{22})>$
- $D.C_{K,R}(m) = <C_{12}(C_{11}(m)), C_{22}(C_{21}((m))>$
- $D.\mathcal{D}_{K,R}(m)$
  $= <\mathcal{D}_{11}(m), C_{11}(m), \mathcal{D}_{12}(C_{11}(m)), \mathcal{D}_{21}(m), C_{21}(m), \mathcal{D}_{22}(C_{21}(m))>$
- $D.\mathcal{V}_K(m, <c_{12}, c_{22}>, <d_{11}, c_{11}, d_{12}, d_{21}, c_{21}, d_{22}>)$
  $= \mathcal{V}_{11}(m) \wedge \mathcal{V}_{21}(m) \wedge \mathcal{V}_{12}(c_{11}) \wedge \mathcal{V}_{22}(c_{21})$

The D construction is quite efficient in computation times (each operation requires one operation from each of the four candidate commitment schemes), and in the size of the commit and decommit strings (commit size is twice that of the candidate commitment schemes, and decommit size consist of four decommitments plus two commitments). In particular, in the size of the commit and decommit strings, it substantially improves upon the sc construction; this may be important for many applications.

However, the D construction has one significant drawback: it uses four component commitment schemes for 1-tolerance, while sc requires only three candidate schemes for 1-tolerance. We can fix this by using only three commitment schemes, but using each of them twice, by connecting in parallel three cascades of two schemes each; we call this the E construction. The definition is omitted for lack of space (and since it should be obvious - especially after reading the next subsection). We next state the tolerance of the D and E constructions. The proof is given in the next subsection.

Lemma 10. *The D (E) construction of $2t+2$ (respectively, $2t+1$) commitment schemes is t-tolerant for* Binding, Relaxed-Binding *and* Hiding *specifications.*

□

5.2 Composing Arbitrary Constructions

We now generalize the idea of combining multiple constructions, as in the previous subsection, to arbitrary constructions and specifications. We still limit our attention to compositions of two constructions. Such compositions accept as input two constructions c and c' and produce a composite construction denoted $c' \circ_I c$, where I is a mapping of the 'candidate functions' to the constructions. We present few simple, and useful, compositions. First, we need to define the relevant mappings I and the composition for given I.

Let c be a construction of plurality p over F which is t-tolerant for $s \to s'$, and let c' be a construction of plurality p' over F which is t'-tolerant for $s' \to s''$. Let p° denote the plurality of the composition of c and c'; namely the input to the composite construction is an ordered set f of p° functions, $f[i] \in F$. The composite construction $c' \circ_I c$ first applies c to p' sets of p functions each, and then applies c' to the p' resulting functions. The composition is defined by the selection of the p functions input to each of the p' applications of the c construction, namely by a mapping $I : \{1, \ldots, p\} \times \{1, \ldots, p'\} \to \{1, \ldots, p^\circ\}$, where $I_i[j]$ identifies the jth function input to the ith c construction. Given I, the I-composition of c' and c, denoted $c' \circ_I c$, is

$$c' \circ_I c\left(f[1], \ldots, f[p^\circ]\right) = c'\left(c\left(f[I_1(1)], \ldots, f[I_1(p)]\right), \ldots, c\left(f[I_{p'}(1)], \ldots, f[I_{p'}(p)]\right)\right)$$

Consider cascade compositions of threshold-tolerant constructions. The following lemma shows that the security of the I-composition of two threshold-tolerant constructions, depends on a simple combinatorial property of mappings I. Consider mapping $I : \{1, \ldots, p\} \times \{1, \ldots, p'\} \to \{1, \ldots, p^\circ\}$ and some set $T \subseteq \{1, \ldots, p^\circ\}$ (of 'weak input functions'). Let $G_i(I, T) = \{I_i[j] | j = 1, \ldots, p\} - T$, i.e. values $I_i[j]$, for some j, which are *not* in T; think of $G_i(I, T)$ as the 'good selections' of I_i. Let $G(I, T)[t] = \{i \text{ s.t. } |G_i(I, T)| \geq p - t\}$. We say that I is a (good) (t, t', t°)-*threshold-composition-structure* if for every $T \subseteq 1, \ldots, p^\circ$ s.t. $|T| \leq t^\circ$ holds: $|G(I, T)[t]| \geq p' - t'$.

Lemma 11. *Let $I : \{1, \ldots, p\} \times \{1, \ldots, p'\} \to \{1, \ldots, p^\circ\}$ be a (good) (t, t', t°)-threshold-composition-structure. Let c be a construction of plurality p over F which is t-tolerant for $s \to s'$. Let c' be a construction of plurality p' over F which is t'-tolerant for $s' \to s''$. Then $c' \circ_I c$, is a construction of plurality p° over F which is t°-tolerant for $s \to s''$.*

Proof: Consider any set f of p° functions, $f[i] \in F$, and assume that $p^\circ - t$ of them satisfy specification s. Namely, for some set $\{i_j\}$ of $p^\circ - t$ indexes holds $s(f[i_j]) = 1$. We need to prove that for every choice $T \subseteq \{1, \ldots, p^\circ\}$ of up to t° functions in f which do not satisfy s, the function resulting from applying composed construction $c \circ_I c'$ to $\{f[1], \ldots, f[p^\circ]\}$ satisfies $s"$. Namely, we need to prove that $s"(c' \circ_I c(f[1], \ldots, f[p^\circ])) = 1$. Let $f'[1], \ldots, f'[p']$ denote the p' intermediate functions, i.e. $f'[i] = c(f[I_i(1)], \ldots, f[I_i(p)])$; hence $c' \circ_I c(f[1], \ldots, f[p^\circ]) = c'(f'[1], \ldots, f'[p'])$.

If $i \in G(I,T)[t]$, namely $|G_i(I,T)| \geq p - t$, then for at least $p - t$ of the functions $f[I_i(1)], \ldots, f[I_i(p)]$ holds $s(f[I_i(j)]) = True$. Since c is t-tolerant for $s \to s'$, it follows that $s'(f'[i]) = True$, for every $i \in G(I,T)$. Since c' is t'-tolerant for $s' \to s"$, it follows that: $s"(c' \circ_I c(f[1], \ldots, f[p^\circ])) = s"(c'(f'[1], \ldots, f'[p'])) = 1$. □

We now present two simple threshold cascade compositions derived from the above lemma, by presenting two simple composition structures:

- Composition structure $D : 0, 1 \times 0, 1 \to 0, 1, 2, 3$ defined as $D_i[j] = 2i + j$ for $i, j \in 0, 1$.
- Composition structure $E : 0, 1 \times 0, 1, 2 \to 0, 1, 2$ defined as $E_i[j] = i + j \mod 3$ for $i \in 0, 1$, $j \in 0, 1, 2$.

By simply checking the combinatorial definition of (t, t', t°)-threshold-composition-structure we get:

Lemma 12. *D and E are both (good) $(0, 1, 1)$ and $(1, 0, 1)$ threshold-composition-structures.* □

From the two Lemmas, we get:

Lemma 13. *Let c, c'_D, c'_E be constructions of plurality 2, 2 and 3 respectively. If c is t-tolerant for $s \to s'$ where $t \in 0, 1$, and c'_D, c'_E are $(1 - t)$-tolerant for $s' \to s"$, then $c'_D \circ_D c$ and $c'_E \circ_E c$ are both 1-tolerant for $s \to s"$.* □

We can now prove Lemma 10, for $t = 1$ (proof for $t > 1$ is similar).

Proof of Lemma 10 Let c be c_c, i.e. c is the cascade construction; and let c'_D, c'_E be $c_{||}$, i.e. the copy-parallel-concat construction, both for commitment schemes. Notice that $D = c_{D'} \circ_D c$, $E = c_{E'} \circ_E c$. The claim follows immediately from Lemmas 4, 7 and 13. □

6 Conclusions and Open Questions

In this work we presented simple, efficient and practical tolerant constructions for some of the most important and practical cryptographic mechanisms, including encryption, signature/MAC and commitment schemes. For encryption, MAC and signature schemes, we simply proved the security of the (very efficient) 'folklore' constructions; for commitment schemes, we present new constructions which are simple compositions of the folklore cascade and parallel (specifically, copy-parallel-concat) constructions. We also present definitions for tolerant constructions and compositions, and some basic yet useful results regarding compositions of constructions.

We believe that efficient tolerant constructions are an important requirement from practical cryptographic primitives; put differently, we should prefer specifications with an efficient tolerant construction. We presented efficient tolerant constructions for several of the important primitives (and specifications) of modern cryptography. However, for others, we did not find (yet?) a (reasonably efficient) tolerant construction. This calls for additional research, to distinguish between specifications with efficient tolerant design, vs. specifications that do not have an efficient tolerant design (and possibly, to find alternate specifications which are sufficient for most applications/scenarios). For example, XOR-parallel-concat is tolerant for $AONT$, but has substantial loss in parameters (instead of s secret bits out of which l must remain secret, we need $2s$ secret bits out of which $s+l$ must remain secret). We hope follow-up works will investigate efficient tolerant constructions for $AONT$, and for other mechanisms not covered by our results. Similarly, the construction in [14] provides tolerance for encryption schemes under $CCA2\text{-}IND$ attacks, but inefficiently; is there a practical, reasonably efficient construction?

Acknowledgment

I wish to thank Mihir Bellare, Ran Canetti, Shai Halevi, Kath Knobe, Boaz Patt-Shamir, Avi Wigderson and anonymous referees for helpful comments and discussions. This work was supported in part by Israeli Science Foundation grant ISF 298/03-10.5.

References

1. C. A. Asmuth and G. R. Blakley. An efficient algorithm for constructing a cryptosystem which is harder to break than two other cryptosystems. Comp. and Maths. with Appls., 7:447-450, 1981.
2. B. Aiello, M. Bellare, G. Di Crescenzo, and R. Venkatesan, Security amplification by construction: the case of doubly-iterated, ideal ciphers, Proc. of CRYPTO 98.
3. Jee Hea An, Yevgeniy Dodis and Tal Rabin, On the Security of Joint Signature and Encryption, in Theory and Application of Cryptographic Techniques, pp. 83-107, 2002. Also in Advances in Cryptology - EUROCRYPT 2002, volume 2332 of Lecture Notes in Computer Science, pages 83-107. Springer-Verlag, 2002.
4. Ross Anderson, Roger Needham. Robustness Principles for Public Key Protocols. In Proceedings of Int'l. Conference on Advances in Cryptology (CRYPTO 95), Vol. 963 of Lecture Notes in Computer Science, pp. 236-247, Springer-Verlag, 1995.
5. Martin Abadi, Roger Needham. Prudent Engineering Practice for Cryptographic Protocols. IEEE Transactions on Software Engineering, 22, 1 (Jan.), 1996, pp. 6-15.
6. Daniel Bleichenbacher. Chosen ciphertext attacks against protocols based on the RSA encryption standard PKCS#1. In Advances in Cryptology - CRYPTO '98, LNCS 1462, pages 1-12. Springer, 1998.

7. M.Bellare, A.Desai, E.Jokipii, P.Rogaway: A Concrete Security Treatment of Symmetric Encryption, Proceedings of the 38th IEEE Symposium on Foundations of Computer Science (FOCS), pp. 394-403, 1997. Revised version at http://www-cse.ucsd.edu/users/mihir/papers/sym-enc.html.
8. Mihir Bellare, Joe Kilian and Phil Rogaway, The security of cipher block chaining, Journal of Computer and System Sciences, Vol. 61, No. 3, Dec 2000, pp. 362-399. Extended abstract in Advances in Cryptology - Crypto 94 Proceedings, Lecture Notes in Computer Science Vol. 839, Y. Desmedt ed, Springer-Verlag, 1994.
9. Mihir Bellare and Chanathip Namprempre. Authenticated encryption: Relations among notions and analysis of the generic construction paradigm. In T. Okamoto, editor, Asiacrypt 2000, volume 1976 of LNCS, pages 531-545. Springer-Verlag, Berlin Germany, Dec. 2000.
10. Mihir Bellare and Phillip Rogaway, Collision-Resistant Hashing: Towards Making UOWHFs Practical, Extended abstract was in Advances in Cryptology- Crypto 97 Proceedings, Lecture Notes in Computer Science Vol. 1294, B. Kaliski ed, Springer-Verlag, 1997. Full paper available at http://www.cs.ucsd.edu/users/mihir/papers/tcr-hash.html.
11. Joonsang Baek, Ron Steinfeld, and Yuliang Zheng. Formal proofs for the security of signcryption. In David Naccache and Pascal Pailler, editors, 5th International Workshop on Practice and Theory in Public Key Cryptosystems - PKC 2002, pp. 80-98, LNCS Vol. 2274, 2002.
12. Danny Dolev, Cynthia Dwork, and Moni Naor. Non-malleable cryptography. In Proceedings of the 23rd Symposium on Theory of Computing, ACM STOC, 1991.
13. Ivan B. Damgård, Lars Ramkilde Knudsen. Enhancing the Strength of Conventional Cryptosystems, BRICS report RS-94-38, November 1994.
14. Yevgeniy Dodis and Jonathan Katz, Chosen Ciphertext Security of Multiple Encryption, Manuscript, December 2003.
15. Ivan B. Damgård, Torben P. Pedersen, Birgit Pfitzmann: On the Existence of Statistically Hiding Bit Commitment Schemes and Fail-Stop Signatures; Crypto '93, LNCS 773, Springer-Verlag, Berlin 1994, 250-265.
16. Ivan B. Damgård, Torben P. Pedersen, Birgit Pfitzmann: Statistical Secrecy and Multi-Bit Commitments; IEEE Transactions on Information Theory 44/3 (1998) 1143-1151.
17. Shimon Even and Oded Goldreich, On the Power of Cascade Ciphers, ACM Transactions on Computer Systems, Vol. 3, 1985, pp. 108-116.
18. National Institute of Standards and Technology, Federal Information Processing Standards Publication, FIPS Pub 180-1: Secure Hash Standard (SHA-1), April 17, (1995), 14 pages.
19. Oded Goldreich, The Foundations of Cryptography, Volume 1 (Basic Tools), ISBN 0-521-79172-3, Cambridge University Press, June 2001.
20. Oded Goldreich, Fragments of a Chapter on Encryptions Schemes, Extracts from working drafts of Volume 2, The Foundations of Cryptography.
21. Oded Goldreich and Shafi Goldwasser and Silvio Micali "How to Construct Random Functions" Journal of the ACM, 33(4), 1984, 792-807.
22. Oded Goldreich, R. Impagliazzo, L. Levin, R. Venkatesen, D. Zuckerman. "Security preserving amplification of randomness", 31st Annual Symposium on Foundations of Computer Science, IEEE Computer Society Press, (1990), 318-326.
23. Shafi Goldwasser and Silvio Micali. "Probabilistic Encryption," JCSS (28), 1984, 270-299.

24. Philippe Golle, Markus Jakobsson, Ari Juels, and Paul Syverson. Universal re-encryption for mixnets. In Tatsuaki Okamoto, editor, RSA Conference Cryptographers' Track, volume 2964 of LNCS, Springer-Verlag, pages 163–178, San Francisco, California, USA, February 2004.
25. Shai Halevi and Silvio Micali, "Practical and Provably-Secure Commitment Schemes from Collision Free Hashing", in Advances in Cryptology - CRYPTO96, Lecture Notes in Computer Science 1109, Springer-Verlag, 1996, pp. 201-215.
26. Johan Hastad, Rudich Impagliazzo, Leonid A. Levin, and Mike Luby, Construction of a Pseudorandom Generator from any One-Way Function. SIAM Journal on Computing, Vol. 28, No. 4, pp. 1364-1396, 1999.
27. Amir Herzberg and Mike Luby, "Public Randomness in Cryptography", proceedings of CRYPTO 1992, ICSI technical report TR-92-068, October, 1992.
28. Amir Herzberg and Shlomit Pinter, "Composite Ciphers", EE Pub. no. 576, Dept of Electrical Engineering, Technion, Haifa, Israel, Feb. 1986.
29. Hugo Krawczyk, "The Order of Encryption and Authentication for Protecting Communications (or: How Secure Is SSL?)," In Crypto '01, pp. 310-331, LNCS Vol. 2139, J. Kilian ed., Springer-Verlag, 2001.
30. Hugo Krawczyk and Tal Rabin. Chameleon signatures. In Network and Distributed System Security Symposium, pages 143-154. The Internet Society, 2000.
31. Arjen K. Lenstra and Eric R. Verheul. Selecting Cryptographic Key Sizes. Journal of Cryptology: The Journal of the International Association for Cryptologic Research, 14(4):255–293, September 2001.
32. U.M. Maurer and J.L. Massey, Cascade ciphers: the importance of being first, Journal of Cryptology, Vol. 6, No. 1, pp. 55-61, 1993.
33. Alfred J. Menezes, Paul C. van Oorschot, Scott A. Vanstone, Handbook of Applied Cryptography, Section 9.2.6, CRC Press, ISBN 0-8493-8523-7, October 1996. Available online at http://www.cacr.math.uwaterloo.ca/hac/.
34. Moni Naor and Moti Yung, Universal one-way hash functions and their cryptographic applications, Proc. 21st Annual ACM Symposium on Theory of Computing (STOC), 1989, pp. 33–43.
35. T. Dierks, C. Allen, The TLS Protocol: Version 1.0, Network Working Group, Internet Engineering Task Force (IETF). Available online at http://www.ietf.org/rfc/rfc2246.txt.
36. Eric Rescorla. SSL and TLS: Designing and Building Secure Systems. Addison-Wesley, 2000.
37. Adi Shamir, How to share a secret, Comm. of the ACM, 22(11):612-613, 1979.
38. Bruce Schneier, Applied Cryptography, John Wiley and Sons, 1996.
39. Victor Shoup, Using hash functions as a hedge against chosen ciphertext attacks, Adv. in Cryptology – Proc. of Eurocrypt '2000, LNCS 1807, pp. 275-288.
40. Adi Shamir and Yael Tauman. Improved online/online signature schemes. In Joe Killian, editor, Proceedings of Crypto 01, volume 2139 of LNCS, pages 355–367. Springer-Verlag, August 2001.
41. Yuliang Zheng, Digital signcryption or how to achieve cost(signature+encryption) << cost(signature)+cost(encryption), in Advances in Cryptology - CRYPTO'97, Berlin, New York, Tokyo, 1997, vol. 1294 of Lecture Notes in Computer Science, pp. 165–179, Springer-Verlag.
42. Phil R. Zimmerman. The Official PGP User's Guide. MIT Press, Boston, 1995.

Simple Password-Based Encrypted Key Exchange Protocols

Michel Abdalla and David Pointcheval

Departement d'Informatique,
École Normale Supérieure,
45 Rue d'Ulm, 75230 Paris Cedex 05, France
{Michel.Abdalla,David.Pointcheval}@ens.fr
http://www.di.ens.fr/users/{mabdalla,pointche}

Abstract. Password-based encrypted key exchange are protocols that are designed to provide pair of users communicating over an unreliable channel with a secure session key even when the secret key or password shared between two users is drawn from a small set of values. In this paper, we present two simple password-based encrypted key exchange protocols based on that of Bellovin and Merritt. While one protocol is more suitable to scenarios in which the password is shared across several servers, the other enjoys better security properties. Both protocols are as efficient, if not better, as any of the existing encrypted key exchange protocols in the literature, and yet they only require a single random oracle instance. The proof of security for both protocols is in the random oracle model and based on hardness of the computational Diffie-Hellman problem. However, some of the techniques that we use are quite different from the usual ones and make use of new variants of the Diffie-Hellman problem, which are of independent interest. We also provide concrete relations between the new variants and the standard Diffie-Hellman problem.

Keywords: Password, encrypted key exchange, Diffie-Hellman assumptions.

1 Introduction

Background. Keys exchange protocols are cryptographic primitives used to provide a pair of users communicating over a public unreliable channel with a secure session key. In practice, one can find several flavors of key exchange protocols, each with its own benefits and drawbacks. An example of a popular one is the SIGMA protocol [19] used as the basis for the signature-based modes of the Internet Key Exchange (IKE) protocol. The setting in which we are interested in this paper is the 2-party symmetric one, in which every pair of users share a secret key. In particular, we consider the scenario in which the secret key is a password.

PASSWORD-BASED KEY EXCHANGE. Password-based key exchange protocols assume a more realistic scenario in which secret keys are not uniformly distributed over a large space, but rather chosen from a small set of possible values (a four-digit pin, for example). They also seem more convenient since human-memorable passwords are simpler to use than, for example, having additional cryptographic devices capable of storing high-entropy secret keys. The vast majority of protocols found in practice do not account, however, for such scenario and are often subject to so-called *dictionary* attacks. Dictionary attacks are attacks in which an adversary tries to break the security of a scheme by a brute-force method, in which it tries all possible combinations of secret keys in a given small set of values (i.e., the dictionary). Even though these attacks are not very effective in the case of high-entropy keys, they can be very damaging when the secret key is a password since the attacker has a non-negligible chance of winning.

To address this problem, several protocols have been designed to be secure even when the secret key is a password. The goal of these protocols is to restrict the adversary's success to on-line guessing attacks only. In these attacks, the adversary must be present and interact with the system in order to be able to verify whether its guess is correct. The security in these systems usually relies on a policy of invalidating or blocking the use of a password if a certain number of failed attempts has occurred.

ENCRYPTED KEY EXCHANGE. The seminal work in the area of password-based key exchange is the encrypted key exchange (EKE) protocol of Bellovin and Merritt [8]. In their protocol, two users execute an encrypted version of the Diffie-Hellman key exchange protocol, in which each flow is encrypted using the password shared between these two users as the symmetric key. Due to the simplicity of their protocol, several other protocols were proposed in the literature based on it [7, 10, 11, 18, 21], each with its own instantiation of the encryption function. Our protocol is also a variation of their EKE protocol.

MINIMIZING THE USE OF RANDOM ORACLES. One of our main goals is to provide schemes that are simple and efficient, but relying as little as possible on random oracles. Ideally, one would want to completely eliminate the need of random oracles as done in the KOY protocol [17]. However, such protocols tend to be less efficient than those based on the EKE protocol of Bellovin and Merritt [8].

Public information: G, g, p, M, N
Secret information: $pw \in Z_p$

User A User B

$x \stackrel{R}{\leftarrow} Z_p \,;\, X \leftarrow g^x$ $y \stackrel{R}{\leftarrow} Z_p \,;\, Y \leftarrow g^y$
$X^\star \leftarrow X \cdot M^{pw}$ $Y^\star \leftarrow Y \cdot N^{pw}$

$\xrightarrow{X^\star}$
$\xleftarrow{Y^\star}$

$SK_A \leftarrow (Y^\star/N^{pw})^x$ $SK_B \leftarrow (X^\star/M^{pw})^y$

Fig. 1. An insecure password-based key exchange protocol.

To understand the difficulties involved in the design of protocols with few random oracles, let us consider the extreme case of the protocol in Figure 1 in which no random oracles are used. Despite being secure against passive attacks, this protocol can be easily broken by an active adversary performing a man-in-the-middle attack. Such an adversary can easily create two different sessions whose session keys are related in a predictable manner. For instance, an adversary can do so by multiplying $X^\star$ by g^r for a known value r. The relation between the underlying session keys SK_A and SK_B is $SK_B = SK_A \cdot Y^r$. Hence, if the adversary learns the value of these two keys, it can perform an off-line dictionary attack using $Y = (SK_B/SK_A)^{-r}$ and $Y^\star$ to recover the password. Moreover, since the adversary can use arbitrary values for r, we cannot detect such attacks.

PROTECTING AGAINST RELATED KEY ATTACKS. In order to fix the problem in the protocol presented in Figure 1 and prevent the adversary from altering the messages, one may be tempted to use message authentication code (MAC) algorithms for key derivation (e.g., by making the session key equal to $\text{MAC}_{SK_A}(A, B, X^\star, Y^\star, 0)$) or key confirmation (e.g., by computing tags via $\text{MAC}_{SK_A}(A, B, X^\star, Y^\star, 1)$). In fact, this is the approach used by Kobara and Imai in the construction of their password-authenticated key exchange protocol [18]. Unfortunately, this approach does not quite work.

Let us now explain the main problems with using MACs. First, the standard notion of security for MACs does not imply security against related key attacks. Hence, new and stronger security notions are required. Second, such new security notions may have to consider adversaries which are given access to a related-key tag-generation oracle. These are oracles that are capable of generating tags on messages under related keys, where the related key function is also passed as a parameter. This is actually the approach used in [18]. However, it is not clear whether such MACs can even be built. Such security notion, for instance, may completely rule out the possibility of using block-cipher-based MAC algorithms since similar security requirements in the context of block ciphers are known to be impossible to achieve [3]. Perhaps, hash-based MAC algorithms may be able to meet these goals, but that does not seem likely without resorting to random oracles, which would defeat the purpose of using MACs in the first place.

SIMPLE CONSTRUCTIONS. In this paper, we deal with the problem of related-key attacks by using a single instance of a random oracle in the key derivation process. We present two simple constructions, whose only difference to one another is the presence of the password in the key derivation function. The presence of the password in the key derivation function is an important aspect, for example, when one wants to consider extensions to the distributed case, where each server only holds a share of password (see [12]).

Surprisingly, the techniques that we use to prove the security of these two constructions are quite different and so are the exact security results. While we are able to provide a tight security reduction for the scheme which includes the password in the key derivation phase, the same cannot be said about the other scheme, for which we can only prove its security in the non-concurrent scenario. However, the techniques that we use to prove the security of the latter are quite interesting and make use of new variants of the Diffie-Hellman problem.

NEW DIFFIE-HELLMAN ASSUMPTIONS. The new variants of the Diffie-Hellman problem that we introduce are called Chosen-Basis Diffie-Hellman assumptions due to the adversary's capability to choose some of the bases used in the definition of the problem. These assumptions are particular interesting when considered in the context of password-based protocols and we do expect to find applications for them other than the ones in this paper. Despite being apparently stronger than the standard Diffie-Hellman assumptions, we prove that this is *not* the case by providing concrete reductions to the computational Diffie-Hellman problem.

Contributions. In this paper, we address the issue of constructing efficient password-based encrypted key exchange protocols. Our main contributions are as follows.

SIMPLE AND EFFICIENT CONSTRUCTIONS IN RANDOM ORACLE MODEL. In this paper, we propose two new password-based encrypted key exchange protocols, called SPAKE1 and SPAKE2, both of which can be proven secure based on the hardness of the computational Diffie-Hellman problem in the random oracle model. Both protocols are comparable in efficiency to any of the existing EKE protocols, if not more efficient, and they only require one random oracle instance. This is contrast with existing EKE constructions, which require either a larger number of random oracle instances or additional ideal models, such as the ideal cipher model. Moreover, neither SPAKE1 nor SPAKE2 requires full domain hash functions or ideal ciphers onto a group, which are hard to implement efficiently. While one protocol is more suitable to scenarios in which the password is shared across several servers, the other enjoys better security properties.

NEW DIFFIE-HELLMAN ASSUMPTIONS. In proving the security of our protocols, we make use of new variations of the computational Diffie-Hellman assumption, called chosen-basis computational Diffie-Hellman assumptions. These new assumptions are of independent interest and we do expect to find new applications for it other than the ones in this paper. Reductions between the problems underlying the new assumptions and the standard computational Diffie-Hellman assumption are also provided.

Related Work. Password-based authenticated key exchange has been extensively studied in the last few years [4, 9–11, 13–15, 17, 20, 22, 12] with the majority of them being submitted for inclusion in the IEEE P1363.2 standard [16], a standard dealing with the issues of password-authenticated key agreement (e.g. EKE) and password-authenticated key retrieval. With the exception of [13, 14, 17], all of these protocols are only proven secure in the random oracle model.

Perhaps, the related work that is closest to ours is the pretty-simple password-authenticated key exchange protocol of Kobara and Imai [18], whose proof of security is claimed to be in the "standard" model. Their protocol consists of EKE phase that is similar to the one used in our protocols followed by an authentication phase based on message authentication code (MAC) algorithms. However, the security model which they use is different from the standard one and hence their result only applies to their specific model. Moreover, as we pointed out above, their protocol needs a stronger security notion for the MAC algorithm and it is not clear whether such MACs can be built without resorting to random oracles, which would contradict their claims.

Organization. In Section 2, we recall the security model for password-based authenticated key exchange. Next, in Section 3, we present our new variants of the Diffie-Hellman problem and their relations to the computational Diffie-Hellman problem. Section 4 then introduces the first of our password-based encrypted key exchange protocols, called SPAKE1, along with its proof of security. SPAKE1 is in fact based on one of the variants of the Diffie-Hellman problem introduced in Section 3. Our second protocol, SPAKE2, is then presented in Section 4 along with its security claims. The proof of security for SPAKE2 can be found in the full version of this paper [2].

2 Security Model for Password-Based Key Exchange

We now recall the security model for password-based authenticated key exchange of Bellare et al. [4].

PROTOCOL PARTICIPANTS. Each participant in the password-based key exchange is either a client $C \in \mathcal{C}$ or a server $S \in \mathcal{S}$. The set of all users or participants $\mathcal{U}$ is the union $\mathcal{C} \cup \mathcal{S}$.

LONG-LIVED KEYS. Each client $C \in \mathcal{C}$ holds a password pw_C. Each server $S \in \mathcal{S}$ holds a vector $pw_S = \langle pw_S[C] \rangle_{C \in \mathcal{C}}$ with an entry for each client, where $pw_S[C]$ is the transformed-password, as defined in [4]. In this paper, we only consider the symmetric model, in which $pw_S[C] = pw_C$, but they may be different in general. pw_C and pw_S are also called the long-lived keys of client C and server S.

PROTOCOL EXECUTION. The interaction between an adversary $\mathcal{A}$ and the protocol participants occurs only via oracle queries, which model the adversary capabilities in a real attack. During the execution, the adversary may create several instances of a participant. While in a concurrent model, several instances may be active at any given time, only one active user instance is allowed for a given intended partner and password in a non-concurrent model. Let U^i denote the instance i of a participant U and let b be a bit chosen uniformly at random. The query types available to the adversary are as follows:

- *Execute*(C^i, S^j): This query models passive attacks in which the attacker eavesdrops on honest executions between a client instance C^i and a server instance S^j. The output of this query consists of the messages that were exchanged during the honest execution of the protocol.
- *Send*(U^i, m): This query models an active attack, in which the adversary may tamper with the message being sent over the public channel. The output of this query is the message that the participant instance U^i would generate upon receipt of message m.
- *Reveal*(U^i): This query models the misuse of session keys by a user. If a session key is not defined for instance U^i or if a *Test* query was asked to either U^i or to its partner, then return $\perp$. Otherwise, return the session key held by the instance U^i.
- *Test*(U^i): This query tries to capture the adversary's ability to tell apart a real session key from a random one. If no session key for instance U^i is defined, then return the undefined symbol $\perp$. Otherwise, return the session key for instance U^i if $b = 1$ or a random key of the same size if $b = 0$.

NOTATION. Following [5, 6], an instance U^i is said to be *opened* if a query *Reveal*(U^i) has been made by the adversary. We say an instance U^i is *unopened* if it is not *opened*. We say an instance U^i has *accepted* if it goes into an accept mode after receiving the last expected protocol message.

PARTNERING. The definition of partnering uses the notion of session identifications (*sid*). More specifically, two instances U_1^i and U_2^j are said to be partners if the following conditions are met: (1) Both U_1^i and U_2^j accept; (2) Both U_1^i and U_2^j share the same session identifications; (3) The partner identification for U_1^i is U_2^j and vice-versa; and (4) No instance other than U_1^i and U_2^j accepts with a partner identification equal to U_1^i or U_2^j. In practice, the *sid* could be taken to be the partial transcript of the conversation between the client and the server instances before the acceptance.

FRESHNESS. The notion of freshness is defined to avoid cases in which adversary can trivially break the security of the scheme. The goal is to only allow the adversary to ask *Test* queries to *fresh* oracle instances. More specifically, we say an instance U^i is *fresh* if it has *accepted* and if both U^i and its partner are *unopened*.

SEMANTIC SECURITY. Consider an execution of the key exchange protocol P by an adversary $\mathcal{A}$, in which the latter is given access to the *Reveal*, *Execute*, *Send*, and *Test* oracles and asks a single *Test* query to a *fresh* instance, and outputs a guess bit b'. Such an adversary is said to win the experiment defining the semantic security if $b' = b$, where b is the hidden bit used by the *Test* oracle.

Let SUCC denote the event in which the adversary is successful. The ake-**advantage** of an adversary $\mathcal{A}$ in violating the semantic security of the protocol P and the **advantage function** of the protocol P, when passwords are drawn from a dictionary $\mathcal{D}$, are respectively

$$\mathbf{Adv}_{P,\mathcal{D}}^{\mathrm{ake}}(\mathcal{A}) = 2 \cdot \Pr[\,\mathrm{SUCC}\,] - 1 \quad \text{and} \quad \mathbf{Adv}_{P,\mathcal{D}}^{\mathrm{ake}}(t, R) = \max_{\mathcal{A}}\{\, \mathbf{Adv}_{P,\mathcal{D}}^{\mathrm{ake}}(\mathcal{A}) \,\},$$

where maximum is over all $\mathcal{A}$ with time-complexity at most t and using resources at most R (such as the number of queries to its oracles). The definition of time-complexity that we use henceforth is the usual one, which includes the maximum of all execution times in the experiments defining the security plus the code size [1].

3 Diffie-Hellman Assumptions

In this section, we recall the definitions for the computational Diffie-Hellman assumption and introduce some new variants of it, which we use in the proof of security of simple password-based encrypted key exchange protocols. We also present some relations between these assumptions. In doing so, we borrow some of the notations in [1].

3.1 Definitions

NOTATION. In the following, we assume a finite cyclic group G of prime order p generated by an element g. We also call the tuple $\mathbb{G} = (G, g, p)$ the represented group.

Computational Diffie-Hellman Assumption: CDH. The CDH assumption states that given g^u and g^v, where u and v were drawn at random from Z_p, it is hard to compute g^{uv}. Under the computational Diffie-Hellman assumption it might be possible for the adversary to compute something interesting about g^{uv} given g^u and g^v.

This can be defined more precisely by considering an Experiment $\mathbf{Exp}_\mathbb{G}^{\text{cdh}}(\mathcal{A})$, in which we select two values u and v in $\mathbb{Z}_p$, compute $U = g^u$, and $V = g^v$, and then give both U and V to an adversary $\mathcal{A}$. Let Z be the output of $\mathcal{A}$. Then, the Experiment $\mathbf{Exp}_\mathbb{G}^{\text{cdh}}(\mathcal{A})$ outputs 1 if $Z = g^{uv}$ and 0 otherwise. Then, we define *advantage* of $\mathcal{A}$ in violating the CDH assumption as $\mathbf{Adv}_\mathbb{G}^{\text{cdh}}(\mathcal{A}) = \Pr[\mathbf{Exp}_\mathbb{G}^{\text{cdh}}(\mathcal{A}) = 1]$ and the *advantage function* of the group, $\mathbf{Adv}_\mathbb{G}^{\text{cdh}}(t)$, as the maximum value of $\mathbf{Adv}_\mathbb{G}^{\text{cdh}}(\mathcal{A})$ over all $\mathcal{A}$ with time-complexity at most t.

Chosen-Basis Computational Diffie-Hellman Assumption: CCDH. The chosen-basis computational Diffie-Hellman problem is a variation of the CDH problem. It considers an adversary that is given three random elements M, N and X in G and whose goal is to find a triple of values (Y, u, v) such that $u = \text{CDH}(X, Y)$ and $v = \text{CDH}(X/M, Y/N)$. The idea behind this assumption is that the adversary may be able to successfully compute either u (e.g., by choosing $Y = g$ and $u = X$) or v (e.g., by choosing $Y = g \cdot N$ and $v = X/M$), but not both. In fact, as we prove later, solving this problem is equivalent to solving the underlying computational Diffie-Hellman problem in $\mathbb{G}$. We now proceed with the formal definition.

Definition 1 (CCDH). *Let $\mathbb{G} = (G, g, p)$ be a represented group and let $\mathcal{A}$ be an adversary. Consider the following experiment, where M, N and X are elements in G,*

> *Experiment* $\mathbf{Exp}_\mathbb{G}^{\text{ccdh}}(\mathcal{A}, M, N, X)$
> $(Y, u, v) \leftarrow \mathcal{A}(M, N, X)$
> $u' \leftarrow \text{CDH}(X, Y)$
> $v' \leftarrow \text{CDH}(X/M, Y/N)$
> *if $u = u'$ and $v = v'$ then $b \leftarrow 1$ else $b \leftarrow 0$*
> *return b*

Now define the advantage of $\mathcal{A}$ in violating the CCDH assumption with respect to (M, N, X), the advantage of $\mathcal{A}$, and the advantage function of the group, respectively, as follows:

$$\mathbf{Adv}_\mathbb{G}^{\text{ccdh}}(\mathcal{A}, M, N, X) = \Pr[\mathbf{Exp}_\mathbb{G}^{\text{ccdh}}(\mathcal{A}, M, N, X) = 1]$$
$$\mathbf{Adv}_\mathbb{G}^{\text{ccdh}}(\mathcal{A}) = \Pr_{M,N,X}\left[\mathbf{Adv}_\mathbb{G}^{\text{ccdh}}(\mathcal{A}, M, N, X)\right]$$
$$\mathbf{Adv}_\mathbb{G}^{\text{ccdh}}(t) = \max_{\mathcal{A}}\{\mathbf{Adv}_\mathbb{G}^{\text{ccdh}}(\mathcal{A})\},$$

where the maximum is over all $\mathcal{A}$ with time-complexity at most t. ◇

Password-Based Chosen-Basis Computational Diffie-Hellman Assumption: PCCDH. The password-based chosen-basis computational Diffie-Hellman problem is a variation of the chosen-basis computational Diffie-Hellman described above that is more appropriate to the password-based setting. The inputs to the problem and the adversarial goals are also somewhat different in this case so let us explain it.

Let $\mathcal{D} = \{1, \ldots, n\}$ be a dictionary containing n equally likely password values and let $\mathcal{P}$ be a public injective map $\mathcal{P}$ from $\{1, \ldots, n\}$ into $\mathbb{Z}_p$. An example of an admissible map $\mathcal{P}$ is the one in which $\{1, \ldots, n\}$ is mapped into the subset $\{1, \ldots, n\}$ of

Z_p. Now let us consider an adversary that runs in two stages. In the first stage, the adversary is given as input three random elements M, N and X in G as well as the public injective map $\mathcal{P}$ and it outputs a value Y in G. Next, we choose a random password $k \in \{1, \ldots, n\}$ and give it to the adversary. We also compute the mapping $r = \mathcal{P}(k)$ of the password k. The goal of the adversary in this second stage is to output a value K such that $K = \text{CDH}(X/M^r, Y/N^r)$.

Note that an adversary that correctly guesses the password k in its first stage can easily solve this problem by computing $r = \mathcal{P}(k)$ and making, for instance, $Y = g \cdot N^r$ and $K = X/M^r$. Since we assume k to be chosen uniformly at random from the dictionary $\{1, \ldots, n\}$, an adversary that chooses to guess the password and follow this strategy can succeed with probability $1/n$.

The idea behind the password-based chosen-basis computational Diffie-Hellman assumption is that no adversary can do much better than the adversary described above. In fact, as we later prove, this should be the case as long as the computational Diffie-Hellman problem is hard in $\mathbb{G}$. We now proceed with the formal definition.

Definition 2 (PCCDH). *Let $\mathbb{G} = (G, g, p)$ be a represented group and let $\mathcal{A}$ be an adversary. Consider the following experiment, where M and N are elements in G, and $\mathcal{P}$ is a public injective map from $\{1, \ldots, n\}$ into Z_p.*

$$\textbf{Experiment } \text{Exp}_{\mathbb{G},n}^{\text{pccdh}}(\mathcal{A}, M, N, X', \mathcal{P})$$
$$(Y', st) \leftarrow \mathcal{A}(\mathsf{find}, M, N, X', \mathcal{P})$$
$$k \stackrel{R}{\leftarrow} \{1, \ldots, n\} \; ; \; r \leftarrow \mathcal{P}(k)$$
$$(K) \leftarrow \mathcal{A}(\mathsf{guess}, st, k)$$
$$X \leftarrow X'/M^r \; ; \; Y \leftarrow Y'/N^r$$
$$\textit{if } K = \text{CDH}(X, Y) \textit{ then } b \leftarrow 1 \textit{ else } b \leftarrow 0$$
$$\textbf{return } b$$

Now define the advantage of $\mathcal{A}$ in violating the PCCDH assumption with respect to $(M, N, X', \mathcal{P})$, the advantage of $\mathcal{A}$, and the advantage function of the group, respectively, as follows:

$$\textbf{Adv}_{\mathbb{G},n}^{\text{pccdh}}(\mathcal{A}, M, N, X', \mathcal{P}) = \Pr[\,\textbf{Exp}_{\mathbb{G},n}^{\text{pccdh}}(\mathcal{A}, M, N, X', \mathcal{P}) = 1\,]$$
$$\textbf{Adv}_{\mathbb{G},n}^{\text{pccdh}}(\mathcal{A}, \mathcal{P}) = \Pr_{M,N,X'}\left[\,\textbf{Adv}_{\mathbb{G},n}^{\text{pccdh}}(\mathcal{A}, M, N, X', \mathcal{P})\,\right]$$
$$\textbf{Adv}_{\mathbb{G},n}^{\text{pccdh}}(t, \mathcal{P}) = \max_{\mathcal{A}}\{\,\textbf{Adv}_{\mathbb{G},n}^{\text{pccdh}}(\mathcal{A}, \mathcal{P})\,\},$$

where the maximum is over all $\mathcal{A}$ with time-complexity at most t. ◇

Set Password-Based Chosen-Basis Computational Diffie-Hellman: S-PCCDH. The set password-based chosen-basis computational Diffie-Hellman problem (S-PCCDH) is a multidimensional variation of the password-based chosen-basis computational Diffie-Hellman problem described above, in which the adversary is allowed to return not one key but a list of keys at the end of the second stage. In this case, the adversary is considered successful if the list of keys contains the correct value. We now proceed with the formal definition.

Definition 3 (S-PCCDH). *Let $\mathbb{G} = (G, g, p)$ be a represented group and let $\mathcal{A}$ be an adversary. Consider the following experiment, where M and N are elements in G, and $\mathcal{P}$ is a public injective map from $\{1, \ldots, n\}$ into $\mathbb{Z}_p$,*

Experiment $\mathbf{Exp}_{\mathbb{G},n,s}^{\text{s-pccdh}}(\mathcal{A}, M, N, X', \mathcal{P})$
$(Y', st) \leftarrow \mathcal{A}(\text{find}, M, N, X', \mathcal{P})$
$k \stackrel{R}{\leftarrow} \{1, \ldots, n\}$; $r \leftarrow \mathcal{P}(k)$
$(\mathcal{S}) \leftarrow \mathcal{A}(\text{guess}, st, k)$
$X \leftarrow X'/M^r$; $Y \leftarrow Y'/N^r$
if $CDH(X, Y) \in \mathcal{S}$ **and** $|\mathcal{S}| \leq s$ **then** $b \leftarrow 1$ **else** $b \leftarrow 0$
return b

As above, we define the advantage of $\mathcal{A}$ in violating the S-PCCDH assumption with respect to $(M, N, X', \mathcal{P})$, the advantage of $\mathcal{A}$, and the advantage function of the group, respectively, as follows:

$$\mathbf{Adv}_{\mathbb{G},n,s}^{\text{s-pccdh}}(\mathcal{A}, M, N, X', \mathcal{P}) = \Pr[\mathbf{Exp}_{\mathbb{G},n,s}^{\text{s-pccdh}}(\mathcal{A}, M, N, X', \mathcal{P}) = 1]$$

$$\mathbf{Adv}_{\mathbb{G},n,s}^{\text{s-pccdh}}(\mathcal{A}, \mathcal{P}) = \Pr_{M,N,X'}\left[\mathbf{Adv}_{\mathbb{G},n,s}^{\text{s-pccdh}}(\mathcal{A}, M, N, X', \mathcal{P})\right]$$

$$\mathbf{Adv}_{\mathbb{G},n,s}^{\text{s-pccdh}}(t, \mathcal{P}) = \max_{\mathcal{A}}\{\mathbf{Adv}_{\mathbb{G},n,s}^{\text{s-pccdh}}(\mathcal{A}, \mathcal{P})\},$$

where the maximum is over all $\mathcal{A}$ with time-complexity at most t. ◇

3.2 Some Relations

In this section, we first provide two relations between the above problems. The first result is meaningful for small n (polynomially bounded in the asymptotic framework). The second one considers larger dictionaries. Then, we show that these assumptions are implied by the classical computational Diffie-Hellman assumption. Finally, we also prove that the most general assumption is also implied by the classical computational Diffie-Hellman assumption.

Relations Between the PCCDH and CCDH Problems. The following two lemmas present relations between the PCCDH and CCDH problems. The first lemma, whose proof can be found in the full version of this paper [2], is oriented to the case of small dictionaries, for which n is polynomially-bounded. However, if n is large, superpolynomial in the asymptotic framework, or more concretely $n \geq 8/\epsilon$, then one should use the second lemma, whose proof can be easily derived from the proof of the first lemma (see [2]).

Lemma 1. *Let $\mathbb{G} = (G, g, p)$ be a represented group, let n be an integer, and let $\mathcal{P}$ be a public injective map from $\{1, \ldots, n\}$ into $\mathbb{Z}_p$.*

$$\frac{2}{n} \geq \mathbf{Adv}_{\mathbb{G},n}^{\text{pccdh}}(t, \mathcal{P}) \geq \frac{1}{n} + \epsilon \implies \mathbf{Adv}_{\mathbb{G}}^{\text{ccdh}}(2t + 3\tau) \geq \frac{n}{128} \times \epsilon^3.$$

Lemma 2. *Let $\mathbb{G} = (G, g, p)$ be a represented group, let n be an integer, and let $\mathcal{P}$ be a public injective map from $\{1, \ldots, n\}$ into $\mathbb{Z}_p$.*

$$\mathbf{Adv}_{\mathbb{G},n}^{\text{pccdh}}(t, \mathcal{P}) \geq \epsilon \geq \frac{8}{n} \implies \mathbf{Adv}_{\mathbb{G}}^{\text{ccdh}}(2t + 3\tau) \geq \frac{\epsilon^2}{32},$$

where τ denotes the time for an exponentiation in $\mathbb{G}$.

Relation Between the CCDH and CDH Problems. The following lemma, whose proof can be found in the full version of this paper [2], shows that the CCDH and CDH problems are indeed equivalent.

Lemma 3. *Let* $\mathbb{G} = (G, g, p)$ *be a represented group.*

$$\mathbf{Adv}_{\mathbb{G}}^{\mathrm{ccdh}}(t) \leq \mathbf{Adv}_{\mathbb{G}}^{\mathrm{cdh}}(t + 2\tau),$$

where τ denotes the time for an exponentiation in $\mathbb{G}$.

Relation Between the S-PCCDH and CDH Problems. The following lemma, whose proof can be found in the full version of this paper [2], gives a precise relation between the S-PCCDH and CDH problems.

Lemma 4. *Let* $\mathbb{G} = (G, g, p)$ *be a represented group, let n and s be integers, and let $\mathcal{P}$ be a public injective map from $\{1, \ldots, n\}$ into $\mathbb{Z}_p$.*

$$\mathbf{Adv}_{\mathbb{G},n,s}^{\mathrm{s-pccdh}}(t, \mathcal{P}) \geq \frac{1}{n} + \epsilon \Longrightarrow \mathbf{Adv}_{\mathbb{G}}^{\mathrm{cdh}}(t') \geq \frac{n^2 \epsilon^6}{2^{14}} - \frac{2s^4}{p},$$

where $t' = 4t + (18 + 2s)\tau$ and τ denotes the time for an exponentiation in $\mathbb{G}$. More concretely,

$$\mathbf{Adv}_{\mathbb{G},n,s}^{\mathrm{s-pccdh}}(t, \mathcal{P}) \geq \frac{1}{n} + \epsilon \geq \frac{1}{n} \times \left(1 + \frac{8(ns)^{2/3}}{p^{1/6}}\right) \Longrightarrow \mathbf{Adv}_{\mathbb{G}}^{\mathrm{cdh}}(t') \geq \frac{n^2 \epsilon^6}{2^{15}}.$$

4 SPAKE1: A Simple Non-concurrent Password-Based Encrypted Key Exchange

We now introduce our first protocol, SPAKE1, which is a *non-concurrent* password-based encrypted key exchange protocol, based on the multi-dimensional version of password-based chosen-basis computational Diffie-Hellman problem, S-PCCDH.

4.1 Description

SPAKE1 is a variation of the password-based encrypted key exchange protocol of Bellovin and Merritt[8], in which we replace the encryption function $\mathcal{E}_{pw}(.)$ with a simple one-time pad function. More specifically, whenever a user A wants to send the encryption of a value $X \in G$ to a user B, it does so by computing $X \cdot M^{pw}$, where M is an element in G associated with user A and the password pw is assumed to be in $\mathbb{Z}_p$. The session identification is defined as the transcript of the conversation between A and B, and the session key is set to be the hash (random oracle) of the session identification, the user identities, and the Diffie-Hellman key. The password pw is *not* an input to the hash function. The full description of SPAKE1 is given in Figure 2.

CORRECTNESS. The correctness of our protocol follows from the fact that, in an honest execution of the protocol, $K_A = K_B = g^{xy}$.

Public information: G, g, p, M, N, H
Secret information: $pw \in \mathbb{Z}_p$

User A User B

$x \xleftarrow{R} \mathbb{Z}_p ;\; X \leftarrow g^x$ $y \xleftarrow{R} \mathbb{Z}_p ;\; Y \leftarrow g^y$
$X^\star \leftarrow X \cdot M^{pw}$ $Y^\star \leftarrow Y \cdot N^{pw}$

$\xrightarrow{X^\star}$
$\xleftarrow{Y^\star}$

$K_A \leftarrow (Y^\star/N^{pw})^x$ $K_B \leftarrow (X^\star/M^{pw})^y$
$SK_A \leftarrow H(A,B,X^\star,Y^\star,K_A)$ $SK_B \leftarrow H(A,B,X^\star,Y^\star,K_B)$

Fig. 2. SPAKE1: a simple non-concurrent password-based key exchange protocol.

4.2 Security

As Theorem 1 states, our non-concurrent password-based key exchange protocol is secure in the random oracle model as long as we believe that the S-PCCDH problem is hard in $\mathbb{G}$.

Theorem 1. *Let $\mathbb{G}$ be a represent group and let $\mathcal{D}$ be a uniformly distributed dictionary of size $|\mathcal{D}|$. Let SPAKE1 describe the password-based encrypted key exchange protocol associated with these primitives as defined in Figure 2. Then, for any numbers t, q_{start}, q_{send}^A, q_{send}^B, q_H, q_{exe},*

$$\mathbf{Adv}^{\text{ake}}_{\text{SPAKE},\mathcal{D}}(t, q_{\text{start}}, q_{\text{send}}^A, q_{\text{send}}^B, q_H, q_{\text{exe}})$$
$$\leq 2 \cdot (q_{\text{send}}^A + q_{\text{send}}^B) \cdot \mathbf{Adv}^{\text{s-pccdh}}_{\mathbb{G},|\mathcal{D}|,q_H}(t', \mathcal{P}) +$$
$$2 \cdot \left(\frac{(q_{\text{exe}} + q_{\text{send}})^2}{2p} + q_H \, \mathbf{Adv}^{\text{cdh}}_{\mathbb{G}}(t + 2q_{\text{exe}}\tau + 3\tau) \right),$$

where q_H represents the number of queries to the H oracle; q_{exe} represents the number of queries to the Execute oracle; q_{start} and q_{send}^A represent the number of queries to the Send oracle with respect to the initiator A; q_{send}^B represents the number of queries to the Send oracle with respect to the responder B; $q_{\text{send}} = q_{\text{send}}^A + q_{\text{send}}^B + q_{\text{start}}$; $t' = t + O(q_{\text{start}}\tau)$; and τ is the time to compute one exponentiation in $\mathbb{G}$.

Since the S-PCCDH problem can be reduced to the CDH problem according to Lemma 4, it follows that SPAKE1 is a secure non-concurrent password-based key exchange protocol in the random oracle model as long as the CDH problem is hard in $\mathbb{G}$, as stated in Corollary 1.

Corollary 1. *Let $\mathbb{G}$ be a represent group and let $\mathcal{D}$ be a uniformly distributed dictionary of size $|\mathcal{D}|$. Let SPAKE1 describe the password-based encrypted key exchange protocol associated with these primitives as defined in Figure 2. Then, for any numbers t, q_{start}, q_{send}^A, q_{send}^B, q_H, q_{exe},*

$$\mathbf{Adv}^{\text{ake}}_{\text{SPAKE},\mathcal{D}}(t, q_{\text{start}}, q^A_{\text{send}}, q^B_{\text{send}}, q_H, q_{\text{exe}})$$

$$\leq 2 \cdot \left(\frac{q^A_{\text{send}} + q^B_{\text{send}}}{|\mathcal{D}|} + \sqrt[6]{\frac{2^{14}}{|\mathcal{D}|^2} \mathbf{Adv}^{\text{cdh}}_{\mathbb{G}}(t') + \frac{2^{15} q^4_H}{|\mathcal{D}|^2 p}} \right) +$$

$$2 \cdot \left(\frac{(q_{\text{exe}} + q_{\text{send}})^2}{2p} + q_H \, \mathbf{Adv}^{\text{cdh}}_{\mathbb{G}}(t + 2q_{\text{exe}}\tau + 3\tau) \right),$$

where $t' = 4t + O((q_{\text{start}} + q_H)\tau)$ and the other parameters are defined as in Theorem 1.

Proof Idea. Let $\mathcal{A}$ be an adversary against the semantic security of SPAKE. The idea is to use $\mathcal{A}$ to build adversaries for each of the underlying primitives in such a way that if $\mathcal{A}$ succeeds in breaking the semantic security of SPAKE, then at least one of these adversaries succeeds in breaking the security of an underlying primitive. Our proof consists of a sequence of hybrid experiments, starting with the real attack and ending in an experiment in which the adversary's advantage is 0, and for which we can bound the difference in the adversary's advantage between any two consecutive experiments.

Proof of Theorem 1. Our proof uses a sequence of hybrid experiments, the first of which corresponds to the actual attack. For each experiment $\mathbf{Exp}_n$, we define an event Succ_n corresponding to the case in which the adversary correctly guesses the bit b involved in the *Test* query.

Experiment $\mathbf{Exp}_0$. This experiment corresponds to the real attack, which starts by choosing a random password pw. By definition, we have

$$\mathbf{Adv}^{\text{ake}}_{\text{SPAKE}}(\mathcal{A}) = 2 \cdot \Pr[\text{Succ}_0] - 1 \tag{1}$$

Experiment $\mathbf{Exp}_1$. In this experiment, we simulate the *Execute*, *Reveal*, and *Send* oracles as in the real attack (see Figure 4 and Figure 5), after having chosen a random password pw. One can easily see that this experiment is perfectly indistinguishable from the real experiment. Hence,

$$\Pr[\text{Succ}_1] = \Pr[\text{Succ}_0] \tag{2}$$

Experiment $\mathbf{Exp}_2$. In this experiment, we simulate all oracles as in Experiment $\mathbf{Exp}_1$, except that we halt all executions in which a collision occurs in the transcript $((A, X^\star), (B, Y^\star))$. Since either $X^\star$ or $Y^\star$ was simulated and thus chosen uniformly at random, the probability of collisions in the transcripts is at most $(q_{\text{send}} + q_{\text{exe}})^2/(2p)$, according to the birthday paradox. Consequently,

$$|\Pr[\text{Succ}_2] - \Pr[\text{Succ}_1]| \leq \frac{(q_{\text{exe}} + q_{\text{send}})^2}{2p} \tag{3}$$

H oracle — On hash query $H(q)$ (resp. $H'(q)$) for which there exists a record (q, r) in the list Λ_H (resp. Λ_H), return r. Otherwise, choose an element $r \in \{0,1\}^{l_k}$, add the record (q, r) to the list Λ_H (resp. Λ_H), and return r.

Fig. 3. Simulation of random oracles H and H'.

| Send queries | - On a query $Send(A^i, \textbf{start})$, assuming A^i is in the correct state, we proceed as follows:
    if ActiveSessionIndex $\neq 0$ **then abort** $A^{\text{ActiveSessionIndex}}$
    ActiveSessionIndex $= i$
    $\theta \stackrel{R}{\leftarrow} Z_p$; $\Theta \leftarrow g^\theta$; $\Theta^\star \leftarrow \Theta \cdot M^{pw}$
    return $(A, \Theta^\star)$
- On a query $Send(B^i, (A, \Theta^\star))$, assuming B^i is in the correct state, we proceed as follows:
    $\phi \stackrel{R}{\leftarrow} Z_p$; $\Phi \leftarrow g^\phi$; $\Phi^\star \leftarrow \Phi \cdot N^{pw}$
    $K \leftarrow (\Theta^\star/M^{pw})^\phi$
    $SK \leftarrow H(A, B, \Theta^\star, \Phi^\star, K)$
    return $(B, \Phi^\star)$
- On a query $Send(A^i, (B, \Phi^\star))$, assuming A^i is in the correct state, we proceed as follows:
    $K \leftarrow (\Phi^\star/N^{pw})^\theta$
    $SK \leftarrow H(A, B, \Theta^\star, \Phi^\star, K)$
    ActiveSessionIndex $= 0$ |

Fig. 4. Simulation of *Send* oracle query.

Experiment Exp$_3$. In this experiment, we replace the random oracle H by a secret one, for computing SK_A and SK_B for all sessions generated via an *Execute* oracle query. As the following lemma shows, the difference between the current experiment and the previous one is negligible as long as the CDH assumption holds. More precisely, we use a private random oracle H', and in the *Execute* oracle queries, one gets $SK_A, SK_B \leftarrow H'(A, B, \Theta^\star, \Phi^\star)$.

Lemma 5. $|\Pr[\text{SUCC}_3] - \Pr[\text{SUCC}_2]| \leq q_H \cdot \mathbf{Adv}_{\mathbb{G}}^{\text{cdh}}(t + 2q_{\text{exe}}\tau + 3\tau)$.

Proof. The proof of Lemma 5 uses the random self-reducibility of the Diffie-Hellman problem. Indeed, the only way for an execution to be altered by the above modification is if the adversary directly asks for $H(A, B, \Theta^\star, \Phi^\star, K)$, which will output something different from $H'(A, B, \Theta^\star, \Phi^\star)$, the answer of a *Reveal* query. But let us simulate the *Execute* oracle with a Diffie-Hellman instance (A, B), and thus $\Theta \leftarrow A \cdot g^\theta$ and $\Phi \leftarrow B \cdot g^\phi$. As a consequence, the above event means that $K = \text{CDH}(\Theta, \Phi) = \text{CDH}(A, B) \times A^\phi \times B^\theta \times g^{\phi\theta}$ is in the list of the queries asked to H: a random guess leads to $\text{CDH}(A, B)$.

Experiment Exp$_4$. The goal of this experiment is to bound the advantage of the adversary during active attacks, in which the adversary has possibly generated the input of a *Send* oracle. To achieve this goal, we change the simulation of the *Send* oracle so that its output is chosen uniformly at random and independently of the password. The session key associated with each oracle is a bit string of appropriate length chosen uniformly at random and independently of input being provided to the *Send* oracle. The exact simulation of the *Send* oracle is as follows:

- On a query of type $(A^i, \textbf{start})$, we reply with $(A, X^\star = g^{x^\star})$ for a random $x^\star \in Z_p$, if A^i is in the correct state. If another concurrent session already exists for user A, then we also terminate that session.

```
┌─────────────────────────────────────────────────────────────────────┐
│ – On query Reveal(U^i), proceed as follows:                         │
│     if session key SK is defined for instance U^i                   │
│     then return SK,                                                 │
│     else return ⊥.                                                  │
│ – On query Execute(A^i, B^j), proceed as follows:                   │
│     θ ←$ Z_p ;  Θ ← g^θ ;  Θ* ← Θ · M^{pw}                          │
│     φ ←$ Z_p ;  Φ ← g^φ ;  Φ* ← Φ · N^{pw}                          │
│     K ← Θ^φ                                                         │
│     SK_A ← H(A, B, Θ*, Φ*, K) ;  SK_B ← SK_A                        │
│     return ((A, Θ*), (B, Φ*))                                       │
│ – On query Test(U^i), proceed as follows:                           │
│     SK ← Reveal(U^i)                                                │
│     if SK = ⊥ then return ⊥                                         │
│     else                                                            │
│        b ←$ {0,1}                                                   │
│        if b = 0 then SK' ← SK else SK' ←$ {0,1}^{l_k}               │
│        return SK'                                                   │
└─────────────────────────────────────────────────────────────────────┘
```

Fig. 5. Simulation of *Execute*, *Reveal* and *Test* queries.

- On a query of type $(B^i, (A, X^\star))$, we reply with $(B, Y^\star = g^{y^\star})$ for a random $y^\star \in Z_p$ and we set the session key SK_B to $H'(A, B, X^\star, Y^\star)$, if B^i is in the correct state.
- On a query of type $(A^i, (B, Y^\star))$, we set the session key SK_A to $H'(A, B, X^\star, Y^\star)$, if A^i is in the correct state.

As the following lemma shows, the adversary cannot do much better than simply guessing the password when distinguishing the current experiment from the previous one.

Lemma 6. $|\Pr[\text{Succ}_4] - \Pr[\text{Succ}_3]| \leq (q_{\text{send}}^A + q_{\text{send}}^B) \cdot \mathbf{Adv}_{G,|\mathcal{D}|,q_H}^{\text{s-pccdh}}(t', \mathcal{P})$, where $t' = t + O(q_{\text{start}}\tau)$.

Proof. The proof of this lemma is based on a sequence of $q_{\text{send}}^A + q_{\text{send}}^B + 1$ hybrid experiments $\mathbf{Hybrid}_{3,j}$, where j is an index between 0 and $q_{AB} = q_{\text{send}}^A + q_{\text{send}}^B$. Let i be a counter for number of queries of the form $(B^k, (A, X^\star))$ or $(A^k, (B, Y^\star))$. That is, we do not count start queries (we do not increment this counter). We define Experiment $\mathbf{Hybrid}_{3,j}$ as follows:

- If $i \leq j$, then we processes the current *Send* query as in Experiment $\mathbf{Exp}_4$.
- If $i > j$, then we processes the current *Send* query as in Experiment $\mathbf{Exp}_3$.

It is clear from the above definition that experiments $\mathbf{Hybrid}_{3,0}$ and $\mathbf{Hybrid}_{3,q_{AB}}$ are equivalent to experiments $\mathbf{Exp}_3$ and $\mathbf{Exp}_4$, respectively. Now let P_j denote the probability of event Succ in Experiment $\mathbf{Hybrid}_{3,j}$. It follows that $\Pr[\text{Succ}_3] = P_0$ and $\Pr[\text{Succ}_4] = P_{q_{AB}}$. Moreover,

$$|\Pr[\text{Succ}_4] - \Pr[\text{Succ}_3]| \leq \sum_{j=1}^{q_{AB}} |P_j - P_{j-1}|.$$

The lemma will follow easily from bounding $|P_j - P_{j-1}|$. In order to so, consider the following algorithm D_j for the S-PCCDH problem in $\mathbb{G}$.

ALGORITHM D_j. Let $U = g^u$, $V = g^v$, and $W = g^w$ be random elements in G and let $\mathcal{P}$ be any injective map from $\{1, \ldots, n\}$ into Z_p. D_j starts running $\mathcal{A}$, simulating all its oracles. The *Reveal*, *Execute*, and *Test* oracles are simulated as in Experiment $\mathbf{Exp}_3$. The *Send* oracle is simulated as follows, Let i be the index of the current *Send* query.

- If the *Send* query is of the form $(A^k, \mathbf{start})$,
  - if $i \leq j$, then D_j replies with $(A, X^\star = Wg^{x^\star})$ for a random $x^\star \in \mathsf{Z}_p$, if A^k is in the correct state. If another concurrent session already exists for user A, then D_j also terminates that session.
  - if $i > j$, then D_j processes it as in Experiment $\mathbf{Exp}_3$.
- If the query is of the form $(B^k, (A, X^\star))$,
  - if $i < j$, then D_j processes it as in Experiment $\mathbf{Exp}_4$.
  - if $i = j$, then D_j replies with $(B, Y^\star = W)$. It also returns $(st, Y' = X^\star)$ as the output of its find stage and waits for the input (st, k) of the guess stage. It then sets the password pw shared between A and B to $\mathcal{P}(k)$ and the session key SK_B to $H(A, B, X^\star, Y^\star, K_B)$, where $K_B = (X^\star/V^{pw})^{w-u\,pw}$. We note that st should contain all the necessary information for D_j to continue the execution of $\mathcal{A}$ and the simulation of its oracles in the guess stage. Let this be *Case B*.
  - if $i > j$, then D_j processes it as in Experiment $\mathbf{Exp}_3$.
- If the *Send* query is of the form $(A^k, (B, Y^\star))$,
  - if $i < j$, then D_j processes it as in Experiment $\mathbf{Exp}_4$.
  - if $i = j$, and A^k is in the correct state, then it returns $(st, Y' = Y^\star)$ as the output of its find stage and waits for the input (st, k) of the guess stage. Then, it sets the password pw shared between A and B to $\mathcal{P}(k)$ and the session key SK_A to $H(A, B, X^\star, Y^\star, K_A)$, where $K_A = (Y^\star/V^{pw})^{w+x^\star-u\,pw}$. Let this be *Case A*.
  - if $i > j$, then D_j processes it as in Experiment $\mathbf{Exp}_3$.

Let K be the part of the input of H that is not present in H' and let $K_1, \ldots, K_{q_H}$ be the list of all such elements. When in Case A, D_j sets $K'_i = K_i/(Y'/V^{pw})^{x^\star}$ for $i = 1, \ldots, q_H$, where $x^\star$ is the value used to compute $X^\star$ in the crucial query. When in Case B, D_j simply sets $K'_i = K_i$. Finally, D_j outputs $K'_1, \ldots, K'_{q_H}$.

We note that in the above, the password is only defined at the j-th step and it is not used before that. Due to the non-concurrency, we do not need to know the password for simulating flows in Experiment $\mathbf{Exp}_4$. We only need it in Experiment $\mathbf{Exp}_3$.

Using the knowledge of u, v, and w in the above, it is clear that the processing of the *Send* queries matches that of Experiment $\mathbf{Hybrid}_{3,j-1}$. However, in the actual description of the S-PCCDH problem, we do not have access to these values. For this reason, the actual algorithm D_j replaces the random oracle H by a secret random oracle H' in the computation of SK_A and SK_B during the processing of the j-th *Send* query. More precisely, it computes SK_A and SK_B as $H'(A, B, X^\star, Y^\star)$. Moreover, we note that in this new scenario, the processing of the *Send* queries matches that of Experiment $\mathbf{Hybrid}_{3,j}$.

PROBABILITY ANALYSIS. Let ASKH represent the event in which the adversary asks for $H(A, B, X^\star, Y^\star, K)$, where $K = \text{CDH}(X^\star/U^{pw}, Y^\star/V^{pw})$ and either $X^\star$ or $Y^\star$ is involved in the crucial j-th query. We first observe that experiments $\mathbf{Hybrid}_{3,j-1}$ and $\mathbf{Hybrid}_{3,j}$ are identical if event ASKH does not happen. Therefore, it follows that the probability difference $|P_j - P_{j-1}|$ is at most $\Pr[\text{ASKH}]$.

However, whenever event ASKH happens, we know that the list of queries asked to H contains the key $K = \text{CDH}(X^\star/U^{pw}, Y^\star/V^{pw})$ involved in the crucial query, and thus D_j will be able to successfully use $\mathcal{A}$ to help it solve the S-PCCDH problem. This is because K_A (Case A) or K_B (Case B) can be used to compute the solution $\text{CDH}(W/U^{pw}, Y'/V^{pw})$ for the S-PCCDH problem as follows:

$$K_A = \text{CDH}(Y^\star/V^{pw}, Wg^{x^\star}/U^{pw}) = \text{CDH}(W/U^{pw}, Y'/V^{pw}) \times (Y'/V^{pw})^{x^\star}$$
$$K_B = \text{CDH}(X^\star/V^{pw}, W/U^{pw}) = \text{CDH}(W/U^{pw}, Y'/V^{pw})$$

Therefore, the list of candidates $K'_1, \ldots, K'_{q_H}$ that D_j outputs should contain the solution for the S-PCCDH problem whenever ASKH happens. Hence, $\Pr[\text{ASKH}]$ is less than or equal to the success probability of D_j. The lemma follows easily from the fact that D_j has time-complexity at most t'. □

5 SPAKE2: A Simple Concurrent Password-Based Encrypted Key Exchange

We now introduce our second protocol, SPAKE2, which is a *concurrent* password-based encrypted key exchange protocol, based on the computational Diffie-Hellman problem, CDH.

5.1 Description

SPAKE2 is a also variation of the password-based encrypted key exchange protocol of Bellovin and Merritt[8] and is almost exactly like SPAKE1. The only difference between the two is in the key derivation function, which also includes the password pw. More specifically, the session key in SPAKE2 is set to be the hash (random oracle) of the session identification, the user identities, the Diffie-Hellman key, and the password. In other words, $SK \leftarrow H(A, B, X^\star, Y^\star, pw, K)$. The session identification is still defined as the transcript of the conversation between A and B.

5.2 Security

As the following theorem states, our concurrent password-based key exchange protocol is secure in the random oracle model as long as the CDH problem is hard in G. The proof of Theorem 2 can be found in the full version of this paper [2].

Theorem 2. *Let $\mathbb{G}$ be a represent group and let $\mathcal{D}$ be a uniformly distributed dictionary of size $|\mathcal{D}|$. Let SPAKE2 describe the password-based encrypted key exchange protocol associated with these primitives as defined in Section 5.1. Then, for any numbers t, q_{start}, q_{send}^A, q_{send}^B, q_H, q_{exe},*

$$\mathbf{Adv}^{\text{ake}}_{\text{SPAKE2},\mathcal{D}}(t, q_{\text{start}}, q^A_{\text{send}}, q^B_{\text{send}}, q_H, q_{\text{exe}})$$
$$\leq 2 \cdot \left(\frac{q^A_{\text{send}} + q^B_{\text{send}}}{n} + \frac{(q_{\text{exe}} + q_{\text{send}})^2}{2p} \right) +$$
$$2 \cdot \left(q_H \, \mathbf{Adv}^{\text{cdh}}_{\mathbb{G}}(t + 2q_{\text{exe}}\tau + 3\tau) + q_H^2 \, \mathbf{Adv}^{\text{cdh}}_{\mathbb{G}}(t + 3\tau) \right),$$

where the parameters are defined as in Theorem 1.

Acknowledgments

The work described in this document has been supported in part by the European Commission through the IST Programme under Contract IST-2002-507932 ECRYPT. The information in this document reflects only the author's views, is provided as is and no guarantee or warranty is given that the information is fit for any particular purpose. The user thereof uses the information at its sole risk and liability

References

1. M. Abdalla, M. Bellare, and P. Rogaway. The oracle Diffie-Hellman assumptions and an analysis of DHIES. In D. Naccache, editor, *CT-RSA 2001*, volume 2020 of *LNCS*, pages 143–158. Springer-Verlag, Apr. 2001.
2. M. Abdalla and D. Pointcheval. Simple password-based encrypted key exchange protocols. Full version of current paper. Available from authors' web pages.
3. M. Bellare and T. Kohno. A theoretical treatment of related-key attacks: RKA-PRPs, RKA-PRFs, and applications. In E. Biham, editor, *EUROCRYPT 2003*, volume 2656 of *LNCS*, pages 491–506. Springer-Verlag, May 2003.
4. M. Bellare, D. Pointcheval, and P. Rogaway. Authenticated key exchange secure against dictionary attacks. In B. Preneel, editor, *EUROCRYPT 2000*, volume 1807 of *LNCS*, pages 139–155. Springer-Verlag, May 2000.
5. M. Bellare and P. Rogaway. Entity authentication and key distribution. In D. R. Stinson, editor, *CRYPTO'93*, volume 773 of *LNCS*. Springer-Verlag, Aug. 1994.
6. M. Bellare and P. Rogaway. Provably secure session key distribution — the three party case. In *28th ACM STOC*, pages 57–66. ACM Press, May 1996.
7. M. Bellare and P. Rogaway. The AuthA protocol for password-based authenticated key exchange. Contributions to IEEE P1363, Mar. 2000.
8. S. M. Bellovin and M. Merritt. Encrypted key exchange: Password-based protocols secure against dictionary attacks. In *1992 IEEE Symposium on Security and Privacy*, pages 72–84. IEEE Computer Society Press, May 1992.
9. V. Boyko, P. MacKenzie, and S. Patel. Provably secure password-authenticated key exchange using Diffie-Hellman. In B. Preneel, editor, *EUROCRYPT 2000*, volume 1807 of *LNCS*, pages 156–171. Springer-Verlag, May 2000.
10. E. Bresson, O. Chevassut, and D. Pointcheval. Security proofs for an efficient password-based key exchange. In *ACM CCS 03*. ACM Press, Oct. 2003.
11. E. Bresson, O. Chevassut, and D. Pointcheval. New security results on encrypted key exchange. In F. Bao, R. Deng, and J. Zhou, editors, *PKC 2004*, volume 2947 of *LNCS*, pages 145–158. Springer-Verlag, Mar. 2004.
12. M. Di Raimondo and R. Gennaro. Provably secure threshold password-authenticated key exchange. In E. Biham, editor, *EUROCRYPT 2003*, volume 2656 of *LNCS*, pages 507–523. Springer-Verlag, May 2003.

13. R. Gennaro and Y. Lindell. A framework for password-based authenticated key exchange. In E. Biham, editor, *EUROCRYPT 2003*, volume 2656 of *LNCS*, pages 524–543. Springer-Verlag, May 2003. http://eprint.iacr.org/2003/032.ps.gz.
14. O. Goldreich and Y. Lindell. Session-key generation using human passwords only. In J. Kilian, editor, *CRYPTO 2001*, volume 2139 of *LNCS*, pages 408–432. Springer-Verlag, Aug. 2001. http://eprint.iacr.org/2000/057.ps.gz.
15. S. Halevi and H. Krawczyk. Public-key cryptography and password protocols. In *ACM Transactions on Information and System Security*, pages 524–543. ACM, 1999.
16. IEEE draft standard P1363.2. Password-based public key cryptography. http://grouper.ieee.org/groups/1363/passwdPK, May 2004. Draft Version 15.
17. J. Katz, R. Ostrovsky, and M. Yung. Efficient password-authenticated key exchange using human-memorable passwords. In B. Pfitzmann, editor, *EUROCRYPT 2001*, volume 2045 of *LNCS*, pages 475–494. Springer-Verlag, May 2001.
18. K. Kobara and H. Imai. Pretty-simple password-authenticated key-exchange under standard assumptions. *IEICE Transactions*, E85-A(10):2229–2237, Oct. 2002. Also available at http://eprint.iacr.org/2003/038/.
19. H. Krawczyk. SIGMA: The "SIGn-and-MAc" approach to authenticated Diffie-Hellman and its use in the IKE protocols. In D. Boneh, editor, *CRYPTO 2003*, volume 2729 of *LNCS*, pages 400–425. Springer-Verlag, Aug. 2003.
20. P. MacKenzie, S. Patel, and R. Swaminathan. Password-authenticated key exchange based on RSA. In T. Okamoto, editor, *ASIACRYPT 2000*, volume 1976 of *LNCS*, pages 599–613. Springer-Verlag, Dec. 2000.
21. P. D. MacKenzie. The PAK suite: Protocols for password-authenticated key exchange. Contributions to IEEE P1363.2, 2002.
22. P. D. MacKenzie, T. Shrimpton, and M. Jakobsson. Threshold password-authenticated key exchange. In M. Yung, editor, *CRYPTO 2002*, volume 2442 of *LNCS*, pages 385–400. Springer-Verlag, Aug. 2002.

Hard Bits of the Discrete Log
with Applications to Password Authentication

Philip Mackenzie* and Sarvar Patel

Bell Laboratories, Lucent Technologies
{philmac,sarvar}@lucent.com

Abstract. Assuming the intractability of solving the discrete logarithm with short exponent problem, it was recently shown that the trailing $n - \omega(\log n)$ bits of the discrete logarithm modulo an n-bit safe prime p are simultaneously hard. However, the question of hardness of the leading bits was left open. In this paper we show that the leading $n - \omega(\log n)$ bits are also simultaneously hard, or equivalently that the distribution of $g^s \bmod p$, where g is a generator of $\mathbb{Z}_p^*$ and s is a uniformly chosen short exponent of $\omega(\log n)$ bits, is indistinguishable from the uniform distribution on $\mathbb{Z}_p^*$. We further show that this result implies the security of a short exponent version of PAK, a password-authenticated key exchange protocol that protects against offline dictionary attacks.

1 Introduction

Informally, a *hard bit* of a one-way function is a bit which is as hard to compute as it is to invert the one way function. Equivalently, for a one-way function f, we say that a bit of f^{-1} is hard if computing that bit is as hard as computing f^{-1} itself. Blum and Micali [5] introduced the concept of hard bits of a one-way function and showed that for the discrete logarithm modulo a prime p, the unbiased most significant bit (i.e., the bit denoting whether the discrete log is greater or less than $\frac{p-1}{2}$) is a hard bit. The question of hard bits is not only theoretically interesting, but is also interesting in practice, for example, in speeding up pseudorandom bit generation.

Our knowledge about hard bits of one-way functions has steadily grown, both for hard bits of arbitrary one-way functions and hard bits of specific one-way functions. Hard bits of arbitrary one-way functions [9, 19, 20] have the advantage that they can be used without considering the specifics of the one-way function. However, hard bits of specific functions have the advantage that they may be easier to compute and/or use. For instance, they may simply be some of the bits of the argument to the one-way function. Informally, a group of bits in the argument of the one way function are simultaneously hard if computing any information about them given the output of the one-way function is as hard as inverting the one-way function. For arbitrary one-way functions it is not possible to have more than $O(\log n)$ simultaneous hard bits, while for specific one-way

---

* Current Address: DoCoMo USA Labs, San Jose, CA, philmac@docomolabs-usa.com

A.J. Menezes (Ed.): CT-RSA 2005, LNCS 3376, pp. 209–226, 2005.
© Springer-Verlag Berlin Heidelberg 2005

functions it is possible (see [27] for a survey of results). In regards to simultaneous hard bits of specific one-way functions, two questions have been investigated: (1) how many bits does the one way function hide and (2) where are they.

It has been shown that the discrete logarithm modulo a prime hides at least $O(\log n)$ bits. Peralta [24] showed that the $O(\log n)$ least significant bits are simultaneously hard, and Long and Wigderson [16] showed that the $O(\log n)$ unbiased most significant bits are simultaneously hard. For almost all primes p, Hastad and Naslund [11] were able to show the individual hardness of all bits, as well as the simultaneous hardness of blocks of $O(\log n)$ bits. For the discrete logarithm modulo a composite, Hastad, Schrift and Shamir [12] showed that $\frac{n}{2} + O(\log n)$ bits are simultaneously hard, and that both the lower half and the upper half of the bits are simultaneously hard, assuming factoring Blum integers is hard.

For the discrete logarithm modulo a safe prime[1] p, Patel and Sundaram [23] showed that the trailing[2] $n - \omega(\log n)$ bits are simultaneously hard assuming the intractability of solving the discrete log with short exponent (DLSE) problem. Obviously this is the maximum possible number of simultaneously hard bits. However, they do not say anything about the simultaneous hardness of the leading bits, and in fact a direct use of their proof technique does not seem to lead to a similar hardness result on the leading bits.

Our Contributions I: In the first part of this paper we are able to show that the leading $n - \omega(\log n)$ bits of the discrete logarithm modulo a safe prime are simultaneously hard assuming the intractability of the DLSE problem. We prove this result by proceeding in two steps. In the first step we prove that showing the simultaneous hardness of the leading bits is equivalent to showing that the distribution of $g^s \bmod p$, where s is a uniformly chosen short exponent of $\omega(\log n)$ bits, is indistinguishable from the uniform distribution on $\mathbb{Z}_p^*$. In the second step we show that a distinguisher between the two distributions above can be turned into a DLSE solver. Since we assume that solving DLSE is intractable, there cannot be such a distinguisher, and hence the leading $n - \omega(\log n)$ bits are simultaneously hard.

The second step is similar to a result in Gennaro [8], except that the *leading* $n - \omega(\log n)$ bits are zero in our case, whereas the *trailing* $n - \omega(\log n)$ bits are zero in his case (except for the lowest bit). Both cases deal with exponents that contain only a small number of non-zero bits, and thus, in some sense, both results deal with "short exponents" (although his short exponents are biased since they contain the leading bits). In his case the result follows by a relatively straightforward reduction to the hardness result of [23]. On the other hand, our case is more difficult, since our exponent s is unbiased, and can cause overflow in the exponent if it is shifted to the leading bits.

We note that analogously to [8], our result can be naturally translated into a simple PRG where at each step the leading bits are output (as opposed to the

---

[1] A prime p is a *safe prime* if $p = 2q + 1$ where q is also prime.
[2] We use the term *trailing bits* to denote the least significant bits and *leading bits* to denote the most significant bits.

trailing bits in [8]). However, in our case one has to take care of the bias of the leading bits, either by dropping them or using a technique such as, e.g., [13] to convert them into unbiased bits.

Application: Password Authenticated Key Exchange: Classical entity authentication and key exchange protocols (e.g., [2]) are not necessarily secure when the two communicating entities only share a password and not a strong secret key. This is because although the data the entities transmit cannot be used to determine their shared strong secret key, it may be used to verify a guess at their key, and thus the protocol would be susceptible to an off-line dictionary attack. In particular, humans are prone to choose passwords with low entropy, and hence it is possible for an attacker to compile a dictionary of likely passwords and do a off-line search by testing candidate passwords from the dictionary against the overheard data.

Lomas et al. [15] presented the first protocols which protected against off-line dictionary attacks, but they were not strictly password only protocols because they assumed that the client had the server's public key. The EKE protocol [3] was the first password-only authenticated key exchange protocol that tried to protect against off-line dictionary attacks. Subsequently, many password only protocols were presented with heuristic arguments, and a large number of them were shown to be insecure. Recently, some protocols were formally proven secure [1, 6, 10, 18].

In particular, the PAK protocol [6] was proven to be as secure as Diffie-Hellman [7] in the random oracle model. The proof of security fails, however, if the parties in the protocol use short exponents, even if one assumes the hardness of the Diffie-Hellman with short exponents (DHSE) problem. (The basic reason is that the proof relies on the fact that the values transmitted by the parties have exactly the same distribution for any password, which is not the case if short exponents are used.) However, there are existing systems which use unauthenticated Diffie-Hellman with short exponents to perform key exchange [22], and a short exponent version of PAK could be an attractive enhancement to these protocols since (1) this would not create a change in the underlying messaging, (2) this would cause the least possible change in the code, and, most importantly, (3) this would not cause any significant increase in the computational complexity.

Our Contributions II: In the second part of this paper we show that we can use our result above on simultaneous hard bits (and more specifically, the result that the distributions $\mathbb{Z}_p^*$ and $\mathbb{Z}_p^*$ restricted to short exponents are indistinguishable assuming the intractability of the DLSE problem) to prove the security of a short exponent version of PAK based on the hardness of the DHSE problem. Initially, one may think this would follow directly. However, for technical reasons this does not seem to be the case, and in fact, for our proof of security to go through the exponents used in PAK need to be chosen from a larger range than the range specified in the DHSE problem. As a concrete example, assuming the DHSE problem is intractable for 256-bit exponents, we would need to use 384-bit exponents in PAK for the proof of security to go through (with a roughly equivalent level of security). Still, this is the most efficient version of PAK (over $\mathbb{Z}_p^*$) that has been proven secure.

2 Preliminaries

2.1 Notation

The integer x can be represented as the bit string $x = x_n, \ldots, x_1$ where $x = x_n \cdot 2^{n-1} + x_{n-1} \cdot 2^{n-2} + \ldots + x_2 \cdot 2^1 + x_1 \cdot 2^0$. $\{0,1\}^n$ denotes the set of n-bit strings and 0^n denotes the string of n zeroes. x_i^j is the binary substring $x_j, \ldots, x_i$. $x \xleftarrow{R} S$ denotes randomly choosing an element x from the set S.

2.2 Discrete Logarithm

For n-bit prime p and generator $g \in \mathbb{Z}_p^*$, the modular exponentiation function $f_{p,g}(x) = g^x \bmod p$ is believed to be a one-way function. It can be computed in polynomial time but it is believed that there is no deterministic or randomized algorithm which can compute the discrete logarithm in time polynomial in n. The index calculus method is the best algorithm known to compute the discrete logarithm, however, even this is infeasible today if p is moderately large, say 1024 bits. The complexity of index calculus is subexponential but not polynomial in n.

Now we define some notation. Let p be an n-bit prime and g be a generator of $\mathbb{Z}_p^*$. We define the discrete log function DL as follows: $\mathsf{DL}(p,g,y) = x$ iff $g^x \equiv y \bmod p$, where $y \in \mathbb{Z}_p^*$. We will define the Diffie-Hellman function DH as follows: $\mathsf{DH}(p,g,x,y) = z$ iff $x = g^a \bmod p$, $y = g^b \bmod p$, and $z = g^{ab} \bmod p$. (For the remainder of the paper, we will omit "mod p" when it is obvious from context. We will also omit p and g from the input to the functions DL and DH when they are obvious from context.)

Discrete Logarithm with Short Exponent Problem. Although performing modular exponentiation can be done in polynomial time, it is still reasonably costly today in terms of computational complexity, especially on computationally bounded devices such as PDAs or cell phones. Even exponentiation modulo a 1024-bit prime can be demanding on such devices. Thus in some protocols that use modular exponentiations, the exponent is restricted to be c bits (e.g., $c = 160$ bits) since this requires fewer multiplications [21]. The security of the protocols then is restricted by the hardness of the discrete logarithm with short exponent (DLSE) problem [23]. The best algorithms known to solve the DLSE problem are square root time algorithms requiring $2^{c/2}$ steps due to Shanks [14] and Pollard [25]. (The index calculus method does not depend on the size of the exponent and its efficiency is related to the size of the whole group.) Then, for instance, to provide 80 bits of security, c should be at least 160 bits.

The DLSE problem has been investigated in the context of the Diffie-Hellman key agreement by van Oorschot and Wiener [21], and they give attacks to discover information about the exponent if random primes are used. They suggest that using safe primes is sufficient to block such attacks. For this reason, we will only consider safe primes for the remainder of the paper (as in [23]).

Now we define some notation. Let g be a generator of $\mathbb{Z}_p^*$. Let $\mathsf{SExp}_{p,g,c} = \{g^x : x \in \{0,1\}^c\}$. We define the short exponent discrete log function $\mathsf{DLSE}(p,g,y,c) = x$ iff $\mathsf{DL}(p,g,y) = x$, but where the domain of y is restricted to $\mathsf{SExp}_{p,g,c}$. We define the short exponent Diffie-Hellman function $\mathsf{DHSE}(p,g,c,x,y) = z$ iff there are values $a,b \in \{0,1\}^c$ such that $\mathsf{DL}(p,g,x) = a$, $\mathsf{DL}(p,g,y) = b$ and $\mathsf{DL}(p,g,z) = ab$.

Let $\mathcal{G}(1^n)$ output an n-bit safe prime p and a generator g for $\mathbb{Z}_p^*$. Then for any algorithm $\mathcal{A}$, let

$$\mathsf{Succ}^{\mathsf{DLSE}}_{p,g,c}(\mathcal{A}) \stackrel{\mathrm{def}}{=} \Pr[y \stackrel{R}{\leftarrow} \mathsf{SExp}_{p,g,c}; x \leftarrow \mathcal{A}(p,g,c,y) : y = g^x]$$

Let $\mathsf{Succ}^{\mathsf{DLSE}}_{n,c}(\mathcal{A}) \stackrel{\mathrm{def}}{=} \mathrm{Ex}[(p,g) \leftarrow \mathcal{G}(1^n) : \mathsf{Succ}^{\mathsf{DLSE}}_{p,g,c}(\mathcal{A})]$. The DLSE_c assumption states that for any probabilistic polynomial-time algorithm $\mathcal{A}$, $\mathsf{Succ}^{\mathsf{DLSE}}_{n,c}(\mathcal{A})$ is negligible.

For any algorithm $\mathcal{A}$, let

$$\mathsf{Succ}^{\mathsf{DHSE}}_{p,g,c}(\mathcal{A}) \stackrel{\mathrm{def}}{=} \Pr[x,y \stackrel{R}{\leftarrow} \mathsf{SExp}_{p,g,c}; z \in \mathcal{A}(p,g,c,x,y) : z = \mathsf{DH}(x,y)]$$

Let $\mathsf{Succ}^{\mathsf{DHSE}}_{n,c}(\mathcal{A}) \stackrel{\mathrm{def}}{=} \mathrm{Ex}[(p,g) \leftarrow \mathcal{G}(1^n) : \mathsf{Succ}^{\mathsf{DHSE}}_{p,g,c}(\mathcal{A})]$. The DHSE_c assumption states that for any probabilistic polynomial-time algorithm $\mathcal{A}$, $\mathsf{Succ}^{\mathsf{DHSE}}_{n,c}(\mathcal{A})$ is negligible.

Let

$$\mathsf{SE}_{c,n} = \{(p,g) \leftarrow \mathcal{G}(1^n); v \stackrel{R}{\leftarrow} \mathsf{SExp}_{p,g,c} : v\}$$

and

$$\mathsf{U}_n = \{(p,g) \leftarrow \mathcal{G}(1^n); v \stackrel{R}{\leftarrow} \mathbb{Z}_p^* : v\}.$$

Let SE_c and U be their respective distribution ensembles, parameterized by n. We define SEdist_c as the problem of distinguishing SE_c from U. For any algorithm $\mathcal{A}$, let

$$\mathsf{Adv}^{\mathsf{SEdist}}_{p,g,c}(\mathcal{A}) \stackrel{\mathrm{def}}{=} |\Pr[x \stackrel{R}{\leftarrow} \mathsf{SExp}_{p,g,c}; 1 = \mathcal{A}(p,g,c,x)] - \Pr[x \stackrel{R}{\leftarrow} \mathbb{Z}_p^*; 1 = \mathcal{A}(p,g,c,x)]|$$

Let $\mathsf{Adv}^{\mathsf{SEdist}}_{n,c}(\mathcal{A}) \stackrel{\mathrm{def}}{=} \mathrm{Ex}[(p,g) \leftarrow \mathcal{G}(1^n) : \mathsf{Adv}^{\mathsf{SEdist}}_{p,g,c}(\mathcal{A})]$. The SEdist_c assumption states that for any probabilistic polynomial-time algorithm $\mathcal{A}$, $\mathsf{Adv}^{\mathsf{SEdist}}_{n,c}(\mathcal{A})$ is negligible.

3 Simultaneous Hardness of the Leading $n - \omega(\log n)$ Bits of the Discrete Log

Let f be a function. Informally, we say that a subset of bits of f^{-1} are simultaneously hard if no probabilistic polynomial-time algorithm can find any information about them given only the output of the function f. A more precise way of stating this is that this subset of bits is indistinguishable from a randomly chosen group of bits with the same distribution given the output of f. Also, for another function h, we say that a subset of bits of f^{-1} are simultaneously h^{-1}-hard if they are simultaneously hard assuming h is one-way.

Using this terminology, we may now precisely state the result of Patel and Sundaram [23] on the simultaneous hardness of the trailing bits of the discrete log.

Theorem 1 ([23]). *For $c = \omega(\log n)$, the trailing $n - c$ bits (except the lsb) of DL modulo an n-bit safe prime p are simultaneously DLSE_c-hard.*

However, this result does not imply anything about the simultaneous hardness of the leading bits, which is what we examine here.

Let $\mathsf{LBits}_{p,g,c} = \{x \xleftarrow{R} \mathbb{Z}_{p-1} : (p,g,g^x, x_{c+1}^n)\}$ and $\mathsf{RLBits}_{p,g,c} = \{x, y \xleftarrow{R} \mathbb{Z}_{p-1} : (p,g,g^x, y_{c+1}^n)\}$. Informally, $\mathsf{LBits}_{p,g,c}$ is a distribution consisting of elements of $\mathbb{Z}_p^*$ along with the leading bits of their discrete logs, and $\mathsf{RLBits}_{p,g,c}$ is a distribution consisting of elements of $\mathbb{Z}_p^*$ along with the leading bits of discrete logs of random elements. Let

$$\mathsf{LB}_{c,n} = \{(p,g) \leftarrow \mathcal{G}(1^n); v \leftarrow \mathsf{LBits}_{p,g,c} : v\}$$

and

$$\mathsf{RLB}_{c,n} = \{(p,g) \leftarrow \mathcal{G}(1^n); v \leftarrow \mathsf{RLBits}_{p,g,c} : v\}$$

Let LB_c and RLB_c be their respective distribution ensembles, parameterized by n.

In a similar way, we will define the corresponding distributions for trailing bits. Let $\mathsf{TBits}_{p,g,c} = \{x \xleftarrow{R} \mathbb{Z}_{p-1} : (p,g,g^x, x_2^{n-c})\}$ and $\mathsf{RTBits}_{p,g,c} = \{x, y \xleftarrow{R} \mathbb{Z}_{p-1} : (p,g,g^x, y_2^{n-c})\}$. Informally, $\mathsf{TBits}_{p,g,c}$ is a distribution consisting of elements of $\mathbb{Z}_p^*$ along with the trailing bits of their discrete logs (except for the lsb), and $\mathsf{RTBits}_{p,g,c}$ is a distribution consisting of elements of $\mathbb{Z}_p^*$ along with the trailing bits of discrete logs of random elements (except for the lsb)[3]. Let

$$\mathsf{TB}_{c,n} = \{(p,g) \leftarrow \mathcal{G}(1^n); v \leftarrow \mathsf{TBits}_{p,g,c} : v\}$$

and

$$\mathsf{RTB}_{c,n} = \{(p,g) \leftarrow \mathcal{G}(1^n); v \leftarrow \mathsf{RTBits}_{p,g,c} : v\}$$

Let TB_c and RTB_c be their respective distribution ensembles, parameterized by n.

Definition 1. *The $n - c$ leading bits of DL are simultaneously hard if LB_c is polynomially indistinguishable from RLB_c.*

Now we are ready to state our main theorem.

Theorem 2. *For $c = \omega(\log n)$, the $n - c$ leading bits of DL modulo an n-bit safe prime are simultaneously DLSE_c-hard.*

---

[3] The trailing bits of discrete logs of random elements are statistically indistinguishable from random bits, but the distributions are not quite identical - see below.

We break the proof of Theorem 2 into two parts. In the first part we prove in Theorem 3 that SE_c is indistinguishable from U if and only if LB_c is indistinguishable from RLB_c. In the second part we prove in Theorem 4 that a distinguisher between SE_c and U can be turned in to a DLSE_c solver. It follows from these two results and the hardness of DLSE_c that the leading $n - c$ bits of DL are simultaneously DLSE_c-hard.

Theorem 3. SE_c *is indistinguishable from* U *iff* LB_c *is indistinguishable from* RLB_c.

Proof. First we show that if one can distinguish $\mathsf{LBits}_{p,g,c}$ from $\mathsf{RLBits}_{p,g,c}$ in time T with probability ϵ, then one can distinguish the uniform distribution on $\mathsf{SExp}_{p,g,c}$ from the uniform distribution on $\mathbb{Z}_p^*$ in time $T + T'$ with probability $\epsilon - \delta$, where T' is polynomial and δ is negligible. (For the remainder of the paper, when it is obvious from context we will use the name of a set to denote the uniform distribution on that set. For instance, above we could say "...if one can distinguish $\mathsf{SE}_{p,g,c}$ from $\mathbb{Z}_p^*$ in time...")

Let D be a distinguisher that distinguishes $\mathsf{LBits}_{p,g,c}$ from $\mathsf{RLBits}_{p,g,c}$. Then we construct a distinguisher D' that distinguishes $\mathsf{SExp}_{p,g,c}$ from $\mathbb{Z}_p^*$.

Let $\mathsf{upper}(p, c) = 2^c \lfloor p2^{-c} \rfloor$, i.e., the result of zeroing the c trailing bits of p.

$D'(y)$
{
$\quad z \xleftarrow{R} \mathbb{Z}_{p-1}$
$\quad z_1^c \leftarrow 0^c$
$\quad y' \leftarrow y \cdot g^z$
$\quad \text{return } D(y', z_{c+1}^n)$
}

If the input y is drawn uniformly from $\mathsf{SExp}_{p,g,c}$, and if $z < \mathsf{upper}(p, c)$, then adding z to the exponent (with the c trailing bits of z set to zero) makes the result y' appear to be picked uniformly from a set $S \subseteq \mathbb{Z}_p^*$ with z as its leading $n - c$ bits, where $S = \{g^x : x \in \mathbb{Z}_{\mathsf{upper}(p,c)}\}$. Hence the statistical difference between $D'(y)$ for $y \xleftarrow{R} \mathsf{SExp}_{p,g,c}$ and $D(v)$ for $v \leftarrow \mathsf{LBits}_{p,g,c}$ is at most $\frac{2^c}{p}$, which is negligible.

On the other hand, if the input y is drawn uniformly from $\mathbb{Z}_p^*$, and if $z < \mathsf{upper}(p, c)$, then adding z to the exponent (with the c trailing bits of z set to zero) makes the result y' appear to be picked uniformly from $\mathbb{Z}_p^*$ with z as random leading $n - c$ bits. Hence the statistical difference between $D'(y)$ for $y \xleftarrow{R} \mathbb{Z}_p^*$ and $D(v)$ for $v \leftarrow \mathsf{RLBits}_{p,g,c}$ is zero.

Next we show that if one can distinguish the uniform distribution on $\mathsf{SExp}_{p,g,c}$ from the uniform distribution on $\mathbb{Z}_p^*$ in time T with probability ϵ, then one can distinguish $\mathsf{LBits}_{p,g,c}$ from $\mathsf{RLBits}_{p,g,c}$ in time $T + T'$ with probability $\epsilon - \delta$, where T' is polynomial and δ is negligible.

Let D be a distinguisher that distinguishes $\mathsf{SExp}_{p,g,c}$ from $\mathbb{Z}_p^*$. Then we construct a distinguisher D' that distinguishes $\mathsf{LBits}_{p,g,c}$ from $\mathsf{RLBits}_{p,g,c}$.

$D'(v)$
{
    Say $v = (y, z_1^{n-c})$
    $w_{c+1}^n \leftarrow z_1^{n-c}$
    $w_1^c \leftarrow 0^c$
    $y' \leftarrow y \cdot g^{-w}$
    return $D(y')$
}

Say $y = g^x$. If v is drawn from $\mathsf{LBits}_{p,g,c}$, then $z_1^{n-c} = x_{c+1}^n$, so $y \cdot g^{-w}$ (for w computed above) sets the leading $n - c$ bits of the discrete log to zero. If $x < \mathsf{upper}(p, c)$ then the result y' appears to be picked uniformly from $\mathsf{SExp}_{p,g,c}$. Hence the statistical difference between $D'(v)$ for $v \leftarrow \mathsf{LBits}_{p,g,c}$ and $D(y)$ for $y \xleftarrow{R} \mathsf{SExp}_{p,g,c}$ is at most $\frac{2^c}{p}$.

If v is drawn from $\mathsf{RLBits}_{p,g,c}$, then z_1^{n-c} is random leading bits, so w is independent from the discrete log of y. Then the result y' appears to be picked uniformly from $\mathbb{Z}_p^*$. Hence the statistical difference between $D'(v)$ for $v \leftarrow \mathsf{RLBits}_{p,g,c}$ and $D(y)$ for $y \xleftarrow{R} \mathbb{Z}_p^*$ is zero.

Before we prove the next theorem, we will need to prove some intermediate lemmas. Let

$$\mathsf{BSExp}_{p,g,c} = \{w \leftarrow \mathbb{Z}_{p-1}; x_{c+1}^n \leftarrow 0^{n-c}; x_2^c \leftarrow w_{n-c+2}^n; x_1^1 \leftarrow w_1^1 : g^x\}.$$

In other words, $\mathsf{BSExp}_{p,g,c}$ are elements of $\mathbb{Z}_p^*$ with "biased" short exponents of size c that are distributed as the $c-1$ leading bits followed by the lsb of random elements of $\mathbb{Z}_p^*$. Let

$$\mathsf{BSE}_{c,n} = \{(p, g) \leftarrow \mathcal{G}(1^n); v \xleftarrow{R} \mathsf{BSExp}_{p,g,c} : v\}.$$

Let BSE_c be the corresponding distribution ensembles, parameterized by n.

Lemma 1. *If $\mathsf{SExp}_{p,g,c}$ can be distinguished from $\mathbb{Z}_p^*$ in time T with probability ϵ then $\mathsf{BSE}_{p,g,c-t}$ can be distinguished from $\mathbb{Z}_p^*$ in time $T + T'$ with probability $\frac{\epsilon}{2}$, where $t = \log \frac{\epsilon}{2}$ and T' is polynomial.*

Proof. Let D distinguish $\mathsf{SExp}_{p,g,c}$ from $\mathbb{Z}_p^*$ in time T with probability ϵ. Then we show how to create D' to distinguish $\mathsf{BSE}_{p,g,c-t}$ from $\mathbb{Z}_p^*$ in time $T + T'$ with probability at least $\frac{\epsilon}{2}$, where $t = \log \frac{\epsilon}{2}$.

$D'(y)$
{
    $s \xleftarrow{R} \{0,1\}^c$
    $y' \leftarrow y \cdot g^s$
    return $D(y')$
}

If y is uniformly drawn from $\mathbb{Z}_p^*$, then adding s to the discrete log will not change the distribution, so y' will be uniform over $\mathbb{Z}_p^*$. Hence the statistical difference between $D'(y)$ for $y \leftarrow \mathbb{Z}_p^*$ and $D(y)$ for $y \xleftarrow{R} \mathbb{Z}_p^*$ is zero.

If y is uniformly drawn from $\mathsf{BSE}_{p,g,c-t}$, then adding s to the discrete log will make the distribution of the result y' to be $\mathsf{MIX}_{p,g,c,t} = \{v \leftarrow \mathsf{SE}_{p,g,c}; y \leftarrow \mathsf{BSE}_{p,g,c-t} : v \cdot y\}$. So D' will distinguish $\mathsf{BSE}_{p,g,c-t}$ from $\mathbb{Z}_p^*$ with the same probability as D distinguishes $\mathsf{MIX}_{p,g,c,t}$ from $\mathbb{Z}_p^*$. Here we bound this probability. Let $p_0 = \Pr[y \leftarrow \mathsf{SE}_{p,g,c}; D(y) = 1]$, $p_1 = \Pr[y \leftarrow \mathsf{MIX}_{p,g,c,t}; D(y) = 1]$, and $p_2 = \Pr[y \leftarrow \mathbb{Z}_p^*; D(y) = 1]$. Then

$$|p_1 - p_2| \geq |p_0 - p_2| - |p_0 - p_1|$$
$$= \epsilon - |p_0 - p_1|$$
$$\geq \epsilon - \mathsf{StatDiff}(\mathsf{SE}_{p,g,c}, \mathsf{MIX}_{p,g,c,t}),$$

where $\mathsf{StatDiff}()$ returns the statistical difference between two distributions. The following claim is proven in the appendix.

Claim. $\mathsf{StatDiff}(\mathsf{SE}_{p,g,c}, \mathsf{MIX}_{p,g,c,t}) = \frac{\epsilon}{2}$.

Hence $|p_1 - p_2| \geq \frac{\epsilon}{2}$.

Lemma 2. *If $\mathsf{BSExp}_{p,g,c}$ can be distinguished from $\mathbb{Z}_p^*$ in time T with probability ϵ then we can distinguish $\mathsf{TBits}_{p,g,c}$ from $\mathsf{RTBits}_{p,g,c}$ in time $T + T'$ with probability ϵ.*

Proof. Let D distinguish $\mathsf{BSExp}_{p,g,c}$ from $\mathbb{Z}_p^*$ with probability ϵ. Then we show how to create D' to distinguish $\mathsf{TBits}_{p,g,c}$ from $\mathsf{RTBits}_{p,g,c}$ with probability ϵ.

In the following protocol $\mathsf{IsbExp}(y)$ is an efficient algorithm that returns the lsb of the exponent of y, and $\mathsf{IsbZeroSqrt}(y)$ is an efficient algorithm that computes the two square roots of y and returns the one with lsb equal to zero.

$D'(v)$
{
    Say $v = (y, z_1^{n-c-1})$
    $y' \leftarrow y \cdot g^{-2z}$
    $b \leftarrow \mathsf{IsbExp}(y')$
    $y' \leftarrow y' \cdot g^{-b}$
    For $(n - c - 1$ times)
        $y' \leftarrow \mathsf{IsbZeroSqrt}(y')$
    $y' \leftarrow y' \cdot g^b$
    $D(y')$
}

Say $y = g^x$. If v is drawn from $\mathsf{TBits}_{p,g,c}$, then $z_1^{n-c-1} = x_2^{n-c}$, so $y \cdot g^{-2z}$ sets the trailing bits (except for the lsb) of the discrete log to zero. The square roots server to shift the $c - 1$ bits of the discrete log down to the trailing bits (again, except for the lsb), so the final result y' appears to be picked uniformly from

$\mathsf{BSExp}_{p,g,c}$. Hence the statistical difference between $D'(v)$ for $v \leftarrow \mathsf{TBits}_{p,g,c}$ and $D(y)$ for $y \xleftarrow{R} \mathsf{BSExp}_{p,g,c}$ is zero.

If v is drawn from $\mathsf{RTBits}_{p,g,c}$, then $z_1^{n-c+t-1}$ is random trailing bits independent from the discrete log of y. Then the result y' of multiplying by g^{-2z} is uniform in $\mathbb{Z}_p^*$. (The same idea was used by Gennaro [8] to create a more efficient discrete logarithm based PRG.) Now we just argue that the remainder of the computation on y' does not change this distribution. First we store the lsb, and set the lsb of y' to be zero. Therefore the distribution of y' is uniform over the quadratic residues. Then we perform square root operations, where we take the square root that is a quadratic residue. This is a permutation on quadratic residues, so the distribution of y' after this is also uniform over the quadratic residues. Finally we add the lsb, which is random, so y' is uniform over $\mathbb{Z}_p^*$. Hence the statistical difference between $D'(v)$ for $v \leftarrow \mathsf{RTBits}_{p,g,c}$ and $D(y)$ for $y \xleftarrow{R} \mathbb{Z}_p^*$ is zero.

Theorem 4. *If SE_c is distinguishable from U, then DLSE_c can be solved in polynomial time.*

Proof. Let D distinguish $\mathsf{SExp}_{p,g,c}$ from $\mathbb{Z}_p^*$ with probability ϵ.
Then we show how to create A to compute $\mathsf{DLSE}(p,g,y,c)$

$A(p,g,y,c)$
{
    For ($guess = 0$ to $2^t - 1$)
        $y' \leftarrow y \cdot (g^{guess \cdot 2^{c-t}})^{-1}$
        $x \leftarrow B(p,g,y',c-t,D^*())$
        if $y = g^x$ return $x + guess \cdot 2^{c-t}$
}

$D^*()$ is a distinguisher between $\mathsf{TBits}_{p,g,c-t}$ and $\mathsf{RTBits}_{p,g,c-t}$ that can be constructed from the previous two lemmas and based on a distinguisher between $\mathsf{SExp}_{p,g,c}$ and $\mathbb{Z}_p^*$. $B(p,g,y,c',D())$ is the Patel-Sundaram [23] polynomial-time algorithm for computing $\mathsf{DLSE}(p,g,y,c')$ given a distinguisher $D()$ between $\mathsf{TBits}_{p,g,c'}$ and $\mathsf{RTBits}_{p,g,c'}$.

When $A(p,g,y,c)$ correctly guesses the t most significant bits of the c-bit exponent of y, it sets them to zero, so that $y' \in \mathsf{SExp}_{p,g,c-t}$. In this case $B(p,g,y',c-t,D^*())$ will return the correct discrete log of y', and then $A()$ will return the correct discrete log of y.

A closer look at the Patel-Sundaram algorithm reveals that the time to solve the DLSE is $O(\frac{(n-c-1)^3 c^2 T(n)}{\epsilon^3})$ where $T(n)$ is the time taken by the distinguisher and ϵ is the distinguisher's advantage; the constant in the expression is small. In our solver, our distinguisher has advantage $\frac{\epsilon}{2}$ and the algorithm $B()$ is called 2^t times. Since $t = \log \frac{2}{\epsilon}$, $B()$ is called $\frac{2}{\epsilon}$ times. Also the size of the exponent in $B()$ is changed from c bits to $c-t$ bits. Thus the time to solve the DLSE using A is $O(\frac{(n-c+t-1)^3 (c-t)^2 T(n)}{\epsilon^4})$. If ϵ is non-negligible, then A runs in polynomial-time.

4 Application: Short Exponent PAK

In this section we show an application of our results to password-authenticated key exchange protocols, and in particular, to the PAK protocols [6]. Although one of the main technical points in the security proof for PAK is that all values seen by the adversary are uniform in the group used, we show that the protocol can be modified so that only short exponents are used (and thus elements are no longer uniform in the whole subgroup), and that this new protocol can be proven secure based on the hardness of DHSE_c, for $c = \omega(\log n)$. As an intermediate step, we show it is based on the hardness of DHSE_c, as well as the hardness of SEdist_c. Using the results of the previous section, this reduces to the hardness of DHSE_c.

The protocol we will modify to use short exponents is the basic implicit authentication version of PAK called PPK. Other variations such as PAK and PAK-Z could be modified and proven secure similarly. Much of our notation and proof techniques will be modeled on MacKenzie [17]. All of these results are in the random oracle model.

4.1 Model

For our proofs of security we use the model of [1]. This model is designed for the problem of authenticated key exchange (ake) between two parties, a client and a server, that share a secret. The goal is for them to engage in a protocol such that after the protocol is completed, they each hold a session key that is known to nobody but the two of them. We assume that secret keys are drawn from $\{0,1\}^\kappa$.

In the following, we will assume some familiarity with the model of [1].

Protocol Participants. Let ID be a nonempty set of principals, each of which is either a client or a server. Thus $ID \stackrel{\text{def}}{=} \mathit{Clients} \cup \mathit{Servers}$, where $\mathit{Clients}$ and $\mathit{Servers}$ are finite, disjoint, nonempty sets. We assume each principal $U \in ID$ is labeled by a string, and we simply use U to denote this string.

Each client $C \in \mathit{Clients}$ has a secret password π_C and each server $S \in \mathit{Servers}$ has a vector $\pi_S = \langle \pi_S[C] \rangle_{C \in \mathit{Clients}}$. Entry $\pi_S[C]$ is the *password record*. In the balanced case we will have $\pi_S[C] = f(\pi_C)$, for some deterministic function f. Let $\mathit{Password}_C$ be a (possibly small) set from which passwords for client C are selected. We will assume that $\pi_C \stackrel{R}{\leftarrow} \mathit{Password}_C$ (but our results easily extend to other password distributions). Clients and servers are modeled as probabilistic poly-time algorithms with an input tape and an output tape.

Execution of the Protocol. A protocol P is an algorithm that determines how principals behave in response to inputs from their environment. In the real world, each principal is able to execute P multiple times with different partners, and we model this by allowing unlimited number of *instances* of each principal. Instance i of principal $U \in ID$ is denoted Π_i^U.

To describe the security of the protocol, we assume there is an adversary $\mathcal{A}$ that has complete control over the environment (mainly, the network), and thus provides the inputs to instances of principals. Formally, the adversary is a probabilistic algorithm with a distinguished query tape. Queries written to this tape are responded to by principals according to P; the allowed queries are formally defined in [1] and summarized here:

Send (U, i, M): Causes message M to be sent to instance Π_i^U. The instance computes what the protocol says to, state is updated, and the output of the computation is given to $\mathcal{A}$. If this query causes Π_i^U to accept or terminate, this will also be shown to $\mathcal{A}$ [4]. To initiate a session between client C and server S, the adversary should send a message containing the server name S to an unused instance of C.

Execute (C, i, S, j): Causes P to be executed to completion between Π_i^C (where $C \in Clients$) and Π_j^S (where $S \in Servers$), and outputs the transcript of the execution. This query captures the intuition of a passive adversary who simply eavesdrops on the execution of P.

Reveal (U, i): Causes the output of the session key held by Π_i^U.

Test (U, i): Causes Π_i^U to randomly select a bit b. If $b = 1$ the session key sk_U^i is output; otherwise, a string is drawn uniformly from the space of session keys and output. A Test query may be asked at any time during the execution of P, but may only be asked once.

Corrupt (U): If $U \in Servers$, this returns $\langle \pi_U[C] \rangle_{C \in Clients}$, and otherwise returns π_U [5].

Partnering. A client or server instance that accepts holds a partner-id pid, session-id sid, and a session key sk. Then instances Π_i^C (with $C \in Clients$) and Π_j^S (with $S \in Servers$) are said to be *partnered* if both accept, they hold (pid, sid, sk) and (pid', sid', sk'), respectively, with $pid = S$, $pid' = C$, $sid = sid'$, and $sk = sk'$, and no other instance accepts with session-id equal to sid.

Freshness. We define the notion of freshness without forward secrecy [1]. An instance Π_i^U is nfs-fresh (fresh with no requirement for forward secrecy) unless either (1) a Reveal (U, i) query occurs, (2) a Reveal (U', j) query occurs where $\Pi_{U'}^j$ is the partner of Π_i^U, or (3) a Corrupt (U') query occurs. (For convenience, we simply disallow Corrupt queries.)

Advantage of the Adversary. We now formally define the authenticated key exchange (ake) advantage of the adversary against protocol P. Let $\text{Succ}_P^{\text{ake}}(\mathcal{A})$ be the event that $\mathcal{A}$ makes a single Test query directed to some fresh instance Π_i^U that has terminated, and eventually outputs a bit b', where $b' = b$ for the bit b that was selected in the Test query. The ake advantage of $\mathcal{A}$ attacking P is defined to be

$$\text{Adv}_P^{\text{ake}}(\mathcal{A}) \stackrel{\text{def}}{=} 2\Pr\left[\text{Succ}_P^{\text{ake}}(\mathcal{A})\right] - 1.$$

The following fact is easily verified.

Fact 1 $\Pr(\text{Succ}_P^{\text{ake}}(\mathcal{A})) = \Pr(\text{Succ}_{P'}^{\text{ake}}(\mathcal{A})) + \epsilon \iff \text{Adv}_P^{\text{ake}}(\mathcal{A}) = \text{Adv}_{P'}^{\text{ake}}(\mathcal{A}) + 2\epsilon$.

---

[4] Recall that accepting implies generating a triple (pid, sid, sk), terminating implies accepting and no more messages will be output. If the protocol aborts, then no more messages will be output, but it does not accept or terminate according to our definitions.

[5] This is the weak corruption model of [1].

4.2 SEPPK Protocol

Let n be the size of p, and c and c' be auxiliary parameters that we will discuss later. The short exponent PPK (SEPPK) protocol is shown in Figure 1, where all arithmetic operations are performed in $\mathbb{Z}_p^*$. The only difference between this protocol and the PPK protocol in [17] is in how x and y are chosen; in PPK they are chosen from $\mathbb{Z}_q$, where $q = p - 1$.

Recall that we assume the random oracle model, and thus we assume the existence of public random functions H from $\{0,1\}^*$ to $\{0,1\}^\infty$, and as shown in [4], this sequence of bits may be used to define the output of H in a specific set. In particular we will assume that we can specify that the output of a random oracle H be interpreted as a (random) element of $\mathbb{Z}_p^*$.

To define hash functions that output elements of $\mathbb{Z}_p^*$, (H_1 and H_2 in PPK), we could, for instance, use a random oracle $H : \{0,1\}^* \to \{0,1\}^{|p|+c}$, and set $H_i(x) \leftarrow H(\text{BYTE}(i) \parallel x) \bmod p$. To simulate H_i, we can simulate H in the following way. So that $H_i(x)$ outputs X (with high probability), we set $H(\text{BYTE}(i) \parallel x)$ as follows, with $w = 2^{|p|+c}$: Generate $r \xleftarrow{R} \mathbb{Z}_w$. If $r > w - (w \bmod p)$ output r, else output $X + p\lfloor r/p \rfloor$.

Security and the Size of the Exponents. Initially, one may think that the security of SEPPK would follow directly from the hardness of SEdist_c, even if the exponents chosen by the players were c bits. In particular, one may attempt to use the hardness of SEdist_c to reduce the security of the SEPPK protocol directly to the security of the standard PPK protocol, since the only difference is in the size of the exponents of the players Diffie-Hellman values. However, the simulation argument in this reduction seems to break down since there will be a session for which we will not know the discrete log of a simulated players Diffie-Hellman value, and for this session we may not be able to determine a correct password guess by the adversary.

Our proof below follows a different approach, using the hardness of SEdist_c to reduce the security of the SEPPK protocol to one where the hash functions return values from $\text{SExp}_{p,g,c}$ instead of $\mathbb{Z}_p^*$, and then using the security of DHSE_c to complete the proof, analogous to the proof of security of the standard PPK protocol, but with an extra padding of c' bits used to make some simulated values indistinguishable from real values.

Example 1. Consider $n = 1024$. Let $c = 256$ and $c' = 128$. That is, we assume the hardness of DHSE_{256} and SEdist_{256}, or equivalently, 128-bit security. Then the exponents in the SEPPK protocol must be 384 bits. Note that this is more efficient than the PPK protocol using similar parameters, since in the PPK protocol, the client needs to perform two 256-bit exponentiations and one extra 768-bit exponentiation.

4.3 Security of SEPPK

Here we prove that the SEPPK protocol is secure, in the sense that an adversary attacking the system cannot determine session keys of fresh instances with greater advantage than that of an online dictionary attack. For concreteness, we

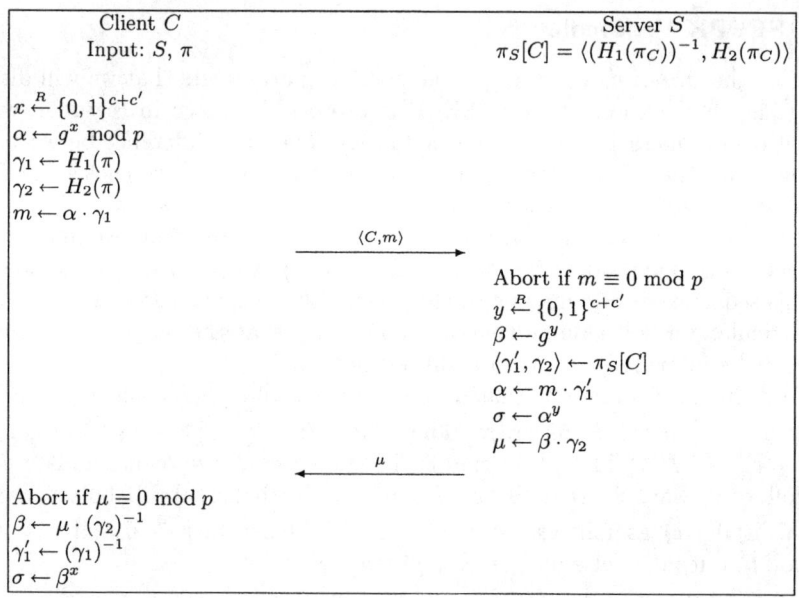

Fig. 1. PPK Protocol over $\mathbb{Z}_p^*$ with generator g. Session ID is $sid = C \parallel S \parallel m \parallel \mu$. Partner ID for C is $pid_C = S$, and partner ID for S is $pid_S = C$. Shared session key is $sk = H_3(\langle C, S, m, \mu, \sigma, \gamma_1' \rangle)$.

prove our theorem based on the hardness of both DHSE_c and SEdist_c. However, by using the result of Theorem 4, our proof below implies that if an adversary can break SEPPK with non-negligible probability, DHSE_c can be solved in polynomial time. Therefore, the security of SEPPK actually relies only on the hardness of DHSE_c.

Let $\mathsf{Succ}_{p,g,c}^{\mathsf{DHSE}}(t,k) = \max_{\mathcal{A}} \left\{ \mathsf{Succ}_{p,g,c}^{\mathsf{DHSE}}(\mathcal{A}) \right\}$, where the maximum is taken over all adversaries of time complexity at most t that output a list containing at most k elements of $\mathbb{Z}_p^*$.

Let $\mathsf{Adv}_{p,g,c}^{\mathsf{SEdist}}(t) = \max_{\mathcal{A}} \left\{ \mathsf{Adv}_{p,g,c}^{\mathsf{SEdist}}(\mathcal{A}) \right\}$, where the maximum is taken over all adversaries of time complexity at most t.

Let t_{exp} be the time required to perform an exponentiation in $\mathbb{Z}_p^*$.

Theorem 5. *Let P be the SEPPK protocol described in Figure 1 except that the exponents x and y are drawn uniformly from $\{0,1\}^{c+c'}$. Assume passwords are drawn uniformly from a dictionary of size N. Fix an adversary $\mathcal{A}$ that runs in time t, and makes $q_{\mathsf{se}}, q_{\mathsf{ex}}, q_{\mathsf{re}}$ queries of type Send, Execute, Reveal, respectively, and q_{ro} queries to the random oracles. Then for $t' = O(t + (q_{\mathsf{ro}} + q_{\mathsf{se}} + q_{\mathsf{ex}})t_{\mathsf{exp}})$:*

$$\mathsf{Adv}_P^{\mathsf{ake}}(\mathcal{A}) = \frac{q_{\mathsf{se}}}{N} + O\left(q_{\mathsf{ro}} \mathsf{Succ}_{p,q,c}^{\mathsf{DHSE}}(t', (q_{\mathsf{ro}})^2) + q_{\mathsf{ro}} \mathsf{Adv}_{p,q,c}^{\mathsf{SEdist}}(t') + \right.$$
$$\left. (q_{\mathsf{se}} + q_{\mathsf{ex}})(q_{\mathsf{ro}} + q_{\mathsf{se}} + q_{\mathsf{ex}})2^{-(c+c')} + q_{\mathsf{ro}} 2^{-c} + q_{\mathsf{se}} 2^{-c'} \right).$$

A proof sketch is given in the appendix.

References

1. M. Bellare, D. Pointcheval, and P. Rogaway. Authenticated key exchange secure against dictionary attacks. In *EUROCRYPT 2000* (LNCS 1807), pp. 139–155, 2000.
2. M. Bellare and P. Rogaway. Entity authentication and key distribution. In *CRYPTO '93* (LNCS 773), pp. 232–249, 1993.
3. S. M. Bellovin and M. Merritt. Encrypted key exchange: Password-based protocols secure against dictionary attacks. In *IEEE Symposium on Research in Security and Privacy*, pages 72–84, 1992.
4. M. Bellare and P. Rogaway. Random oracles are practical: A paradigm for designing efficient protocols. In 1st *ACM Conference on Computer and Communications Security*, pages 62–73, November 1993.
5. M. Blum and S. Micali. How to Generate Cryptographically Strong Sequences of Pseudo-Random Bits. In *SIAM Journal of Coumputing* 13(4):850–864, November 1984.
6. V. Boyko, P. MacKenzie, and S. Patel. Provably secure password authentication and key exchange using Diffie-Hellman. In *EUROCRYPT 2000* (LNCS 1807), pp. 156–171, 2000.
7. W. Diffie and M. Hellman. New directions in cryptography. *IEEE Trans. Info. Theory*, 22(6):644–654, 1976.
8. R. Gennaro. An Improved Pseudo-Random Generator Based on Discrete Log. In *CRYPTO 2000* (LNCS 1880), pp. 469–481, 2000.
9. O. Goldreich and L. Levin. A hard core predicate for any one way function. *21st ACM Symposium on the Theory of Computing*, pp. 25-32, 1989.
10. O. Goldreich and Y. Lindell. Session-Key Generation using Human Passwords Only. In *CRYPTO 2001* (LNCS 2139), pp. 408–432, 2001.
11. J. Hastad and M. Naslund. The Security of all RSA and discrete log bits. Manuscript, 1999. (Preliminary version appears in 39th FOCS, 1998, pp. 510–519.)
12. J. Hastad, A. Schrift, and A. Shamir. The discrete logarithm modulo a composite hides $O(n)$ bits. *Journal of Computer and System Sciences*, 47:376–404, 1993.
13. A. Juels, M. Jakobsson, E. Shriver, and B. Hillyer. How to turn loaded dice into fair coins. *IEEE Transactions on Information Theory*, 46(3): 911-921, 2000.
14. *The Art of Computer Programming (vol 3): Sorting and Searching*, Addison, Wesley, 1973.
15. T. Lomas, L .Gong, J. Saltzer, and R. Needham. Reducing risks from poorly chosen keys. *ACM Operating Systems Review*, 23(5):14–18, Dec. 1989. Proceedings of the 12th ACM Symposium on Operation System Principles.
16. D. Long and A. Wigderson. The discrete log hides $O(\log n)$ bits. *SIAM Journal of Computing* 17, 413–420, 1988.
17. P. MacKenzie. The PAK suite: Protocols for password-authenticated key exchange. DIMACS Technical Report 2002-46, October 2002.
18. P. MacKenzie, S. Patel, and R. Swaminathan. Password authenticated key exchange based on RSA. In *ASIACRYPT 2000* (LNCS 1976), pp. 599–613, 2000.
19. M. Naslund. Universal hash functions and hard core bits. In *EUROCRYPT 1995* (LNCS 921), pp. 356–366, 1995.
20. M. Naslund. All bits in ax+b are hard. In *CRYPTO 1996* (LNCS 1109), pp. 114–128, 1996.
21. P. van Oorschot and M. Wiener. On Diffie-Hellman key agreement with short exponents. *EUROCRYPT'96* (LNCS 1070), pp. 332–343, 1996.

22. TIA/EIA/IS-683-C. Over-the-Air service provisioining of mobile stations in spread spectrum systems.
23. S. Patel and G. Sundaram. An efficient discrete log pseudo random generator. In *CRYPTO '98* (LNCS 1462), pp. 304-317, 1998.
24. R. Peralta. Simultaneous security of bits in the discrete log. In *EUROCRYPT 1985* (LNCS 219), pp. 62–72, 1985.
25. J. M. Pollard. Monte Carlo methods for index computation (mod p). *Mathematics of Computation*, 32, No. 143:918–924, 1978.
26. D. R. Stinson. Universal hash families and the leftover hash lemma, and applications to cryptography and computing. *J. Combin. Math. Combin. Comput.*, 42, (2002), 3-31.
27. M. I. Gonzalez Vasco and M. Naslund. A survey of hard core functions. In Proceedings of the *Workshop on Comp. Number Theory and Cryptography*, Singapore 1999, Birkhauser, 2001, 227-256.

A Proofs

Proof (of Claim 3). Let $X = \mathsf{SE}_{p,g,c}$ and $Y = \mathsf{MIX}_{p,g,c,t}$. Note the following:

1. For $z \in \{g^x : 0 \leq x < 2^{c-t} - 1\}$, $2^{-c} = \Pr(X = z) \geq \Pr(Y = z)$.
2. For $z \in \{g^x : 2^{c-t} - 1 \leq x < 2^c - 1\}$, $2^{-c} = \Pr(X = z) = \Pr(Y = z)$.
3. For $z \in \{g^x : 2^c \leq x < 2^c + 2^{c-t} - 1\}$, $2^{-c} \geq \Pr(Y = z) \geq \Pr(X = z)$.

Then $\mathsf{StatDiff}(X,Y) = \frac{1}{2}\sum_z |\Pr(X = z) - \Pr(Y = z)| = \sum_z \max\{\Pr(X = z), \Pr(Y = z)\} - 1 \leq 2^{-c}(2^{c-t} - 1) + 2^{-c}(2^c - 2^{c-t}) + 2^{-c}(2^{c-t} - 1) - 1 \leq 2^{-t} \leq \frac{\epsilon}{2}$, where a proof for the second equality can be found in Stinson [26].

Proof (of Theorem 5 – Sketch). The proof of this theorem closely follows the proof of security for PPK given in [17], which we call the *PPK proof*. We will sketch this proof, focusing on the changes that need to be made to handle short exponents.

As in the PPK proof, our proof will proceed by introducing a series of protocols $P_0, P_{0.5}, P_1, \ldots, P_7$ related to P, with $P_0 = P$. In P_7, $\mathcal{A}$ will be reduced to a simple online guessing attack that will admit a straightforward analysis. Note that we add an addition protocol $P_{0.5}$ in our analysis. We describe these protocols informally in Figure 2. Except for $P_{0.5}$ and P_2, these descriptions are identical to those in the PPK proof. Note that we do not have any sort of random self-reducibility in $\mathsf{SExp}_{p,g,c}$, which somewhat reduces the strength of our security reductions. For each i from 1 to 7, we will bound the decrease in advantage of $\mathcal{A}$ attacking protocol P_i compared to $\mathcal{A}$ attacking protocol P_{i-1}.

$P_0 \to P_1$: First we bound the decrease in advantage of $\mathcal{A}$ attacking protocol $P_{0.5}$ compared to $\mathcal{A}$ attacking protocol P_0. This is by a straightforward hybrid argument on the calls to $H_1(\cdot)$ and $H_2(\cdot)$, so the decrease in advantage of $\mathcal{A}$ is at most $2q_{\mathrm{ro}}\mathsf{Adv}^{\mathsf{SEdist}}_{p,g,c}(t')$.
Next we bound the decrease in advantage of $\mathcal{A}$ attacking protocol P_1 compared to $\mathcal{A}$ attacking protocol $P_{0.5}$. This is by the same argument as the PPK proof, except that since the hash functions output values in a different range, the q^{-1} factor is replaced by $2^{-c+c'}$, so the decrease in advantage of $\mathcal{A}$ is at most $O((q_{\mathrm{se}} + q_{\mathrm{ex}})(q_{\mathrm{ro}} + q_{\mathrm{se}} + q_{\mathrm{ex}}))2^{-(c+c')}$.

P_0	The original protocol P.
$P_{0.5}$	All values returned by $H_1(\cdot)$ and $H_2(\cdot)$ are drawn from $\mathsf{SExp}_{p,g,c}$.
P_1	If honest parties randomly choose m or μ values seen previously in the execution of the protocol, the protocol halts and the adversary fails.
P_2	The protocol answers Send and Execute queries using m and μ drawn randomly from $\mathsf{SExp}_{p,g,c+c'}$, without making any random oracle queries. Subsequent random oracle queries by the adversary are backpatched, as much as possible, to be consistent with the responses to the Send and Execute queries. (This is a standard technique for proofs involving random oracles.)
P_3	If an $H_3(\cdot)$ query is made, it is not checked for consistency against Execute queries. That is, instead of backpatching to maintain consistency with an Execute query, the protocol responds with a random output.
P_4	If a correct password guess is made against a client instance or server instance (determined by an $H_3(\cdot)$ query using the correct inputs to compute a session key), the protocol halts and the adversary automatically succeeds.
P_5	If the adversary makes two password guesses against the same server instance, the protocol halts and the adversary fails.
P_6	If the adversary makes two password guesses against the same client instance, the protocol halts and the adversary fails.
P_7	The protocol uses an internal password oracle that holds all passwords and only accepts simple queries that test whether a given password is the correct password for a given client/server pair. The test for correct password guesses (from P_4) is changed so that whenever the adversary makes a password guess, a query is submitted to the oracle to determine if it is correct.

Fig. 2. Informal description of protocols P_0 through P_7.

$P_1 \to P_2$: Using m and μ drawn randomly from $\mathsf{SExp}_{p,g,c+c'}$ introduces a statistical distinguishability factor of $O(q_{\text{se}})2^{-c'}$. Then as in the PPK proof, the remaining possibility of distinguishing the protocols is when $\mathcal{A}$ makes an $H_3(\langle C, S, \cdot, \cdot, \cdot, \gamma' \rangle)$ query where $\gamma' = H_1(\pi_C)$, but $\mathcal{A}$ has not actually made the $H_1(\pi_C)$ query. (That is, the adversary "guesses" the correct output of an $H_1(\cdot)$ query.) However, in our proof the $H_1(\cdot)$ function outputs values in a different range, so the q^{-1} factor is replaced by $2^{-c+c'}$, and thus the distinguishing factor becomes $O(q_{\text{ro}})2^{-c}$.

$P_2 \to P_3$: Similar to the PPK proof (with DHSE replacing CDH), this can be shown using a standard reduction from DHSE. The decrease in advantage of the adversary is at most $2\mathsf{Succ}^{\mathsf{DHSE}}_{p,g,c}(t', q_{\text{ro}}) + O(q_{\text{ex}})2^{-c'}$.

$P_3 \to P_4$: As in the PPK proof, there is no decrease in the advantage of the adversary.

$P_4 \to P_5$: Similar to the PPK proof (with DHSE replacing CDH), this can be shown using a standard reduction from DHSE. However, as opposed to the PPK proof, one must make a guess at the $H_1(\cdot)$ query involved in the adversary's double password guess, since we do not have random self-reducibility in $\mathsf{SExp}_{p,g,c}$. The decrease in advantage of the adversary is at most $2q_{\text{ro}}\mathsf{Succ}^{\mathsf{DHSE}}_{p,g,c}(t', (q_{\text{ro}})^2) + O(q_{\text{se}})2^{-c'}$.

$P_5 \to P_6$: This is a reduction from DHSE, similar to the previous one. The decrease in advantage of the adversary is at most $2q_{\text{ro}}\text{Succ}^{\text{DHSE}}_{p,g,c}(t',(q_{\text{ro}})^2) + O(q_{\text{se}})2^{-c'}$.

$P_6 \to P_7$ As in the PPK proof, by inspection one can see that these two protocols are indistinguishable.

The probability of $\mathcal{A}$ succeeding in P_7 is now at most the probability of $\mathcal{A}$ guessing the correct password, plus the probability of succeeding while not guessing the correct password. Since there are at most q_{se} queries to the password oracle, and passwords are chosen uniformly from a dictionary of size N, the probability of $\mathcal{A}$ guessing the correct password is at most $\frac{q_{\text{se}}}{N}$. Then the probability of $\mathcal{A}$ succeeding in P_7 without guessing the password is the probability of $\mathcal{A}$ making a Test query and succeeding in guessing the bit used in that query. But as in the PPK proof, the view of $\mathcal{A}$ can be seen to be independent of the unrevealed session keys, and thus the probability of succeeding in this way is exactly $\frac{1}{2}$. Thus the advantage of $\mathcal{A}$ in P_7 is at most $\frac{q_{\text{se}}}{N}$, and the theorem follows.

Proofs for Two-Server Password Authentication

Michael Szydlo and Burton Kaliski

RSA Laboratories,
Bedford, MA 01730, USA
{mszydlo,bkaliski}@rsasecurity.com

Abstract. Traditional password-based authentication and key-exchange protocols suffer from the simple fact that a single server stores the sensitive user password. In practice, when such a server is compromised, a large number of user passwords, (usually password hashes) are exposed at once. A natural solution involves splitting password between two or more servers. This work formally models the basic security requirement for two-server password authentication protocols, and in this framework provides concrete security proofs for two protocols. The first protocol considered [7] appeared at USENIX'03, but contained no security proof. For this protocol, we provide a concrete reduction to the computational Diffie-Hellman problem in the random oracle model. Next we present a second protocol, based on the same hard problem, but which is simpler, and has an easier, tighter reduction proof.

Keywords: password authentication, secret sharing, concrete security reduction

1 Introduction

Passwords remains the most widespread method of user authentication to date, despite their inherent weaknesses. For example, user passwords, or password hashes are often stored in a server database, and the user authenticates by sending the password back using a server-side SSL authenticated channel. Of course, all password systems permit an attacker to make some number of guesses before the server locks the account down. However, a much more serious vulnerability exists: in case of a server compromise, an attacker may obtain *all* user passwords, or password hashes in the database at once.

Strengthening Passwords: The convenience of user-chosen password authentication protocols has caused them to be widely deployed. Among the weaker protocols one finds passwords sent in the clear, reusable low entropy PINs, and hash based challenge response techniques. A commonly used, and better, approach is to send a password or password hash over a server-side SSL- authenticated connection. Conceptually, these approaches still suffer from the fact that a user may be tricked into revealing his password to a server who does not know it. Starting with *Encrypted Key Exchange*[2] of Bellovin and Merritt, the benefits of a zero-knowledge based approach were realized. The goal of such protocols is to provide an authentication procedure which does not reveal a user password to any party who does not already have it. This line of research continued in several directions [8, 14, 4, 12], and represent a significant improvement in client-server protocols.

Multiple Server Use: Despite the improvements described above, single server password based authentication protocols do not protect from server compromise in a satisfactory way. Typically, an attacker who breaches a server will be able to obtain a very large number of user passwords, perhaps after running a dictionary attack (salt merely slows this). The natural approach to addressing this vulnerability is the use of multiple servers. In such schemes, the capability of verifying a password is split among two or more machines, and more than a certain threshold number of servers need to collude to recover the password. Starting with the work of Ford and Kaliski[6], various zero-knowledge multiple server password protocols have been proposed [9, 13, 7]. Multi-server protocols should provide basic username-password authentication to the collection of servers, without using special hardware or long-term client side key storage. Even for low-entropy passwords, an attacker should not be able to improve upon the naive guessing strategy without corrupting a threshold number of servers. On the other hand, these protocols do not pretend to have unrealistic goals of preventing denial of service or protecting user passwords in the case of client compromise.

Our focus, the two-server approach, is appealing for several reasons. With just two servers, the largest risk of wholesale password theft is greatly diminished. Also, deploying a large number of independently run servers appears logistically challenging, whereas the addition of a second server may be feasible in practice.

Provable Security: Increasingly, it has been realized that the proposal of a cryptographic scheme is only as valuable as its accompanying provable security analysis. The security proof techniques based on complexity theoretic foundations, including the copious general results on secure multi-party computation and threshold cryptography, provide tools for analyzing the kinds of protocols we are interested in. Typically, this framework is used to present asymptotic security definitions and security proofs[1]. However, for a protocol which is to be deployed, a *concrete* security analysis is required.

Our Contributions: In this work, we describe an appropriate framework for analyzing the concrete security of two-server password based authentication schemes. For a (random oracle) variant of [7], we provide a concrete reduction to the computational Diffie-Hellman problem. We also present a second protocol, based on the same hard problem, whose security proof is tighter, and simpler. All of our the security proofs are presented as explicit reduction algorithms which relate the difficulty of two computational problems. This approach allows for a more transparent concrete security analysis. Given that difficult security proofs, are sometimes left unread, we hope that our explicit approach is helpful.

---

[1] The somewhat misleading identification of the term "polynomial time" with "efficient" is due to the notion's stability among different models of computation.

1.1 Organization

The rest of this paper is organized as follows. In Section 2, we discuss the structure and desired security properties of a two-server password authentication protocol. In Section 3, we describe a general framework for concrete security reductions. In Section 4, we recall the protocol from [7], and present a new protocol. In Section 5, we define the concrete adversarial experiments appropriate for our two party authentication protocols. In Section 6 and 7 we provide the explicit reduction algorithm and state the concrete security result for these schemes. We summarize the results and conclude in Section 9. In the appendix, we discuss the unmodified scheme of [7], and the use of a Decisional Diffe-Hellman oracle.

2 Communication Framework and Desired Security

A two server authentication protocol involves a *Client* and two servers. Following [7], the two servers will be denoted *Blue* and *Red*. During an enrollment phase, the user chooses a password, which is processed by the client to produce registration messages for each server. Later, when a claimant enters a password, the client prepares and sends authentication messages to each server. After the two servers complete a verification protocol, the claimant is notified of the result by one or both servers.

To model a scenario in which the identities of the Blue and Red servers are easily ascertained, we assume that all parties employ a *secure channel* to Blue and Red. In practice, this can be realized with SSL. Architecturally, it may be desirable for the client to communicate with a single server, and this is easily accomplished by treating one server (say Blue) as a router.

The reader will easily verify that the protocols we describe are *complete*; a claimant with correct password will always authenticate correctly. More difficult is to show the soundness property: that an adversary can not do much better than password guessing.

Password Privacy: In this paper, we are interested in measuring if the two-party password protocols optimally protects the sensitive password data in the event that one server is compromised, and that compromising one server does not help an adversary authenticate to the other server. An experiment to test this should be designed so that an adversary

1. Tries to authenticate as a user who has previously enrolled.
2. May compromise one server, gaining the ability to impersonate messages.
3. May pose as the user and interact in some number of rounds, (denoted T).
4. May prompt the actual user to authenticate with the correct password.
5. Is allowed some bounded number of random oracle calls, (denoted Q)[2].
6. Is compared with an ideal-world adversary, allowed T password guesses.

---

[2] Artificially counting random oracle calls this way is a feature of random oracle security proofs. When arguing that the security carries over to protocols using a hash function, Q is usually set to be proportional to the adversary's running time.

To simplify the formal experiments which follow, we make an additional assumption on the protocol. Namely we assume the two servers employ a session management technique which precludes simultaneous authentication attempts for the same username, and also eliminate attacks which confuse messages corresponding to distinct usernames[3]. This means that the adversary will not gain any advantage by interleaving messages among concurrent sessions.

Limitations of the Model: For concreteness, we set the adversarial goal to be persuading the non-corrupted server to authenticate the adversary as user *username*. It is straightforward to alter the exact experiments described below for the goal of correctly guessing the password. This can be a more natural goal, for instance, when one server (Blue) is granting access to some service, and another (Red) is present to elliminate a single point of password compromise. Then, the natural goal of an adversary compromising Blue is to learn user passwords.

The adversarial capabilities described above do not measure the potential advantage an adversary might gain from *keying error*. Since the adversary is only allowed to activate the client on the *correct* password, the model does not capture the potential advantage for an adversary who observes a client launching the authentication sequence with a incorrect but related password. Although it is clumsy to model, it is conceivable that an adversary might benefit from this.

3 Formal Security Model

3.1 Adversarial Experiment Background

Parties and Experiments: All cryptographic parties are modeled as known stateful, probabalistic algorithms, whose inputs and outputs are interpreted as messages. The adversary, denoted *Adv*, is modeled by an arbitrary, unspecified algorithm. *Experiments*, or *hard problems* are algorithms which call one or more parties (black box) and output a bit $\{0,1\}$, and are used to describe joint properties of the parties. For each adversary *Adv*, we denote the *running-time* of *Adv* in experiment *Exp* by $Time_{Exp}(Adv)$, and the *success* by $Succ_{Exp}(Adv) = Prob[Exp^{Adv}() = 1]$.

Adversarial Capabilities: A *security property* of a protocol is defined in terms of an experiment which specifies both the *adversarial capabilities* (limiting the number and order of messages sent), and the *adversarial objective*, or success criteria. These experiments are designed to measure the ability of an attacker to disrupt normal protocol flow, or to learn a secret.

Concrete Reduction: A *concrete reduction* from hard problem Exp_{Re} to Exp_{Hp} is a black box conversion of adversary Adv_{Re} for the first to an adversary Adv_{Hp} for the second. More specifically, it consists of:

1. An algorithm *Reduce*, defining $Adv_{Hp} = Reduce^{Adv_{Re}}$.
2. A formula for $Succ_{Exp_{Hp}}(Adv_{Hp})$ in terms of $Succ_{Exp_{Re}}(Adv_{Re})$.
3. A formula for $Time_{Exp_{Hp}}(Adv_{Hp})$ in terms of $Time_{Exp_{Re}}(Adv_{Re})$.

---

[3] The username is included as input to the each random oracle (hash function) call.

To be meaningful, Exp_{Hp} should represent a well studied computational problem, such as factoring, or the computational Diffe-Hellman problem[4].

A *comparative concrete reduction* from hard problem Exp_{Re} to Exp_{Id} and Exp_{Hp}, also includes an algorithm $ReduceId$, defining $Adv_{Id} = ReduceId^{Adv_{Re}}$, and a formula for $Succ_{Exp_{Hp}}(Adv_{Hp})$ in terms of the real-ideal world advantage:

$$Ad = Succ_{Exp_{Re}}(Adv_{Re}) - Succ_{Exp_{Id}}(Adv_{Id}). \quad (1)$$

No Complexity Assumptions: In the concrete framework, no notion of computational indistinguishability is required, and complexity assumptions play a reduced role[5]. Although not required by this framework, a *security parameter k* can be used to calibrate the scheme so that the underlying hard problem is more difficult, and thus the attacker's task more costly.

Random Oracle Disclaimer: Unfortunately, our protocols involve hash functions, yet our security statements do not reduce to the associated hard problems of inverting a hash function on a random input or of finding a collision. Instead, our analysis pretends that the hash functions are replaced with "truly random" functions[1]. As with all random oracle proofs, the security statements we prove, describe most directly properties of a *related* protocol in which all parties must query a distinct cryptographic **trusted third party** to evaluate the hash function. This party, easily implemented with a sorted table, chooses the random function in stages by replying to queries randomly and consistently[6].

4 Two Protocols

Secret Sharing Basis: We now describe a slightly modified version of the protocol[7], and describe our new protocol. Before we begin, we provide the basic intuition and notation common to each. During the registration phase the client splits the password *pass* into shares by choosing a random pad R as the first share, and setting the second share $P = R \oplus pass'$. Later, during the authentication phase, a claimant using password *pass'*, selects a distinct random pad, R', and sets $P' = R' \oplus pass'$. The Blue server computes $\hat{A} = P \oplus P'$, while the Red server obtains $\hat{B} = R \oplus R'$. Clearly $\hat{A} = \hat{B} \iff pass = pass'$.

Relationship with Password Key Exchange: The problem of comparing two values in zero knowledge is known as the *socialist millionaire's problem* [5, 10, 11, 3]. Password based key exchange protocols must solve this problem. The authentication protocols we consider are not key exchange protocols, *per se*, as they already utilize SSL session keys, but the interaction between the two servers also follows a solution to the socialist millionaire's problem. The reader

---
[4] Reduction arguments relying on stronger assumptions such as the Decisional Diffe-Hellman assumption, or oracle "gap-assumptions" are somewhat less compelling.
[5] Common complexity-theoretic notions of "negligible function", and "computational indistinguishability" may distract from the focus of a concrete security analysis.
[6] We prefer to explicitly describe the random oracle as a trusted third party so as not to overstate the security implications of a random oracle proof.

is encouraged to compare messages exchanged between Blue and Red (especially in the new scheme) with zero-knowledge key-exchange protocols such as PAK[4].

We also remark that the security of the *three-party protocols* we consider does not follow automatically from a solution to the socialist millionaire's problem. In general, deducing security properties of composed protocol instances is difficult; furthermore, in our particular case, the adversary can influence both $\hat{A}$ and $\hat{B}$.

Notation: The first protocol is from [7] (modifications discussed below), and the new protocol follows the same framework. For easy comparison, we use the same notation and display the two side by side. Let k be an integer, $\mathcal{G}$ be a cyclic group of order q with generator G, and $gp.gen(k)$ an algorithm which generates (a description of) $\mathcal{G}$.[7] Let $pass, pass', R, P, R', P' \in \{0,1\}^k$, e, f be integers in $[1, q]$, $A, B, Y_0, Y_1, Z_0, Z_1 \in \mathcal{G}$; $H_0, H_1 \in \{0,1\}^k$, and $ok_0, ok_1 \in \{0,1\}$. Let w be a function $\{0,1\}^* \to \mathcal{G}$, and let h be a function $\{0,1\}^* \to \{0,1\}^k$, implemented by random oracles W, and H. The symbol $\xleftarrow{\$}$ denotes a random assignment.

Modified BJKS(k)	New Scheme(k)
Parameters	**Parameters**
$(\mathcal{G}, q, G) \leftarrow gp.gen(k)$	$(\mathcal{G}, q, G) \leftarrow gp.gen(k)$
Registration	**Registration**
$pass \leftarrow passgen()$	$pass \leftarrow passgen()$
$R \xleftarrow{\$} \{0,1\}^k$	$R \xleftarrow{\$} \{0,1\}^k$
$P \leftarrow R \oplus pass$	$P \leftarrow R \oplus pass$
Authentication	**Authentication**
Client.auth1$(pass', U)$:	Client.auth1$(pass', U)$:
$R' \xleftarrow{\$} \{0,1\}^k$	$R' \xleftarrow{\$} \{0,1\}^k$
$P' \leftarrow R' \oplus pass'$	$P' \leftarrow R' \oplus pass'$
Blue.auth1(P', U):	Blue.auth1(P', U):
$e \xleftarrow{\$} [1, q]$	$e \xleftarrow{\$} [1, q]$
$A \leftarrow w(P \oplus P', U)$	$A \leftarrow w(P \oplus P', U)$
$Y_0 \leftarrow AG^e$	$Y_0 \leftarrow AG^e$
Red.auth1(R', Y_0, U):	Red.auth1(R', Y_0, U):
$f \xleftarrow{\$} [1, q]$	$f \xleftarrow{\$} [1, q]$
$\star$ $B \leftarrow w(R \oplus R', U)$	$B \leftarrow w(R \oplus R', U)$
$Y_1 \leftarrow BG^f$	$\star$ $Y_1 \leftarrow G^f$
$Z_1 \leftarrow (Y_0/B)^f$	$Z_1 \leftarrow (Y_0/B)^f$
$\star$ $H_1 \leftarrow h(Y_0, Y_1, Z_1, R \oplus R', U)$	$H_1 \leftarrow h(Y_0, Y_1, Z_1, R \oplus R', U)$
Blue.auth2(Y_1, H_1):	Blue.auth2(Y_1, H_1):
$Z_0 \leftarrow (Y_1/A)^e$	$\star$ $Z_0 \leftarrow Y_1^e$
$\star$ $H_0 \leftarrow h(H_1, Z_0, P \oplus P', U)$	$\star$ $ok_0 \leftarrow H_1 \stackrel{?}{=} h(Y_0, Y_1, Z_0, P \oplus P', U)$
Red.auth2(H_0):	$\star$ if(ok_0), $conf \leftarrow P \oplus P'$ else $\emptyset$
$\star$ $ok_1 \leftarrow H_0 \stackrel{?}{=} h(H_1, Z_1, R \oplus R', U)$	$\star$ Red.auth2$(ok_0, conf)$:
Blue.auth3():	$\star$ $ok_1 \leftarrow R \oplus R' \stackrel{?}{=} conf$
$\star$ $ok_0 \leftarrow H_1 \stackrel{?}{=} h(Y_0, Y_1, Z_0, P \oplus P', U)$	$\star$ Blue.auth3():
Client.auth2(ok_0).	Client.auth2(ok_0).

---

[7] This description includes an efficient test of membership for $\mathcal{G}$. E.g., $\mathcal{G} = \mathcal{Z}_p^*$.

Modifications: The first scheme described above differs from [7] in computations marked with the symbol $\star$. The main difference is that $P \oplus P'$ and $R \oplus R'$ have been added as hash function inputs. This allows for a less awkward and more efficient security proof. Also, the $\mathcal{G}$ -membership check for Y_0, Y_1 is not depicted, but implicitly assumed to be part of the message parsing.

Although the new scheme resembles the first in form, it is related more naturally to the Diffie-Hellman problem (and to PAK), and thus has a tighter proof. In the new scheme, Y_1 is set to G^f, instead of BG^f, no H_0, is computed, and instead the Blue server sends $R \oplus R'$ back to the Red server as a confirmation message. These further differences are also marked by the symbol $\star$.

5 Adversarial and Computational Problems

Following the framework of Section 3, we now describe the adversarial experiments corresponding to the actual adversary (Exp_{Re}), the ideal-world adversary (Exp_{Id}), and the underlying hard problem (Exp_{Hp}). The concrete security statements tie together the performance of adversaries for three such experiments, quantifying the informal statement "the adversary can't do significantly better than guessing".

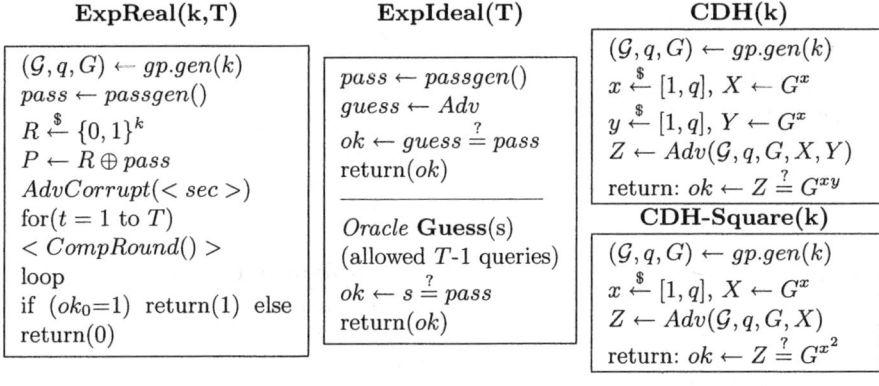

The ideal world adversarial experiment is very simple. It effectively must guess the password in T tries. Clearly, the success probability of any adversary in **ExpIdeal(T)** is less than the sum of the most common T passwords produced by $passgen()$. We denote this probability $GuessProb(T)$.

The hard problem experiment above simply corresponds to the Diffie-Hellman problem for group $\mathcal{G}$. Experiment **CDH-Square(k)** corresponds to the problem of computing $DH(X, X)$ from X, where $DH(G^a, G^b)$ denotes the element G^{ab}.

The real-world experiment is more interesting, and depends on which server is compromised. The secret $< sec >$ revealed during compromise is either P if Blue is compromised, or R if Red is compromised. The per-round interactions ($CompRound()$ above), are described in more detail below with **ExpRealBlue(k,T)** and **ExpRealRed(k,T)**, which reflect the adversary's capabilities listed in Section 2. Separate experiments are given for the new scheme.

5.1 Per Round Interactions

These experiments are designed to follow the framework of in Section 2, except for an additional simplification. Namely, the possibility of the adversary "activating" a user to authenticate with the correct password is treated separately, so that we can assume that the adversary must interact within a session. It is not difficult to see why the adversary will not benefit from this activity. In such a situation $P \oplus P' = R \oplus R'$, and regardless of whether the adversary has compromised Blue or Red, the messages of the other server may be perfectly simulated, so the adversary can learn nothing. Further details are in Appendix D.

CompRedRound

> **Adversary:**
> $P' \leftarrow AdvClient1()$
> **Blue.auth1(P'):**
> $e \stackrel{\$}{\leftarrow} [1, q]$
> $A \leftarrow w(P \oplus P', U)$
> $Y_0 \leftarrow AG^e$
> **Adversary:**
> $Y_1, H_1 \leftarrow AdvRed.auth1(Y_0)$
> **Blue.auth2(Y_1, H_1):**
> $Z_0 \leftarrow (Y_1/A)^e$
> $H_0 \leftarrow h(H_1, Z_0, P \oplus P', U)$
> **Adversary:**
> $AdvRed.auth2(H_0)$
> **Blue.auth3():**
> $ok_0 \leftarrow H_1 \stackrel{?}{=} h(Y_0, Y_1, Z_0, P \oplus P', U)$
> **Adversary:**
> $AdvClient2(ok_0)$

CompRedRoundNew

> **Adversary:**
> $P' \leftarrow AdvClient1()$
> **Blue.auth1(P'):**
> $e \stackrel{\$}{\leftarrow} [1, q]$
> $A \leftarrow w(P \oplus P', U)$
> $Y_0 \leftarrow AG^e$
> **Adversary:**
> $Y_1, H_1 \leftarrow AdvRed1(Y_0)$
> **Blue.auth2(Y_1, H_1):**
> $Z_0 \leftarrow (Y_1/A)^e$
> $ok_0 \leftarrow H_1 \stackrel{?}{=} h(Y_0, Y_1, Z_0, P \oplus P', U)$
> if(ok_0), $conf \leftarrow P \oplus P'$ else $\emptyset$
> **Adversary:**
> $AdvRed2(ok_0, conf)$
> **Blue.auth3():**
> **Adversary:**
> $AdvClient2(ok_0)$

CompBlueRound

> **Adversary:**
> $R' \leftarrow AdvClient1()$
> $Y_0 \leftarrow AdvBlue1()$
> **Red.auth1(R', Y_0, U):**
> $f \stackrel{\$}{\leftarrow} [1, q]$
> $B \leftarrow w(R \oplus R', U)$
> $Y_1 \leftarrow BG^f$
> $Z_1 \leftarrow (Y_0/B)^f$
> $H_1 \leftarrow h(Y_0, Y_1, Z_1, R \oplus R', U)$
> **Adversary:**
> $H_0 \leftarrow AdvBlue2(Y_1, H_1)$
> **Red.auth2(H_0):**
> $ok_1 \leftarrow H_0 \stackrel{?}{=} h(H_1, Z_1, R \oplus R')$
> **Adversary:**
> $AdvClient2()$

CompBlueRoundNew

> **Adversary:**
> $R' \leftarrow AdvClient1()$
> $Y_0 \leftarrow AdvBlue1()$
> **Red.auth1(R', Y_0, U):**
> $f \stackrel{\$}{\leftarrow} [1, q]$
> $B \leftarrow w(R \oplus R', U)$
> $Y_1 \leftarrow G^f$
> $Z_1 \leftarrow (Y_0/B)^f$
> $H_1 \leftarrow h(Y_0, Y_1, Z_1, R \oplus R', U)$
> **Adversary:**
> $ok, conf \leftarrow AdvBlue2(Y_1, H_1)$
> **Red.auth2$(ok_0, conf)$:**
> $ok_1 \leftarrow R \oplus R' \stackrel{?}{=} conf$
> **Adversary:**
> $AdvClient2()$

These per round interactions have been naturally derived from the real protocols by inserting the adversary's where the compromised server would be active.

6 Concrete Reduction Algorithms

We now present the reduction algorithms, which will ultimately convert the real world adversary into algorithm for solving **CDH(k)** or **CDH-Square(k)**.

6.1 Strategy

We consider adversaries which compromise the Blue and Red servers separately, and we prove each concrete reduction statement in two stages.

In Stage 1, we immediately present the reduction algorithm itself: *Reduce*, which yields a **CDH(k)** adversary $Adv_{Hp} = Reduce^{Adv_{Re}}$ for each real-world adversary Adv_{Re}. Viewing the transcript of a real-world experiment $Exp_{Re}^{Adv_{Re}}$ as a sequence of random variables, we will define (separately for each experiment) an event called an *effective guess* which can be loosely interpreted as a password guess, and we say the event E_{OverT} occurs if there are more than T effective guesses. In this stage we also relate the success $Succ_{Exp_{Hp}}(Adv_{Hp})$ to the probability $Prob[E_{OverT}]$.

The goal of Stage 2, is to compare $Prob[E_{OverT}]$ with the success of the real and ideal-world adversaries. To this end we study the interaction of a real world adversary with a perfect simulator that calls a password guessing oracle. By limiting the number of password guessing oracle calls to T, we obtain an auxiliary reduction algorithm *ReduceId*, which yields an ideal-world Adversary $Adv_{Id} = ReduceId^{Adv_{Re}}$ for each real-world adversary Adv_{Re}. As an ideal world adversary, its success must be bounded by $GuessProb(T)$. The simulation is also constructed such that if all effective guesses are incorrect, the success probability is only $T2^{-k}$. Taken together, this implies that $Succ_{Exp_{Re}}(Adv_{Re}) \leq Prob[E_{OverT}] + GuessProb(T) + T2^{-k}$. Rewriting this relation as

$$Prob[E_{OverT}] \geq Succ_{Exp_{Re}}(Adv_{Re}) - GuessProb(T) - T2^{-k}, \qquad (2)$$

and combining it with the bound of stage 1, we finally relate $Succ_{Exp_{Hp}}(Adv_{Hp})$, to the adversarial advantage (Eq. 1).

6.2 Compromise-Red Reduction (Modified BJKS)

For this experiment, we define an **effective guess** on $p\tilde{a}ss$ in round t to be the event that for the Y_0, Y_1, H_1, P' sent in round t, oracle H was called on input $(Y_0, Y_1, \tilde{Z}, \tilde{P}, U)$ or on input $(H_1, \tilde{Z}, \tilde{P}, U)$, where $\tilde{P} = P' \oplus R \oplus p\tilde{a}ss$, $\tilde{A} = W(\tilde{P})$, and $\tilde{Z} = DH(\frac{Y_0}{A}, \frac{Y_1}{A})$. We also let $Q_\dagger = Q + 2T$ denote the maximal number of H oracle queries in the entire experiment.

For the BJKS-modified scheme, our aim is reduce to **CDH-Square(k)**. Thus *Reduce* accepts X as input and attempts to compute $DH(X, X)$. *Reduce* calls Adv_{Re} black box, and employs code from the actual protocol, as well as custom versions, *simulations*, of certain functions, generally following the flow of Exp_{Re}. For this reduction, a custom W-oracle embeds X by responding to each new query with X^r for a random $r \in [1, q]$. Next, for one randomly chosen round

$t_0 \in [1, T]$, the usual adversarial round interaction will be replaced with one in which Blue's normal operation is simulated so that (1) $Y_0 = G^a$ has a random known discrete log a, (2) H_0 is set to be equal to a value coinciding with a random H oracle response in the range of indices $[1, Q_\dagger]$, and (3) ok_0 is chosen randomly from $\{0, 1\}$.

Such a simulation is not perfect, but its transcript distribution follows that of an actual real world adversary with probability at least $1/(2Q_\dagger)$. Since t_0 is random, with probability at least $Prob[E_{OverT}]/(2Q_\dagger T)$, two effective guesses will be made on round t_0. In the final stage, two indices in $[1, Q_\dagger]$ are chosen at random. If these indices correspond to H queries of the two effective guesses, these H oracle queries will include the distinct pairs $(\tilde{P}, Z)$ and $(\tilde{P}', Z')$, such that $Z = DH(\frac{Y_0}{\tilde{A}}, \frac{Y_1}{\tilde{A}})$ for $\tilde{A} = W(\tilde{P})$, and $Z' = DH(\frac{Y_0}{\tilde{A}'}, \frac{Y_1}{\tilde{A}'})$ for $\tilde{A}' = W(\tilde{P}')$. Searching through the maximum of $Q + T + 2$ calls to the W oracle, we locate the two integers r, and r' such that $\tilde{A} = X^r$ and $\tilde{A}' = X^{r'}$. Provided $r, r', r - r' \neq 0$, the assignment

$$D \leftarrow \{[Z(Y_1/\tilde{A})^{-a}]^{1/r} \,/\, [Z'(Y_1/\tilde{A}')^{-a}]^{1/r'}\}^{1/(r-r')} \quad (3)$$

yields $DH(X, X)$. The final chance of success is $Prob[E_{OverT}]R/(2\{Q_\dagger^3 T)$, where

$$R = 1 - (Q + T + 2 + 2)(Q + T + 2 + 1)/2q \quad (4)$$

lower bounds the chance that all the W-oracle results are distinct and nonzero.

ReduceRed(Adv, T,Q)(X,$\mathcal{G}$,q,G)

for $i = 1$ to $Q + T + 2$:
  $r_i \xleftarrow{\$} [1, q]$, $W_i \leftarrow X^{r_i}$
$t_0 \xleftarrow{\$} [1, T]$, $ind_1 \xleftarrow{\$} [1, Q_\dagger]$
$hrand \xleftarrow{\$} \{0, 1\}^k$

$pass \leftarrow passgen()$
$R \xleftarrow{\$} \{0, 1\}^k$, $P \leftarrow R \oplus pass$
$AdvCorrupt(R)$
for($t = 1$ to T):
  if($t = t_0$) SimulatedRound()
  else CompRedRound()

$ind_2, ind_3 \xleftarrow{\$} [1, Q_\dagger]$
$(Z_0, \tilde{P}) \leftarrow Hin(ind_2)$
$(Z_0', \tilde{P}') \leftarrow Hin(ind_3)$
$A \leftarrow W(\tilde{P}), r \leftarrow Wseek(A)$
$A' \leftarrow W(\tilde{P}'), r' \leftarrow Wseek(A')$
$D \leftarrow \{\frac{Z(Y_1/A)^{-a}]^{1/r}}{Z'(Y_1/A')^{-a}]^{1/r'}}\}^{1/(r-r')}$
output(D)

W,H Oracle Simulation

Oracle **W**(s) -Adv gets Q queries.
On i'th new query: $w(s) = W_i$.
Oracle **H**(s) -Adv gets Q queries.
ind_1'th new query: $h(s) = hrand$.
Other new queries: $h(s) \xleftarrow{\$} \{0, 1\}^k$

Simulated Round
Adversary:
$P' \leftarrow AdvClient1()$
Blue.sim1(P'):
$a \xleftarrow{\$} [1, q]$, $Y_0 \leftarrow G^a$
Adversary:
$Y_1, H_1 \leftarrow AdvRed.auth1(Y_0)$
Blue.sim2(Y_1, H_1):
$H_0 \leftarrow hrand$
Adversary:
$AdvRed.auth2(H_0)$
Blue.sim3():
$ok_0 \xleftarrow{\$} \{0, 1\}$
Adversary:
$AdvClient2(ok_0)$

Passing to stage 2, our goal is to find a bound on $Prob[E_{OverT}]$. This is accomplished with another reduction, $Adv_{Id} = ReduceId^{Adv_{Re}}$ creating an ideal world adversary which makes guesses to a password guessing oracle. Algorithm *ReduceId* is constructed from **ExpRealRed(k,T)**, but replaces the algorithms of Blue, H, and W with simulated algorithms, *which do not directly make use of the password.*

ReduceRedId(Adv, T,Q)

Simulated Round	**W,H Oracle Simulation**
Blue.sim1(P'): $a \xleftarrow{\$} [1,q]$, $Y_0 \leftarrow G^a$ **Adversary:** $Y_1, H_1 \leftarrow AdvRed.auth1(Y_0)$ Blue.sim2(Y_1, H_1): for each h_{in} $\quad (valid, p\tilde{a}ss) \leftarrow \textbf{Valid}(h_{in}, t)$ $\quad$ if($valid$ AND **IdealGuess**($p\tilde{a}ss$)) $\quad\quad H_0, H_1' \leftarrow \textbf{Correct}(h_{in})$ $\quad\quad$ return(H_0) $H_0 \leftarrow hrand_t \xleftarrow{\$} \{0,1\}^k$ $H_1' \leftarrow hrand'_t \xleftarrow{\$} \{0,1\}^k$ Blue.sim3(): $ok_0 \leftarrow H_1' \stackrel{?}{=} H_1$	Oracle **W**(s) On i'th new query: $r_i \xleftarrow{\$} [1,q]$ $w(s) = G_i^r$. Oracle **H**(h_{in}) for each completed round t, $\quad (valid, p\tilde{a}ss) \leftarrow \textbf{Valid}(h_{in}, t)$ $\quad$ if($valid$ AND **IdealGuess**($p\tilde{a}ss$)) $\quad\quad h(h_{in}) \leftarrow hrand_t$ $\quad\quad h(h'_{in}) \leftarrow hrand'_t$ Other new queries: $\quad h(h_{in}) \xleftarrow{\$} \{0,1\}^k$ Oracle **IdealGuess**($p\tilde{a}ss$)) return ($p\tilde{a}ss \stackrel{?}{=} pass$)

In contrast with Stage 1, Blue's messages will be simulated for every round. The challenge is to simulate Blue's second and third interactions: **Blue.sim2**(Y_1, H_1), and **Blue.sim3**(). Both H_0 and ok_0 must be consistent with the unknown password. This is accomplished by examining all queries made to H so far, to determine if they corresponds to effective guesses. This is done with a test

$$(valid, pass) \leftarrow \textbf{Valid}(\tilde{Z}, \tilde{P}, t)$$

which returns $valid = 1$, if $\tilde{Z} = DH(\frac{Y_0}{\tilde{A}}, \frac{Y_1}{\tilde{A}})$ for $\tilde{A} = W(\tilde{P})$, and 0 otherwise, and also returns the guessed password $pass = \tilde{P} \oplus P' \oplus R$, when the guess is valid. In fact, this test can be efficiently performed if the discrete log of $\tilde{A}$, and Y_0 are known[8]. The test is used three ways, (1) to ensure each H_0 is consistent with previous H-queries, (2) to ensure each ok_0 is consistent with previous H-queries, (3) to ensure future H-queries are consistent with H_0, ok_0 values of previous rounds.

A slight complication arrises from the fact that an effective guess can correspond to two types of hash calls. To deal with this, the simulator uses the function **Correct**(h_{in}) which looks up the hash query values $\tilde{Z}$, and $\tilde{P}$, and returns both hash values $H_1' = h(Y_0, Y_1, \tilde{Z}, \tilde{P}, U)$ and $H_0 = h(H_1, \tilde{Z}, \tilde{P}, U)$, once the password is discovered. If the H-oracle has not been called on the inputs corresponding to the correct password, the simulator will choose random responses $hrand_t$ and $hrand'_t$ for the two types of hashes. Later H queries are always

---

[8] Actually, the efficiency of Adv_{Id} is not important for our argument.

checked, and if one is found to correspond to an effective guesses for a previous round, both types of hash are answered consistently, with $hrand_t$ and $hrand'_t$.

This simulation is perfect if the number of **IdealGuess()** queries is not limited, and the number of queries is distributed exactly as the number of effective guesses made by Adv_{Re} in Exp_{Re}. Thus by the argument above in Section 6.1, we obtain bound (Eq. 2), on $Prob[E_{OverT}]$, so we can can relate $Succ_{Exp_{Hp}}(Reduce^{Adv_{Re}})$ to the advantage Ad defined in (Eq. 1). Letting R be the constant of (Eq. 4), and $Ad' = Ad - T2^{-k}$, we obtain

$$Succ_{Exp_{Hp}}(Reduce^{Adv_{Re}}) \geq \frac{Ad'R}{2TQ_\dagger^3}. \qquad (5)$$

Furthermore, $Time_{Exp_{Hp}}(Reduce^{Adv_{Re}}) = T_A + T_E + T_R$, where T_E is the time of the experiment Exp_{Re} itself, $T_A = Time_{Exp_{Re}}(Adv_{Re})$, and T_R is the additional time incurred by the reducer itself, which includes a cost proportional to $Qlog(Q)$ to manage the random oracle tables, the $Q + T + 2$ exponentiations for the $W = X^r$ embedding, and the exponentiations required for the final extraction of the candidate $DH(X, X)$ value.

6.3 Compromise-Blue Reduction (BJKS)

This reduction is quite similar to the compromise Red one, except the definition of *effective guess* is somewhat simpler as only one type of H query need be considered. The success of the derived adversary $Reduce^{Adv_{Re}}$ satisfies the same inequality (Eq. 5). Further details on the simulation are given in Appendix A.

6.4 Completing the Reduction to CDH

The reductions, as presented, reduce to the hard problem **CDH − Square(k)**. In order to further reduce to **CDH(k)**, we use a well known trick. The equation

$$DH(X,Y)^2 = \sqrt{DH(XY, XY)}/\sqrt{DH(X/Y, X/Y)} \qquad (6)$$

defines a reduction *Reduce*, such that for all **CDH − Square(k)** solvers S, with success ϵ completing in time τ, $T = Reduce^S$ is a **CDH(k)** solver with success ϵ^2 which completes in time 2τ. Using this approach, a significant loss of success probability results in our reductions. However, when the *decisional* Diffe-Hellman problem is feasible, the situation improves. In this case, we focus on measuring *expected time*, and a **CDH − Square(k)** solver S, with expected time τ can be converted to a **CDH(k)** solver T with expected time 2τ.

7 Reductions for the New Protocol

The reductions for the new protocol are significantly simpler. First, the apparently small change of setting $Y_1 = G^f$ (instead of $Y_1 = BG^f$), enables a direct

CDH(k) reduction in the compromise Blue experiment, rather than indirect approach via the **CDH − Square(k)** problem. Secondly, using the confirmation message $conf$, allows a direct comparison of the real-world adversary to the password guessing adversary, without even mentioning the **CHD** problem. The difference can easily be seen to be related to the chance of an H-oracle collision.

7.1 Compromise-Red (New Scheme)

For this experiment an **effective guess** on $pass$ in round t denotes the event that for the Y_0, Y_1, H_1, P' sent in round t, oracle H was called on input $(Y_0, Y_1, \tilde{Z}, \tilde{P}, U)$ resulting in H_1 where $\tilde{P} = P' \oplus R \oplus pass$, $\tilde{A} = W(\tilde{P})$, and $\tilde{Z} = DH(\frac{Y_0}{\tilde{A}}, Y_1)$. Note the different requirement for $\tilde{Z}$, and the additional H_1 requirement.

For this experiment we do not need the two stage proof, and instead construct an ideal world adversary directly. The construction of $ReduceID$, follows the same strategy described above in Section 6.2. Specifically, the W oracle is programmed with values of known discrete log, a random $Y_0 = G^a$ is sent with known discrete log a, and consistent values of H_0 are produced by the simulator. All of Blue server messages may be perfectly simulated provided that the oracle **IdealGuess()** is called for each effective guess. As above, this simulation implies Inequality (Eq. 2). However, two effective guesses on a single round would imply

$$h(Y_0, Y_1, \tilde{Z}, \tilde{P}, U) = H_1 = h(Y_0, Y_1, \tilde{Z}, \tilde{P}, U),$$

which is a hash collision. Thus $Prob[E_{OverT}] < S$, where

$$S = 1 - Q(Q-1)/2q \tag{7}$$

is a lower bound on the probability that none of the Q H-oracle results coincide. The resulting bound on the adversarial advantage has nothing to do with **CDH(k)**.

$$Succ_{Exp_{Re}}(Adv_{Re}) - GuessProb(T) \leq S + T2^{-k}. \tag{8}$$

7.2 Compromise-Blue (New Scheme)

For this experiment an **effective guess** on $pass$ round t denotes the event that for the Y_0, Y_1, H_1, P' sent in round t, oracle H was called on input $(Y_0, Y_1, \tilde{Z}, \tilde{P}, U)$, where $\tilde{P} = P' \oplus R \oplus pass$, $\tilde{A} = W(\tilde{P})$, and $\tilde{Z} = DH(\frac{Y_0}{\tilde{A}}, Y_1)$. In this experiment, the maximal number of H-queries is $Q_\dagger = Q + T$.

The reduction $ReduceId$, creating the ideal world adversary, follows the strategy of Section 6.2, yielding the usual Inequality (Eq. 2). However, in contrast with the scheme of [7], the main $Reduce$ algorithm, is able to solve **CDH(k)** directly. To define $Reduce$, we let X, Y be the **CDH(k)** problem instance. Following Section 6.2, the W oracle is programmed with values $B = X_i^r$, and for a randomly chosen round $t_0 \in [1, T]$, Y_1 is set equal to Y, and H_1 is set to be equal to the a'th H-oracle response, for a random index a in the interval $[1, Q_\dagger]$.

In the final stage, two indices in $[1, Q_\dagger]$ are chosen at random. If these indices correspond to the H queries of two effective guesses, these H oracle queries will include two pairs $(\tilde{P}, Z)$ and $(\tilde{P}', Z')$, such that $Z = DH(\frac{Y_0}{A}, Y_1)$ for $\tilde{A} = W(\tilde{P})$, and $Z' = DH(\frac{Y_0}{A'}, Y_1)$ for $\tilde{A}' = W(\tilde{P}')$. Searching through the (up to $Q+T+2$) W oracle queries, we find the two integers r, and r' such that $A = X^r$ and $A' = X^{r'}$. Provided $r - r' \neq 0$, the formula

$$D \leftarrow [Z/Z']^{r-r'} \tag{9}$$

yields $DH(X,Y)$. The success is at least $Prob[E_{OverT}]R/(Q_\dagger^3 T)$, where R is as in (Eq. 4), (a slightly smaller R suffices), so combining with (Eq. 2), yields

$$Succ_{Exp_{H_p}}(Reduce^{Adv_{Re}}) \geq \frac{Ad'R}{TQ_\dagger^3}. \tag{10}$$

8 Protocol and Proof Variants

Unmodified BJKS: We were still able to find security proofs for the *unmodified* scheme of [7] without the modifications described in Section 4. This analysis is described in Appendix B. The resulting success bound obtained was $\frac{AdR}{TQ_\dagger^5}$.

Using a Decision Oracle: If there is an efficient Diffe-Hellman decision algorithm for $\mathcal{G}$, it makes sense to exploit this in the reduction proofs. For certain elliptic curves, the Weil pairing provides such an efficient procedure. In addition to the technique described in Section 6.4, a decision oracle can be used to improve the simulations in *Reduce*. These techniques are discussed in Appendix C, and we include the expected running time of the CDH adversaries in the summary.

Red Server Notification: In the protocols we have studied, the Red server does not relay the result ok_1 back to the client. The proof for the protocol variant which includes this message requires some small modifications to the simulations.

9 Conclusions

We summarize the success of the derived **CDH(k)** Adversaries in this table. This allows a comparison of the reduction efficiency. The final columns show how an adversary can make Q_{DDH} DDH queries to obtain a success rate of $Ad'H/T$, and thus can solve **CDH(k)** by with probability near one, by repeating the algorithm *Reps* times. Notice that the difference in reduction "tightness" is significantly more pronounced when a decisional oracle is not available.

Scheme	Corrupted	Time	Success	Q_{DDH}	Reps
BJKS	Blue/Red	$2(T_E + T_A + T_R)$	$(Ad'R/2T)^2/Q_\dagger^{10}$	Q^2	$2T/Ad'R$
BJKS-M	Blue/Red	$2(T_E + T_A + T_R)$	$(Ad'R/2T)^2/Q_\dagger^6$	Q	$2T/Ad'R$
New	Blue	$T_E + T_A + T_R$	$Ad'R/2TQ_\dagger^3$	Q	$T/Ad'R$
	Red	-	-		

The revised scheme presented in this paper is also preferable from several other viewpoints: its security proof is more transparent, the message flows are more directly related to the usual Diffe-Hellman problem, and the server portion of the three party protocol (except the confirmation message) closely relates to the well studied PAK key exchange protocol. The second protocol was also designed to have a security proof, whereas the security proof for scheme [7] was found after its introduction. The preferred approach is to design the protocols concurrently with the proofs.

This work has presented a framework which may be useful for proving *concrete security statements* concerning two-party password-based authentication protocols. More generally, we hope that the approach of presenting explicit *reduction algorithms* instead of reduction proofs will be perceived as adding higher level of transparency to security proofs. Additionally, we hope our approach also illustrates that meaningful concrete security statements can be stated and proved independently of any traditional complexity-theory based foundations[9]. While very useful for feasibility results, complexity theory certainly does not encompass all of cryptography!

Acknowledgments

The authors would like to thank Phil MacKenzie for useful discussions, and the anonymous reviewers for comments and corrections.

References

1. M. Bellare and P. Rogaway. Random oracles are practical: A paradigm for designing efficient protocols. In *1st ACM Conference on Computer and Communications Security*, pages 62–73. ACM Press, 1993.
2. S. M. Bellovin and M. Merritt. Encrypted key exchange: Password-based protocols secure against dictionary attacks. In *IEEE Computer Society Symposium on Research in Security and Privacy*, pages 72–84. IEEE Press, 1992.
3. F. Boudot, B. Schoenmakers, and J. Traoré. A fair and efficient solution to the socialist millionaires' problem. *Discrete Applied Mathematics*, 111(1-2):23–36, 2001.
4. V. Boyko, P. MacKenzie, and S.Patel. Provably secure password-authenticated key exchange using diffie-hellman. In B. Preneel, editor, *Advances in Cryptology - Eurocrypt '00*, pages 156–, Berlin, 2000. Springer-Verlag. LNCS No. 1807.
5. R. Fagin, M. Naor, and P. Winkler. Comparing information without leaking it. *CACM*, 39(5):77–85, May 1996.
6. W. Ford and B. S. Kaliski Jr. Server-assisted generation of a strong secret from a password. In *Proceedings of the IEEE 9th International Workshop on Enabling Technologies (WETICE)*. IEEE Press, 2000.
7. B. Kaliski and M. Szydlo J. Brainard, A. Juels. Nightingale: A new two-server approach for authentication with short secrets. In *Proceedings of the 12th USENIX Workshop on Security*, pages 1–2. IEEE Computer Society, 2003.
8. D. P. Jablon. Research papers on strong password authentication, 2002. URL: www.integritysciences.com/links.html.

---

[9] **Corollary**: Assuming computational Diffe-Hellman, we have $Lim_{k\to\infty} Ad(T,k) = 0$.

9. D.P. Jablon. Password authentication using multiple servers. In David Naccache, editor, *Topics in Cryptology - CT-RSA 2001*, pages 344–360. Springer-Verlag, 2001. LNCS no. 2020.
10. M. Jakobsson and A. Juels. Mix and match: Secure function evaluation via ciphertexts. In T. Okamoto, editor, *ASIACRYPT 2000*, pages 162–177. Springer-Verlag, 2000. LNCS no. 1976.
11. M. Jakobsson and M. Yung. Proving without knowing: On oblivious, agnostic, and blindfolded provers. In *CRYPTO '96*, pages 186–200, 1996. LNCS no. 1109.
12. P. MacKenzie, S. Patel, and R. Swaminathan. Password-authenticated key exchange based on rsa. In T. Okamoto, editor, *Advances in Cryptology - Asiacrypt '00*, pages 599–, Berlin, 2000. Springer-Verlag. LNCS No. 1976.
13. P. Mackenzie, T. Shrimpton, and M. Jakobsson. Threshold password-authenticated key exchange. In M. Yung, editor, *CRYPTO 2002*, pages 385–400. Springer-Verlag, 2002. LNCS no. 2442.
14. M.Bellare, D.Pointcheval, and P. Rogaway. Authenticated key exchange secure against dictionary attacks. In B. Preneel, editor, *Advances in Cryptology - Eurocrypt '00*, pages 139–, Berlin, 2000. Springer-Verlag. LNCS No. 1807.

A Compromise Blue Details (BJKS)

This reduction algorithm takes the same form as that of Section 6.2, and we just include it for completeness. For this experiment an **effective guess** on *pass* in round t denotes the event that for the Y_0, Y_1, R' sent in round t, oracle H was called on input $(Y_0, Y_1, \tilde{Z}, \tilde{R}, U)$ where $\tilde{R} = R' \oplus P \oplus pass$, $\tilde{B} = W(\tilde{R})$, and $\tilde{Z} = DH(Y_0/\tilde{B}, Y_1/\tilde{B})$. The problem instance X is embedded via the W oracle, and the Red server messages are simulated for a randomly chosen round t_0. In this one round, the simulation sets $Y_1 = G^a$ to be a random group element with known discrete log a, and imperfectly simulates H_1 setting it to equal one of the H oracle responses. This round's the Red server simulation is shown below. Once all T rounds complete, the reducer chooses two indices at random from $[1, Q_\dagger]$ and obtains the pairs (Z, B) and (Z', B') where $B = X^r$, and $B' = X^{r'}$.

With probability $Prob[E_{OverT}]/Q_\dagger T$, there are two effective guesses in round t_0. If additionally, the two chosen indices correctly correspond to the effective guesses, the desired $DH(X, X)$ may be algebraically derived from Z_0, Z'_0, B, B', r, r', a, (see Eq. 3)), provided $r, r', r - r'$ are all non-zero. We conclude that the success is at least $Prob[E_{OverT}]/(\{Q_\dagger^3 T R)$.

Simulated Round - ReduceBlue	Simulated Round - ReduceBlueID
Red.sim1(R', Y_0):	**Red.sim1**(R', Y_0):
$a \xleftarrow{\$} [1, q], Y_1 \leftarrow G^a$	$a \xleftarrow{\$} [1, q], Y_1 \leftarrow G^a$
$H_1 \leftarrow hrand$	for each $\tilde{H} = H(h_{in})$
Red.sim2(H_0):	$\quad (valid, \tilde{pass}) \leftarrow \textbf{Valid}(h_{in}, t)$
(nothing)	$\quad \text{if}(valid \text{ AND } \textbf{IdealGuess}(\tilde{pass}))$
	$\quad\quad H_1 \leftarrow \tilde{H}$, return
	$H_1 \leftarrow hrand_t \xleftarrow{\$} \{0,1\}^k$
	Red.sim2(H_0):
	(nothing)

The *ReduceId* algorithm uses the same strategy as Section 6.2, although the simpler definition of effective guess makes the simulation easier (details below). By the same argument above in Section 6.1, we obtain bound (Eq. 2), on $Prob[E_{OverT}]$, so we can can relate $Succ_{Exp_{H_p}}(Reduce^{Adv_{Re}})$ to the advantage Ad (Eq. 1), obtaining for R of (Eq. 4),

$$Succ_{Exp_{H_p}}(Reduce^{Adv_{Re}}) \geq \frac{Ad'R}{TQ_\dagger^3}. \tag{11}$$

B Unmodified BJKS Scheme

Our modification of the scheme from [7] simply added the values $P \oplus P'$ and $R \oplus R'$ into the hash function inputs. Since the reducer was able to examine the list of H queries, to obtain the $P \oplus P'$ as well as Z_0, the candidate $A = W(P \oplus P')$ values could be computed.

We were still able to find security proofs for the original scheme, without this modification. In this case, the strategy employed by the reducer is to guess the correct W oracle values at random from the entire list of oracle calls, thus obtaining the required $A = G^r$ and $A' = G'^r$, albeit with reduced probability.

The second stage, the comparison with the ideal strategy, was also somewhat more difficult. Following the approach of Section 6.2, the simulator was not able to directly connect the H queries to the password. Instead, the simulation can be made by searching through the list of W oracle queries, and for each image A compute an associated Z. Each Z is compared with the H oracle query list to determine for which passwords "an effective guess" has been made. The resulting simulation is not perfect, since at a later point in the execution, the W oracle may coincidentally produce a random A which corresponds to an effective guess. In such a case, the simulated H_0 or ok_0 may have been inconsistent with the $\{H_0, ok_0\}$ corresponding to the correct password. This imperfect simulation results in a small error term in the inequality comparing the $Prob[E_{OverT}]$, $Succ_{Re}$, and $ExpIdeal(T)$.

An alternate approach to the proof, which enumerates every possible password, may be more efficient for small password dictionaries.

C Using a Decisional Diffe Hellman Oracle

In our proofs, a DDH oracle can be used to check whether a call to the H-oracle is an *effective guess*. This is useful in two ways. First, a *perfect* simulation in *Reduce* can be achieved at a cost of $Q_\dagger$ DDH-queries. One query is made for each call to the H-oracle with respect to the chosen round t_0. (Optionally, all T rounds can be used), then the approach of setting H_0 or H_1 randomly is replaced with a strategy which checks all previous H calls for effective guesses.

The second place where the DDH oracle is useful is in the final stage of *Reduce*, where the two Z values are selected. By knowing which H queries are effective guesses, the correct pair can be found when it exists.

Having implemented these changes to *Reduce*, the success can be improved to $Ad'R/T$. Thus, the expected number of repeats of the whole experiment required to solve the hard problem with high probability is approximately $T/Ad'H$ times.

This approaches works for the modified BJKS scheme as well as the new scheme, and for the modified BJKS scheme, passing from **CDH − Square(k)** to **CDH(k)** incurs a mere doubling of expected time. However, for the *unmodified* BJKS scheme, it is more difficult to determine an effective guess. By checking each W and each H call, effective guesses may still be determined, but the required number of DDH-queries is now $Q_{\dagger}^2$. In this case, we also note that the simulation required for *ReduceId* has an extra, small error term.

D Partially Passive Adversaries

To simplify the argument in Section 5.1, we had deferred consideration of adversaries which are active at the time that the valid client authenticates. In some models, the adversary is allowed to "trigger" the client some number of times. We now provide further justification for the claim that this activity will not help the adversary. For convenience, we note the details of the per-round interactions in the cases that Blue or Red is compromised.

CompRedRoundPassive	CompBlueRoundPassive
Client.auth1$(pass, U)$:	**Client.auth1**$(pass, U)$:
$R' \xleftarrow{\$} \{0,1\}^k$	$R' \xleftarrow{\$} \{0,1\}^k$
$P' \leftarrow R' \oplus pass'$	$P' \leftarrow R' \oplus pass'$
Blue.auth1(P'):	**Adversary:**
$e \xleftarrow{\$} [1, q]$	$Y_0 \leftarrow AdvBlue1(P')$
$A \leftarrow w(P \oplus P', U)$	**Red.auth1**(R', Y_0, U):
$Y_0 \leftarrow AG^e$	$f \xleftarrow{\$} [1, q]$
Adversary:	$B \leftarrow w(R \oplus R', U)$
$Y_1, H_1 \leftarrow AdvRed.auth1(Y_0, R')$	$Y_1 \leftarrow BG^f$
Blue.auth2(Y_1, H_1):	....
....	Continues as in Section 5.
Continues as in Section 5.	

Unlike in Section 5.1, the adversary who corrupts Red is presented with an R' generated by the honest client. As a result, the adversary knows $R \oplus R' = P \oplus P'$, and both the client and Blue server can be perfectly simulated, for example, **Client.sim1**:$R' \xleftarrow{\$} \{0,1\}^k$, **Blue.sim1**: $e \xleftarrow{\$} [1, q]$; $A \leftarrow w(R \oplus R', U)$; $Y_0 \leftarrow AG^e$. This simulation applies to the both the BJKS and new protocols.

Similarly, the adversary who corrupts Blue is presented with an P' generated by the honest client, so this adversary also knows $P \oplus P' = R \oplus R'$, and both the client and Blue server can be perfectly simulated, for example, **Client.sim1**:$P' \xleftarrow{\$} \{0,1\}^k$, **Blue.sim1**: $f \xleftarrow{\$} [1, q]$; $B \leftarrow w(P \oplus P', U)$; $Y_1 \leftarrow BG^e$. Thus, the adversary who prompts the client to enter the correct password, will not learn anything from this round of interaction.

Design and Analysis of Password-Based Key Derivation Functions

Frances F. Yao[1] and Yiqun Lisa Yin[2]

[1] Department of Computer Science,
City University of Hong Kong,
Kowloon, Hong Kong
csfyao@cityu.edu.hk

[2] Princeton Architecture Laboratory for Multimedia and Security,
Princeton University,
Princeton, NJ 08544
yyin@princeton.edu

Abstract. A password-based key derivation function (KDF) – a function that derives cryptographic keys from a password – is necessary in many security applications. Like any password-based schemes, such KDFs are subject to key search attacks (often called dictionary attacks). Salt and iteration count are used in practice to significantly increase the workload of such attacks. These techniques have also been specified in widely adopted industry standards such as PKCS and IETF. Despite the importance and wide-spread usage, there has been no formal security analysis on existing constructions. In this paper, we propose a general security framework for password-based KDFs and introduce two security definitions each capturing a different attacking scenario. We study the most commonly used construction $H^{(c)}(p\|s)$ and prove that the iteration count c, when fixed, does have an effect of stretching the password p by $\log_2 c$ bits. We then analyze the two standardized KDFs in PKCS#5. We show that both are secure if the adversary cannot influence the parameters but subject to attacks otherwise. Finally, we propose a new password-based KDF that is provably secure even when the adversary has full control of the parameters.

1 Introduction

1.1 Background and Motivation

Cryptographic keys are essential in virtually all security application. In practice, however, inputs to an application are typically raw key materials, such as passwords, that are not yet in the form to be used as keys. Therefore, a key derivation function (KDF) – a function that derives cryptographic keys from keying materials – is often a necessary component in all security applications.

There are many usage scenarios for key derivation functions depending on the form of the input. For example, the input can be a user password, a random seed value from some entropy source, or an output value from a cryptographic operation such as Diffie-Hellman key agreement. The second scenario is typically handled by a pseudorandom number generator (e.g., the FIPS PRNG in [6]), and the third scenario is handled by hashing the long output down to the required key length (e.g., the KDF1 in [8]). In both scenarios, there is usually enough entropy in the key materials.

In this paper, we focus our study on password-based key derivation functions. Unlike the other two scenarios mentioned above, passwords, in particular those chosen by a user, often are short or have low entropy. Therefore, special treatment is required in key derivation to defend against exhaustive key search attacks.

One basic approach for designing a password-based key derivation function is to derive the key from the password p and a random known value s (called salt), by applying a function H (such as hash, keyed hash, or block cipher) for a number of iterations c (called iteration count). For example, the following is a typical construction[1]:

$$key = H^{(c)}(p\|s).$$

Intuitively, the salt s serves the purpose of creating a large set of possible keys corresponding to a given password, among which one key is selected according to the salt used in each execution of the KDF. The iteration count c serves the purpose of increasing the cost of deriving each key, thereby significantly increasing the workload of key search attacks. These two techniques have been commonly used in practice and also specified in widely adopted industry standards, including PKCS [13] and IETF [4].

In a more general setting, we can view a password-based KDF as a method for stretching any short keys (not necessarily passwords) into longer keys. An efficient key stretching method can be useful in strengthening security without any architectural or policy changes for complex systems (e.g., a credit-card system). In addition, password-based KDF can be used in a straightforward way to define password-based encryption schemes and message authentication schemes.

Despite their importance and wide-spread usage, there has been no formal security analysis of existing password-based KDFs. Furthermore, there has been no general security framework for analyzing such KDFs. It is possible that the above popular KDF construction is insecure against some sophisticated key-search attacks even though the parameters are chosen large enough and the underlying hash function H is sound (e.g., it can be considered as a random oracle).

1.2 Our Framework

In this paper, we propose a general security framework for studying password-based key derivation functions. Our framework aims at capturing various types of key-search attacks and allowing concrete analysis of attacker's success probability relating to its available computational resource for launching key-search attacks. We also model salt and iteration count in a way as they are used in practice so that their impact on the overall security of the scheme can be quantified.

---

[1] Throughout the paper, we use $H^{(c)}$ to denote that H is applied c times and use $\|$ to denote the concatenation of two strings.

Key search attacks on password-based KDFs can be quite sophisticated [10], but they generally target at the *construction* of the key derivation function F and treat the underlying function H as a black-box transformation. This motivates us to model the underlying primitive H as a random oracle [2] and hence its internal structure is ignored in the analysis.

We define two levels of security for the KDF depending on the capability of the adversary A: – in the *weak* model A can only observe the output of F while in the *strong* model A can query F on inputs of its choice. In both models, A can make queries to H, and the maximum number of such queries captures A's available computational power to launch key search attacks. Roughly, the construction of F is secure under each model if A with the given capability cannot distinguish the output of F for an unknown password p from a random string.

1.3 Main Results

Using the proposed framework, we first study the security of the iterative construction. We prove that if an adversary A makes at most t queries to H, then the success probability Adv that it can distinguish the derived key for an unknown p from a random string of the same length n satisfies

$$\frac{\lfloor t/c \rfloor}{|PW|} < Adv < \frac{\lfloor t/c \rfloor}{|PW|} + \frac{t^2}{2^n}.$$

The upper bound is dominated by the first term, since the second term is negligible in practical settings. The above result implies that, for a fixed iteration count, there is no short-cut in the key search other than computing H iteratively for each password. In other words, the iteration count c increases the workload of exhaustive key search by a factor of c. So the iteration construction effectively stretches a k-bit key to a $(k + \log_2 c)$-bit key.

We then focus our attention on practical password-based KDFs and analyze the two KDFs in PKCS#5 [13] – the *de facto* standard for password-based cryptography. Our analysis on the iterative construction implies that the two KDFs are weakly secure as long as the computational resource available to the attacker is much less than $c|PW|$ (although it can be much larger than $|PW|$). We also show that neither KDF is secure in the strong model and discuss how such security weakness may be exploited to mount attacks in practical scenarios.

Based on the insight gained from our earlier analysis, we propose a new password-based key derivation function with enhanced security. The main idea in our construction is to include iteration count explicitly in the input to the derivation function to prevent it from being manipulated by the attacker. We show that the new KDF is secure in the strong model.

1.4 Related Work

Rigorous analysis of password-based key derivation schemes seems to have received relatively little attention compared to other cryptographic schemes. In

[10], the term *key-stretching* is used for conversion of low-entropy keys into longer keys by mechanisms such as iterated hash. A connection was made between the cost of computing $H^{(c)}$ and the cost of finding a collision for H. Roughly speaking, if $H^{(c)}$ could be computed on average with fewer than $c/2$ calls to H, then this would lead to a collision search for H faster than the naive birthday attack. Although it would be hard to translate the result into a standard concrete security model, it is certainly of practical interest. Our results on $H^{(c)}$ are well quantified; moreover, the framework for KDF is rigorously defined and the effects of salt and iteration are studied in a standard *distinguish-from-random* model with respect to specific query types.

In [14], the UNIX password hashing algorithm is analyzed. The core of the algorithm is roughly $f(p) = DES_p^{(25)}(0)$, that is, to encrypt the value zero 25 times using the password p as the key. So the algorithm has an iterative structure somewhat similar to the password-based KDFs considered here. It is proved that the algorithm is a secure hashing function if DES is a secure block cipher. However, as pointed out by the authors, their analysis only implies that iteration does not harm security, but they are not able to show that iteration actually enhances security as one would intuitively expect.

Key derivation functions in a general sense share some similarity in their design, such as the use of hash functions to process the raw key materials. For instance, the pseudorandom number generator (PRNG) defined in FIPS [6] is hash-based and can derive long keys from a random seed. The exact construction, however, is quite different: in the password-based KDF the hash is applied iteratively while in the PRNG the hash outputs are concatenated to produce the key. Therefore, security analysis of PRNG type of key derivation [5] does not directly apply to password-based key derivation.

Another related subject is password-based authentication and key exchange protocols, and the goal of such protocols is to authenticate two parties who share a common password. Existing protocols use various public-key techniques such as RSA or Diffie-Hellman in a way that the messages exchanged between the parties provide little or no help for an attacker to guess the password. A survey of well-analyzed protocols can be found in [9].

2 Security Framework and Definitions

2.1 KDF Model

We denote a password-based key derivation function as

$$y = F(p, s, c)$$

where p is password, s is salt, c is iteration count, and y is derived key of length n. Let the set of passwords, salts, and iteration counts be PW, S, and C. For a fixed $p \in PW$, we can view $y = F_p(s, c)$ as a function with input (s, c) and output y. We interchangeably write $F(p, s, c)$ or $F_p(s, c)$ depending on the context.

For ease of analysis, we make some assumptions on the sets PW, S, and C. We assume that $PW = \{0,1\}^l$, $S = \{0,1\}^s$, and $C = [c_*, c^*]$ for some integers $0 < c_* < c^*$. The upper limit ensures that the iteration count is not too large, since otherwise the KDF becomes too slow and useless. In addition, we assume that the length of $p\|s\|c$ is at most n, although our analysis can be extended to the more general case (see Section 4.3 for further discussions).

We denote the underlying primitive that is used to construct F as H. For example, H can be instantiated using practical hash functions as building blocks. Since key search attacks typically treats H as a black-box transformation without exploiting its internal structure, we model H as a random function from $\mathcal{R}_n$, the set of all functions from $\{0,1\}^n$ to $\{0,1\}^n$. That is, we focus our analysis on how F is constructed based on H rather than the structure of H itself.

2.2 Attack Model

Consider a typical usage scenario of a password-based KDF in which two users Alice and Bob share a password p. To encrypt a message, Alice sets s, c and derives a key $y = F(p, s, c)$. She then uses y to encrypt and obtain the ciphertext z. Alice sends (z, s, c) in the clear to Bob. Bob derives the key y by computing $y = F(p, s, c)$ and uses y to decrypt z.

From the attacker's point of view, the salt s and the count c are both *known*. The attacker usually does not have control of s or c, but in certain scenarios they can be *chosen*. The derived key y may be hidden from the attacker, but it can become *known* for various reasons (e.g., it was leaked out due to system security holes), in which case the attacker obtains a tuple (y, s, c) corresponding to some unknown password p.

In our attack model, we assume that s and c can be either known or chosen, and the derived key y is always known. An attacker A to a password-based KDF is a polynomial-time algorithm that may use the following two types of oracle queries:

- H *Query:* Query the underlying function H on input x and obtain $H(x)$. That is, A has access to oracle $H(.)$.
- F *Query:* For an unknown password p, query the key derivation function F on input (s, c) and obtain the derived key $y = F_p(s, c)$. That is, A has access to oracle $F_p(.,.)$ for some unknown p.

In practice, the number of H queries can be quite large, since it is determined by the adversary's available computational resource for performing an offline key search attack. In contrast, the number of F queries is very limited, since it is usually determined by the security design and policy of the system, not the adversary.

2.3 Security Definition

We introduce two security definitions – weakly secure and strongly secure – depending on attacker's capability. In the weak model, we assume that the attacker

A can make only queries to H, while in the strong model, we assume that A can make queries to both H and F. The goal of the attacker A is to distinguish the derived key $y = F_p(s,c)$ for $p \xleftarrow{r} PW$ from a random string of the same length[2].

Definition 1. Weakly Secure KDF. Let $y = F_p(s,c)$ be a password-based KDF. Let $b \in \{0,1\}$. We consider the following experiment depending on b:

Experiment E_b
$p_0 \xleftarrow{r} PW$ // password is generated at random
$H \xleftarrow{r} \mathcal{R}_n$ // H is generated at random
$s_0 \leftarrow S, c_0 \leftarrow C$ // salt and count are fixed and known
If $b = 0$, then $y_0 \leftarrow F_{p_0}(s_0, c_0)$, else $y_0 \xleftarrow{r} \{0,1\}^n$
$i \leftarrow 0$
repeat
$\quad i \leftarrow i + 1$
$\quad A$ chooses x_i and is given $H(x_i)$
until A reaches the maximum number of queries
A outputs either 0 or 1

The success probability of A is defined as

$$Adv_A(t) = Pr_{E_1}[A = 1] - Pr_{E_0}[A = 1].$$

where t denote the maximum number of queries to H. The maximum success probability achievable by any adversary A is denoted by $Adv(t)$.

Definition 2. Strongly Secure KDF. Let $y = F_p(s,c)$ be a password-based KDF. Let $b \in \{0,1\}$. We consider the following experiment depending on b:

Experiment E_b
$p_0 \xleftarrow{r} PW$ // password is generated at random
$H \xleftarrow{r} \mathcal{R}_n$ // H is generated at random
$s_0 \leftarrow S, c_0 \leftarrow C$ // salt and count are fixed and known
If $b = 0$, then $y_0 \leftarrow F_{p_0}(s_0, c_0)$, else $y_0 \xleftarrow{r} \{0,1\}^n$
$i \leftarrow 0$
repeat
$\quad i \leftarrow i + 1$
$\quad A$ first decides which type of queries
$\quad$ If H query, A chooses x_i and is given $H(x_i)$
$\quad$ If F query, A chooses $(s_i, c_i) \neq (s_0, c_0)$
$\qquad$ and is given $y_i = F_{p_0}(s_i, c_i)$
until A reaches the maximum number of queries
A outputs either 0 or 1

The success probability of A is defined as

$$Adv_A(t,m) = Pr_{E_1}[A = 1] - Pr_{E_0}[A = 1],$$

---

[2] If W is a set, than $w \xleftarrow{r} W$ denotes selecting w uniformly at random from W.

where t and m denote the maximum number of H and F queries, respectively. The maximum success probability achievable by any adversary A in is denoted by $Adv(t, m)$.

3 Password-Based KDFs in Practice

Password-based key derivation functions are commonly used in practice. They are also specified in industry standards such as PKCS#5, PKCS#12, IETF, and openPGP. Here we describe the KDFs in PKCS#5, which is considered as the de facto standard for password-based cryptography. KDFs in other standards mostly follow similar designs.

Two password-based KDFs are specified in PKCS#5 v2.0 [13]: PBKDF1 and PBKDF2. Some recommendations are given regarding the use of salt and iteration count. For example, S should be at least 64 bits, and c should be at least 1000.

The underlying function in PBKDF1 is a hash function $H()$ such as MD2, MD5, or SHA1. The derived key is defined as $y = H^{(c)}(p\|s)$. Most other standards or implementations use this construction.

PBKDF2 was intended to provide more security. The underlying function in PBKDF2 is a keyed hash function $H_k()$, such as HMAC [1]. The password p is used as the key k in each invocation of $H_k()$. The derived key is defined as $y = U_1 \oplus U_2 \oplus ... \oplus U_c$, where $U_i = H_p^{(i)}(s)$ for $i = 1, .., c$. The exclusive-ors adds an extra layer of protection, but at the core of the construction is still the iterative application of H_p.

4 Effects of Iteration Count

In this section, we focus our analysis on iteration count and quantify its effect on the security of KDF. A KDF function with an iterative structure is of the form
$$y = H^{(c)}(p, s) = H^{(c)}(p\|s).$$
We will show that this construction is secure as long as the adversary only has access to H and its computational resource is significantly less than $c|PW|$, which is formally stated in the following theorem.

Theorem 1. *In the weakly secure model for KDF, if the adversary makes at most t H queries, then the maximum success probability $Adv(t)$ satisfies*

$$\frac{\lfloor t/c_0 \rfloor}{|PW|} < Adv(t) < \frac{\lfloor t/c_0 \rfloor}{|PW|} + \frac{t^2}{2^n}.$$

Before going into the proof details, we first try to understand the result by considering a practical scenario. Let
$$|PW| = 2^{40}, n = 128, c = 2^{16}, t = 2^{44}.$$

Setting $c = 2^{16}$ adds little overhead at the user end for deriving a single key[3], but the workload of a straightforward dictionary attack increases to $c|PW| = 2^{56}$ from 2^{40} (when $c = 1$). With $t = 2^{44}$ queries to H, the attacker can certainly correctly compute a fraction of $\frac{t/c}{|PW|} = 2^{-12}$ of the derived keys. Our result shows that this is indeed the *best* the attacker can do, since the probability for correctly computing more than 2^{-12} of the derived keys is at most $\frac{t^2}{2^n} = 2^{-40}$. Effectively, the iteration count stretches a 40-bit password into a $40 + \log_2 c = 56$-bit key.

4.1 Graph Representation of H

For the purpose of the proof, we set up a graph to represent a random function H and the adversary's query process for H. This graph-based approach allows us to visualize the adversary's knowledge gained in the query process, and makes the proof more intuitive.

Let G_H be a directed graph on the vertex set $\{0,1\}^n$; a directed edge (x,y) exists in G_H if and only if $H(x) = y$. Hence every vertex has out-degree 1 and G_H contains 2^n edges. The adversary, by probing a sequence of t edge "$H(x) =?$", discovers a subgraph Q_H of G_H which is referred to as the *query graph*. Since the same query graph Q_H can arise from different functions H, it is sometimes convenient to write Q without referring to a specific H.

4.2 Analysis of Probabilities

We start by defining two games R (for "random") and K (for KDF) which correspond to the two experiments E_1 and E_0, respectively. For each game, we specify how to simulate the oracle H upon adversary's queries. In the game specification, there are some extra computing steps – they are hidden to A and hence do not affect the behavior of A, but they will help our analysis.

We note that the two games are very similar, and the only difference is in Step 4 which is shown by the underline. Two flags bad_1 and bad_2 are set when certain "bad" event occurs. The set Y contains all distinct values of $H(x)$ for which x has been queried[4].

In Game R, the answers seen by the adversary A are exactly the same as in E_1. The difference is that the game contains two extra steps – Step 2 for detecting collisions and Step 3 for detecting whether $H^{(c_0)}(p_0 \| s_0)$ has been computed. So the success probabilities of A in game R and experiment E_1 are the same, which is stated in the following lemma.

Lemma 1. $Pr_R[A = 1] = Pr_{E_1}[A = 1]$.

In Game K, the answers seen by the adversary A are almost the same as in experiment E_0 with possible exception on queries $H(u_i)$. We will show in the next lemma that this apparent difference will not affect A's success probability.

---

[3] On a Pentium 4 running at 2.1GHz, 2^{16} SHA-1 operations take less than 0.02 second according to the benchmarks for Wei Dai's CRYPTO++ Library.

[4] We also include $u_0 = p_0 \| s_0$ in Y is for detecting the event that u_0 is not the *first* vertex of a path. It is not necessary to do so, but makes later analysis easier.

Initially, $H(.)$ is undefined. Choose $p_0 \xleftarrow{r} PW$ and $y_0 \xleftarrow{r} \{0,1\}^n$.
Set $i \leftarrow 0, u_0 \leftarrow p_0 \| s_0, Y \leftarrow \{u_0, y_0\}$.

On oracle query $H(x)$:

1. Choose $y \xleftarrow{r} \{0,1\}^n$.
2. If $y \notin Y$, set $Y \leftarrow Y \cup \{y\}$.
   Else if $y \in Y$, set bad_1.
3. If $x = u_i$ and $i < c_0$, set $i \leftarrow i + 1$ and $u_i \leftarrow y$.
   Else if $x = u_i$ and $i = c_0$, set bad_2.
4. Define $H(x) = y$ and return y.

Fig. 1. Game R.

Initially, $H(.)$ is undefined. Choose $p_0 \xleftarrow{r} PW$ and $y_0 \xleftarrow{r} \{0,1\}^n$.
Set $i \leftarrow 0, u_0 \leftarrow p_0 \| s_0, Y \leftarrow \{u_0, y_0\}$.

On oracle query $H(x)$:

1. Choose $y \xleftarrow{r} \{0,1\}^n$.
2. If $y \notin Y$, set $Y \leftarrow Y \cup \{y\}$.
   Else if $y \in Y$, set bad_1.
3. If $x = u_i$ and $i < c_0$, set $i = i + 1$ and $u_i \leftarrow y$.
   Else if $x = u_i$ and $i = c_0$, set $y \leftarrow y_0$. Set bad_2.
4. Define $H(x) = y$ and return y.

Fig. 2. Game K.

Lemma 2. $Pr_K[A = 1] = Pr_{E_0}[A = 1]$.

Proof: In experiment E_0, H is chosen randomly at the beginning and y_0 is then set to be $y_0 = H^{(c_0)}(u_0)$. Therefore, for $i = 0, 1, ... c_0 - 1$, each value $H(u_i)$ is chosen at random *before* the experiment starts. In Game K, for $i = 0, 1, ... c_0 - 2$, each value $H(u_i)$ is chosen at random as the game *proceeds*. Only the last value $H(u_{c_0-1})$ is chosen at random (to be y_0) *before* the game starts.

Since all these values are chosen at random and they are all independent of each other, there is no difference from the adversary's point of view. Hence the success probability of A is the same. **QED**

Using the above lemmas, we have $Adv_A(t) = Pr_R[A = 1] - Pr_K[A = 1]$. So we now consider the relation between Game R and Game K. Let BAD_1 be the event that flag bad_1 gets set, and similarly for BAD2. Let $BAD = BAD_1 \cup BAD_2$. It is easy to see that the answers seen by A are exactly the same if neither bad event occurs. Furthermore, each bad event occurs with the same probability in the two games.

Lemma 3. *(1)* $Pr_R[A = 1|\overline{\text{BAD}}] = Pr_K[A = 1|\overline{\text{BAD}}]$.
(2) $Pr_R[\text{BAD}] = Pr_K[\text{BAD}]$.

Following a standard probability argument (such as that in [11]), we have $Adv_A(t) < Pr_R[\text{BAD}]$. So we only need to derive an upper bound on $Pr_R[\text{BAD}]$ for proving the theorem.

Proof of Theorem 1. For simplicity, we omit the R in the subscript.

$$\begin{aligned} Pr[\text{BAD}] &= Pr[\text{BAD}_1 \cup \text{BAD}_2] \\ &= Pr[\text{BAD}_1] + Pr[\text{BAD}_2|\overline{\text{BAD}_1}] \cdot Pr[\overline{\text{BAD}_1}] \\ &\leq Pr[\text{BAD}_1] + Pr[\text{BAD}_2|\overline{\text{BAD}_1}]. \end{aligned}$$

It is easy to see that $Pr[\text{BAD}_1] < (t^2/2 + 2t)/2^n$, since the probability of a collision within t queries is at most $(t^2/2)/2^n$, and the probability that any of the t values of H collide with u_0 or y_0 is at most $2t/2^n$. Assuming $t \geq 4$, we have $Pr[\text{BAD}_1] < t^2/2^n$.

We next bound the second term $Pr[\text{BAD}_2|\overline{\text{BAD}_1}]$. If the event BAD_1 doesn't occur, then the query graph consists of a set of disjoint paths on which u_0 can only appear as the *first* vertex. Note that BAD_2 is the event that there is a path of length at least c_0 starting from vertex u_0. With t edges, there are at most $\lfloor t/c_0 \rfloor$ such paths. For each path, with probability at most $1/|PW|$, the first vertex is $u_0 = p_0 \| s_0$. Hence $Pr[\text{BAD}_2|\overline{\text{BAD}_1}] < \frac{\lfloor t/c_0 \rfloor}{|PW|}$. Combining the two upper bounds, we obtain that $Adv(t) < Pr[\text{BAD}] \leq \frac{2t^2}{2^n} + \frac{\lfloor t/c_0 \rfloor}{|PW|}$.

The lower bound can be achieved easily by computing full paths of length c_0 for $\lfloor t/c_0 \rfloor$ passwords in PW. **QED**

4.3 Discussions on c-th Iterate of a Random Function

It may be helpful to review some mathematical background on the c-th iterate of a random function. Although random functions have been studied extensively in the literature, the c-th iterate function $H^{(c)}$ has received relatively little attention. For example, it is well known that the image of a random function has size $(1 - e^{-1})2^n$. What about the image of $H^{(c)}$? At what rate does the image size decrease with c? In a 1990 paper by Flajolet and Odlyzko [7] they provided answers to these questions by deriving the following recurrence.

Image Size of $H^{(c)}$. *The image size of $H^{(c)}$ is $(1 - \tau_c)2^n$, where τ_c satisfies the recurrence $\tau_0 = 0, \tau_{c+1} = e^{-1+\tau_c}$. Furthermore, asymptotically $1 - \tau_c = 2/c$.*

In other words, the image size of $H^{(c)}$ decreases arithmetically with c (not geometrically as one might guess at first). This illustrates that $H^{(c)}$ is significantly different from a random function as c gets larger, and its analysis is a nontrivial matter. It is also interesting to note that, although the image size of $H^{(c)}$ goes down by a factor of $2/c$ compared with that of H, yet Theorem 1 shows that the attacker's workload must increase by a factor of c.

Theorem 1 proves tight bounds for the password space. When $t < c$, the lower bound on $Adv(t)$ in the theorem becomes zero. Below, we derive a nontrivial lower bound by describing a strategy of the attacker. Let $d(x)$ denote the number of divisors of x, and define $d[x_1, x_2] = \max d(x), x_1 \le x \le x_2$.

Lemma 4. *With $t < c$ queries, the attacker can achieve $Adv_A(t) > \frac{t^2}{2^{n+1}} + (1 - \frac{t^2}{2^{n+1}}) \frac{d[c, c-t]}{2^n}$.*

Proof. The attacker simply computes $u, H(u), H^{(2)}(u), \ldots$ iteratively and hopes that the sequence becomes periodic, i.e., a repeated value occurs before $H^{(t)}(u)$ so that $H^{(c)}(u)$ is determined. The probability for this to happen is $t(t+1)/2^{n+1}$. In the case the chain $(u, H(u), \ldots, H^{(t)}(u))$ is not periodic, the adversary will select a vertex on the path $v = H^{(\delta)}(u)$ where δ is chosen to maximize the number of divisors $d(c - \delta)$ for $0 \le \delta \le t$. The probability of success in this case is at least $d[c, c-t]/2^n$ as stated. **QED**

We note that, somewhat surprisingly, the lower bound gets better with larger c, as the attacker may find a number $c - \delta$ in the range $[c - t, c]$ with a large number of divisors so that it is more likely to have $H^{(c)}(u) = H^{(\delta)}(u)$ through periodicity.

In our modelling of the password space, we assumed that all passwords are of the same length ℓ and are chosen by users with equal probability. It is not difficult to extend our analysis to obtain similar security bounds when these assumptions are removed. For example, let a set of passwords of arbitrary length be added to PW. After one iteration of H these points are distributed randomly in the domain $\{0,1\}^n$, and additional iterations (adjusting c to be $c-1$) would behave just as analyzed before. Similarly, even if the passwords have different probabilities originally, after one iteration this will have no effect on the collision probabilities which depend only on the fact $prob[H(x) = y] = \frac{1}{2^n}$ in the domain $\{0,1\}^n$.

5 Security Analysis of KDFs in PKCS#5

In the preceding section, we analyzed the basic iterative construction $H^{(c)}$. The analysis implies that the two KDFs in PKCS#5 are weakly secure as long as the adversary's computational resource is far less than $c|PW|$, even though it can be much larger than $|PW|$.

In what follows, we analyze the security of the two KDFs under the strongly secure model – that is, the adversary is allowed a few queries to F. We show that neither KDF is secure under this model and we also explore how such security weakness can be exploited to launch attacks in practical settings.

5.1 PBKDF1

The attack on PBKDF1 is based on an obvious relation between keys derived using the *same* salt. For any salt s and two iteration counts $c_0 < c_1$, let $y_i = F(p, s, c_i) = H^{(c_i)}(p\|s)$. Then, it is easy to see that $y_1 = H^{(c_1 - c_0)}(y_0)$. This relation allows an attacker to distinguish y_0 from a random function with one F query (s, c_1) and $(c_1 - c_0)$ H queries.

Note that if the key $y_0 = H^{(c_0)}(p\|s)$ were ever compromised for some reason, then any key derived using the same salt s and an iteration count *larger than* c_0 would all be compromised. This might happen in practice if the user (or the security administrator of the system) decides to increment the iteration count. Therefore, it is a good practice in general to use different salt values in deriving different keys.

5.2 PBKDF2

The derived keys in PBKDF2 also suffer from non-randomness, although the relations among keys are slightly more complicated. Let s be any salt value and let c_1, c_2, c_3 be three consecutive iteration counts. For $i = 1, 2, 3$, define $y_i = F(p, s, c_i) = U_1 \oplus ... \oplus U_{c_i}$. Then, we have $y_1 \oplus y_2 = U_{c_2}$ and $y_2 \oplus y_3 = U_{c_3}$. This yields the following relation among the three keys: $(y_2 \oplus y_3) = H_p(y_1 \oplus y_2)$, where $H_p()$ is the underlying function HMAC.

This relationship among keys opens the door to dictionary attacks. The attacker simply computes the HMAC function $H_p(U_{c_2})$ for all possible passwords p, and the password that gives $H_p(U_{c_2}) = U_{c_3}$ is very likely to be the correct password used in the scheme. Once p is known, it is easy to distinguish the derived key from a random string. We remark that the workload of the above attack is $|PW|$, no matter what c is. This implies the iteration count does not add much (or any) protection in PBKDF2 against dictionary attacks.

6 Effects of Salt

A salt serves the purpose of creating a large set of possible long keys corresponding to a password p. If the salt is s bits long, then the number of possible long keys can be as large as 2^s. Each time the KDF is executed with a salt, either selected by the user or generated at random, one of the 2^s long keys is selected.

One natural question is the following: Suppose that an adversary has computed the long keys correspond to all the passwords $p \in PW$ for a salt s_1. That is, the adversary has a table of size $|PW|$ in which each entry contains the value $(p, H^{(c)}(p\|s_1))$ for some $p \in PW$. Does this table provide the adversary some shortcuts to derive long keys using a *different* salt $s_2 \neq s_1$?

The answer is certainly "No", which is well-known in practice. Using the graph-based approach, we can show that the set of paths corresponding to s_1 and the set of paths corresponding to s_2 are all disjoint with high probability, and hence the table for s_1 provides essentially no information for derived keys using s_2. The detailed analysis is similar to that for Theorem 1 and thus omitted here.

7 New Proposal for Strongly Secure KDF

In this section, we present a new proposal for strongly secure KDF based on our study on the effects of iteration counts and salt, as well as the analysis on existing KDFs.

We first note that the graph-based analysis provides insights on the exact way that each parameter contributes to the overall security: The computation process for deriving a key corresponds to a path in the query graph. So choosing a larger iteration count forces the attacker to traverse a *longer* path for deriving each key, while choosing a different salt value forces the attacker to traverse a *different* path in the computation. It is also easy to see why the KDFs in PKCS#5 are not strongly secure using the query graph. For example, in the case on PBKDF1, the F queries (s, c_0) and (s, c_1) correspond to two paths that overlap. This extra information allows the attacker to distinguish the derived key from a random string.

Based on the above discussion, we can see that a strongly secure KDF should be constructed in a way that the values of $y = F(p, s, c)$, for different p, s and c, are nearly independent of each other. Certainly, there are various ways of achieving this goal. Here we propose a simple construction that maintains the same efficiency as the KDFs in PKCS#5. The idea is to include iteration count explicitly as an input to the hash function H. More specifically, the new KDF is

$$y = F^*(p, s, c) = H^{(c)}(p\|s\|c).$$

In what follows, we prove that the above KDF is strongly secure – secure even when the adversary can choose (s, c) and make F queries. We assume that there are lower and upper limits to the c_i acceptable in queries to F, that is, $c_* < c_i < c^*$. Indeed, without a lower limit, the adversary can always set $c = 1$ in the F query and then perform an offline key search attack with complexity $O(|PW|)$.

Theorem 2. *In the strongly secure model for KDF, if the adversary makes at most t queries to H and at most m queries to F, then the maximum success probability $Adv(t, m)$ satisfies*

$$\frac{\max(\lfloor (t - c^*)/c_* \rfloor, m)}{|PW|} < Adv(t, m) \frac{\lfloor t/c_* \rfloor + 2m}{|PW|} + \frac{(t + m)^2}{2^n}.$$

Proof. We provide a sketch of the proof here, and the details are given in the appendix. The upper bound proof uses the same type of arguments as that of Theorem 1. More specifically, we define two games R' and K' which the adversary might play. Since there are now two types of queries to deal with, the simulator needs to maintain some extra information during the course of the game to make sure that its answers to oracle queries F^* and H are consistent. Then following a similar analysis, we only need to bound the probability of some bad events to obtain the upper bound in the theorem.

For the lower bound, we describe two strategies. In strategy A, the adversary computes separate paths of length c_* using queries to H and makes only one F query. In strategy B, the adversary constructs a single path of length t and picks m appropriate vertices to make m queries to F. The success probability is the maximum of the two as stated in the theorem. **QED**

8 Conclusions

Password-based key derivation functions are necessary in many security applications. Despite their importance and wide-spread usage, rigorous analysis of such functions seems to have received relatively little attention in the literature compared with many other cryptographic schemes.

In this paper, we define a general security framework for password-based key derivation functions where salt and iteration count are included as parameters. Under this framework, we focus on the most commonly used construction $H^{(c)}(p\|s)$ and prove that the iteration count c, when fixed, does have an effect of stretching the password by $\log_2 c$ bits. Our analysis is done using a random functional graph representing H, conditioned upon a query graph representing information revealed to the attacker. It provides insights on the exact way that each parameter contributes to the overall security.

We then analyze two widely deployed KDFs defined in PKCS#5. We show that both are secure the adversary cannot influence the parameters, but are subject to attacks otherwise. We also consider how such security weaknesses can be exploited in practice.

Finally, based on the insight gained from our earlier analysis, we propose a new password-based key derivation that is provably secure even when the attacker has full control of the salt and iteration count. The new proposal achieves stronger security while preserving the same efficiency as existing KDFs. We expect that the new proposal will find its application in practical implementations.

Acknowledgements

We would like to thank the anonymous referees for many helpful comments.

References

1. M. Bellare, R. Canetti and H. Krawczyk. Keyed Hash Functions for Message Authentication. In *Advances in Cryptology – Crypto '96*, Springer-Verlag, 1996. Crypto'96.
2. M. Bellare and P. Rogaway. Random Oracles are practical: A Paradigm For Designing Efficient Protocols. In *First ACM Conference on Computer and Communications Security*, 1993.
3. S. Bellovin and M. Merritt. Encrypted Key Exchange: Password-Based Protocols Secure Against Dictionary Attacks. In *Proceedings of the IEEE Symposium on Research in Security and Privacy*, 1992.
4. T. Dierks and C. Allen. The TLS Protocol Version 1.0. IETF RFC 2246, *Internet Request for Comments*, January 1999.
5. A. Hevia, A. Desai, and Y. L. Yin. A Practical-Oriented Treatment of Pseudorandom Number Generators. In *Advances in Cryptology – Eurocrypt '02*, Springer-Verlag, 2002.
6. FIPS PUB 186-2. Digital Signature Standard. National Institute of Standards and Technologies, 1994.

7. P. Flajolet and A. M. Odlyzko. Random mapping statistics. In *Advances in Cryptology - EUROCRYPT '89*, Springer-Verlag, 1990.
8. IEEE Std 1363-2000: Standard Specifications for Public-Key Cryptography. *IEEE Computer Society*, 2000.
9. IEEE P1363.2: Standard Specifications for Password-Based Public-Key Cryptographic Techniques. Draft D15. May 2004.
http://grouper.ieee.org/groups/1363/passwdPK/draft.html.
10. J. Kelsey, B. Schneier, C. Hall, and D. Wagner. Secure Applications of Low-Entropy Keys. In *Proceedings of the First International Workshop ISW '97*, Springer-Verlag, 1998.
11. J. Killian and P. Rogaway. How To Protect DES Against Exhaustive Key Search Attacks. In *Advances in Cryptology - CRYPTO '96*, Springer-Verlag, 1996.
12. A. M. Odlyzko, private communication. 2003.
13. RSA Laboratories' PKCS#5 v2.0: Password-Based Cryptography Standard. 1999.
14. D. Wagner and I. Goldberg. Proofs of Security For The UNIX Password Hashing Algorithm. In *Advances in Cryptology - Asiacrypt '00*, Springer-Verlag, 2000.

Proof of Theorem 2

Upper Bound

The upper bound proof uses the same type of arguments as that of Theorem 1. We start by specifying two games R' and K' (see Figures 3 and 4). Since there are now both H and F^* queries to deal with, the simulator needs to maintain some necessary information during the course of the game to make sure that its answers to both types of oracle queries are consistent.

Before diving into the detailed descriptions of the two games, it is instructional to compare at a high level how oracle query H is handled in Game K' and Game K. The main difference is the additional Step 4 (marked as *new*) in Game K', which is for updating the necessary information maintained by the simulator.

In both games, the simulator keeps track of all the F^* queries as well as the H queries starting at $p_0\|s\|c$. More precisely, it maintains a set

$$L = \{(s_k, c_k, y_k, u_k, i_k)\}$$

where each item in L is a 5-tuple such that either the query (s_k, c_k) has been made to F^* or the query $x = p_0\|s_k\|c_k$ has been made to H. The other three entries are defined as follows:

- If the query to F^* has been made, then $y_k = H^{(c_k)}(x)$ Otherwise, $y_k = *$ meaning it is still undefined.
- If the query to H has been made, then i_k is the number of consecutive queries to H made thus far starting at x, and u_k is the last answer.

It also maintains the set of all "starting points" in L, that is, a set $X = \{x_k\}$ where $x_k = p_0\|s_k\|c_k$.

Following similar analysis as that of Theorem 1, we have that $Adv_A(t, m) < Pr[\mathsf{BAD}] \le Pr[\mathsf{BAD}_1] + Pr[\mathsf{BAD}_2|\overline{\mathsf{BAD}_1}]$. So we only need to derive an upper

Initially, $H(.)$ and $F_{p_0}(.,.)$ are both undefined.
Choose $p_0 \xleftarrow{r} PW$ and $y_0 \xleftarrow{r} \{0,1\}^n$.
Set $i_0 \leftarrow 0, x_0 \leftarrow p_0\|s_0\|c_0, Y \leftarrow \{y_0\}$.
Set $X \leftarrow \{x_0\}, L \leftarrow \{(s_0, c_0, y_0, u_0, i_0)\}$.
Set $j \leftarrow 0$.

On oracle query $H(x)$:

1. Choose $y \xleftarrow{r} \{0,1\}^n$.
2. If $y \notin Y$, set $Y \leftarrow Y \cup \{y\}$.
   Else if $y \in Y$, set bad_1.
3. If $x = x_k \in X$ and $i_k < c_k$, set $i_k \leftarrow i_k + 1$ and $u_k \leftarrow y$.
   Else if $x = x_k \in X$ and $i_k = c_k$, set $\underline{y \leftarrow y_k}$. Set bad_2.
4. (*new step compared with Game K*)
   If $x \notin X$ and $x = p_0\|s\|c$, then $X \leftarrow X \cup x$ and add a new item in L:
   $j \leftarrow j+1, s_j \leftarrow s, c_j \leftarrow c, y_j \leftarrow *, u_j \leftarrow x, i_j \leftarrow 1$
5. Define $H(x) = y$ and return y.

On oracle query $F_{p_0}^*(s,c)$:

1. Choose $y \xleftarrow{r} \{0,1\}^n$.
2. If $y \notin Y$, set $Y \leftarrow Y \cup \{y\}$.
   Else if $y \in Y$, set bad_1.
3. Let $x = p_0\|s\|c$.
   If $x = x_k \in X$ and $i_k < c_k$, set $y_k \leftarrow y$.
   Else if $x = x_k \in X$ and $i_k = c_k$, set $\underline{y \leftarrow y_k}$. Set bad_2.
4. If $x \notin X$, then $X = X \cup x$ and add a new item in L:
   $j \leftarrow j+1, s_j \leftarrow s, c_j \leftarrow c, y_j \leftarrow y, u_j \leftarrow x, i_j \leftarrow 0$
5. Define $F_{p_0}^*(s,c) = y$ and return y.

Fig. 3. Game K'.

Game R' is the same as Game K', except that the execution of the underlined step ($y \leftarrow y_k$) is removed.

Fig. 4. Game R'.

bound on the probability of each bad event. Analyzing the first term is straightforward. Since there are $t + m$ queries in total, $Pr[\text{BAD}_1] < (t+m)^2/2^n$.

Analyzing the second term $Pr[\text{BAD}_2|\overline{\text{BAD}_1}]$ is somewhat more complex. First, let Q_1 be the query graph corresponding to the t H queries. If the attacker uses only Q_1, its success probability is bounded by $\frac{\lfloor t/c_* \rfloor}{|PW|}$ as shown in Theorem 1, except that c_0 is replaced with its lower limit c_*. Next, we consider the effect of F^* queries. We observe that (unlike in the proof of Theorem 1) $x_j = p_0\|s_j\|c_j$ doesn't have to be the *first* vertex of a path, since the adversary is allowed to choose (s_k, c_k) and make a query to F^*, and this provides the adversary more chances of success. To quantify this advantage, we consider the number of vertices of the form $\{p\|s_i\|c_i, 1 \leq i \leq m\}$, denoted by m'. The success probability using F^* queries is bounded by $m'/|PW|$. Note that $m' = m + q$ where q is the expected number of collisions in $s\|c$ among all the vertices $p\|s\|c$ in Q_1. Since the expected value of q is $t|PW|/2^n \ll 1$, it can be shown that the probability that $m' = m + q \geq m + m = 2m$ is negligible.

Combining all the probabilities, we prove that $Adv(t, m)$ is bounded by $\frac{\lfloor t/c_* \rfloor + 2m}{|PW|} + \frac{(t+m)^2}{2^n})$ as stated.

Lower Bound

For the lower bound, we describe two strategies from which the adversary can pick the one yielding better success probability depending on the parameters. In strategy A, the adversary computes separate paths of length c_* for $\lfloor (t - c^*)/c_* \rfloor$ passwords $p_i\|s\|c_*$ using $t - c^*$ queries to H. He then makes a F query asking for $y = F_p^*(s, c_*)$. With probability $\lfloor (t - c^*)/c_* \rfloor /|PW|$, vertex y coincides with the endpoint of one of the paths, thus revealing the password p_0. In such an event the adversary then makes c_0 more H queries to compute $y'_0 = H^{(c_0)}(p_0\|s_0\|c_0)$ and answers 1 if $y'_0 = y_0$. All together the adversary used at most t queries to H and one F queries to achieve success probability of $\lfloor (t - c^*)/c_* \rfloor /|PW|$.

In strategy B, the adversary constructs Q_1 to be a single path of length t starting from an arbitrary $p\|s\|c$. With probability $1 - O(t^2/2^n)$, the path will be cycle-free. Its first $t - c^*$ vertices $p_i\|s_i\|c_i$ have their full paths $T_{p_i\|s_i\|c_i}$ completely contained in Q_1. Assuming m to be much smaller than $t - c^*$, the adversary can pick m vertices $p_i\|s_i\|c_i$ along the path with distinct p_i and make at most m queries to F with the corresponding (s_i, c_i)'s. With probability $m/|PW|$, it can identify the password. This completes the proof of Theorem 2. **QED**

A New Two-Party Identity-Based Authenticated Key Agreement

Noel McCullagh[1,*] and Paulo S.L.M. Barreto[2]

[1] School of Computing,
Dublin City University,
Glasnevin, Dublin 9, Ireland
noel.mccullagh@computing.dcu.ie

[2] Escola Politécnica, Universidade de São Paulo,
Av. Prof. Luciano Gualberto, tr. 3, 158,
BR 05508-900 São Paulo(SP), Brazil
pbarreto@larc.usp.br

Abstract. We present a new two-party identity-based key agreement that is more efficient than previously proposed schemes. It is inspired on a new identity-based key pair derivation algorithm first proposed by Sakai and Kasahara. We show how this key agreement can be used in either escrowed or escrowless mode. We also describe conditions under which users of different Key Generation Centres can agree on a shared secret key. We give an overview of existing two-party key agreement protocols, and compare our new scheme with existing ones in terms of computational cost and storage requirements.

Keywords: authenticated key agreement, identity-based cryptography, bilinear maps, Tate pairing.

1 Introduction

In this paper we propose a new two-party authenticated identity-based key agreement from bilinear maps. The basic idea behind an identity-based cryptosystem is that end users can choose an arbitrary string, for example email addresses or other online identifiers, as their public key. This eliminates much of the overhead associated with key management. In traditional PKI settings, key agreement protocols relies on the parties obtaining each other's certificates, extracting each other's public keys, checking certificate chains (which may involve many signature verifications) and finally generating a shared secret. The technique of identity-based encryption (IBE) greatly simplifies this process. This idea was first proposed by Shamir [19] in 1984, made viable by Cocks [8] and Boneh and Franklin [4] in 2001, further streamlined by Sakai and Kasahara [16] in 2003, and is currently an area of very active research (see e.g. [9] for a survey).

---

* This author wishes to thank Enterprise Ireland for their support with this research under grant IF/2002/0312/N.

There are many key agreement protocols based on bilinear maps, and most have subsequently been broken. One of the first applications of pairing based cryptography was a tripartite key agreement protocol by Joux [12]. Although this key agreement does not authenticate the users, and thus is susceptible to the man-in-the-middle attack, it was a significant step in the development of pairing based cryptography. This original scheme was not identity-based. Many key agreements from bilinear maps have been since proposed. Scott [17], Smart [24], and Chen and Kudla [6] have proposed two-party key agreement protocols, none of which have been broken. All of these schemes require that all parties involved in the key agreement are clients of the same Key Generation Centre (KGC). Nalla proposes a tripartite identity-based key agreement in [13], and Nalla and Reddy propose a scheme in [14], but both have been broken [7, 21]. Shim presents two key agreements [23, 22], but both these schemes have been broken by Sun and Hsieh [25]. Another authenticated tripartite key agreement proposed by Al-Riyami and Patterson [1] was broken by Shim [20].

Most identity-based key agreement protocols have the property of *key escrow*: the trusted authority that issues private keys can recover the agreed session key. This feature is either acceptable, unacceptable, or desired depending on the circumstances. For example, escrow is essential in situations where confidentiality as well as an audit trail is a legal requirement, as in confidential communication in the health care profession. There are other examples, such as personal communications, where it would be advantageous to turn escrow off.

The two-party key agreements proposed by Smart and by Chen and Kudla are escrowed schemes by default. A modification suggested by Chen and Kudla [6] to remove escrow can also be applied to Smart's scheme. However, this modification creates additional computational overhead. Scott's scheme does not allow escrow, and there seems no obvious way to introduce this feature – bar one party to the protocol sending a third party a copy of the agreed key.

Chen and Kudla also suggest a modification that allows two parties that have their public keys generated by two different Key Generation Centre's to communicate. We say that these parties are members of *different domains*. Most key agreements require both parties to be from the same domain. This, for example, might mean that two workers from the same company would be able to generate a shared secret, however employees from two different companies would not be able to generate such a shared secret. We suggest a protocol that, without pairing precomputation, is twice as efficient as the scheme suggested in [6].

We suggest key agreement between domains is an important property of this scheme as, from a commercial viewpoint, identity-based cryptography (IBC) seems particularly well suited to encrypted telephony and encrypted VoIP. For encrypted VoIP to work on a global scale there simply must be compatibility between networks, and therefore key agreement between different networks is important.

Our contributions in this paper are:

- An efficient identity-based authenticated key agreement protocol that can be instantiated in either escrowed or escrowless mode without imposing extra computational steps.
- An efficient key agreement that allows users who have their private keys generated by distinct Key Generation Centres to establish a shared secret without additional overhead, provided standardised curve parameters are used.

This paper is organised as follows. Section 2 introduces basic mathematical concepts. Section 3 describes our proposed authenticated key agreement with escrow, and section 4 introduces our proposed escrowless scheme. In section 5 we present a key agreement protocol for members of distinct key generation domains. We discuss efficiency issues in section 6 and security issues in section 7. Finally, we draw our conclusions in section 8.

2 Mathematical Preliminaries

An elliptic curve $\mathbb{E}(\mathbb{F}_{q^k})$ is the set of solutions (x,y) over the field $\mathbb{F}_{q^k}$ to an equation of the form $y^2 = x^3 + Ax + B$, together with an additional point at infinity, denoted O. There exists an Abelian group law on $\mathbb{E}$, with explicit formulas for computing the coordinates of a point $P_3 = P_1 + P_2$ from the coordinates of P_1 and P_2. Scalar multiplication of a point is defined as the repeated addition of a point to itself n, e.g. $3P_1 = P_1 + P_1 + P_1$. O is the identity element.

The number of points of an elliptic curve $\mathbb{E}(\mathbb{F}_{q^k})$ is called the order of the curve over the field $\mathbb{F}_{q^k}$. A point P has order r if $rP = O$ for the smallest possible $r > 0$. The set of r-torsion points on $\mathbb{E}$ is the set $\mathbb{E}[r] = \{P \in \mathbb{E} \mid rP = O\}$. The order of a point always divides the curve order. There is an operation on a point in the extension field that will reduce that point to a point in the base field; this is called the trace map, and is denoted as $\mathrm{Tr}(P)$. One of these cyclic subgroups is called the *trace zero* subgroup, $\mathcal{T} = \{P \in \mathbb{E} \mid \mathrm{Tr}(P) = O\}$. A subgroup $\mathbb{G}$ of an elliptic curve is said to have embedding degree k if its order r divides $q^k - 1$ for the smallest possible k. We assume $k > 1$. We let $\mathbb{G}_0$ be the group of order r defined over $\mathbb{F}_q$ and $\mathbb{G}_1$ be the trace zero group, again of order r. The results of Weil and Tate pairing operations equate to one of the r-th roots of unity. Again this is a group of order r, we call this group $\mathbb{G}_2$ [11].

The modified Tate pairing over supersingular curves [5] denoted $\hat{t}(P,Q)$ is $t(P, \psi(P))$ where $t : \mathbb{G}_0 \times \mathbb{G}_1 \to \mathbb{G}_2$ is the Tate pairing and $\psi : \mathbb{G}_0 \to \mathbb{G}_1$ is an efficiently computable distortion map [11]. It is an example of a bilinear map of the form $\hat{t} : \mathbb{G}_0 \times \mathbb{G}_0 \to \mathbb{G}_2$ where $\mathbb{G}_0$ and $\mathbb{G}_2$ are groups of order r.

The possibility of exploiting differences between the pairings $\hat{t}(P,P)$ and $t(P,Q)$ to implement protocols with different properties has occurred to other authors [10, 17]. We use the modified Tate pairing in the escrowed system, and the Tate pairing in the escrowless system.

3 An Authenticated Key Agreement with Escrow

As with all other identity-based cryptosystems we assume the existence of a trusted Key Generation Centre (KGC) that is responsible for the creation and secure distribution of users private keys. This agreement algorithm can be implemented using the modified Tate pairing.

Setup: The KGC inputs a security parameter κ into a BDH parameter generator $\mathcal{B}_{mt}$ which returns two groups $\mathbb{G}_0$ and $\mathbb{G}_2$, both of prime order r, a suitable bilinear map $\hat{t} : \mathbb{G}_0 \times \mathbb{G}_0 \to \mathbb{G}_2$ (which can be implemented as the modified Tate pairing), a generator element P such that $\langle P \rangle = \mathbb{G}_0$, and a random oracle $\mathcal{H} : \{0,1\}^* \to \mathbb{Z}_r^*$. The KGC randomly generates a master secret $s \in_R \mathbb{Z}_r^*$, and calculates a master public key sP. The parameters and master public key are distributed to the users of the system through a secure authenticated channel. We assume that the number of users is polynomial in κ.

Extract: The KGC checks that a user has a claim to a particular online identifier. If they do, the KGC generates their private key and communicates it privately to them. Let Alice's online identifier map to $a \in \mathbb{Z}_r^*$ by means of the random oracle $\mathcal{H}$. Alice's public key is $(a+s)P$, which can be computed as $aP + sP$. The KGC computes Alice's private key as $A_{pri} = (a+s)^{-1}P$. While it may be argued that this key pair derivation is not as elegant as that in the Boneh-Franklin IBE [4], since the public key no longer relies on the user's identity alone, most key agreements, except Scott's and Ryu et al.'s [15], also use the KGC's master secret in the key agreement stage.

Key Agreement: Assume that Alice and Bob have private keys issued by the same KGC, respectively A_{pri} and B_{pri}. Alice and Bob each generate one unique random nonce $x_a, x_b \in_R \mathbb{Z}_r^*$, respectively.

$$
\begin{array}{ll}
\text{Alice} & \text{Bob} \\
A_{KA} = x_a(bP + sP) & \rightleftarrows \quad B_{KA} = x_b(aP + sP) \\
key_a = \hat{t}(B_{KA}, A_{pri})^{x_a} & key_b = \hat{t}(A_{KA}, B_{pri})^{x_b}
\end{array}
$$

This scheme is consistent because:

$$
\begin{aligned}
key_a &= \hat{t}(B_{KA}, A_{pri})^{x_a} \\
&= \hat{t}(P, P)^{x_a x_b} \\
&= \hat{t}(A_{KA}, B_{pri})^{x_b} \\
&= key_b.
\end{aligned}
$$

The escrow property derives from the KGC's ability to recover the shared session key by computing:

$$
\begin{aligned}
x_a P &= (s+b)^{-1} A_{KA}, \\
x_b P &= (s+a)^{-1} B_{KA}, \\
key &= \hat{t}(x_a P, x_b P).
\end{aligned}
$$

Our scheme is *role symmetric*, with each party performing the same operations and thus incurring the same computational cost.

4 An Authenticated Key Agreement Without Escrow

The key agreement without escrow differs only slightly from the algorithm given in section 3. Again there are three algorithms, **Setup**, **Extract** and **Key Agreement**. This key agreement protocol can be implemented using the conventional Tate pairing, not the modified Tate pairing as in the escrowed scheme.

Setup: The KGC inputs a security parameter κ into a BDH parameter generator $\mathcal{B}_t$ which returns three groups $\mathbb{G}_0, \mathbb{G}_1$ and $\mathbb{G}_2$, $\mathbb{G}_0$ and $\mathbb{G}_2$ being groups of prime order r, a suitable bilinear map $t : \mathbb{G}_0 \times \mathbb{G}_1 \to \mathbb{G}_2$ (which can be implemented as the Tate pairing), two generator elements P and Q such that $\langle P \rangle = \mathbb{G}_0$ and $\langle Q \rangle = \mathbb{G}_1$, and a random oracle $\mathcal{H} : \{0,1\}^* \to \mathbb{Z}_r^*$. It is important that the discrete logarithm between $\psi(P)$ and Q is unknown[1]. This can be achieved by obtaining P and Q as the output of random oracles $\mathcal{H}_0 : \{0,1\}^* \to \mathbb{G}_0$ and $\mathcal{H}_1 : \{0,1\}^* \to \mathbb{G}_1$ evaluated on publicly known constant strings cs_0 and cs_1 (cs_0 and cs_1 may be the same string). The KGC randomly generates a master secret $s \in_R \mathbb{Z}_r^*$, and calculates a master public key sP. The parameters, master public key and the constant strings used in the derivation of P and Q are distributed to the users of the system through a secure authenticated channel. We assume that the number of users is polynomial in κ.

Extract: The KGC checks that a user has a claim to a particular online identifier. If they do, the KGC generates their private key and communicates it privately to them. Let Alice's online identifier map to $a \in \mathbb{Z}_r^*$ by means of the random oracle $\mathcal{H}$. Alice's public key is $A_{pub} = (a+s)P$, which can be computed as $aP + sP$. Alice's private key is generated as $A_{pri} = (a+s)^{-1}Q$. End user Alice is encouraged to check that the KGC has used the correct Q in the construction of her private key by checking the following:

$$P \leftarrow \mathcal{H}_0(cs_0)$$
$$Q \leftarrow \mathcal{H}_1(cs_1)$$
$$t(A_{pub}, A_{pri}) \stackrel{?}{=} t(P, Q)$$

Key Agreement: Assume that Alice and Bob have private keys issued by the same KGC, respectively A_{pri} and B_{pri}. Alice and Bob each generate one unique random nonce $x_a, x_b \in \mathbb{Z}_r^*$, respectively.

Alice		Bob
$A_{KA} = x_a(bP + sP)$	$\rightleftarrows$	$B_{KA} = x_b(aP + sP)$
$key_a = t(B_{KA}, A_{pri})^{x_a}$		$key_b = t(A_{KA}, B_{pri})^{x_b}$

---

[1] If the KGC knows λ such that $\psi(P) = \lambda Q$, it can use the distortion map to get a representation in $\langle Q \rangle$ of A_{KA} or B_{KA} and then recover the session key using the technique outlined in the previous section. On non-supersingular curves no efficiently computable distortion map exists [26] and this attack does not apply.

This scheme is consistent because

$$\begin{aligned}key_a &= t(B_{KA}, A_{pri})^{x_a} \\ &= t(P, Q)^{x_a x_b} \\ &= t(A_{KA}, B_{pri})^{x_b} \\ &= key_b.\end{aligned}$$

We also note that, although the KGC has the ability to generate the private keys of both users in the protocol, it is not able to obtain the shared session key for any particular run of the protocol. The KGC can, in this instance, easily compute $t(P,Q)^{x_a}$ and $t(P,Q)^{x_b}$, but calculating the key from these values involves solving the Computational Diffie-Hellman Problem (CDHP) over the group $\mathbb{G}_2$ [27].

5 Key Agreement Between Members of Distinct Domains

We now look at key agreements between members of separate domains. This idea was first suggested in [6]. We suggest a scheme that is twice as efficient as their scheme without precomputation, whilst being similar with precomputation. Again this protocol can be instantiated in escrowed or escrowless mode.

For key agreement to be possible between members of different groups all that is needed is for the points P, Q in the case of the escrowless system, or just P in the case of the escrowed system, and the curve description to be the same (standardised). Elliptic curves, suitable group generator points and other cryptographic tools have been standardised for non-IBE applications, for example in the NIST FIPS standards. It is reasonable, therefore, to assume the availability of standard pairing-friendly curves as well.

Once these group generator points and curves have been agreed upon, each KGC can generate their own random master secret.

Alice's private key is generated by KGC_1 with a master secret s_1. Bob's private key is generated by KGC_2 with a master secret s_2. Alice's public key is $(a+s_1)P$ and her private key is $A_{pri} = (a+s_1)^{-1}P$. Likewise, Bob's public key is $(b+s_2)P$ and his private key is $B_{pri} = (b+s_2)^{-1}P$. Notice that now Alice must obtain s_2P (the master public key of Bob's KGC) and vice-versa; it is critical that the master public keys are obtained in an authenticated manner, as with any IBC scheme.

Alice and Bob now perform the authenticated key agreement:

$$\begin{array}{ccc} \textbf{Alice} & & \textbf{Bob} \\ A_{KA} = x_a(bP + s_2 P) & \rightleftarrows & B_{KA} = x_b(aP + s_1 P) \\ key_a = \hat{t}(B_{KA}, A_{pri})^{x_a} & & key_b = \hat{t}(A_{KA}, B_{pri})^{x_b} \end{array}$$

This scheme is consistent because

$$key_a = \hat{t}(B_{KA}, A_{pri})^{x_a}$$
$$= \hat{t}(P, P)^{x_a x_b}$$
$$= \hat{t}(A_{KA}, B_{pri})^{x_b}$$
$$= key_b.$$

6 Efficiency

Smart's protocol [24] requires each party to perform 2 point scalar multiplications and 2 pairing evaluations. One of these pairings can be partially precomputed, reducing the cost to 1 point scalar multiplication, 1 pairing evaluation and 1 pairing exponentiation per party at an additional storage cost of one pairing per recipient. Our new scheme achieves the same efficiency without incurring the extra storage requirements.

The Chen-Kudla authenticated key agreement protocol [6] requires 2 elliptic curve point scalar multiplications, 1 point addition and 1 pairing evaluation.

Scott's key agreement [17], using the pairing as a SPEKE generator, only requires two pairing exponentiations when precomputation is used. Again it restricts all users to having private keys generated by the same KGC.

The scheme proposed here requires 1 point scalar multiplication, 1 pairing exponentiation and one 1 pairing evaluation. We note that a pairing exponentiation is quicker than a point scalar multiplication.

We also note that the method of generating public keys from identities – namely, by mapping identities to integer coefficients and performing a scalar multiplication – is faster than the technique used in Boneh-Franklin key pair generation. Their technique involves mapping the identifier to a coordinate, solving the curve equation and then multiplying by a large cofactor to generate a point of order r. Public keys in our system will always be points of order r.

In Smart's protocol the recipient's public key is used either explicitly or implicitly (if pairings are precomputed) to complete the protocol. In our scheme, public keys of form $uP + sP$ may be stored to save one scalar multiplication, with the advantage that such values require a much smaller storage space than pairing values, namely, a fraction[2] $1/k$ where k is the embedding degree of the curve $E(\mathbb{F}_q)$.

We leave public key generation out of the following complexity analysis as it is only slightly faster for our system – and can be precomputed in all IBE systems. We also leave out $E(\mathbb{F}_{q^k})$ multiplication, point addition and hashing as they are fast to compute compared to the other principle operations.

Key: $p = $ *pairing evaluation, $e = E(\mathbb{F}_{q^k})$ (pairing) exponentiation, $m = $ scalar multiplication, $n = $ number of recipients, $s = $ storage space per pairing evaluation, rac $= $ requires additional computation (two point multiplications).*

---

[2] If pairing compression techniques as described in [18] are used, the fraction is $2/k$ in general or $3/k$ in a special case.

	Proposed	Smart	Chen-Kudla	Scott
No Precomp	**1p+1e+1m**	2p+1m	1p+2m	1p+2e
Precomp	**1p+1e+1m**	1p+1e+1m+ns	1p+1e+1m+ns	2e
Escrow	**Yes / No**	Yes / No (rac)	Yes / No (rac)	No
Between Domains	**Yes**	No	No	No

7 Security of the Proposed Scheme

The proof of security of the above algorithm relies on the conjectured intractability of a problem which Zhang et al. [28] call the **Bilinear Inverse Diffie-Hellman Problem**: For $\alpha, \beta \in \mathbb{Z}_r^*$, given $P, \alpha P, \beta P$, compute $v = \hat{t}(P,P)^{\alpha^{-1}\beta}$.

7.1 The Security of the Authentication Mechanism

Assuming that the BIDHP is hard (with respect to the security parameter κ), we now show the security of the above protocols.

We adopt the security model proposed by Bellare and Rogaway [2], modified by Blake-Wilson et al. [3], and used in proving the security of the key agreement protocol introduced in [6] and others.

The model includes a set of parties, each modelled by an oracle. We use the notation $\prod_{i,j}^n$, meaning a participant/oracle i believing that it is participating in the n-th run of the protocol with j. Oracles keep transcripts of all communications in which they have been involved. Each oracle has a secret private key, issued by a KGC, which has run a BDH parameter generator $\mathcal{B}$ and published groups $\mathbb{G}_0$ and $\mathbb{G}_2$, a bilinear map of the form $e : \mathbb{G}_0 \times \mathbb{G}_0 \to \mathbb{G}_2$, a group generator P of $\mathbb{G}_0$, and a master public key sP.

The model contains and adversary E which has access to all message flows in the system. E is not a user or KGC. All oracles only communicate with each other via E. E can replay, modify, delay, interleave or delete messages. E is benign if it acts like a wire and does not modify communication between oracles. From [2], if two oracles receive, via the adversary, property formatted messages that have been generated exclusively by the other oracle, and both oracles accept, we say that these two oracles have had a matching conversation.

The adversary at any time can make the following queries:

Create: E sets up a new oracle in the system that has public key ID, of E's choosing. E has access to the identity / public key of the oracle. The private key is obtained from the KGC.
Send: E sends a message of his choice to an oracle i, $\prod_{i,j}^n$, in which case i assumed that the message came from j. E can also instruct the actual oracle j to start a new run of the protocol with i by sending a λ. Using the terminology of [6] an oracle is an *initiator oracle* if the first message that it receives is a λ, otherwise it is a *responder oracle*.
Reveal: E receives the session key that is currently being held by a particular oracle.

Corrupt: E receives the long term asymmetric private key being held by a particular oracle.

Test: E receives either the session key or a random value from a particular oracle. Specifically, to answer the query the oracle flips a fair coin $c \in \{0, 1\}$; if the answer is 0 it outputs the agreed session key, and if the answer is 1 it outputs a random element of $\mathbb{G}_2$. E then must decide whether c is 0 or 1; call this prediction c'. E's advantage in distinguishing the actual session key held by an uncorrupted party from a key sampled at random from $\mathbb{G}_2$ in this game, with respect to the security parameter κ, is given by:

$$Advantage^E(\kappa) = |Pr[c' = c] - 1/2|$$

The Test query can be performed only once, against an oracle that is in the Accepted state (see below), and which has not previously been asked a Reveal or Corrupt query.

An oracle may be in one of the following states (it cannot be in more than one state).

Accepted: If the oracle decides to accept a session key, after receipt of properly formated messages.

Rejected: If the oracle decides to not to accept and aborts the run of the protocol.

*: If the oracle has yet to decide whether to accept to reject for this run of the protocol. We assume that there is some time out on this state.

Opened: If a Reveal query has been performed against this oracle for its last run of the protocol (its current session key is revealed).

Corrupted: If a Corrupt query has ever been performed against this oracle.

Definition 1.
A protocol is an AK protocol if:

- *In the presence of the benign adversary on $\prod_{i,j}^n$ and $\prod_{j,i}^t$, both oracles always accept holding the same session key, and this key is distributed uniformly at random on $\mathbb{G}_2$; if for every adversary E:*
- *If uncorrupted oracles $\prod_{i,j}^n$ and $\prod_{j,i}^t$, have matching conversations then both oracles accept and hold the same session key;*
- *$Advantage^E(\kappa)$ is negligible.*

Theorem 1. *The proposed key agreement protocol is a secure AK protocol.*

Proof. Condition 1 holds as follows: Both oracles accept holding the same session key as a direct result of the commutativity of exponentiation of members of the group $\mathbb{G}_2$. The session key is distributed uniformly at random by the fact that both oracles generate truly random $x \in \mathbb{Z}$. Therefore the product of these elements will also be random. Since the exponent is random, and $e(P, P)$ is a generator of the group $\mathbb{G}_2$, the session key will be uniformly distributed over $\mathbb{G}_2$.

Condition 2 holds by the fact that if they have matching conversations then the communication was generated entirely by the two oracle's. Therefore, by the bilinearity of the pairing and the commutativity of exponentiation they accept and hold the same session key.

Condition 3 holds as follows: Consider by contradiction that $Advantage^E(\kappa)$ is non-negligible. Then we can construct from E an algorithm $\mathcal{F}$ that solves the BIDHP with non-negligible advantage. $\mathcal{F}$ is given as input the output of the BDH generator $\mathcal{B}$. $\mathcal{F}$'s task is to solve the BIDHP, namely, given P, αP and βP, compute $v = \hat{t}(P, P)^{\alpha^{-1}\beta}$.

All queries by the adversary E now pass through $\mathcal{F}$.

Create: For each oracle $\mathcal{F}$ chooses $y_i \in_R \mathbb{Z}_p^*$, creates a public key as $u_i P = (y_i P - sP)$, and computes the private key as $y_i^{-1} P$. Obviously $y_i P = u_i P + sP$. However, for the j-th oracle $\mathcal{F}$ answers αP. Since $\mathcal{F}$ does not know α, it cannot calculate $\alpha^{-1} P$, the correct private key for this oracle.

Corrupt: $\mathcal{F}$ answers Corrupt queries in the usual way, revealing the private key of the oracle being queried. However, $\mathcal{F}$ does not know the private key for oracle j; if E asks a Corrupt query on oracle j, $\mathcal{F}$ gives up.

Send: $\mathcal{F}$ answers all send queries in the usual way, except if E asks Send $\prod_{i,j}^n$, $\mathcal{F}$ answers $x_i P$, for a known x_i, which is, from E's perspective, indistinguishable from $x_t(\alpha P)$ for a random $x_t \in_R \mathbb{Z}_q^*$. In response it will get a value from j, this is set as the value βP – this is a genuine value from j and $\mathcal{F}$ does not influence it.

Test: At some point E will ask a single Test query of some oracle, which we assume is oracle j; if it is not, $\mathcal{F}$ aborts. The chance of $\mathcal{F}$ picking j is $\xi = 1/n$ where n is the number of oracles (Create queries). Since it is picked it must have accepted and it must be holding a session key of the form $e(P, P)^{x_i x_j}$. However, $\mathcal{F}$ cannot compute this key and hence cannot simulate the query, so it simply outputs a random element of the group $\mathbb{G}_2$.

If $\mathcal{F}$ does not abort and E does not detect $\mathcal{F}$'s inconsistency in answering the Test query then its advantage in predicting the correct session key is $Advantage^E(\kappa)$ as before. For this to be non-negligible, E must have some advantage in calculating $e(P, P)^{x_i x_j}$, given $\beta P = x_j \alpha P$ as input from j.

If E does not detect any inconsistencies in $\mathcal{F}$'s responses, then $\mathcal{F}$ must have non-negligible advantage $A(\kappa)$ in calculating $e(P, P)^{x_i x_j}$, but, since $\mathcal{F}$ does not know j's private key the session key was calculated as $e(P, P)^{\alpha^{-1}\beta x_i}$. Provided that $\mathcal{F}$ is able to calculate $e(P, P)^{\alpha^{-1}\beta x_i}$, it can calculate $e(P, P)^{\alpha^{-1}\beta}$ since it knows x_i.

We assume that there is some timeout τ_s on the length of a run of the protocol, including the time spent in the $*$ state. We also assume that some time τ_c is allocated to allow the construction of oracles in the Create query, and time τ_o allocated for each Corrupt query. We assume that n oracles are needed, and that m send queries are needed, and o corrupt queries are needed. The expected time needed to solve the BIDHP is:

$$\frac{(n\tau_c)(m\tau_s)(o\tau_o)\xi}{A(\kappa)}$$

Table 1. Key offset attack.

$$
\begin{array}{lcl}
\text{Alice} & E & \text{Bob} \\
K_A = x_a(s+b)P & \rightarrow K'_A = \epsilon K_A \rightarrow & \\
 & \leftarrow K'_B = \epsilon K_B \leftarrow & K_B = x_b(s+a)P \\
key = e(K'_B, A_{pri})^{x_a} & & key = e(K'_A, B_{pri})^{x_b} \\
\quad = e(P,P)^{x_a x_b \epsilon} & & \quad = e(P,P)^{x_a x_b \epsilon}
\end{array}
$$

We note that our protocol is vulnerable to an attack described by Blake-Wilson et al. [3], namely, that an active adversary can offset the agreed session key by an exponent ϵ unbeknownst to Alice or Bob. Most key agreements without key confirmation are vulnerable to this attack, for example, those by Chen and Kudla, Smart and Scott. The attack is shown below, with E, being an active attacker.

Although this attack (which exists against many key agreements) is interesting, it should be noted that it does not allow the attacker to gain any knowledge of the agreed session key.

7.2 Further Security Considerations

Here we look at the new key agreement using a few security definitions that are often used to judge key agreements. We only consider the basic protocol given in section 3.

Known Key Security: If one session key is compromised this does not mean that any other session keys are compromised. This is from the fact that the agreed session keys rely on random ephemeral keys. A session key as a result is distributed uniformly in $\mathbb{G}_2$ with no connection to other session keys.

Key-Compromise Impersonation: If Alice's private key is exposed, it does not enable an adversary to impersonate Bob to Alice. This stems from the fact that Alice uses Bob's public key in her contribution to the shared secret.

Unknown Key-Share Resilience: Alice cannot be coerced into sharing a key with Charlie thinking she is sharing a key with Bob. Again, this come from the fact that Alice explicitly uses Bob's public key in her contribution to the session key.

Forward Secrecy: Compromise of either Alice's private key or Bob's private key does not appear to allow an attacker to recover any past session keys. On the other hand, compromise of the KGC's master secret in the escrowed scheme allows all past agreed session keys to be recovered.

Key Control: Because both parties have an input into the key, neither entity is able to force the full session key to be a preselected value. However, Bob can set certain bits of the agreed session key by carefully selecting his ephemeral key

x_b until be achieves the desired result. It does not appear possible for Bob to set any substantial number of bits in a reasonable time frame. Again, this key agreement is no less secure in this respect that most other key agreements. As with all key agreements a short timeout on a particular run of the protocol may be advisable.

8 Conclusion

We have presented a new ID-based key agreement protocol inspired on the Sakai-Kasahara key pair generation algorithm. The proposed scheme improves on the performance of the Smart and the Chen-Kudla key agreement protocols, can be instantiated in either escrowed or escrowless mode, and can be carried out by clients of distinct KGC's.

References

1. S. S. Al-Riyami and K. G. Paterson. Tripartite authenticated key agreement protocols from pairings. In *IMA Conference on Cryptography and Coding*, volume 2898 of *Lecture Notes in Computer Science*, pages 332–359. Springer-Verlag, 2003.
2. M. Bellare and P. Rogaway. Entity authentication and key distribution. In *Advances in Cryptology – Crypto'93*, volume 773 of *Lecture Notes in Computer Science*, pages 232–249. Springer-Verlag, 1994.
3. S. Blake-Wilson, D. Johnson, and A. Menezes. Key agreement protocols and their security analysis. In *IMA International Conference on Cryptography and Coding*, volume 1355 of *Lecture Notes in Computer Science*, pages 30–45. Springer-Verlag, 1997.
4. D. Boneh and M. Franklin. Identity-based encryption from the weil pairing. In *Advances in Cryptology – Crypto'2001*, volume 2139 of *Lecture Notes in Computer Science*, pages 213–229. Springer-Verlag, 2001.
5. L. Chen and K. Harrison. Multiple trusted authorities in identifier based cryptography from pairings on elliptic curves. Trusted Systems Laboratory, HP, 2003. http://www.hpl.hp.com/techreports/2003/HPL-2003-48.pdf.
6. L. Chen and C. Kudla. Identity based authenticated key agreement from pairings. Cryptology ePrint Archive, Report 2002/184, 2002. http://eprint.iacr.org/2002/184.
7. Z. Chen. Security analysis on Nalla-Reddy's ID-based tripartite authenticated key agreement protocols. Cryptology ePrint Archive, Report 2003/103, 2003. http://eprint.iacr.org/2003/103.
8. C. Cocks. An identity based encryption scheme based on quadratic residues. In *VIII IMA International Conference on Cryptography and Coding*, volume 2260 of *Lecture Notes in Computer Science*, pages 360–363. Springer-Verlag, 2001. "http://www.cesg.gov.uk/site/ast/idpkc/media/ciren.pdf.
9. R. Dutta, R. Barua, and P. Sarkar. Pairing-based cryptography : A survey. Cryptology ePrint Archive, Report 2004/064, 2004. http://eprint.iacr.org/2004/064.
10. S. Galbraith. Personal communication, 2004.
11. S. Galbraith and V. Rotger. Easy decision-diffie-hellman groups. Cryptology ePrint Archive, Report 2004/070, 2004. http://eprint.iacr.org/2004/070.

12. A. Joux. A one round protocol for tripartite Diffie-Hellman. In *Proceedings of Algorithmic Number Theory Symposium*, volume 1838 of *Lecture Notes in Computer Science*, pages 385–394. Springer-Verlag, 2000.
13. D. Nalla. ID-based tripartite key agreement with signatures. Cryptology ePrint Archive, Report 2003/144, 2003. http://eprint.iacr.org/2003/144.
14. D. Nalla and K. C. Reddy. ID-based tripartite authenticated key agreement protocols from pairings. Cryptology ePrint Archive, Report 2003/004, 2003. http://eprint.iacr.org/2003/004.
15. Eun-Kyung Ryu, Eun-Yoon, and Kee-Young Yoo. An efficient ID-based autenticated key agreement protocol from pairings. In *NETWORKING 2004*, volume 3042 of *Lecture Notes in Computer Science*, pages 1458–1463. Springer-Verlag, 2004.
16. R. Sakai and M. Kasahara. ID based cryptosystems with pairing on elliptic curve. In *2003 Symposium on Cryptography and Information Security – SCIS'2003*, Hamamatsu, Japan, 2003. http://eprint.iacr.org/2003/054.
17. M. Scott. Authenticated ID-based key exchange and remote log-in with insecure token and PIN number. Cryptology ePrint Archive, Report 2002/164, 2002. http://eprint.iacr.org/2002/164/.
18. M. Scott and P. S. L. M. Barreto. Compressed pairings. In *Advances in Cryptology – Crypto'2004*, volume 3152 of *Lecture Notes in Computer Science*. Springer-Verlag, 2004. to appear.
19. A. Shamir. Identity based cryptosystems and signature schemes. In *Advances in Cryptology – Crypto'84*, volume 0196 of *Lecture Notes in Computer Science*, pages 47–53. Springer-Verlag, 1984.
20. K. Shim. Cryptanalysis of Al-Riyami-Paterson's authenticated three party key agreement protocols. Cryptology ePrint Archive, Report 2003/122, 2003. http://eprint.iacr.org/2003/122.
21. K. Shim. Cryptanalysis of ID-based tripartite authenticated key agreement protocols. Cryptology ePrint Archive, Report 2003/115, 2003. http://eprint.iacr.org/2003/115.
22. K. Shim. Efficient ID-based authenticated key agreement protocol based on Weil pairing. *Electronics Letters*, 39(8):653–654, 2003.
23. K. Shim. Efficient one round tripartite authenticated key agreement protocol from Weil pairing, 2003.
24. N. P. Smart. An identity based authenticated key agreement protocol based on the Weil pairing. *Electronics Letters*, 38:630–632, 2002.
25. H.-M. Sun and B.-T. Hsieh. Security analysis of Shim's authenticated key agreement protocols from pairings. Cryptology ePrint Archive, Report 2003/113, 2003. http://eprint.iacr.org/2003/113.
26. E. Verheul. Evidence that XTR is more secure than supersingular elliptic curve cryptosystems. In *Advances in Cryptology – Eurocrypt'2001*, volume 2045 of *Lecture Notes in Computer Science*, pages 195–210. Springer-Verlag, 2001.
27. Y. Yacobi. A note on the bilinear Diffie-Hellman assumption. Cryptology ePrint Archive, Report 2002/113, 2002. http://eprint.iacr.org/2002/113.
28. F. Zhang, R. Safavi-Naini, and W. Susilo. An efficient signature scheme from bilinear pairings and its applications. In *International Workshop on Practice and Theory in Public Key Cryptography – PKC'2004*, Lecture Notes in Computer Science, pages 277–290. Springer-Verlag, 2004.

Accumulators from Bilinear Pairings and Applications

Lan Nguyen

Centre for Information Security,
University of Wollongong, Wollongong 2522, Australia
ldn01@uow.edu.au

Abstract. We propose a dynamic accumulator scheme from bilinear pairings and use it to construct an identity-based (ID-based) ring signature scheme with constant-size signatures and to provide membership revocation to group signature schemes, identity escrow schemes and anonymous credential systems. The ID-based ring signature scheme and the group signature scheme have very short signature sizes. The size of our group signatures with membership revocation is only half the size of those in the well-known ACJT00 scheme, which does not provide membership revocation. The schemes do not require trapdoor, so system parameters can be shared by multiple groups belonging to different organizations. All schemes are provably secure in formal models. We generalize the definition of accumulators and provide formal models for ID-based ad-hoc anonymous identification schemes and identity escrow schemes with membership revocation.

Keywords: Dynamic accumulators, ID-based, ring signatures, group signatures, identity escrow, membership revocation, privacy, anonymity.

1 Introduction

An *accumulator* scheme, introduced by Benaloh and de Mare [5] and further developed by Baric and Pfitzmann [3], allows aggregation of a large set of inputs into one constant-size value. For a given element, there is a *witness* that the element was included into the accumulated value whereas it is not possible to compute a witness for an element that is not accumulated. Camenisch and Lysyanskaya [11] extended the concept to *dynamic* accumulators, that means the costs of adding or deleting elements and updating individual witnesses do not depend on the number of elements aggregated. Accumulators have been found in a number of privacy-enhancing applications, including *ad-hoc anonymous identification, ring signatures* [13], *identity escrow* and *group signature* schemes with *membership revocation* [11].

Ring signature schemes, introduced by Rivest et al. [19] and further studied in [9], allows a user to form an ad-hoc group without a central authority and sign messages on behalf of the group. A user might not even know that he has been included in a group and even a party with unlimited computing resources can not identify the signer. Zhang and Kim [23] extended the concept to *ID-based* ring signature schemes, where the group is formed by using members' identities rather than their public keys. ID-base cryptography was introduced by Shamir [20] to simplify key management in public key primitives. In any ID-based system, there is a central authority, called *Private Key Generator* (PKG), to extract private keys from identities. In ID-based ring signature schemes, to comply with the ad-hoc property, the involvement of a central authority is limited to only setting up initial public parameters and generating private keys from identities, and not for forming groups.

While having simple group formation set up as an advantage, the size of ring signatures linearly depends on the group size, as the verifier needs to know at least the group description. However, as pointed out in [13], in many scenarios, the group does not change for a long time or has a short description. So an appropriate measurement of ring signature sizes does not need to include the group description and it is a good direction to find constant-size ring signatures without the group description part. A ring signature scheme (DKNS04) with such a property has been proposed by Dodis et al. [13]. They provide an ad-hoc anonymous identification scheme, where a user can form ad-hoc groups and anonymously prove membership in such groups, and use the Fiat-Shamir heuristics [14] to convert it into the ring signature scheme. The DKNS04 scheme requires user public keys to be primes, that does not seem to allow an ID-based extension. This paper provides the first ID-based ring signature scheme with constant-size signatures (without counting the list of identities to be included in the ring).

The notion of ring signatures is originated from the notion of group signatures, which was introduced by Chaum and Van Heyst [12]. A group signature scheme allows a group member to sign a message on behalf of the group without revealing his identity, and without allowing the message to be linkable to other signed messages that are verifiable under the same public key. The main difference with ring signature schemes lies in the role of a *group manager*. The group manager registers new users by issuing membership certificates that contains registration details, and in case of dispute revokes anonymity of a signed message by 'opening' the signature. In some schemes the functions of the group manager can be split between two managers: an *issuer* and an *opener*. An identity escrow system [15] can be converted into a group signature scheme using the Fiat-Shamir heuristic [14], and group signatures have been used as building blocks for *anonymous credential* systems [2]. A formal model (BSZ04) of group signature schemes was proposed by Bellare et al. [4] with four security requirements (correctness, anonymity, traceability and non-frameability). In Crypto 2000, Ateniese et al. (ACJT00) [1] proposed an efficient group signature scheme with very short length and low computation cost. Ateniese and de Medeiros later proposed an efficient group signature scheme (AdM03) [2] that is 'without trapdoor' in the sense that none of the parties in the system, including the group manager, need to know the trapdoor for generating system parameters. They also outline the importance of this property in real-world applications.

Providing efficient fully dynamic group signature schemes, where users can be revoked from the group, has been a serious challenge. The most notable scheme (CL02) was proposed in [11]; and Tsudik and Xu [22] later proposed another scheme (TX03), which requires less exponentiations in some operations. Both schemes use dynamic accumulators as the key for efficiency improvements and this method can provide membership revocation for other primitives, such as identity escrow and anonymous credential systems. Using this approach to extend AdM03 scheme to be a trapdoor-free group signature scheme with membership revocation does not seem to be easy, as the user certificates are not suitable to be accumulated by the dynamic accumulator used in CL02 and TX03 schemes. Based on this method, but using a different dynamic accumulator, we construct a trapdoor-free group signature scheme with membership revocation.

Our Contribution

In this paper, we propose a new dynamic accumulator scheme and its provably secure applications with a number of attractive properties. The applications are an ID-based ring signature scheme, a group signature scheme with membership revocation and their interactive counterparts, i.e. an ID-based ad-hoc anonymous identification scheme and an identity escrow scheme with membership revocation. The dynamic accumulator can also be used to provide membership revocation for anonymous credential systems. We also generalize the model of accumulators and provide formal models of ID-based ad-hoc anonymous identification schemes and identity escrow schemes with membership revocation, based on the models in [13, 4].

The schemes have a number of attractive properties. Both signature schemes provide the shortest signature sizes compared to corresponding schemes previously proposed. For example, at a comparable level of security when the CL02 and ACJT00 schemes use 1024 bit composite modulus and our group signature scheme with membership revocation uses elliptic curve groups of order 160 bit prime, the signature size in our scheme is just nearly one fourth and one half of the size of an CL02 signature and an ACJT00 signature, respectively. For higher security levels this ratio will be smaller, and ACJT00 scheme does not provide membership revocation. Like CL02 scheme, no procedure in our scheme linearly depends either on the current group size or the total number of revoked members. Our ID-based ring signature scheme is the first one providing signatures with fixed size. All previous normal ring signature schemes, except for the one in [13], have signature sizes linearly dependent on the group size. When using elliptic curve groups of order 160 bit prime, our ring signature size is only about 200 bytes.

Our schemes are completely trapdoor-free. Though being trapdoor-free, the AdM03 scheme uses a trapdoor in the initialization of the system and assumes that the initializing party "safely forgets" the trapdoor. Besides, the AdM03 scheme does not provide membership revocation. Finally in our group signature scheme, the interactive protocol underlying the signature scheme achieves perfect zero-knowledge without any computational assumption whereas in many previous schemes, including ACJT00 and CL02 schemes, the corresponding protocols achieve statistical zero-knowledge under the Strong RSA assumption. We note that all these zero-knowledge proofs including ours, is in honest verifier model.

The organization of the paper is as follows. We recall some background knowledge in section 2 and present the models of dynamic accumulators, ID-based ad-hoc anonymous identification, ID-based ring signature and identity escrow with membership revocation schemes in section 3. Section 4 and 5 give descriptions of our dynamic accumulator scheme, an ID-based ad-hoc anonymous identification scheme, an ID-based ring signature scheme and their security proofs. Section 6 exemplifies the application of our dynamic accumulator to membership revocation by providing an identity escrow scheme with membership revocation and section 7 provides efficiency comparison.

2 Preliminaries

Notation. Let $\mathbb{N}$ be the set of positive integers. For a function $f : \mathbb{N} \to \mathbb{R}^+$, if for every positive number α, there exists a positive integer l_0 such that for every integer $l > l_0$, it holds that $f(l) < l^{-\alpha}$, then f is said to be *negligible*. If there exists a positive number α_0 such that for every positive integer l, it holds that $f(l) < l^{\alpha_0}$, then f is said to be *polynomial-bound*. Let PT denote polynomial-time, PPT denote probabilistic PT and DPT denote deterministic PT. An adversary is an interactive Turing machine. For a PT algorithm $\mathcal{A}(\cdot)$, "$x \leftarrow \mathcal{A}(\cdot)$" denotes an output from the algorithm. For a set $\mathbf{X}$, "$x \leftarrow \mathbf{X}$" denotes an element uniformly chosen from $\mathbf{X}$. For interactive Turing machines $\mathcal{A}(\cdot)$ and $\mathcal{B}(\cdot)$, "$(a \leftarrow \mathcal{A}(\cdot) \leftrightarrow \mathcal{B}(\cdot) \to b)$" denotes that a and b are random variables corresponding to outputs of the joint computation between $\mathcal{A}(\cdot)$ and $\mathcal{B}(\cdot)$. Finally, "$\Pr[Procedures|Predicate]$" denotes the probability that *Predicate* is true after executing the *Procedures*.

2.1 Bilinear Pairings

Let $\mathbb{G}_1, \mathbb{G}_2$ be cyclic additive groups generated by P_1 and P_2, respectively, whose orders are a prime p, and $\mathbb{G}_M$ be a cyclic multiplicative group with the same order p. Suppose there is an isomorphism $\psi : \mathbb{G}_2 \to \mathbb{G}_1$ such that $\psi(P_2) = P_1$. Let $e : \mathbb{G}_1 \times \mathbb{G}_2 \to \mathbb{G}_M$ be a bilinear pairing with the following properties:

1. **Bilinearity:** $e(aP, bQ) = e(P, Q)^{ab}$ for all $P \in \mathbb{G}_1, Q \in \mathbb{G}_2, a, b \in \mathbb{Z}_p$
2. **Non-degeneracy:** $e(P_1, P_2) \neq 1$
3. **Computability:** There is an efficient algorithm to compute $e(P, Q)$ for all $P \in \mathbb{G}_1, Q \in \mathbb{G}_2$

For simplicity, hereafter, we set $\mathbb{G}_1 = \mathbb{G}_2$ and $P_1 = P_2$. But our schemes can be easily modified for the general case when $\mathbb{G}_1 \neq \mathbb{G}_2$. For a group $\mathbb{G}$ of prime order, hereafter, we denote the set $\mathbb{G}^* = \mathbb{G}\backslash\{\mathcal{O}\}$ where $\mathcal{O}$ is the identity element of the group.

We define a Bilinear Pairing Instance Generator as a PPT algorithm $\mathcal{G}$ that takes as input a security parameter 1^l and returns a uniformly random tuple

$\mathbf{t} = (p, \mathbb{G}_1, \mathbb{G}_M, e, P)$ of bilinear pairing parameters, including a prime number p of size l, a cyclic additive group $\mathbb{G}_1$ of order p, a multiplicative group $\mathbb{G}_M$ of order p, a bilinear map $e : \mathbb{G}_1 \times \mathbb{G}_1 \to \mathbb{G}_M$ and a generator P of $\mathbb{G}_1$.

2.2 Complexity Assumptions

The q-SDH assumption originates from a weaker assumption introduced by Mitsunari et al. [16] to construct traitor tracing schemes [21] before being proposed by Boneh and Boyen [6]. It intuitively means that there is no PPT algorithm that can compute a pair $(c, \frac{1}{s+c}P)$, where $c \in \mathbb{Z}_p$, from a tuple $(P, sP, \ldots, s^q P)$, where $s \in_R \mathbb{Z}_p^*$.

q-Strong Diffie-Hellman (q-SDH) Assumption. *For every PPT algorithm $\mathcal{A}$, the following function $\mathrm{Adv}_\mathcal{A}^{q\text{-}SDH}(l)$ is negligible.*

$$\mathrm{Adv}_\mathcal{A}^{q\text{-}SDH}(l) = \Pr[(\mathcal{A}(\mathbf{t}, P, sP, \ldots, s^q P) = (c, \frac{1}{s+c}P)) \wedge (c \in \mathbb{Z}_p)]$$

where $\mathbf{t} = (p, \mathbb{G}_1, \mathbb{G}_M, e, P) \leftarrow \mathcal{G}(1^l)$ and $s \leftarrow \mathbb{Z}_p^*$.

Intuitively, the DBDH assumption [7] states that there is no PPT algorithm that can distinguish between a tuple $(aP, bP, cP, e(P, P)^{abc})$ and a tuple (aP, bP, cP, Γ), where $\Gamma \in_R \mathbb{G}_M^*$ (i.e., chosen uniformly random from $\mathbb{G}_M^*$) and $a, b, c \in_R \mathbb{Z}_p^*$. It is defined as follows.

Decisional Bilinear Diffie-Hellman (DBDH) Assumption. *For every PPT algorithm $\mathcal{A}$, the following function $\mathrm{Adv}_\mathcal{A}^{DBDH}(l)$ is negligible.*

$$\mathrm{Adv}_\mathcal{A}^{DBDH}(l) = |\Pr[\mathcal{A}(\mathbf{t}, aP, bP, cP, e(P, P)^{abc}) = 1] - \Pr[\mathcal{A}(\mathbf{t}, aP, bP, cP, \Gamma) = 1]|$$

where $\mathbf{t} = (p, \mathbb{G}_1, \mathbb{G}_M, e, P) \leftarrow \mathcal{G}(1^l)$, $\Gamma \leftarrow \mathbb{G}_M^*$ and $a, b, c \leftarrow \mathbb{Z}_p^*$.

3 Models

3.1 Accumulators

We generalize definitions of accumulators provided in [11, 13] as follows (in [11, 13], $\mathbf{U}_f = \mathbf{U}_g$ and the bijective function g is the identity function $g(x) = x$).

Definition 1. *An accumulator is a tuple $(\{\mathbf{X}_l\}_{l \in \mathbb{N}}, \{\mathbf{F}_l\}_{l \in \mathbb{N}})$, where $\{\mathbf{X}_l\}_{l \in \mathbb{N}}$ is called the* value domain *of the accumulator; and $\{\mathbf{F}_l\}_{l \in \mathbb{N}}$ is a sequence of families of pairs of functions such that each $(f, g) \in \mathbf{F}_l$ is defined as $f : \mathbf{U}_f \times \mathbf{X}_f^{ext} \to \mathbf{U}_f$ for some $\mathbf{X}_f^{ext} \supseteq \mathbf{X}_l$, and $g : \mathbf{U}_f \to \mathbf{U}_g$ is a bijective function. In addition, the following properties are satisfied:*

- *(Efficient Generation) There exists an efficient algorithm $\mathcal{G}$ that takes as input a security parameter 1^l and outputs a random element $(f, g) \in_R \mathbf{F}_l$, possibly together with some auxiliary information a_f.*

- (Quasi Commutativity) *For every* $l \in \mathbb{N}$, $(f,g) \in \mathbf{F}_l$, $u \in \mathbf{U}_f$, $x_1, x_2 \in \mathbf{X}_l$: $f(f(u,x_1),x_2) = f(f(u,x_2),x_1)$. *For any* $l \in \mathbb{N}$, $(f,g) \in \mathbf{F}_l$, *and* $\mathbf{X} = \{x_1,...,x_q\} \subset \mathbf{X}_l$, *we call* $g(f(...f(u,x_1)...,x_1))$ *the accumulated value of the set* $\mathbf{X}$ *over* u. *Due to quasi commutativity, the value* $f(...f(u,x_1)...,x_1)$ *is independent of the order of the* x_i's *and is denoted by* $f(u,\mathbf{X})$.
- (Efficient Evaluation) *For every* $(f,g) \in \mathbf{F}_l$, $u \in \mathbf{U}_f$ *and* $\mathbf{X} \subset \mathbf{X}_l$ *with polynomial-bound size:* $g(f(u,\mathbf{X}))$ *is computable in time polynomial in* l, *even without the knowledge of* a_f.

Definition 2. *(Collision Resistant Accumulator). An accumulator is defined as* collision resistant *if for every PPT algorithm* $\mathcal{A}$, *the following function* $Adv_{\mathcal{A}}^{col.acc}(l)$ *is negligible.*

$$Adv_{\mathcal{A}}^{col.acc}(l) = Pr[(f,g) \leftarrow \mathbf{F}_l; u \leftarrow \mathbf{U}_f; (x,w,\mathbf{X}) \leftarrow \mathcal{A}(f,g,\mathbf{U}_f,u)|$$
$$(\mathbf{X} \subseteq \mathbf{X}_l) \wedge (w \in \mathbf{U}_g) \wedge (x \in \mathbf{X}_f^{ext}\setminus\mathbf{X})$$
$$\wedge(f(g^{-1}(w),x) = f(u,\mathbf{X}))]$$

We say that w *is a* witness *for the fact that* $x \in \mathbf{X}_l$ *has been accumulated in* $v \in \mathbf{U}_g$ *whenever* $g(f(g^{-1}(w),x)) = v$. *The notion of witness for a set of values* $\mathbf{X} \subseteq \mathbf{X}_l$ *can be defined similarly.*

Definition 3. *(Dynamic Accumulator). A* dynamic accumulator *is defined as a collision resistant accumulator with the following properties:*

- (Efficient Addition) *there exist PT algorithms* $\mathcal{D}_a, \mathcal{W}_a$ *such that, if* $v = g(f(u,\mathbf{X}))$, $x \in \mathbf{X}$, $x' \notin \mathbf{X}$ *and* $g(f(g^{-1}(w),x)) = v$, *then (i)* $\mathcal{D}_a(a_f,v,x') = v'$ *such that* $v' = g(f(u,\mathbf{X}\cup\{x'\}))$; *and (ii)* $\mathcal{W}_a(f,g,v,v',x,x',w) = w'$ *such that* $g(f(g^{-1}(w'),x)) = v'$.
- (Efficient Deletion) *there exist PT algorithms* $\mathcal{D}_d, \mathcal{W}_d$ *such that, if* $v = g(f(u,\mathbf{X}))$, $x, x' \in \mathbf{X}$, $x \neq x'$ *and* $g(f(g^{-1}(w),x)) = v$, *then (i)* $\mathcal{D}_d(a_f,v,x') = v'$ *such that* $v' = g(f(u,\mathbf{X}\setminus\{x'\}))$; *and (ii)* $\mathcal{W}_d(f,g,v,v',x,x',w) = w'$ *such that* $g(f(g^{-1}(w'),x)) = v'$.

Similar to Theorem 2 in [11], we can easily prove the following theorem about security of dynamic accumulators against adaptive attacks.

Theorem 1. *Suppose* $\mathcal{DA}$ *is a dynamic accumulator and* $\mathcal{O}$ *is an interactive Turing machine, which operates as an oracle as follows. It receives input* (f, g, a_f, u), *where* $(f,g) \in \mathbf{F}_l$ *and* $u \in \mathbf{U}_f$. *It maintains a list of values* $\mathbf{X}$ *which is initially empty, and the current accumulated value,* v, *which is initially* $g(u)$. *It responds to two types of messages: when receiving the* (add, x) *message, it checks that* $x \in \mathbf{X}_l$, *and if so, adds* x *to the list* $\mathbf{X}$ *and updating the accumulated value (using efficient addition* $\mathcal{D}_a$), *it then sends back this updated value; similarly, when receiving the* (delete, x) *message, it checks that* $x \in \mathbf{X}$, *and if so, deletes it from the list and updates* v *(using efficient deletion* $\mathcal{D}_d$) *and sends back the updated value. In the end of the computation,* $\mathcal{O}$ *returns the current values for* $\mathbf{X}$ *and* v. *Let* $\mathbf{U}_f^{ext} \times \mathbf{X}_f^{ext}$ *denote the domains for which the computational* procedure for function f is defined. For every PPT adversary $\mathcal{A}$, the following function $Adv_{\mathcal{A}}^{adap.col}(l)$ is negligible.

$$Adv_{\mathcal{A}}^{adap.col}(l) = Pr[(f,g) \leftarrow \mathbf{F}_l; u \leftarrow \mathbf{U}_f; (x,w) \leftarrow \mathcal{A}(f,g,\mathbf{U}_f,u) \leftrightarrow \\ \mathcal{O}(f,g,a_f,u) \rightarrow (\mathbf{X},v) - (\mathbf{X} \subseteq \mathbf{X}_l) \wedge (w \in \mathbf{U}_g) \wedge \\ (x \in \mathbf{X}_f^{ext}\backslash \mathbf{X}) \wedge (f(g^{-1}(w),x) = f(u,\mathbf{X}))]$$

3.2 Identity-Based Ad-Hoc Anonymous Identification Schemes

Syntax. The following definition is quite the same as the definition of an ad-hoc anonymous identification scheme in [13] except for some ID-based-related features: a KeyGen algorithm replaces the Register algorithm and the Setup does not maintain a database of users' public keys.

An identity-based ad-hoc anonymous identification scheme is defined as a tuple $\mathcal{IA}$ =(Setup, KeyGen, Make-GPK, Make-GSK, IAID$_P$, IAID$_V$) of PT algorithms, which are described as follows.

- Setup takes as input a security parameter 1^l and returns the public parameters *params* and a master key *mk*. The master key is only known to the Private Key Generator (PKG).
- KeyGen, run by the PKG, takes as input *params*, *mk* and an arbitrary identity of an user and outputs a private key for the user. The identity is used as the corresponding public key.
- Make-GPK takes as input *params* and a set of identities and deterministically outputs a single group public key which is used in the identification protocol IAID described below. Its cost linearly depends on the number of identities being aggregated. The algorithm is *order invariant* that means the order of aggregating the identities does not matter.
- Make-GSK takes as input *params*, a set of identities and a pair of an identity and the corresponding private key and deterministically outputs a single group secret key which is used in the identification protocol IAID described below. Its cost linearly depends on the number of identities being aggregated. It can be observed that a group secret key $gsk \leftarrow$ Make-GSK$(params, \mathbf{S}', (s_{id}, id))$ corresponds to a group public key $gpk \leftarrow$ Make-GPK$(params, \mathbf{S})$ if and only if $\mathbf{S} = \mathbf{S}' \cup \{id\}$. And more than one group secret key might corresponds to the same group public key.
- IAID = (IAID$_P$, IAID$_V$) is the two party identification protocol, which allows the prover (IAID$_P$) to anonymously show his membership in a group of identities he constructed by himself. Both of the prover and the verifier (IAID$_V$) takes as input *params* and a group public key; IAID$_P$ is also given a corresponding group secret key; and IAID$_V$ finally outputs 0 (reject) or 1 (accept). The cost of the protocol is independent from the number of identities that were aggregated in the group public key.

Security Requirements. The requirements are quite the same as thoses for ad-hoc anonymous identification schemes in [13], including Correctness, Soundness and Unconditional Anonymity, which are described in the full version of this paper [17].

3.3 ID-Based Ring Signature Schemes

Based on the model in [23], an ID-based ring signature scheme is as a tuple $\mathcal{IR}$ =(RSetup, RKeyGen, RSign, RVerify) of PT algorithms. RSetup and RKeyGen are defined the same as Setup and KeyGen in ID-based ad-hoc anonymous identification schemes. The PPT algorithm RSign takes as input the public parameter $params$, a user private key s_{id}, a set of identities, which includes the identity corresponding to s_{id}, and a message m; and outputs a signature for m. The DPT algorithm RVerify takes as input a set of identities, a message and a ring signature; and outputs either accept or reject.

There are three security requirements for ID-based ring signature schemes: Correctness, Unforgeability against Chosen Message, Group and Signer Attacks (UNF-CMGSA), and Unconditional Anonymity. Correctness intuitively requires that if RSign is given a valid private key correponding to an identity in the input set of identities, then its output signature is accepted by RVerify with overwhelming probability. UNF-CMGSA intuitively requires an adversary, who can adaptively play a chosen message-group-signer attack many times, can not forge a new ring signature with non-negligible probability. The chosen message-group-signer attack allows the adversary to adaptively choose a message, a group of identities, specify a signer in that group and query RSign for the corresponding signature. Unconditional Anonymity intuitively requires that given a ring signature, the adversary cannot tell the identity of the signer with a probability non-negligibly larger than a random guess, even assuming that the adversary has unlimited computing resources.

An ID-based ad-hoc anonymous identification scheme $\mathcal{IA}$ can be converted to an ID-based ring signature scheme $\mathcal{IR}$ by applying the Fiat-Shamir heuristics. Based on arguments similar to those in [13], we have the following lemma.

Lemma 1. *If $\mathcal{IA}$ provides Correctness, Soundness and Unconditional Anonymity, then the non-interactive dual $\mathcal{IR}$ provides Correctness, UNF-CMGSA (in the random oracle model), and Unconditional Anonymity.*

3.4 Identity Escrow Schemes with Membership Revocation

Based on the BSZ04 formal model for group signature schemes, we propose a formal model for identity escrow schemes with membership revocation. The model can be used for many existent schemes, such as ones in [11,22], where some public information needs to be updated after each addition or deletion of group members. The main extensions from the BSZ04 formal model are as follows.

- A public archive arc records history of the public information that needs to be updated. After each addition or deletion of group members, the issuer needs to add new information to arc.

- The issuer, with access to arc and reg, uses an algorithm Revoke to remove a specified member from the group by updating arc.
- Apart from the unchanged *membership secret key* (private signing key in the BSZ04 model), each group member also keeps a *membership witness*. Based on information in the public archive, each group member can run an algorithm Update to update the membership witness.
- There is an algorithm CheckArchive, that can be run by any party after each change in the public archive. This algorithm checks if the issuer updates the archive arc correctly. With such an algorithm, we can assume arc is always updated correctly.

An identity escrow scheme with membership revocation is a tuple $\mathcal{IE}$ =(GKg, UKg, Join, Iss, IEID$_P$, IEID$_V$, Open, Judge, Revoke, Update, CheckArchive) of PT algorithms, where GKg generates public parameters and secret keys, UKg generates personal public and private keys (different from membership secret keys) for users, the protocol (Join, Iss) allows a user to join the group and get a membership secret key and a membership witness, the protocol IEID=(IEID$_P$, IEID$_V$) allows a group member to anonymously prove his membership, Open revokes a IEID transcript to find the prover and Judge decides if the Open finds the right prover. The security requirements are Correctness, Anonymity, Traceability and Non-frameability. More details about the model are provided in the full version of this paper [17].

4 A Dynamic Accumulator from Bilinear Pairings

We propose a dynamic accumulator $\mathcal{DA}1 = (\{\mathbf{X}_l\}_{l\in\mathbb{N}}, \{\mathbf{F}_l\}_{l\in\mathbb{N}})$ from Bilinear Pairings as follows.

- Efficient Generation: To generate an instance of the accumulator from a security parameter l, use $\mathcal{BPG}$ to generate a tuple $\mathbf{t} = (p, \mathbb{G}_1, \mathbb{G}_M, e, P)$ and $s \in_R \mathbb{Z}_p^*$. Compute a tuple $\mathbf{t}' = (P, sP, \ldots, s^q P)$, where q is the upper bound on the number of elements to be accumulated by the accumulator. The corresponding functions (f, g) for $\mathbf{t}, \mathbf{t}'$ are defined as:

$f: \mathbb{Z}_p \times \mathbb{Z}_p \to \mathbb{Z}_p$ and $g: \mathbb{Z}_p \to \mathbb{G}_1$
$f: (u, x) \mapsto (x+s)u$ $g: u \mapsto uP$

The corresponding domain for elements to be accumulated is $\mathbb{Z}_p\backslash\{-s\}$ and the auxiliary information is $a_f = s$. The tuple $\mathbf{t}' = (P^{(0)} = P, P^{(1)} = sP, \ldots, P^{(q)} = s^q P)$ can be distributively constructed by many parties so that all of them need to cooperate to find s.
- Quasi Commutativity: It holds that: $f(f(u, x_1), x_2) = f(u, \{x_1, x_2\}) = (x_1 + s)(x_2 + s)u$.
- Efficient Evaluation: For $u \in \mathbb{Z}_p$ and a set $\mathbf{X} = \{x_1, ..., x_k\} \subset \mathbb{Z}_p\backslash\{-s\}$, where $k \leq q$, the value $g(f(u, \mathbf{X})) = \prod_{i=1}^{k}(x_i + s)uP$ is computable in time polynomial in l from the tuple $\mathbf{t}' = (P, sP, \ldots, s^q P)$ and without the knowledge of the auxiliary information s.

- Efficient Addition: Suppose $V = g(f(u, \mathbf{X}))$, $x \in \mathbf{X}$, $x' \notin \mathbf{X}$ and $g(f(g^{-1}(W), x)) = V$, then $V' = g(f(u, \mathbf{X} \cup \{x'\}))$ can be computed as $V' = (x' + s)V$. And the value W' such that $g(f(g^{-1}(W'), x)) = V'$ can be computed as $W' = V + (x' - x)W$.
- Efficient Deletion: Suppose $V = g(f(u, \mathbf{X}))$, $x, x' \in \mathbf{X}$, $x \neq x'$ and $g(f(g^{-1}(W), x)) = V$, then $V' = g(f(u, \mathbf{X} \setminus \{x'\}))$ can be computed as $V' = 1/(x' + s)V$. And the value W' such that $g(f(g^{-1}(W'), x)) = V'$ can be computed as $W' = (1/(x' - x))(W - V')$.

Theorem 2 states the collision resistant property of $\mathcal{DA}1$ based on the Strong Diffie Hellman assumption.

Theorem 2. *The accumulator $\mathcal{DA}1$ provides Collision Resistance if the q-SDH assumption holds, where q is the upper bound on the number of elements to be accumulated by the accumulator.*

Proof. Suppose there is a PPT adversary $\mathcal{A}$ that can break Collision-Resistance property of $\mathcal{DA}1$, we show a construction of a PPT adversary $\mathcal{B}$ that can break the q-SDH assumption. Suppose a tuple $challenge = (P, zP, \ldots, z^q P)$ is given, where $z \in_R \mathbb{Z}_p^*$, we show that $\mathcal{B}$ can compute $(c, 1/(z+c)P)$, where $c \in \mathbb{Z}_p$ with non-negligible probability. Let $u \in_R \mathbb{Z}_p^*$, as $\mathcal{A}$ breaks Collision-Resistance property of $\mathcal{DA}1$, he can output $\mathbf{X} = \{x_1, \ldots, x_k\} \subset \mathbb{Z}_p \setminus \{-z\}$, $x \in \mathbb{Z}_p \setminus (\{-z\} \cup \mathbf{X})$ and $W \in \mathbb{G}_1$ such that $k \leq q$ and $(x+z)W = \prod_{i=1}^{k}(x_i+z)uP$. From this equation and the tuple $challenge$, $(1/(x+z))P$ can be computed and hence the q-SDH assumption is broken.

5 An ID-Based Ad-Hoc Anonymous Identification Scheme

5.1 Descriptions

As defined in the formal model, our scheme is a tuple $\mathcal{IA}1$ =(Setup, KeyGen, Make-GPK, Make-GSK, IAID$_P$, IAID$_V$) of PT algorithms, which are described as follows.

Setup, on a security parameter l, generates an instance of the accumulator above, including functions (f, g) and tuples $\mathbf{t} = (p, \mathbb{G}_1, \mathbb{G}_M, e, P)$ and $\mathbf{t}' = (P, P_{pub} = sP, \ldots, s^q P)$, where $s \in_R \mathbb{Z}_p^*$ and q is the upper bound on the number of identities to be aggregated. The auxiliary information s can be safely deleted, as it will never be used later. It also generates $Q \in_R \mathbb{G}_1^*$, $u, s_m \in_R \mathbb{Z}_p^*$ and computes $Q_{pub} = s_m Q$. Let $\mathbb{I}$ be the set of all possible identities, choose a collision-free function $\mathcal{H} : \mathbb{I} \to \mathbb{Z}_p \setminus \{-s\}$, that means the probability that an PPT adversary can find $id_0 \neq id_1$ so that $\mathcal{H}(id_0) = \mathcal{H}(id_1)$ is negligible. $\mathcal{H}$ does not have to be a hash or one-way function, but using a collision-free hash function $\mathcal{H} : \{0,1\}^* \to \mathbb{Z}_p$ is also acceptable. Then, the public parameters are $params = (l, \mathbf{t}, \mathbf{t}', f, g, Q, Q_{pub}, u, \mathcal{H})$ and the master key is $mk = s_m$.

KeyGen extracts a private key $s_{id} = R_{id}$ for an identity id as $R_{id} = 1/(\mathcal{H}(id) + s_m)Q$. The user can verifies the private key by checking $e(\mathcal{H}(id)Q + Q_{pub}, R_{id}) \stackrel{?}{=} e(Q, Q)$.

Make-GPK, given a set of identities $\{id_i\}_{i=1}^k$, computes the set $\mathbf{X} = \{\mathcal{H}(id_i)\}_{i=1}^k$ and generates the group public key for the set $gpk = V = g(f(u, \mathbf{X}))$.

Make-GSK generates the group secret key gsk for a user id and a set of identities $\{id_i\}_{i=1}^k$ by computing the set $\mathbf{X'} = \{\mathcal{H}(id_i)\}_{i=1}^k$, $h_{id} = \mathcal{H}(id)$ and the witness $W = g(f(u, \mathbf{X'}))$. The group secret key is $gsk = (h_{id}, s_{id}, W)$.

(IAID$_P$, IAID$_V$). This protocol IAID has the common input $params$ and gpk and the prover (user id) also has gsk. It is a combination of the proof that an identity is accumulated and a proof of knowledge of the user private key correponding to that identity. The protocol proves the knowledge of (h_{id}, R_{id}, W) satisfying equations $e(h_{id}Q + Q_{pub}, R_{id}) = e(Q, Q)$ and $e(h_{id}P + P_{pub}, W) = e(P, V)$.

1. IAID$_P$ generates $r_1, r_2, r_3, r_4, k_1, k_2, k_3, k_4 \in_R \mathbb{Z}_p$ and computes

    $U_1 = r_1(h_{id}Q + Q_{pub}); U_2 = r_1^{-1} R_{id}; U_3 = r_2 U_1 + r_3 H$
    $S_1 = r_1 r_2 (h_{id}P + P_{pub}); S_2 = (r_1 r_2)^{-1} W$
    $T_1 = k_1 Q + k_2 Q_{pub} + k_3 H; T_2 = k_4 U_1 + k_3 H; T_3 = k_1 P + k_2 P_{pub}$

2. IAID$_P \longrightarrow$ IAID$_V$: $U_1, U_2, U_3, S_1, S_2, T_1, T_2, T_3$
3. IAID$_V$ verifies that $e(U_1, U_2) \stackrel{?}{=} e(Q, Q)$ and $e(S_1, S_2) \stackrel{?}{=} e(P, V)$ (This can be done concurrently with next rounds).
4. IAID$_P \longleftarrow$ IAID$_V$: $c \in_R \mathbb{Z}_p$
5. IAID$_P$ computes $s_1 = k_1 + cr_1 r_2 h_{id}$, $s_2 = k_2 + cr_1 r_2$, $s_3 = k_3 + cr_3$, $s_4 = k_4 + cr_2$
6. IAID$_P \longrightarrow$ IAID$_V$: s_1, s_2, s_3, s_4
7. IAID$_V$ verifies that $T_1 \stackrel{?}{=} s_1 Q + s_2 Q_{pub} + s_3 H - cU_3$; $T_2 \stackrel{?}{=} s_4 U_1 + s_3 H - cU_3$; $T_3 \stackrel{?}{=} s_1 P + s_2 P_{pub} - cS_1$

5.2 Security

Theorem 3. *The ID-based ad-hoc anonymous identification scheme $\mathcal{IA}1$ provides Correctness and Unconditional Anonymity. The scheme $\mathcal{IA}1$ provides Soundness if the q-Strong Diffie-Hellman assumption holds, where q is the upper bound of the group size.*

Theorem 3 can be easily concluded from the zero-knowledge property of IAID that is stated in Lemma 2. Correctness and Unconditional Anonymity is based on the completeness and perfect zero-knowledge properties of the IAID protocol, respectively. Soundness of $\mathcal{IA}1$ is based on the soundness property of the IAID protocol, the collision-resistance property of the accumulator $\mathcal{DA}1$ and the fact that: if a PPT adversary $\mathcal{A}$ can compute a new pair of hashed identity and private key $(h_{id}^*, R_{id}^* = 1/(h_{id}^* + s_m)Q)$ from a set of $\{(h_{id}^{(i)}, R_{id}^{(i)} = 1/(h_{id}^{(i)} + s_m)Q)\}_{i=1}^q$, then $\mathcal{A}$ can break the q-SDH assumption (see [6] for a proof of this).

Lemma 2. *The IAID protocol is a (honest-verifier) perfect zero-knowledge proof of knowledge* (h_{id}, R_{id}, W) *satisfying equations* $e(h_{id}Q + Q_{pub}, R_{id}) = e(Q, Q)$ *and* $e(h_{id}P + P_{pub}, W) = e(P, V)$.

Proof. As the proof for completeness is straightforward, we present the proofs for Soundness and Zero-knowledge property only, as follows.

Soundness: If the protocol accepts with non-negligible probability, we show that the prover must have the knowledge of (h_{id}, R_{id}, W) with the relations stated in the theorem. Suppose the protocol accepts for the same commitment $(U_1, U_2, U_3, S_1, S_2, T_1, T_2, T_3)$ with two different pairs of challenges and responses $(c, s_1, ...s_4)$ and $(c', s'_1, ..., s'_4)$. Let $f_i = \frac{s_i - s'_i}{c - c'}, i = 1, ..., 4$, then

$$U_3 = f_1 Q + f_2 Q_{pub} + f_3 H; \ U_3 = f_4 U_1 + f_3 H; \ S_1 = f_1 P + f_2 P_{pub}$$

so $U_1 = f_1 f_4^{-1} Q + f_2 f_4^{-1} Q_{pub}$.

Let $h_{id} = f_1 f_2^{-1}$, $R_{id} = f_2 f_4^{-1} U_2$ and $W = f_2 S_2$, then $e(h_{id}Q + Q_{pub}, R_{id}) = e(Q, Q)$ and $e(h_{id}P + P_{pub}, W) = e(P, V)$, as $e(U_1, U_2) = e(Q, Q)$ and $e(S_1, S_2) = e(P, V)$. So the prover have the knowledge of (h_{id}, R_{id}, W) satisfying the relations.

Zero-Knowledge: The simulator chooses $c, s_1, ...s_4 \in_R \mathbb{Z}_p$, $b_1, b_2 \in_R \mathbb{Z}_p^*$, $U_3 \in_R \mathbb{G}_1$ and compute $U_1 = b_1 Q$, $U_2 = b_1^{-1} Q$, $S_1 = b_2 P$, $S_2 = b_2^{-1} V$, $T_1 = s_1 Q + s_2 Q_{pub} + s_3 H - c U_3$, $T_2 = s_4 U_1 + s_3 H - c U_3$ and $T_3 = s_1 P + s_2 P_{pub} - c S_1$. We can see that the distribution of the simulation is the same as the distribution of the real transcript.

5.3 Constant-Size Identity-Based Ring Signatures

Applying the Fiat-Shamir heuristics to the ID-based ad-hoc anonymous identification scheme $\mathcal{IA}1$ results in an ID-based ring signature scheme $\mathcal{IR}1$ with constant-size signatures. More specifically, each signature contains $(U_1, U_2, U_3, S_1, S_2, c, s_1, s_2, s_3, s_4)$, where c is computed from a hash function (a random oracle). Both the signer and the verifier only need to perform a computation proportional to the ring size once, and get some constant-size information (the group secret key and the group public key, respectively), on which they can produce/verify many subsequent signatures in constant time. The security of the scheme is stated in Theorem 4, which is based on results in Theorem 3 and Lemma 1.

Theorem 4. *The ID-based ring signature scheme* $\mathcal{IR}1$ *provides Correctness and Unconditional Anonymity. It also provides UNF-CMGSA in the random oracle model under the q-SDH assumption, where q is the upper bound of the group size.*

6 Application to Membership Revocation

We show how dynamic accumulators can be used to achieve membership revocation for group signature, identity escrow and anonymous credential systems.

In particular, we provide membership revocation to an identity escrow scheme (NS04) proposed in [18], and prove its security in the formal model above. The scheme can be easily converted to a group signature scheme (using Fiat-Shamir heuristics) or extended to an anonymous credential system; all of them provide membership revocation.

6.1 An Identity Escrow Scheme with Membership Revocation

As defined in the formal model, our identity escrow scheme involves a trusted party for initial set-up, two group managers (the issuer and the opener), and users, each with a unique identity $i \in \mathbb{N}$, that may become group members. The scheme is a tuple $\mathcal{IE}1$ =(GKg, UKg, Join, Iss, IEID$_P$, IEID$_V$, Open, Judge, Revoke, Update, CheckArchive) of PT algorithms which are defined as follows. We assume that the group size and the number of queries asked by the adversary are polynomial-bounded.

GKg: Suppose l is a security parameter and the generator $\mathcal{BPG}$ generates a tuple of bilinear pairing parameters $\mathbf{t} = (p, \mathbb{G}_1, \mathbb{G}_M, e, P) \leftarrow \mathcal{G}(1^l)$, that is also the publicly shared parameters. Choose a hash function $\mathcal{H} : \{0,1\}^* \to \mathbb{Z}_p$, which is assumed to be a random oracle in the security proofs. Choose $P_0, G, H \in_R \mathbb{G}_1$, $x, x' \in_R \mathbb{Z}_p^*$ and compute $P_{pub} = xP$, $\Theta = e(G, G)^{x'}$.

An instance of the dynamic accumulator $\mathcal{DA}1$ is also generated by choosing $Q \in_R \mathbb{G}_1$, $s \in_R \mathbb{Z}_p^*$, computing $Q_{pub} = sQ$ and defining functions (f, g), correponding to the domain $\mathbb{Z}_p \backslash \{-s\}$ for elements to be accumulated and the auxiliary information $a_f = s$, as:

$$f : \mathbb{Z}_p \times \mathbb{Z}_p \to \mathbb{Z}_p \quad \text{and} \quad g : \mathbb{Z}_p \to \mathbb{G}_1$$
$$f : (u, a) \mapsto (a+s)u \qquad g : u \mapsto uQ$$

Note that unlike the definition of $\mathcal{DA}1$, the tuple $\mathbf{t}' = (Q, sQ, \ldots, s^q Q)$ is not needed to be generated here. The reason is that the evaluation of the accumulated value can be done by the issuer with the knowledge of the auxiliary information s; and the efficient addition and efficient deletion properties allow witnesses to be updated without the knowledge of the tuple $\mathbf{t}'$.

Besides tables reg and upk, there is also a *public archive*, as a table arc. Each entry j (row j^{th}) on the table will have three attributes, the first attribute contains a certificate part of an user, who was added to or deleted from the group. The second attribute is just one bit, to indicate whether the user was added (1) or deleted (0). The third attribute contains the *group accumulated value* V_j (more description of this value will be given) after adding or deleting that user.

Initially, the public archive is empty, a $u \in_R \mathbb{Z}_p^*$ is generated and the group accumulated value is set to $V_0 = uQ$. The group public key is gpk =$(u, Q, Q_{pub}, P, P_0, P_{pub}, H, G, \Theta)$, the issuing key is $ik = (s, x)$, and the opening key is $ok = x'$.

UKg: This algorithm generates keys that provide authenticity for messages sent by the user in the (Join, Iss) protocol. This algorithm is the key generation algorithm K_S of any digital signature scheme $(K_S, Sign, Ver)$ that is unforgeable against chosen message attacks (UNF-CMA). A user i runs the UKg algorithm that takes as input a security parameter 1^l and outputs a personal public and private signature key pair $(upk[i], usk[i])$. Public Key Infrastructure (PKI) can be used here. Although any UNF-CMA signature scheme can be used, but using schemes whose security is based on DBDH or SDH assumptions, will reduce the underlying assumptions of our group signature scheme.

(Join, Iss): In this protocol, an user i and the issuer first generate a value $x_i \in \mathbb{Z}_p^*$ so that its randomization is contributed by both parties, but its value is only known by the user. The issuer then generates (a_i, S_i) for the user so that $e(a_i P + P_{pub}, S_i) = e(P, x_i P + P_0)$. The user uses $usk[i]$ to sign his messages in the protocol. Suppose the current group accumulated value, which is publicly known, is V_j (there have been j entries on the table arc), the issuer computes a new group accumulated value $V_{j+1} = (a_i + s)V_j$ and appends an entry $(a_i, 1, V_{j+1})$ to the table. Note that the formal model assumes the communication to be private and authenticated. In case the user i was revoked and now rejoins the group again ($reg[i]$ has been filled), he and the issuer just need to perform the steps 8, 9, 10 of the protocol. The protocol is as follows.

1. user $i \longrightarrow$ issuer: $I = yP + rH$, where $y, r \in_R \mathbb{Z}_p^*$.
2. user $i \longleftarrow$ issuer: $u, v \in_R \mathbb{Z}_p^*$.
3. The user computes $x_i = uy + v$, $P_i = x_i P$.
4. user $i \longrightarrow$ issuer: P_i and a proof of knowledge of (x_i, r') such that $P_i = x_i P$ and $vP + uI - P_i = r'H$ (see [10] for this proof).
5. The issuer verifies the proof, then chooses $a_i \in_R \mathbb{Z}_p^*$ different from all corresponding elements previously issued, and computes $S_i = \frac{1}{a_i + x}(P_i + P_0)$.
6. user $i \longleftarrow$ issuer: a_i, S_i.
7. The user computes $\Delta_i = e(P, S_i)$, verifies if $e(a_i P + P_{pub}, S_i) = e(P, x_i P + P_0)$, and stores the *membership secret key* $gsk[i] = (x_i, a_i, S_i, \Delta_i)$. Note that only the user knows x_i. The issuer also computes Δ_i and makes an entry in the table reg: $reg[i] = (i, \Delta_i, \langle \text{Join, Iss}\rangle$ transcript so far).
8. Suppose the current group accumulated value is V_j, the issuer computes a new *group accumulated value* $V_{j+1} = (a_i + s)V_j$ and appends $(a_i, 1, V_{j+1})$ to the table arc.
9. user $i \longleftarrow$ issuer: $j + 1, V_{j+1}$
10. The user verifies that $e(a_i Q + Q_{pub}, V_j) = e(Q, V_{j+1})$, then set his current *membership witness* to be $(j+1, W_{i,j+1})$ where $W_{i,j+1} = V_j$.

(IEID$_P$, IEID$_V$): This protocol IEID shows an user i's knowledge of (a_i, S_i) and a secret x_i such that: $e(a_i P + P_{pub}, S_i) = e(P, x_i P + P_0)$ and a_i has been accumulated in the current group accumulated value. The protocol does not reveal any information about his knowledge to anyone, except for the opener, who can only compute Δ_i by decrypting an encryption of that value. Before the protocol is started, user i checks the table arc to find the latest group accumulated value V_j and runs Update algorithm to compute his current membership witness $(j, W_{i,j})$ (or the issuer asks the users to run Update after changes in the table arc). The protocol is then run between user i (as IEID$_P$) and a verifier IEID$_V$ as follows.

1. IEID$_P$ computes $E = tG$, $\Lambda = \Delta_i \Theta^t$
2. The following sub-protocol, which we call the Proving protocol, is performed.
   (a) IEID$_P$ generates $r_1, ..., r_5, k_0, ..., k_7 \in_R \mathbb{Z}_p^*$ and computes: $R_1 = r_4(a_i Q + Q_{pub})$; $R_3 = r_4^{-1} W_{i,j}$; $R_2 = r_1 r_2 R_1 + r_5 H$; $U_1 = r_1(a_i P + P_{pub})$; $U_3 = r_2 r_4 S_i$; $U_2 = r_2 r_4 U_1 + r_3 H$; $X = r_1 r_2 r_4 (x_i P + P_0)$; $T_1 = k_1 P + k_2 P_{pub} + k_0 H$; $T_2 = k_3 P + k_2 P_0$; $T_4 = k_5 U_1 + k_0 H$; $T_3 = k_1 Q + k_2 Q_{pub} + k_4 H$; $T_5 = k_6 R_1 + k_4 H$; $T_6 = k_7 G - k_5 E$; $\Pi = \Theta^{k_7} \Lambda^{-k_5}$.
   (b) IEID$_P$ $\longrightarrow$ IEID$_V$: $E, \Lambda, R_1, R_2, R_3, U_1, U_2, U_3, X, T_1, ..., T_6, \Pi$.
   (c) IEID$_V$ verifies that $e(R_1, R_3) \stackrel{?}{=} e(Q, V_j)$ and $e(U_1, U_3) \stackrel{?}{=} e(P, X)$ (This can be done concurrently with next rounds).
   (d) IEID$_P$ $\longleftarrow$ IEID$_V$: $c \in_R \mathbb{Z}_p$.
   (e) IEID$_P$ computes in $\mathbb{Z}_p$: $s_0 = k_0 + cr_3$; $s_1 = k_1 + cr_1 r_2 r_4 a_i$; $s_2 = k_2 + cr_1 r_2 r_4$; $s_3 = k_3 + cr_1 r_2 r_4 x_i$; $s_4 = k_4 + cr_5$; $s_5 = k_5 + cr_2 r_4$; $s_6 = k_6 + cr_1 r_2$; $s_7 = k_7 + cr_2 r_4 t$.
   (f) IEID$_P$ $\longrightarrow$ IEID$_V$: $s_0, ..., s_7$.
   (g) IEID$_V$ verifies that $T_1 \stackrel{?}{=} s_1 P + s_2 P_{pub} + s_0 H - cU_2$; $T_2 \stackrel{?}{=} s_3 P + s_2 P_0 - cX$; $T_4 \stackrel{?}{=} s_5 U_1 + s_0 H - cU_2$; $T_3 \stackrel{?}{=} s_1 Q + s_2 Q_{pub} + s_4 H - cR_2$; $T_5 \stackrel{?}{=} s_6 R_1 + s_4 H - cR_2$; $T_6 \stackrel{?}{=} s_7 G - s_5 E$; $\Pi \stackrel{?}{=} \Theta^{s_7} \Lambda^{-s_5} e(P, cU_3)$.

Open: To open an IEID transcript $(E, \Lambda, ...)$ to find the prover, the opener computes $\Delta_i = \Lambda e(E, G)^{-x'}$ and a non-interactive zero-knowledge proof ϱ of knowledge of x so that $\Theta = e(G, G)^{x'}$ and $\Lambda/\Delta_i = e(E, G)^{x'}$ (see [10] for this proof); and finds the corresponding entry i in the table reg. If no entry is found, it returns $(0, \Delta_i, \varrho)$. Otherwise, it returns $(reg[i], \varrho)$.

Judge: Anyone can run the Judge algorithm as follows. On an output $(reg[i], \varrho)$ by the Open algorithm for an IEID transcript $(E, \Lambda, ...)$, it returns reject if verification of the proof ϱ rejects. Otherwise, it returns accept. On an output $(0, \Delta_i, \varrho)$ by Open, it returns reject if verification of the proof ϱ rejects; otherwise, it returns accept.

Revoke: To remove an user i from the group, the issuer retrieves the user's a_i from the table reg and the current group accumulated value V_j and computes a new group accumulated value $V_{j+1} = (1/(a_i + s))V_j$. The issuer appends a new entry $(a_i, 0, V_{j+1})$ on the table arc.

Update: Given access to the arc table, which currently has n rows, an user i with a membership witness $(j, W_{i,j})$ computes a new witness as follows. Its cost is about $n - j$ scalar multiplications.

for $(k = j + 1; k++; k \leq n)$ do
   retrieve from row k^{th} of arc the entry (a, b, V_k);
   if $b = 1$, then $W_{i,k} = V_{k-1} + (a - a_i)W_{i,k-1}$
     else $W_{i,k} = (1/(a - a_i))(W_{i,k-1} - V_k)$ end if;
end for;
return $(n, W_{i,n})$;

CheckArchive: Any party, after a change on the public archive, can run this algorithm as follows.

retrieve from the new row of arc the entry (a, b, V_k);
if $(b = 1)$ then return $(e(aQ + Q_{pub}, V_{k-1}) = e(Q, V_k))$
else return $(e(aQ + Q_{pub}, V_k) = e(Q, V_{k-1}))$;

6.2 Security

Theorem 5. *The identity escrow scheme with membership revocation $\mathcal{IE}1$ provides Correctness.*

Theorem 6. *The scheme $\mathcal{IE}1$ provides Anonymity if the Decisional Bilinear Diffie-Hellman assumption holds.*

Theorem 7. *The scheme $\mathcal{IE}1$ provides Traceability if the q-Strong Diffie-Hellman assumption holds, where q is the upper bound of the group size.*

Theorem 8. *The scheme $\mathcal{IE}1$ provides Non-frameability if the Discrete Logarithm assumption holds over the group $\mathbb{G}_1$ and the digital signature scheme $(K_S, Sign, Ver)$ is UNF-CMA.*

Proofs of these theorems can be found in the full version [17]. They are based on the Coalition-Resistance of the NS04 scheme, the Collision-Resistance of $\mathcal{DA}1$ and the Zero-knowledge property of the Proving protocol, which is stated in Lemma 3.

Lemma 3. *The Proving protocol in the IEID protocol is a (honest-verifier) perfect zero-knowledge proof of knowledge of $W_{i,j}$, (a_i, S_i), x_i and t such that $e(a_i Q + Q_{pub}, W_{i,j}) = e(Q, V_j)$, $e(a_i P + P_{pub}, S_i) = e(P, x_i P + P_0)$, $E = tG$ and $\Lambda = e(P, S_i)\Theta^t$.*

Proof. As the proof for completeness is straightforward, we present the proofs for Soundness and Zero-knowledge property only, as follows.

Soundness: If the protocol accepts with non-negligible probability, we show that the prover must have the knowledge of $W_{i,j}$, (a_i, S_i), x_i and t with the relations stated in the theorem. Suppose the protocol accepts for the same commitment $(R_1, R_2, R_3, U_1, U_2, U_3, X, T_1, ..., T_6, \Pi)$ with two different pairs of challenges and responses $(c, s_0, ...s_7)$ and $(c', s'_0, ..., s'_7)$. Let $f_i = \frac{s_i - s'_i}{c - c'}, i = 0, ..., 7$, then

$$U_2 = f_1 P + f_2 P_{pub} + f_0 H; \quad U_2 = f_5 U_1 + f_0 H$$
$$R_2 = f_1 Q + f_2 Q_{pub} + f_4 H; \quad R_2 = f_6 R_1 + f_4 H$$
$$X = f_3 P + f_2 P_0; \quad E = f_7 f_5^{-1} G; \quad \Lambda = e(P, f_5^{-1} U_3) \Theta^{f_7 f_5^{-1}}$$

so $U_1 = f_1 f_5^{-1} P + f_2 f_5^{-1} P_{pub}$ and $R_1 = f_1 f_6^{-1} Q + f_2 f_6^{-1} Q_{pub}$.

Let $a_i = f_1 f_2^{-1}$, $S_i = f_5^{-1} U_3$, $x_i = f_3 f_2^{-1}$, $t = f_7 f_5^{-1}$ and $W_{i,j} = f_2 f_6^{-1} R_3$, then $E = tG$, $\Lambda = e(P, S_i)\Theta^t$, $e(a_i P + P_{pub}, S_i) = e(P, x_i P + P_0)$ (as $e(U_1, U_3) = e(P, X)$) and $e(a_i Q + Q_{pub}, W_{i,j}) = e(Q, V_j)$ (as $e(R_1, R_3) = e(Q, V_j)$). So the prover have the knowledge of $W_{i,j}$, (a_i, S_i), x_i and t satisfying the relations.

Zero-Knowledge: The simulator chooses $c, s_0, ... s_7 \in_R \mathbb{Z}_p, b_1, b_2 \in_R \mathbb{Z}_p^*, U_1, U_2, R_2 \in_R \mathbb{G}_1$ and compute $U_3 = b_1 P$, $X = b_1 U_1$, $R_1 = b_2 Q$, $R_3 = b_2^{-1} V_j$, $T_1 = s_1 P + s_2 P_{pub} + s_0 H - cU_2$, $T_2 = s_3 P + s_2 P_0 - cX$, $T_4 = s_5 U_1 + s_0 H - cU_2$, $T_3 = s_1 Q + s_2 Q_{pub} + s_4 H - cR_2$, $T_5 = s_6 R_1 + s_4 H - cR_2$, $T_6 = s_7 G - s_5 E$, $\Pi = \Theta^{s_7} \Lambda^{-s_5} e(P, cU_3)$. We can see that the distribution of the simulation is the same as the distribution of the real transcript.

7 Efficiency Comparison

Our ID-based ring signature scheme is the first to provide constant-size signatures. Although the tuple $\mathbf{t'}$ is long, users just need to download it once, and they do not need to obtain the whole $\mathbf{t'}$. The signature size is also very much smaller than that of the current state-of-the-art normal ring signature scheme DKNS04. For elliptic curve group of 160-bit prime order, the signature size is only about 200 bytes. In the future, when higher levels of security are required, this difference even grows much larger. The same conclusion can be drawn for the size of our group signatures in comparison with those in the CL02 and TX03 schemes, and even the ACJT00 scheme, which does not have membership revocation. Note that $e(Q,Q)$ and $e(P,V)$ in IAID and $e(Q,V_j)$ in IEID can be pre-computed and published before the executions of the protocol.

We now make a specific comparison of sizes in our new group signature scheme with membership revocation with those in ACJT00 and CL02 schemes. We assume that our scheme is implemented by an elliptic curve or hyperelliptic curve over a finite field. p is a 160-bit prime, $\mathbb{G}_1$ is a subgroup of an elliptic curve group or a Jacobian of a hyperelliptic curve over a finite field with order p and compression techniques are used. $\mathbb{G}_M$ is a subgroup of a finite field of size approximately 2^{1024}. A possible choice of these parameters can be from Boneh et al.'s short signature scheme [8], where $\mathbb{G}_1$ is derived from the curve $E/GF(3^\iota)$ defined by $y^2 = x^3 - x + 1$. In addition, we assume that system parameters in the ACJT00 and CL02 schemes are $\epsilon = 1.1$, $l_p = 512$, $k = 160$, $\lambda_1 = 838$, $\lambda_2 = 600$, $\gamma_1 = 1102$ and $\gamma_2 = 840$. We summarize the result in the following table.

	Signature	gpk	gsk	ik	ok	Membership Revocation
ACJT00	1087	768	370	128	128	No
CL02 scheme	1968	1280	370	256	128	Yes
Our scheme	470	289	188	40	20	Yes

References

1. G. Ateniese, J. Camenisch, M. Joye, and G. Tsudik. A practical and provably secure coalition-resistant group signature scheme. CRYPTO 2000, Springer-Verlag, LNCS 1880, pp. 255-270.
2. G. Ateniese, and B. de Medeiros. Efficient Group Signatures without Trapdoors. ASIACRYPT 2003, Springer-Verlag, LNCS 2894, pp. 246-268.

3. N. Baric and B. Pfitzmann. Collision-free accumulators and fail-stop signature schemes without trees. EUROCRYPT 1997, Springer-Verlag, LNCS 1233, pp. 480-494.
4. M. Bellare, H. Shi, and C. Zhang. Foundations of Group Signatures: The Case of Dynamic Groups. Cryptology ePrint Archive: Report 2004/077.
5. J. Benaloh and M. de Mare. One-way accumulators: A decentralized alternative to digital signatures. EUROCRYPT 1993, Springer-Verlag, LNCS 765, pp. 274-285.
6. D. Boneh, and X. Boyen. Short Signatures Without Random Oracles. EUROCRYPT 2004, Springer-Verlag, LNCS 3027, pp. 56-73.
7. D. Boneh, and X. Boyen. Efficient Selective-ID Secure Identity-Based Encryption Without Random Oracles. EUROCRYPT 2004, Springer-Verlag, LNCS 3027, pp. 223-238.
8. D. Boneh, B. Lynn, and H. Shacham. Short signatures from the Weil pairing. ASIACRYPT 2001, Springer-Verlag, LNCS 2248, pp.514-532.
9. E. Bresson, J. Stern, and M. Szydlo. Threshold ring signatures and applications to ad-hoc groups. CRYPTO 2002, Springer-Verlag, LNCS 2442, pp. 465-480.
10. J. Camenisch, and M. Michels. A group signature scheme with improved efficiency. ASIACRYPT 1998, Springer-Verlag, LNCS 1514.
11. J. Camenisch, and A. Lysyanskaya. Dynamic Accumulators and Application to Efficient Revocation of Anonymous Credentials. CRYPTO 2002, Springer-Verlag, LNCS 2442, pp. 61-76.
12. D. Chaum, and E. van Heyst. Group signatures. CRYPTO 1991, LNCS 547, Springer-Verlag.
13. Y. Dodis, A. Kiayias, A. Nicolosi, and V. Shoup. Anonymous Identification in Ad Hoc Groups. EUROCRYPT 2004, Springer-Verlag, LNCS 3027, pp. 609-626.
14. A. Fiat, and A. Shamir. How to prove yourself: practical solutions to identification and signature problems. CRYPTO 1986, Springer-Verlag, LNCS 263, pp. 186-194.
15. J. Killian, and E. Petrank. Identity escrow. CRYPTO 1998, Springer-Verlag, LNCS 1642, pp. 169-185.
16. S. Mitsunari, R. Sakai, and M. Kasahara. A new traitor tracing. IEICE Trans. Vol. E85-A, No.2, pp.481-484, 2002.
17. L. Nguyen. Accumulators from Bilinear Pairings and Applications. Full version.
18. L. Nguyen, and R. Safavi-Naini. Efficient and Provably Secure Trapdoor-free Group Signature Schemes from Bilinear Pairings. ASIACRYPT 2004, Springer-Verlag, LNCS.
19. R. Rivest, A. Shamir, and Y. Tauman. How to leak a secret. ASIACRYPT 2001, Springer-Verlag, LNCS 2248, pp.552-565.
20. A. Shamir, Identity-based cryptosystems and signature schemes. CRYPTO 1984, LNCS 196, Springer-Verlag, pp. 47-53.
21. V. To, R. Safavi-Naini, and F. Zhang. New traitor tracing schemes using bilinear map. DRM Workshop 2003.
22. G. Tsudik, and S. Xu. Accumulating Composites and Improved Group Signing. ASIACRYPT 2003, Springer-Verlag, LNCS 2894, pp. 269-286.
23. F. Zhang, and K. Kim. ID-Based Blind Signature and Ring Signature from Pairings. ASIACRYPT 2002, Springer-Verlag, LNCS 2501, pp. 533-547.

Computing the Tate Pairing*

Michael Scott

School of Computing,
Dublin City University,
Ballymun, Dublin 9, Ireland
mike@computing.dcu.ie

Abstract. We describe, in detail sufficient for easy implementation, a fast method for calculation of the Tate pairing, as required for pairing-based cryptographic protocols. We point out various optimisations and tricks, and compare timings of a pairing-based Identity Based Encryption scheme with an optimised RSA implementation.

Keywords: Elliptic curves, pairing-based cryptosystems.

1 Introduction

In the fast growing world of pairing-based cryptography (for background see [1]) there are many protocols, many pairings (Tate, Weil, modified Weil etc.) and many choices for the *embedding degree* k, as well as a choice of super- or non-supersingular curves over fields of large or small characteristic. The range of protocols is impressive, many with novel properties [6, 7, 28]. For a recent review see [11]. However so far there are not many reported implementations of the fast algorithms for pairings that have been developed in [2, 4, 13].

Here for the sake of being concrete we will focus exclusively on the Tate Pairing on non-supersingular curves over a field of large prime characteristic. We will also focus on the case $k = 2$ for the following reasons:

- It simplifies the description
- Choosing $k = 2$ makes it easy to pick a group order of the lowest possible Hamming weight which is very efficient.
- Choosing $k = 2$ allows us to implement the Tate pairing based protocols using only $E(\mathbb{F}_p)$ elliptic curves as supported by many cryptographic libraries.
- $k = 2$ permits the important *denominator elimination* optimisation [2].
- It allows for easy times-2 compression of the Tate pairing value [25].
- In protocols elliptic curve point multiplication can often be replaced with faster exponentiation using the identity $e_r(wP, Q) = e_r(P, Q)^w$.
- Elliptic curves suitable for pairing based cryptosystems are, by design, in flagrant breach of the MOV condition, as required for "ordinary" elliptic curves [20]. The ECC community recently got a scare when Semaev [27] suggested that a new index calculus type attack on normal elliptic curves may

---
* Research supported by Enterprise Ireland grant IF/2002/0312/N.

be possible. In the context considered here an index calculus attack is already possible [20], and therefore we need not be too concerned. Nevertheless a choice of a small value of k reduces the impact of any such new attack.
- For a given level of security it is our experience that $k = 2$ is fastest.
- In many protocols it is required to do a point multiplication prior to application of the Tate pairing. Using $k = 2$ this implies a point multiplication only on an $E(\mathbb{F}_p)$ curve, rather than a point multiplication on a curve defined over a higher extension field, which would be computationally more expensive.
- $\mathbb{F}_{p^2}$ arithmetic is particularly easy to implement. This is sometimes called the *quadratic extension field*. If it is assumed in this paper that the prime modulus p is 3 mod 4, then an element in $\mathbb{F}_{p^2}$ can be considered as a "complex number", $a+bi$, $a, b \in \mathbb{F}_p$, where i is $\sqrt{-1}$. Note that -1 is always a quadratic non-residue for a 3 mod 4 prime. There are exactly $(p-1)(p+1)$ elements in the field $\mathbb{F}_{p^2}$. Note that $(a+ib)^p = (a-ib)$, where $a-bi$ is the *conjugate* of $a+ib$. Also an element $\in \mathbb{F}_{p^2}$ can be squared (or multiplied) using just two (or three) $\mathbb{F}_p$ modular multiplications using the identity $(a+bi)^2 = (a+b)(a-b) + 2abi$ and Karatsuba's method respectively. Sometimes we use the notation $[a, b]$ to denote the $\mathbb{F}_{p^2}$ number $a + bi$.
- Using $k = 2$ the time-critical function is 512-bit modular multiplication. This is the same operation as required for 1024-bit RSA decryption using the Chinese Remainder theorem and therefore it is likely to be supported by hardware accelerators and co-processors. Highly optimized code for this common operation may be already supported by cryptographic software libraries.

We do concede that $k = 2$ may not be optimal in some settings such as a short signature scheme, like for example the BLS scheme [7].

In this paper we draw heavily from the theoretical results described by Barreto et al. [4] and [2]. Our results improve a little on those described there using ideas from [25].

2 The Curve

There are many ways proposed to find non-supersingular curves of low embedding degree suitable for pairing-based protocols. See for example [3, 5, 8, 10, 21] and [26]. Using these methods the existance of a suitable elliptic curve is first determined, and then the actual parameters of the curve are found using the method of Complex Multiplication as described in [14] and implemented in [23].

The particular curve we will use (found using the "folklore" method described by Galbraith in Chapter 9 of [5]), is described in the Weierstrass form

$$E : y^2 = x^3 - 3x + B$$

with $B \in \mathbb{F}_p$. If $x, y \in \mathbb{F}_p$, the curve has $\#E(\mathbb{F}_p)$ points on it, where $\#E(\mathbb{F}_p) = p + 1 - t$ and t is the *trace of the Frobenius* [20]. If $x, y \in \mathbb{F}_{p^2}$, it has $\#E(\mathbb{F}_{p^2}) = (p + 1 - t)(p + 1 + t)$ points. A related *twisted* curve $E'(\mathbb{F}_p)$ is

$$E' : y^2 = x^3 - 3x - B$$

If $x, y \in \mathbb{F}_p$, this curve will have $p+1+t$ points on it. For our chosen curve:-

B = 6806165982543682940158586534684000322786886482451629214574812129884878382661217060174101978023037681795174235816499484606521501517622872852116277695499501

p = 11711338024714009669995700965425239711927177698599625717955894184681899877662827977441218356846207573509472307873756662300754437232398452830779100780970303

$\#E$ = 11711338024714009669995700965425239711927177698599625717955894184681899877662611539569996945969293708404400344208273812850399351303651875378098503534075638

t = 216437871221410876913865105071963665482849450355085928746577452680597246894666

Note that p is 512-bits long and is congruent to 3 mod 4. $\#E(\mathbb{F}_p)$ has (by design) a 160-bit prime factor r of low Hamming weight, where $r = 2^{159} + 2^{17} + 1$, a Solinas prime. The group of points on $E(\mathbb{F}_{p^2})$ of order r exhibit the required $k = 2$ embedding degree behaviour [24] – observe that $r \mid p+1$ – and therefore form a suitable setting for calculation of the Tate pairing. Note that the discrete logarithm problem over $\mathbb{F}_{p^2}$ where p is 512 bits is regarded as being approximately as hard as a discrete logarithm problem over $\mathbb{F}_p$ where p is 1024 bits. Therefore this curve satisfies contemporary security requirements. Roughly speaking it will be as difficult to "break" as 1024–bit RSA.

Note that although we have chosen to use a non-supersingular curve here, the method described will also work without modification for a large prime characteristic $k = 2$ supersingular curve, as originally suggested by Boneh & Franklin [6].

3 The Tate Pairing Algorithm

The notation for the Tate pairing is $e_r(P, Q)$, where P and Q are points on the elliptic curve $E(\mathbb{F}_{p^k})$, with P a point of order r. Q may also be of order r, but not necessarily (see below). The Tate pairing evaluates as an element in $\mathbb{F}_{p^k}$.

The algorithm itself consists of two parts; an application of Miller's algorithm, followed by a final exponentiation. For an easy-to-read introduction to the Tate pairing, Miller's algorithm, and its derivation from divisor theory see [18]. See also [29] for a nice discussion on pairings.

Now we describe the optimised BKLS algorithm for the particular case of $k = 2$, with denominator elimination applied [2]. Basically (and very loosely) Miller's algorithm first carries out an implicit multiplication of P by r, using the standard line-and-tangent double-and-add algorithm for elliptic curve point multiplication [20]. The result of this multiplication will (of course) be O, the point at infinity, as P is of order r. If a line or tangent should ever pass through Q then the pairing algorithm will fail, but it can be arranged that this won't ever happen (see below). At each step in the process an $\mathbb{F}_{p^2}$ value is calculated from a distance relationship between the current line or tangent and the point Q. This consists of a numerator (derived from the line or tangent associated with the addition or doubling of a point), and a denominator (derived from a vertical line through the destination point).

This value is multiplicatively accumulated, and its final value is the output of Miller's algorithm.

However this value, an element in $\mathbb{F}_{p^2}$ may not be of order r. To ensure a unique answer of order r the output of Miller's algorithm must be exponentiated to the power of $(p-1)(p+1)/r$. This is then the final result of the Tate pairing.

Observe that this final exponentiation itself can be considered in two parts – an exponentiation to the power of $(p-1)$ followed by an exponentiation to the power of $(p+1)/r$. The first exponentiation to $(p-1)$ ensures that any $\mathbb{F}_p$ component of the output of Miller's algorithm is reduced to 1 (by Fermat's little theorem), and hence can be ignored. An important observation in [2] is that under certain circumstances (that will pertain here) the "denominator" component is always in $\mathbb{F}_p$ and hence can be discarded. So we only deal here with the numerator.

In contrast to P, Q is not involved in any point addition or multiplication. Only the coordinates of Q are actually required.

P and Q could be chosen as linearly-independent points from $E(\mathbb{F}_{p^2})[r]$, and represented in standard (x, y) affine coordinates with $x, y \in \mathbb{F}_{p^2}$. However from an implementation view-point the requirement for point multiplication on $E(\mathbb{F}_{p^2})$ is a little difficult, as most existing cryptographic libraries would not support this. However this problem can be neatly side-stepped by modifying the algorithm to accept P as a point on $E(\mathbb{F}_p)$. If it helps, think of P as an point on $E(\mathbb{F}_{p^2})$ whose coordinates have an imaginary part of zero. Solinas [29] calls this rather nicely *Miller light*. It solves another problem - if Q is truly on $E(\mathbb{F}_{p^2})$ then P and Q can never be linearly dependant, and no line generated in the implicit multiplication of P will ever pass through Q. This *Miller light* algorithm will obviously be much faster.

As already stated Q need not be of order r. In fact it can be any point on the curve. Whatever its value it will be a member of a *coset* which does include exactly one point of order r (or the point at infinity). All points in the same coset are equivalent as far as the Tate pairing is concerned [24]. Our only problem is to ensure that Q is not in a coset associated with a point of order r, which is in the same subgroup as P (otherwise P and Q will be linearly dependent). This condition is met by points of the form $Q([a, 0], [0, d])$. It is not difficult to see that if there are $p + 1 - t$ points of the form $Q([a, 0], [c, 0])$ on $E(\mathbb{F}_p)$ then there will be $p + 1 + t$ points of the form $Q([a, 0], [0, d])$ on $E(\mathbb{F}_{p^2})$. (Simply substitute all possible $a < p$ for x in the curve equation, not forgetting the point at infinity. If the right hand side is a quadratic residue the points are $Q([a, 0], [\pm c, 0])$, otherwise they are $Q([a, 0], [0, \pm d])$). There will always be a subgroup of order r consisting of points of this latter form. Such points stay in this form under point multiplication. Furthermore there is a simple relationship between such points on $E(\mathbb{F}_{p^2})$ and points on the twisted curve $E'(\mathbb{F}_p)$. In fact for every point $Q([-a, 0], [0, d])$ on the curve defined over the quadratic extension field $\mathbb{F}_{p^2}$, there is a point $Q(a, d)$ on the twisted curve defined over $\mathbb{F}_p$ [3]. This *isomorphism* is very convenient, as it means that Q can be treated as a point on $E'(\mathbb{F}_p)$

In many protocols, for example [6], there is a need to hash and map an arbitrary string to a curve point of order r. For a general point on $E(\mathbb{F}_{p^2})$ this requires point multiplication by the large co-factor $(p+1-t)(p+1+t)/r$. However if the string is hashed instead to a point on the twisted curve, then the cofactor is the much smaller $(p+1+t)/r$.

Next we describe the BKLS algorithm in detail. If the current point in the implicit multiplication is $A(x, y)$, and if the next point doubling or addition generates a line of slope λ, then we will require the function

$f(A, \lambda, Q)$
1. $x, y \leftarrow A$
2. $a, d \leftarrow Q$
3. **return** $y - \lambda(a + x) - di$

which calculates and returns an $\mathbb{F}_{p^2}$ value. We also need a function which adds two points (or doubles a point) on the elliptic curve $E(\mathbb{F}_p)$. Assume therefore the existance of a function $A.add(B)$ which adds B to A and returns the line slope λ. Next we need a function to calculate the contribution of the most recent point addition/doubling to the $\mathbb{F}_{p^2}$ *Miller variable m*.

$g(A, B, Q)$
1. $T = A$
2. $\lambda = A.add(B)$
3. **return** $f(T, \lambda, Q)$

Now we are ready for the full BKLS algorithm.

Capital letters indicate an elliptic curve point. The variables r, n, p, and i are all simple integers. The notation n_i refers to the i-th bit of n. The variable λ and the coordinates of A, P and Q are all in $\mathbb{F}_p$. Only the Miller variable m requires support for elementary $\mathbb{F}_{p^2}$ arithmetic. The standard elliptic curve point addition/doubling formula involves the calculation of the line slope so there is no extra work involved in obtaining λ [20]. The conjugate of m is denoted $\bar{m}$.

BKLS(r, p, P, Q)
1. $m = 1$
2. $A = P$
3. $n = r - 1$
4. **for** $i \leftarrow \lfloor \lg(r) \rfloor - 2$ **downto** 0 **do**
5. $m = m^2 \cdot g(A, A, Q)$
6.   **if** $n_i = 1$ **then** $m = m \cdot g(A, P, Q)$
7. **end for**
8. $m = \bar{m}/m$
9. $m = m^{(p+1)/r}$
10. **return** m

As noted by Duursma and Lee [12] the very last step of Miller's algorithm can be ignored, as it does not contribute to the pairing value, so we do not need to process the last bit of r. As a result, with our choice of r as a Solinas prime with a Hamming weight of 3, the condition $n_i = 1$ is met precisely once.

As pointed out in [25], the value of m after line 8 of this algorithm is already *unitary*. Unitary values like $m = u + vi$ have the following useful properties:

- $u^2 + v^2 = 1$
- $(u + vi)^{-1} = (u - vi)$
- $(u + vi)^n = V_n(2u)/2 + U_n(2u)vi$

where V_n and U_n are the well-known Lucas sequences. The first property tells us that given u, then v can be uniquely determined from its sign. If it is clear from the protocol that this sign is not important, then v can be dropped.

This compressed pairing $\varepsilon_r(P,Q)$ as defined in [25] can be calculated in line 9 of the algorithm as $V_{(p+1)/r}(2u)$ and this single $\mathbb{F}_p$ value can be returned in line 10. Fortunately there is a well-known fast and efficient laddering algorithm which calculates $V_n(\cdot)$, and requires very little memory [25]. Note that unitary values remain unitary under any subsequent exponentiation. Therefore any subsequent exponentiation of the compressed pairing value can also be computed using the fast Lucas $V_n(\cdot)$ function.

3.1 Resistance to Simple Power Attack Analysis

We can make the following generalisations about pairing-based cryptographic protocols based on the BKLS tate pairing algorithm described above

- r and p are fixed and public
- secrets may be introduced as unknown points P and/or Q
- secrets may be introduced as exponents of pairings.

In the light of these observations it is of interest to consider the resistance of pairing-based protocols to so-called SPA attacks [17]. Observe first that the path taken through the code in the course of the execution of the BKLS algorithm is independent of any possible secrets. Furthermore, using the compressed pairing, any subsequent exponentiation of a pairing value can be carried out using a laddering algorithm, which is also known to be resistant to SPA [16]. Therefore one might with reasonable confidence expect that the power consumption profile of (and execution time for) such protocols will be constant and independent of any secret values.

To increase resistance to more sophisticated SPA and DPA attacks we suggest the following simple counter-measures

- Exploit bilinearity and calculate $e(P, Q) = e(sP, tQ)^{1/st}$, where s and t are random variables.
- Multiply the Miller variable at any time prior to line 8 in the BKLS algorithm by a random element of F_p. This does not effect the result, as all such contributions are eliminated by the final exponentiation.

4 Optimisations

In this section we will describe various speed-ups and tricks.

4.1 Projective Coordinates

It has been the experience of many that elliptic curve point addition and doubling over $E(\mathbb{F}_p)$ is faster if the points are represented in (x, y, z) *projective* coordinates rather than in (x, y) *affine* coordinates [14]. This is because affine point addition or point doubling requires an expensive modular inversion mod p. This operation is hard to optimise. If a single pairing value is being calculated and if precomputation is not possible, then our experience is that projective coordinates, which do not require a modular inversion, are faster in practice. In the context of pairings the use of projective coordinates is also recommended by Izu and Takagi [15].

The modification to the algorithm above is simple - we just need a projective version of the $A.add(B)$ which will return the line slope as a rational $\lambda = \lambda_n/\lambda_d$, and a new $f(\cdot)$ function. It is assumed that initially P and Q are presented in affine coordinates (with $z=1$).

$f(A, \lambda, Q)$
1. $x, y, z \leftarrow A$
2. $a, d \leftarrow Q$
3. $\lambda_n, \lambda_d \leftarrow \lambda$
4. **return** $y\lambda_d - \lambda_n(az^3 + xz) - dz^3\lambda_d \cdot i$

4.2 Precomputation

For the calculation of $e_r(P, Q)$ in the context of a particular protocol, the parameter P may be fixed. It may for example be an individual's fixed private key, or it may be a system global. Whatever the reason, if P is fixed then it makes sense to calculate the points and slopes that arise in the implicit multiplication of rP just once, and store them, as suggested in [2] and [13]. In this case it makes sense to revert to affine coordinates, as no point addition or doubling will be needed.

The modification to the BKLS algorithm is straightforward. Assume the prior storage of $\{x_j, y_j, \lambda_j\}$ for each point A_j that arises in the point multiplication. Then in line 2 of the $g(\cdot)$ function instead of calculating the next point and slope, simply extract them from this precomputed store. The size of the store will be $3 \cdot (512/8) \cdot 160 = 30720$ bytes.

For supersingular curves we like to use the pairing $\hat{e}_r(P, Q) = e_r(P, \phi(Q))$, where $\phi(\cdot)$ is a distortion map. In this case $\hat{e}_r(P, Q) = \hat{e}_r(Q, P)$, and so a fixed parameter can always be exploited for precomputation. However this is not true for non-supersingular curves as $e_r(P, Q) \neq e_r(Q, P)$.

4.3 Products of Pairings

In some protocols, for example [28], it is necessary to calculate the product of two or more pairings. Consider the calculation of $e_r(P, Q) \cdot e_r(R, S)$. Each pairing requires an implicit point multiplication by r, an application of Miller's algorithm, and a final exponentiation. We suggest three optimisations which apply if the two pairing are calculated simultaneously rather than separately:

- Since the implicit multiplications of P and R occur in lock-step with one another, it makes sense to use affine coordinates in conjunction with Montgomery's trick. This means that just one modular inversion will be required instead of two. Montgomery's trick is based on the simple observation that $1/x = y/xy$ and $1/y = x/xy$.
- Both pairings can share the same *Miller variable* m. This means only a single squaring of m in line 5 will be required.
- Both pairings can share the final exponentiation (as pointed out by Solinas [29]).

The second and third optimisations depend on the observation that the final result will be the product of each pairing's Miller variable. These optimisations apply equally well to the product of multiple pairings. For the product of two pairings it will be about 50% faster than computing two separate pairings.

4.4 Protocol Optimisations

In pairing based protocols the most important property of the pairing is *bilinearity*.

$$e_r(aP, bQ) = e_r(P, Q)^{ab}$$

It is more efficient to calculate $e_r(P, wQ)$ as $e_r(P, Q)^w$, as exponentiation by w in F_{p^2} will be much faster that point multiplication by w on $E(F_p)$ (Note however that this may not be true for values of $k > 2$.) A protocol like the Boneh and Franklin IBE scheme [6] may demand that an arbitrary string be hashed and mapped to the second Tate pairing parameter Q of order r. Since Q is on the twisted curve, the hashing would result in a random point S of order $p+1+t$, and the mapping would require a point multiplication by a large constant cofactor $c = (p+1+t)/r$, so $Q = cS$. However bilinearity applies *even though S is not of order r*. So rather than calculate $e_r(P, cS)$ again its faster to calculate $e_r(P, S)^c$. Depending on the protocol it may in fact be valid to dispense with the co-factor altogether. Alternatively, as for example in the Boneh and Franklin scheme, it may be possible to calculate $e(P, cS) = e(cP, S)$, and the constant c can be permanently combined with a constant P, and thus eliminated completely from the calculation [19].

On occasion a protocol (such as Boneh and Franklin IBE decryption) requires us to check that a curve point is in fact of order r, and then to calculate a pairing. Since the pairing carries out an implicit point multiplication of its first parameter by r, these functions can be combined. First re-organise the protocol if necessary so that the point whose order is to be tested is the first parameter of the pairing. Then in the BKLS algorithm insert the line

7a. **if** $A \neq -P$ **then return** Wrong_Order

5 Case Study – The Sakai and Kasahara Identity Based Encryption Scheme

Here we describe the simplest variant of the Sakai and Kasahara Identity Based Encryption scheme [22], and deploy some of our optimisations, and make use of the compressed pairing. This IBE method is not as well-known as that of Boneh and Franklin [6], but it has its advantages, for example no pairing calculation is required for encryption. It is not secure against a chosen-ciphertext attack, but then neither is unadorned RSA with which we will be comparing it. As for all IBE schemes, it can be described in four stages, Setup, Extract, Encrypt and Decrypt.

- **Setup:** Global parameters are generated by the trusted key-issuing authority. It generates a suitable curve (like ours) and generates random points $P \in E(\mathbb{F}_p)[r]$ and $Q \in E'(\mathbb{F}_p)[r]$, and then generates its own master secret $s \in \mathbb{F}_r$. The authority makes public the values Q, sQ, and $g = \varepsilon_r(P, Q)$. Also made public are two hash functions $h_1 : \{0,1\}^* \to \mathbb{F}_r$ and $h_2 : \mathbb{F}_p \to \{0,1\}^c$
- **Extract:** Each user approaches the trusted authority and is issued with a private key. In the case of Alice, her identifying string is hashed to a value $a = h_1(Alice's\ Identity)$ and she is issued with the private key $D = \frac{1}{s+a}P$
- **Encrypt:** To encrypt a secret session key k and send it to Alice, first find $a = h_1(Alice's\ Identity)$ and then generate a random $w \in \mathbb{F}_r$ and create the ciphertext
$$C_1 = w(sQ + aQ)$$
$$C_2 = k \oplus h_2(V_w(g))$$
- **Decrypt:** Alice recovers the session key as $k = h_2(\varepsilon_r(D, C_1)) \oplus C_2$.

where $V_w(\cdot)$ is the Lucas function and $\varepsilon_r(\cdot)$ is the compressed pairing. The correctness of the algorithm follows immediately from bilinearity. Note that a pairing calculation is only needed for decryption. And for this pairing the first parameter is a constant – Alice's private key. Therefore the precomputation optimisation of section 4.2 is appropriate. For encryption Alice's public key can be regarded as $(sQ + aQ)$, and if multiple messages are to be sent to Alice, then this value can also be precalculated and cached. The calculation of aQ, if required, is carried out on the twisted curve $E'(\mathbb{F}_p)$, as is the subsequent point multiplication by w. If the point to be multiplied is fixed, then the precomputation method of Brickell et al. can be used here to advantage [9].

6 Results

The IBE scheme described above was implemented using a mixture of C++, C and assembly, both with and without precomputation, and compared with a similarly optimised RSA implementation (standard windowing techniques for modular exponentiation are used). We focus on the decryption operation in both cases - clearly RSA encryption will be much faster. The RSA implementation uses as a public key the product of two 512-bit prime factors, and decryption uses the

Chinese Remainder Theorem. So in both cases 512-bit modular multiplication or squaring in $\mathbb{F}_p$ is the time-critical operation. This was implemented using inline unlooped assembly language. We are not claiming that our multi-precision implementation is the fastest possible on the targeted hardware, a 1GHz Pentium III, but it is the same implementation for both IBE and RSA. For the pairing calculation without precomputation projective coordinates are used.

It is possible to calculate exactly the number of $\mathbb{F}_p$ multiplications (called henceforth a *mul*) needed for the pairing; for now we count squarings as multiplications, although squarings will be a little quicker. Using the standard algorithms for point addition and doubling as described in [14], a projective point addition requires 11 muls, and a doubling requires 8. The projective $f(\cdot)$ function needs 8 muls. Hence the $g(\cdot)$ function performs 16 muls for each point doubling and 19 muls for each addition. Line 5 of the main algorithm will require one $\mathbb{F}_{p^2}$ squaring, and one $\mathbb{F}_{p^2}$ multiplication at a cost of 5 muls, plus a call to $g(\cdot)$ for a point doubling, for a total of 21 muls. The loop is repeated 159 times, so line 5's total contribution is $159 \cdot 21 = 3339$ muls. The point addition of line 6 is only carried out once, and adds an extra 22 muls. So by the end of Miller's algorithm the cost so far is 3361 muls. Line 9 calls for a $\mathbb{F}_{p^2}$ modular inversion. Assuming that we are going to compress the pairing, this calls for 5 muls and a single inversion in $\mathbb{F}_p$. The exponent $(p+1)/r$ will be $512 - 160 = 352$ bits, and the Lucas laddering algorithm [25] requires exactly 2 muls per bit. So the grand total for the whole pairing is $3361 + 5 + 352 \cdot 2 = 4070$ muls plus one modular inversion.

Using precomputation the savings are substantial. The cost of each call to the $g(\cdot)$ function now falls to just 1 mul. Repeating the analysis above the total is now reduced to 1667 muls plus one inversion. By contrast a representative run of an RSA decryption program requires 1020 modular squarings and 222 modular multiplications.

Some tests show that, in the environment used, a modular squaring is equivalent to 0.89 of a modular multiplication, and that a modular inversion costs close to 28 muls. Introducing these equivalent costs, we can specify the computation required quite precisely in terms of the number of muls.

Timings are for the number theoretic parts of the algorithm; message padding and hashing are excluded. Our actual timing do not quite live up to those that could be projected from the raw mul counts. The pairing calculation, unlike RSA, involves a large number of modular additions and subtractions which have been ignored in the analysis above. In our implementation elliptic curve addition and subtraction is done in C, whereas the pairing is implemented in C++, which will be a little slower due to the tendency of C++ to create and destroy temporary variables. Finally precomputation may suffer somewhat as the precomputed data plus program variables are too large for the Pentium III 16K byte L1 data cache.

Table 1. Sakai and Kasahara IBE decryption – 1GHz PIII.

algorithm	$\mathbb{F}_p$ Muls	Time (ms)
IBE decryption, w/o precomp.	3992.7	20.0
IBE decryption, with precomp.	1660.1	10.2
RSA decryption	1126.8	4.5

A similarly optimised implementation of the Tate pairing, without precomputation, on a Compaq iPaq 3660 PDA powered by a 206MHz 32-bit StrongARM processor took just 355ms. This indicates that pairing-based cryptography may find application on low powered devices such as PDAs and mobile phones.

7 Conclusions

We have shown that pairing-based cryptography is perhaps not as slow or as difficult as was generally thought. The $k = 2$ case is particularly simple and accessible, and can be quickly implemented using existing crypto resources. Various optimisations have been suggested. An implementation of an IBE scheme shows that pairing-based cryptography can perform nearly as well as long established techniques such as RSA.

Acknowledgments

Thanks to Paulo Barreto for useful comments, and Noel McCullagh for drawing to my attention the Sakai and Kasahara IBE scheme [22] and for help with the implementation on the Compaq PDA.

References

1. P. S. L. M. Barreto. The pairing-based crypto lounge, 2004. http://planeta.terra.com.br/informatica/paulobarreto/pblounge.html.
2. P.S.L.M. Barreto, H.Y. Kim, B. Lynn, and M. Scott. Efficient algorithms for pairing-based cryptosystems. In *Advances in Cryptology – Crypto'2002*, volume 2442 of *Lecture Notes in Computer Science*, pages 354–68. Springer-Verlag, 2002.
3. P.S.L.M. Barreto, B. Lynn, and M. Scott. Constructing elliptic curves with prescribed embedding degrees. In *Security in Communication Networks – SCN'2002*, volume 2576 of *Lecture Notes in Computer Science*, pages 263–273. Springer-Verlag, 2002.
4. P.S.L.M. Barreto, B. Lynn, and M. Scott. On the selection of pairing-friendly groups. In *Selected Areas in Cryptography – SAC 2003*, volume 3006 of *Lecture Notes in Computer Science*, pages 17–25. Springer-Verlag, 2003.
5. I. F. Blake, G. Seroussi, and N. P. Smart, editors. *Advances in Elliptic Curve Cryptography, Volume 2*. Cambridge University Press, 2005.
6. D. Boneh and M. Franklin. Identity-based encryption from the Weil pairing. *SIAM Journal of Computing*, 32(3):586–615, 2003.
7. D. Boneh, B. Lynn, and H. Shacham. Short signatures from the Weil pairing. In *Advances in Cryptology – Asiacrypt'2001*, volume 2248 of *Lecture Notes in Computer Science*, pages 514–532. Springer-Verlag, 2002.
8. F. Brezing and A. Weng. Elliptic curves suitable for pairing based cryptography. Cryptology ePrint Archive, Report 2003/143, 2003. Available from http://eprint.iacr.org/2003/143.
9. E. F. Brickell, D. M. Gordon, K. S. McCurley, and D. B. Wilson. Fast exponentiation with precomputation: Algorithms and lower bounds. In *Advances in Cryptology – Eurocrypt'92*, volume 658 of *Lecture Notes in Computer Science*, pages 200–207. Springer-Verlag, 1993.

10. R. Dupont, A. Enge, and F. Morain. Building curves with arbitrary small MOV degree over finite prime fields. Cryptology ePrint Archive, Report 2002/094, 2002. http://eprint.iacr.org/2002/094.
11. R. Dutta, R. Barua, and P. Sarkar. Pairing-based cryptography : A survey. Cryptology ePrint Archive, Report 2004/064, 2004. http://eprint.iacr.org/2004/064.
12. I. Duursma and H.-S. Lee. Tate-pairing implementations for tripartite key agreement. In *Advances in Cryptology – Asiacrypt 2003*, volume 2894 of *Lecture Notes in Computer Science*, pages 111–123. Springer-Verlag, 2003.
13. S. Galbraith, K. Harrison, and D. Soldera. Implementing the Tate pairing. In *Algorithm Number Theory Symposium – ANTS V*, volume 2369 of *Lecture Notes in Computer Science*, pages 324–337. Springer-Verlag, 2002.
14. IEEE Std 1363-2000. Standard specifications for public-key cryptography. IEEE P1363 Working Group, 2000.
15. T. Izu and T. Takagi. Efficient computations of the Tate pairing for the large MOV degrees. In *ICISC 2002*, volume 2587 of *Lecture Notes in Computer Science*, pages 283–297, 2003.
16. M. Joye and S. Yen. The Montgomery powering ladder. In *Cryptographic Hardware and Embedded Systems - CHES 2002*, volume 2523 of *Lecture Notes in Computer Science*, pages 291–302, Berlin, Germany, 2003. Springer-Verlag.
17. P. Kocher, J. Jaffe, and B. Jun. Introduction to differential power analysis and related attacks, 1998. http://www.cryptography.com/dpa/technical.
18. W. Mao and K. Harrison. Divisors, bilinear pairings, and pairing enabled cryptographic applications, 2003. http://hplbwww.hpl.hp.com/people/wm/research/pairing.pdf.
19. N. McCullagh. Personal Communication, 2004.
20. A. Menezes. *Elliptic Curve Public Key Cryptosystems*. Kluwer Academic Publishers, 1993.
21. A. Miyaji, M. Nakabayashi, and S. Takano. New explicit conditions of elliptic curve traces for FR-reduction. *IEICE Transactions on Fundamentals*, E84-A(5):1234–1243, 2001.
22. R. Sakai and M. Kasahara. ID based cryptosystems with pairing on elliptic curve. Cryptography ePrint Archive, Report 2003/054, 2003. http://eprint.iacr.org/2003/054.
23. M. Scott, 2002. http://ftp.compapp.dcu.ie/pub/crypto/cm.exe.
24. M. Scott, 2002. http://www.computing.dcu.ie/~mike/tate.html.
25. M. Scott and P. Barreto. Compressed pairings. In *Advances in Cryptology – Crypto' 2004*, volume 3152 of *Lecture Notes in Computer Science*, pages 140–156. Springer-Verlag, 2004. Also available from http://eprint.iacr.org/2004/032/.
26. M. Scott and P. Barreto. Generating more MNT elliptic curves. Cryptology ePrint Archive, Report 2004/058, 2004. Available from http://eprint.iacr.org/2004/058/.
27. I. Semaev. Summation polynomials and the discrete logarithm problem on elliptic curves. Cryptography ePrint Archive, Report 2004/031, 2003. http://eprint.iacr.org/2004/031/.
28. N. P. Smart. An identity based authenticated key agreement protocol based on the Weil pairing. *Electronics Letters*, 38:630–632, 2002.
29. J. Solinas. ID-based digital signature algorithms, 2003. http://www.cacr.math.uwaterloo.ca/conferences/2003/ecc2003/solinas.pdf.

Fast and Proven Secure Blind Identity-Based Signcryption from Pairings

Tsz Hon Yuen and Victor K. Wei

Department of Information Engineering,
The Chinese University of Hong Kong,
Shatin, Hong Kong
{thyuen4,kwwei}@ie.cuhk.edu.hk

Abstract. We present the first blind identity-based signcryption (BIBSC). We formulate its security model and define the security notions of blindness and parallel one-more unforgeability (p1m-uf). We present an efficient construction from pairings, then prove a security theorem that reduces its p1m-uf to Schnorr's ROS Problem in the random oracle model plus the generic group and pairing model. The latter model is an extension of the generic group model to add support for pairings, which we introduce in this paper. In the process, we also introduce a new security model for (non-blind) identity-based signcryption (IBSC) which is a strengthening of Boyen's. We construct the first IBSC scheme proven secure in the strengthened model which is also the fastest IBSC in this model or Boyen's model. The shortcomings of several existing IBSC schemes in the strengthened model are shown.

1 Introduction

Identity-based cryptography is a kind of public key cryptography that using recipient's identity as the public key. The identity can be name, email address or any other arbitrary strings that can identify a recipient uniquely. Usually a trusted authority (TA) is needed to generate private keys according to the public keys. The advantage is that distribution of public key in advance is not needed. The concept of identity-based cryptography was firstly proposed by Shamir [21] in 1984. Since then, there are many suggestions for the implementation of identity-based encryption ([12, 23, 16, 10]). However they were not fully satisfactory. In 2001, Boneh and Franklin [4] proposed the first practical identity-based encryption scheme using pairings on elliptic curves. Identity-based encryptions prior to [4] either requires high complexity to compute the key pairs or is insecure against colluders. There are also developments in identity-based signatures [6], authenticated key agreement, etc.

Blind signature was introduced by Chaum [7], which provides anonymity of users in applications such as e-cash. It allows users to get a signature of a message in a way that the signer learns neither the message nor the resulting signature.

Privacy and authenticity are also the basic aims of public key cryptography. We have encryption and signature to achieve these aims. Zheng [27] proposed that encryption and signature can be combined as "signcryption" which can be more efficient in computation than running encryption and signature separately. The security of signcryption is discussed by An et al. [1]

1.1 Contributions

This paper makes the following contributions to the literature:
1. We present the first blind identity-based signcryption (BIBSC). Roughly speaking, BIBSC works as follows: Upon request from Warden, a blind signcryption oracle makes a commitment, then blindly signs and computes the randomness term in the encryption part. Warden deblinds the signature and uses the randomness term returned to produce a signcryption.
2. We formulate the first BIBSC security models to define security notions including blindness and parallel one-more unforgeability (p1m-uf).
3. We construct the first BIBSC scheme from pairings, and prove its security. The blindness of our BIBSC scheme is statistical ZK, and the p1m-uf is reduced to Schnorr's ROS Problem in the random oracle model plus the generic group and pairing model (GGPM).
4. We introduce the generic group and pairing model (GPPM) which is an extension of the generic group model [18, 22, 20] by including support for pairings. We use this model to prove p1m-uf of our BIBSC scheme.
5. We introduce a strengthening of Boyen's [5] security model for identity-based signcryption (IBSC) to add support of authenticated encryption.
6. We construct the first proven secure IBSC scheme in the strengthened model. It is also the fastest and shortest IBSC scheme in our model as well as in Boyen's [5] model.
7. The shortcomings of several existing IBSC schemes in the strengthened model are shown.

1.2 Organization

In Section 2, we define the preliminaries. In Section 3, we define the IBSC and BIBSC security models. In Section 4, we introduce our schemes. In Section 5, we introduce the generic group and pairing model. In Section 6, we compare our IBSC scheme with existing schemes.

2 Preliminaries

2.1 Related Results

Shamir [21] suggested an identity-based signature scheme. Boneh and Franklin [4] proposed an identity-based encryption scheme. There are some papers [15, 5, 13, 11, 9, 14] concerning the combination of identity-based signature and encryption to form IBSC schemes. The most expensive single operation is pairing computations. Schemes of [15, 5, 14] use 5 pairings, while [13, 9] use 6, and [11] uses 4. [5] is proven secure in a stronger model than [15, 13]. [11] has no security proof.

Blind signatures was introduced by Chaum [7]. Some identity-based blind signature schemes was proposed in [24–26].

2.2 Pairings

Our BIBSC and IBSC schemes use bilinear pairings on elliptic curves. We now give a brief revision on the property of pairings and some candidate hard problems from pairings that will be used later.

Let G_1, G_2, G_3 be cyclic groups of prime order q, writing the group action multiplicatively. Let g_1 (resp. g_2) be a generator of G_1 (resp. G_2). There exists ψ which is isomorphism from G_2 to G_1, with $\psi(g_2) = g_1$.

Definition 1. *A map $e : G_1 \times G_2 \to G_3$ is called a bilinear pairing if, for all $x \in G_1, y \in G_2$ and $a, b \in Z$, we have $e(x^a, y^b) = e(x, y)^{ab}$, and $e(g_1, g_2) \neq 1$.*

Definition 2. *(Co-BDH Problem). The co-Bilinear Diffie-Hellman problem is that, given $P, P^\alpha, P^\beta \in G_1$, $Q \in G_2$, for unknown $\alpha, \beta \in Z_q$, to compute $e(P, Q)^{\alpha\beta}$.*

Definition 3. *(Co-CDH Problem). The co-Computational Diffie-Hellman problem is that, given $P, P^\alpha \in G_1$, $Q \in G_2$, for unknown $\alpha \in Z_q$, to compute Q^α.*

2.3 Blind Signatures and Schnorr's ROS Problem

Blind signature is described as follows: Upon request from Warden, a signing oracle makes a commitment, then blindly signs a message for Warden. Warden deblinds the signature such that the signing oracle knows neither the message nor the output signature.

Parallel one-more forgery against blind signature is that an attacker interacts with a signer l times and produces $l + 1$ signatures from these interactions. Schnorr [20] reduced the parallel one-more unforgeability (p1m-uf) of the blind Schnorr signature to the ROS Problem in the random oracle plus generic group model (ROM+GGM). The followings are from Schnorr [20]:

Definition 4. *(ROS Problem). Find an overdetermined, solvable system of linear equations modulo q with random inhomogeneities. Specifically, given an oracle random function $F : Z_q^l \leftarrow Z_q$, find coefficients $a_{k,i} \in Z_q$ and a solvable system of $l + 1$ distinct equations of Eq. (1) in the unknowns $c_1, \ldots, c_l$ over Z_q:*

$$a_{k,1}c_1 + \ldots + a_{k,l}c_l = F(a_{k,1}, \ldots, a_{k,l}) \text{ for } k = 1, \ldots, t. \quad (1)$$

Theorem 1. *[20] Given generator g, public key h and an oracle for H, let a generic adversary $\mathcal{A}$ performs t generic steps and interacts with a signer for l times. If $\mathcal{A}$ succeeds in a parallel attack to produce $l + 1$ signatures with a probability of success better than $\binom{t}{2}/q$, then $\mathcal{A}$ must solve the ROS-problem in ROM+GGM.*

3 BIBSC and Enhanced IBSC Security Model

We define the first security model for BIBSC and also an enhancement of Boyen's security model for IBSC. For logistics, we present the latter first.

Intuitions: Basically, signcryption reuses the randomness in signing as the randomness in encryption, to achieve bandwidth conservation. Lower complexity is also a goal. In blind signcryption, below, the "prover oracle" delivers both the blind signature as well as the intermediate encryption results which reuses the randomness. In comparison, the prover oracle in a blind signature scheme delivers only the signature.

In the naive sign-then-encrypt (StE) instantiation, the recipient can decrypt, and then re-encrypt the (sender) signed plaintext to a third party. The resulting signcryption is a valid signcryption but the signer and the encryptor are distinct. Boyen's *ciphertext unlinkability* [5] extends this basic idea. In the naive encrypt-the-sign (EtS) instantiation, the encryptor and the signer are assured to be the same. The *authenticated encryption* [1] extends this basic idea. Our security model supports both ciphertext unlinkability and authenticated encryption in two different but closely related *dual versions*.

3.1 Enhanced IBSC Security Model

We present an enhancement of Boyen's security model for IBSC. The main addition is to add support for *authenticated encryption*. The signer cannot deny signcrypting the message to the recipient. Boyen's model is restricted to *ciphertext unlinkability* where this assurance is not required. Our model below is capable of supporting authenticated encryption, resp. ciphertext unlinkability.

3.1.1 Primitives.
An IBSC scheme consists of four algorithms: (Setup, Extract, Signcrypt, Unsigncrypt). The algorithms are specified as follows: Setup: On input a security parameter k, the TA generates $\langle \zeta, \pi \rangle$ where ζ is the randomly generated master key, and π is the corresponding public parameter.
Extract: On input ID, the TA computes its corresponding private key S_{ID} (corresponding to $\langle \zeta, \pi \rangle$) and sends back to its owner in a secure channel.
Signcrypt: On input the private key of sender A, S_A, recipient identity ID_B and a message m, outputs a ciphertext σ corresponding to π.
Unsigncrypt: On input private key of recipient B, S_B, and ciphertext σ, decrypt to get sender identity ID_A, message m and signature s corresponding to π. Verify s and verify if encryptor = signer. Output $\top$ for "true" or $\bot$ for "false".

We make the consistency constraint that if $\sigma \leftarrow Signcrypt(S_A, ID_B, m)$, then $m \leftarrow Unsigncrypt(S_B, \sigma)$.

3.1.2 Indistinguishability.
Indistinguishability for IBSC against adaptive chosen ciphertext attack (IND-IBSC-CCA2) is defined as in the following game. The adversary is allowed to query the random oracles, key extraction oracle, signcryption oracle and unsigncryption oracle. The game is defined as follows:

1. The simulator selects the public parameter and sends to the adversary.
2. The adversary performs polynomial number of oracle queries adaptively.
3. The adversary generates m_1, ID_{A1}, ID_{B1}, and sends to the simulator. The adversary knows S_{A1}. The simulator generates m_0, ID_{A0}, ID_{B0}, randomly chooses $b \in_R \{0,1\}$. The simulator delivers the challenge ciphertext $\sigma \leftarrow Signcrypt(S_{Ab}, ID_{Bb}, m_b)$ to the adversary.

4. The adversary performs polynomial number of oracle queries adaptively.
5. The adversary tries to compute b, in the following three sub-games:
   (a) The simulator ensures $B0 = B1$, $m_0 = m_1$.
   (b) The simulator ensures $A0 = A1$, $m_0 = m_1$.
   (c) The simulator ensures $A0 = A1$, $B0 = B1$.

The adversary wins the game if he can guess b correctly. The *advantage* of the adversary is the probability, over half, that he can compute b accurately.

The oracles are defined as follows:

Key Extraction Oracle $\mathcal{KEO}$: Upon input an identity, the key extraction oracle outputs the private key corresponding to this identity.

Signcryption Oracle $\mathcal{SO}$: Upon input m, ID_A, ID_B, the signcryption oracle produces a valid signcryption σ for the triple of input.

Unsigncryption Oracle $\mathcal{UO}$: Upon input ciphertext σ and recipient ID, the unsigncryption oracle outputs the decryption result and the verification outcome.

Oracle query to $\mathcal{KEO}$ with input ID_{B0} or ID_{B1} is not allowed. Oracle query to $\mathcal{SO}$ with input (m_1, ID_{A1}, ID_{B1}) is not allowed. Oracle query to $\mathcal{UO}$ for the challenge ciphertext from the simulator is not allowed.

Definition 5. *(Indistinguishability). An IBSC is* IND-IBSC-CCA2 *secure if no PPT adversary has a non-negligible advantage in any of the sub-games above.*

Our security notion above is a strong one. It incorporates previous security notions including *insider-security* in [1], *indistinguishability* in [15], and *anonymity* in [5].

3.1.3 Existential Unforgeability.
Existential unforgeability against adaptive chosen message attack for IBSC (EU-IBSC-CMA) is defined as in the following game. The adversary is allowed to query the random oracles, $\mathcal{KEO}$, $\mathcal{SO}$ and $\mathcal{UO}$, which are defined in the above section. The game is defined as follows:

1. The simulator selects the public parameter and sends to the adversary.
2. The adversary performs polynomially number of oracle queries adaptively.
3. The adversary delivers a recipient identity ID_B and a ciphertext σ.

The adversary wins the game if he can produce a valid (σ, ID_B) such that σ can be decrypted, under the private key of ID_B, to a message m, sender identity ID_A and a signature s which passes all verification tests.

Oracle query to $\mathcal{KEO}$ with input ID_A is not allowed. The adversary's answer (σ, ID_B) should not be computed by $\mathcal{SO}$ before.

Definition 6. *(Existential Unforgeability). An IBSC is* EU-IBSC-CMA *secure if no PPT adversary has a non-negligible probability in winning the game above.*

The adversary is allowed to get the private key of the recipient in the adversary's answer. This gives us an *insider-security* in [1]. This model for *authenticated encryption* is stronger than Boyen's [5] existential unforgeability in the sense that our model provides non-repudiation for the ciphertext while Boyen's provides non-repudiation for the decrypted signature only. For *ciphertext unlinkability*, we have to add one more restriction for our model. Oracle queries to $\mathcal{SO}$ for (ID_A, m) in the adversary's answer using any recipient identity are not allowed. Then the model changes to non-repudiation for signature only.

3.2 Introducing BIBSC Security Model

We will propose the primitives of blind version of IBSC and then define the security notions for blindness and parallel one-more unforgeability.

3.2.1 Primitives. A BIBSC is a five-tuple (Setup, Extract, BlindSigncrypt, Warden, Unsigncrypt) where Setup, Extract and Unsigncrypt are identical as primitives in IBSC. (BlindSigncrypt, Warden) is a 3-move interactive protocol as follows. Input to BlindSigncrypt is the sender identity ID_A and its private key S_A, and the recipient identity ID_B. Input to Warden is ID_A, ID_B and a message m.

1. BlindSigncrypt sends a commit X to Warden.
2. Warden challenges BlindSigncrypt with h.
3. BlindSigncrypt sends back the response W and V to Warden.

Finally Warden outputs a ciphertext σ.

3.2.2 Blindness. Here we define the blindness of BIBSC. The adversary is allowed to makes q_B queries to blind signcryption oracle $\mathcal{BSO}$, q_H queries to random oracles, q_S queries to $\mathcal{SO}$, and q_U queries to $\mathcal{UO}$. The adversary keeps the transcript $\mathcal{T}$ recording the interaction between BlindSigncrypt and Warden.

Definition 7. *(Blindness) A BIBSC is blind if given a ciphertext σ by Warden, $Prob\{\sigma$ by $Warden\} = Prob\{\sigma$ by $Warden|\mathcal{T}\}$*

3.2.3 Parallel One-More Unforgeability. Parallel one-more unforgeability for BIBSC (p1m-uf) is defined as in the following game. It is similar to the one-more forgery for traditional blind signature schemes [2, 3, 26].

1. The sender identity ID_A is given to the adversary.
2. The adversary makes a total of q_B queries to blind signcryption oracle $\mathcal{BSO}_{ID_k}$, $1 \leq k \leq K$, q_H queries to random oracles, q_K queries to $\mathcal{KEO}$, q_S queries to $\mathcal{SO}$, and q_U queries to $\mathcal{UO}$.
3. The adversary delivers q_B+1 tuples (ID_i, σ_i) to the simulator, $1 \leq i \leq q_B+1$.

The adversary wins the game if he can produce $q_B + 1$ valid distinct tuples (ID_i, σ_i) that can decrypts, under the private key of ID_i, to message m_i, sender identity ID_A, and signature s_i which passes the verification tests. The $\mathcal{SO}, \mathcal{UO}$ and $\mathcal{KEO}$ are same as the one in IBSC. We have the new interactive $\mathcal{BSO}$:

$\mathcal{BSO}_{ID_A}$: Upon input ID_B, it returns a number X. Then input a number h. It produces an output (W, V) based on sender ID_A, recipient ID_B, X and h. It is required that the private key of ID_A is never extracted by $\mathcal{KEO}$.

Definition 8. *(Parallel One-more Unforgeability). A BIBSC is p1m-uf secure if no PPT adversary has a non-negligible probability in winning the above game.*

4 Efficient and Secure BIBSC (resp. IBSC) Schemes

We present our constructions of efficient and secure BIBSC and IBSC schemes from pairings. For logistics of presentation, we present the IBSC scheme first.

4.1 A New Efficient and Secure IBSC Scheme

This IBSC scheme follows the primitives in Section 2. Let G_1, G_2, G_3 be (multiplicative) cyclic groups of order q. The pairing is given as $e : G_1 \times G_2 \to G_3$. Now we define our scheme as follows.

Setup: The setup of TA is similar to [4]. On input a security parameter $n \in N$, a generator $G[1^n]$ generates G_1, G_2, G_3, q and e. The TA chooses a generator $P \in G_1$ and picks a random $s \in Z_q$ as the master key. Then the TA sets $P_{TA} = P^s \in G_1$. After that the TA chooses cryptographic hash functions $H_0 : \{0,1\}^* \to G_2, H_1 : \{0,1\}^* \times G_2 \times \{0,1\}^* \to Z_q, H_2 : G_3 \to \{0,1\}^*, H_3 : G_3 \times \{0,1\}^* \to G_2$. The system parameters are $\langle q, G_1, G_2, G_3, e, P, P_{TA}, H_0, H_1, H_2, H_3 \rangle$.

Extract: Given a user identity string $ID \in \{0,1\}^*$, his public key is $Q_{ID} = H_0(ID) \in G_2$. His private key $S_{ID} = (Q_{ID})^s \in G_2$ is calculated by TA.

Signcrypt: Suppose Alice wants to signcrypt a message m to Bob. Assume Alice's identity is ID_A with public key Q_A and private key S_A. Bob's identity is ID_B.

- Sign: Alice chooses a random $r \in Z_q$ and computes:

$$X = P^r \in G_1$$
$$h = H_1(m, X, ID_B) \in Z_q$$
$$W = S_A{}^h Q_A{}^r \in G_2$$

- Encrypt: Alice computes $Q_B = H_0(ID_B) \in G_2$ and:

$$V = e(P_{TA}{}^r, Q_B) \in G_3$$
$$Y = H_3(V, ID_A) \oplus W \in G_2$$
$$Z = H_2(V) \oplus \langle ID_A, m \rangle \in \{0,1\}^*$$

Alice outputs a ciphertext $\sigma = \langle X, Y, Z \rangle$ and sends to Bob.

Unsigncrypt: Bob receives the ciphertext $\sigma = \langle X, Y, Z \rangle$.

- Decrypt: Assume the private key of Bob is S_B. Bob computes:

$$V' = e(X, S_B)$$
$$\langle ID_A, m \rangle = H_2(V') \oplus Z$$

Output $\langle ID_A, m \rangle$ together with $\langle X, Y, V' \rangle$ to Verify.

- Verify: Bob computes $W' = H_3(V', ID_A) \oplus Y$ and compares if:

$$e(P, W') = e(XP_{TA}{}^h, Q_A) \quad \text{where } h = H_1(m, X, ID_B)$$

Output $\top$ if the above verification is true, or output $\bot$ if false.

In Section 3.1, Unsigncrypt also requires the verification for checking encryptor = signer. It is implicitly done in Decrypt and Verify as both of them use the same X in σ to decrypt and verify.

Finally, we show the consistency constraint is satisfied in Decrypt and Verify. In Decrypt, V can be recovered as: $e(X, S_B) = e(P^r, Q_B{}^s) = e(P_{TA}{}^r, Q_B)$. In Verify, if the signature is valid, both sides should be equivalent because:
$e(P, W) = e(P, S_A{}^h Q_A{}^r) = e(P, Q_A{}^{(sh+r)}) = e(P^{(r+sh)}, Q_A) = e(XP_{TA}{}^h, Q_A)$.

Theorem 2. *Our IBSC scheme is* **IND-IBSC-CCA2** *secure provided the co-BDH Problem is hard in the random oracle model.*

Theorem 3. *Our IBSC scheme is* **EU-IBSC-CMA** *secure provided the co-CDH Problem is hard in the random oracle model.*

Proof sketches of the above two theorems are in Appendix B.

Dual Support of Ciphertext Unlinkability (CU) and Authenticated Encryption (AE): One of the main difference between our IBSC scheme and Boyen's scheme [5] is that our scheme has linkability (AE) while Boyen's scheme has unlinkability (CU). In our original AE version, we include the recipient identity in the signature, such that the adversary cannot reuse the signature s by sender ID_A for other recipients and encrypt s to forge a signcryption from ID_A to the adversary himself.

As unlinkability may also be important in some applications, we provide the CU version of our scheme. The only change is that in Sign change $h = H_1(m, X)$. Other steps remain unchanged. Therefore this CU version is as efficient as the original AE version. Notice that by changing to CU, unforgeability for ciphertext reduces to unforgeability for signature only, as in [5].

4.2 The First BIBSC Scheme

In this BIBSC scheme, Setup, Extract and Unsigncrypt are the same as Section 4.1. We describe the interactive protocol for BlindSigncrypt and Warden below:

BlindSigncrypt	Warden
randomly choose r	randomly choose α, β
send $X = P^r \in G_1 \longrightarrow$	
	compute $\hat{X} = X^\alpha P^\beta \in G_1$, $\hat{h} = H(m, \hat{X}, ID_B) \in Z_q$
	$\longleftarrow$ send $h = \alpha^{-1}\hat{h} \in Z_q$
send $W = S_A{}^h Q_A{}^r \in G_2$	
and $V = e(P_{TA}{}^r, Q_B) \in G_2 \longrightarrow$	
	compute $\hat{W} = W^\alpha Q_A{}^\beta \in G_2$
	compute $\hat{V} = V^\alpha e(P_{TA}{}^\beta, Q_B) \in G_3$
	compute $\hat{Y} = H_3(\hat{V}, ID_A) \oplus \hat{W} \in G_2$
	compute $\hat{Z} = H_2(\hat{V}) \oplus \langle ID_A, m \rangle \in \{0,1\}^*$
	output $\sigma = \langle \hat{X}, \hat{Y}, \hat{Z} \rangle$

Consistency is verified as:

$$e(P, \hat{W}) = e(P, W^\alpha Q_A^\beta) \quad \text{and} \quad \hat{V} = V^\alpha e(P_{TA}{}^\beta, Q_B)$$
$$= e(P, Q_A)^{s\hat{h}+\alpha r+\beta} \qquad\qquad = e(P^{s(r\alpha+\beta)}, Q_B)$$
$$= e(P_{TA}{}^{\hat{h}} X^\alpha P^\beta, Q_A) \qquad\qquad = e(X^\alpha P^\beta, S_B)$$
$$= e(\hat{X} P_{TA}{}^{\hat{h}}, Q_A) \qquad\qquad\qquad = e(\hat{X}, S_B)$$

Theorem 4. *Our BIBSC scheme has blindness.*

Theorem 5. *Our BIBSC scheme is plm-uf secure provided Schnorr's ROS Problem is hard in the ROM+GGPM.*

Proof sketches of the above two theorems are in Appendix B.

Remark: In our proof, we use an alternative representation for $\hat{Y}$ and $\hat{Z}$. Let θ_4 (resp. θ_5) be a bijective mapping from G_2 to G_4 (resp. from $\{0,1\}^*$ to G_5) where G_4 (resp. G_5) is a cyclic group. Change $H_2 : G_3 \to G_5, H_3 : G_3 \times \{0,1\}^* \to G_4$. Then $\hat{Y} = H_3(\hat{V}, ID_A) \oplus \theta_4(\hat{W}) \in G_4$ and $\hat{Z} = H_2(\hat{V}) \oplus \theta_5(\langle ID_A, m \rangle) \in G_5$. In Unsigncrypt, we can use θ_4^{-1} and θ_5^{-1} to recover the message. The efficiency and security of our BIBSC scheme will not be affected.

5 Generic Group and Pairing Model (GGPM)

We briefly introduce the generic group and pairing model (GGPM) by extending the generic group model (GGM) of [18, 22, 19], to include support for the pairing oracle. There are two types of data, namely, group elements in G_1, G_2, and G_3, and non-group data. The group cardinalities are prime numbers q_1, q_2, q_3 respectively, with $q_1 = q_2 = q_3 = q$. Non-group data are integers in Z (or in Z_q depending on convention). The *base elements* of G_3 can be randomly generated, obtained from the blind signcryption oracle, or computed as the pairing of one element from G_1 and one element from G_2. The GGPM consists of:

1. Three GGMs, one for each of G_1, G_2, and G_3. Denote their *encodings* by $\theta_i : G_i \to S_i$, $i = 1, 2, 3$.
2. A pairing oracle, $\hat{e} : S_1 \times S_2 \to S_3$, satisfying bilinear properties.
3. Other oracles in the security model such as $\mathcal{BSO}$, $\mathcal{KEO}$ and random oracle.

The encodings θ_i are that non-group operations are meaningless. Similar to [20] each *generic step* is a computation of one of the following:

mex-1: $Z_q^{d_1} \times G_1^{d_1} \to G_1, (a_1^{(1)}, \cdots, a_{d_1}^{(1)}, g_1^{(1)}, \cdots, g_{d_1}^{(1)}) \mapsto \prod_i (g_i^{(1)})^{a_i^{(1)}}$

mex-2: $Z_q^{d_2} \times G_2^{d_2} \to G_2, (a_1^{(2)}, \cdots, a_{d_2}^{(2)}, g_1^{(2)}, \cdots, g_{d_2}^{(2)}) \mapsto \prod_{i'} (g_{i'}^{(2)})^{a_{i'}^{(2)}}$

mex-3: $Z_q^{d_3+d_1 d_2} \times G_3^{d_3} \times G_1^{d_1} G_2^{d_2} \to G_3,$

$(a_1^{(3)}, \cdots, a_{d_3+d_1 d_2}^{(3)}, g_1^{(3)}, \cdots, g_{d_3}^{(3)}, (g_1^{(1)}, g_1^{(2)}), \cdots, (g_{d_1}^{(1)}, g_{d_2}^{(2)}))$

$\mapsto \prod_{i=1}^{d_3} (g_i^{(3)})^{a_i^{(3)}} \prod_{j=1}^{d_1} \prod_{k=1}^{d_2} e(g_j^{(1)}, g_k^{(2)})^{a_{d_3+d_2(j-1)+k}^{(3)}}$

mex-p: $Z_q^{d_1+d_2} \times G_1^{d_1} \times G_2^{d_2} \to G_3,$

$(a_1^{(4)}, \cdots, a_{d_1}^{(4)}, a_1^{(5)}, \cdots, a_{d_2}^{(5)}, g_1^{(1)}, \cdots, g_{d_1}^{(1)}, g_1^{(2)}, \cdots, g_{d_2}^{(2)})$

$\mapsto \prod_j \prod_k e(g_j^{(1)}, g_k^{(2)})^{a_j^{(4)} a_k^{(5)}}$

The elements $g_i^{(1)}$'s are P, P_{TA}, $\mathcal{BSO}$ commitments X_i's, and randomly generate G_1 elements. The elements $g_i^{(2)}$'s are Q_{ID}'s, S_{ID}'s, $\mathcal{BSO}$ responses W_i's, and randomly generate G_2 elements. The elements $g_i^{(3)}$'s are $\mathcal{BSO}$ responses V_i's, randomly generate G_3 elements, and pairing oracle outputs. Similar to [20], we can omit randomly generated group elements, below, w.l.o.g.

A (non-interactive) **generic algorithm** is a sequence of t_{total} generic steps

1. Inputs are: $f_1^{(u)}, \cdots, f_{t'_u}^{(u)} \in G_u$ for $u = 1, 2, 3$, $1 \leq t'_u < t_{total}$, where $t' = \sum_u t'_u < t_{total}$ and non-group data like Z_q in given ciphertext or signature.
2. Computation steps are: $f_i^{(u)} = \prod_{j=1}^{i-1}(f_j^{(u)})^{a_{i,j}^{(u)}}$, for $i = t'_u+1, \cdots, t_u$, $u = 1, 2$, and $f_i^{(3)} = \prod_{j=1}^{i-1}(f_j^{(3)})^{a_{i,j}^{(3)}} \cdot \prod_{1 \leq k, \ell < t} e(f_k^{(1)}, f_\ell^{(2)})^{b_{i,k,\ell}}$ for $i = t'_3 + 1, \cdots, t_3$, where $t_{total} = t_1 + t_2 + t_3 + t_4$ and exponents $a_{i,j}^{(u)}$ depends arbitrarily on i, j, and non-group inputs.
3. Ouputs are: non-group data and group elements $f_{\sigma_1}^{(u)}, \cdots, f_{\sigma_d}^{(u)}$ where the integers $\sigma_1, \cdots, \sigma_d \in \{1, \cdots, t_u\}$ that depend arbitrarily on the non-group input.

The generic adversary can also perform equality test, if-then-else, looping, and other logical operations. We omit discussions about them here.

In the generic algorithm, each computation step $f_\sigma^{(u)}$ must be represented as the product of powers of group elements $g_i^{(1)}$'s, $g_{i'}^{(2)}$'s, $g_{i''}^{(3)}$'s, and $e(g_k^{(1)}, g_\ell^{(2)})$'s. There are only polynomially many group elements involved in any PPT algorithm. Each step can be represented as a sequence of exponents, and that representation should be unique. A *collision* is when a step can have multiple representations w.r.t. the bases consisting of the prescribed set of group elements. The following lemma shows the *collision* probability for $f_i^{(1)}, f_j^{(2)}, f_k^{(3)}$ are negligible except when involving oracle queries. The proof is similar to Schnorr's Lemma 1 and omitted.

Lemma 6. *In an arbitrary instantiation of the generic groups and the generic pairing, the probability of a PPT generic algorithm being able to compute a collision is negligible, except the collisions obtained via oracle queries. The probability is taken over randomized instantiations of all randomly generated base elements.*

Oracle assisted collisions are obtained from the $\mathcal{BSO}$ which are of the type $e(A, B) = e(C, D)$ in G_3. The $\mathcal{KEO}$ also yields collisions in G_2. The identity-based characteristics need special attention in the proof of this lemma.

Next we elaborate on **interactive generic algorithms**. We count the following generic steps:

- group operations mex-1, mex-2, mex-3, mex-p
- queries to hash oracle H
- queries to key extraction oracle $\mathcal{KEO}$
- interactions with a blind signcryption oracle $\mathcal{BSO}$.

A **generic adversary** is an interactive algorithm that interacts with $\mathcal{BSO}$. The construction is similar to Schnorr's, unless specified below. The *input* consists of generators $g^{(1)}, g^{(2)}, g^{(3)}$, public keys $Q_1, \cdots, Q_K \in G_2$, master public key

$P_{TA} \in G_1$, group order q, pairing $e(\cdot,\cdot)$ and collection of messages, ciphertexts and so on, which can be broken into group elements and non-group data.

$\mathcal{A}$'s *transmission* to $\mathcal{KEO}$ depends arbitrarily on given group elements and non-group data. Notice that key extraction for sender's private key is not allowed.

The *restriction* is that $\mathcal{A}$ can use group elements only for generic group operations, equality tests and for queries to hash oracle and $\mathcal{KEO}$, whereas non-group data can be arbitrarily used without charge. The computed group elements are given as explicit multiplicative combinations of given group elements. Let $X_\ell = g^{(1)^{r_\ell}} \in G_1, W_\ell = Q_A^{r_\ell + sh_\ell} \in G_2, V_\ell = e(X_\ell, S_{B_\ell})$ for $\ell = 1, \cdots, l$ be the group elements that $\mathcal{A}$ gets from $\mathcal{BSO}$ using the sender ID_A and recipient ID_{B_ℓ}. A computed $f_j^{(1)} \in G_1$ is of the form $f_j^{(1)} = P^{a_{j,-1}^{(1)}} P_{TA}^{a_{j,0}^{(1)}} \prod_{\ell=1}^{l} X_\ell^{a_{j,\ell}^{(1)}}$, where the exponents $a_{j,-1}^{(1)}, \cdots, a_{j,l}^{(1)} \in Z_q$ depend arbitrarily on given non-group data. A computed $f_j^{(2)} \in G_2$ is of the form $f_j^{(2)} = Q_A^{a_{j,0}^{(2)}} \prod_{\ell=1}^{l} W_\ell^{a_{j,\ell}^{(2)}}$, where the exponents depend arbitrarily on given non-group data. A computed $f_j^{(3)} \in G_3$ is of the form $f_j^{(3)} = e(P, Q_A)^{a_{j,-1}^{(3)}} e(P_{TA}, Q_A)^{a_{j,0}^{(3)}} \prod_{\ell=1}^{l} V_\ell^{a_{j,\ell}^{(3)}}$.

Powers and Limitations of GGM and GGPM. Because co-CDH and one-more co-CDH are collisions in GGPM, Lemma 6 implies they are hard. The real-world interpretations of this and other GGPM-based results are discussed in Appendix A.

6 Comparisons

The bandwidth and complexity efficiencies of our IBSC scheme is compared against a collection of existing schemes in Table 1.

The computation time includes the number of pairings and exponential computation as they are the most expensive in IBSC scheme. The actual number of computation which cannot be pre-computed (when the recipient identity and the message is not yet known) is shown in bracket.

For fair comparison on ciphertext size, we assume that a message m of length $||m||$ have to cut into k pieces for signcryption, usually with 160-bit for each piece. The 160-bit randomness is reused by multiple 160-bit blocks in the same message. We assume this bandwidth-conserving manoeuvre does not reduce security. We ignore the bandwidth cost of sending the sender identity by assuming it is sent just once, or not sent at all as the recipient is expecting a few senders. $||G_1||$ (resp. $||F_p||$) denotes the size of G_1 (resp. F_p) element, which is about 160 bits for most representative in elliptic curve implementation and signcryption applicatons. In LQ2 [14], δ is 160 bits for ciphertext unlinkability, and 0 bit for ciphertext linkability.

Schemes M, LQ1, NR and CYSC are not IND-A secure, because the unsigncryption requires the knowledge of sender identity in advance. Schemes LQ1 and NR are not IND-B secure because of the following: any adversary who knows the sender identity, private key and the message signcrypted can distinguish the recipient identity. [13] showed M is not IND-CCA2 secure. Schemes LQ1 and NR

Table 1. Comparing bandwidth and complexity efficiencies of IBSC schemes. IND-A (resp. IND-B, IND-C) means sender anonymity (resp. recipient anonymity, message confidentiality). StE (Sign-then-Encrypt) and EtS (Encrypt-then-Sign) use ID-based encryption from [4] and ID-based signature from [6].

Scheme	Security				Ciphertext Size	Signcrypt Time		Unsigncrypt Time	
	IND			EU					
	A	B	C			#pair	#exp	#pair	#exp
EtS	×	√	√	√	$(2k+1)\|\|G_1\|\| + 2\|\|m\|\|$	1	4 (1)	3	1 (1)
StE	√	√	√	×	$(2k+1)\|\|G_1\|\| + 2\|\|m\|\|$	1	4 (1)	3	1 (1)
M [15]	×	√	×	√	$(k+1)\|\|G_1\|\| + \|\|m\|\|$	1	3 (1)	4	1 (1)
LQ1 [13]	×	×	*	√	$k(\|\|G_1\|\| + \|\|F_p\|\|) + \|\|m\|\|$	2	2 (1)	4	1 (1)
NR [11]	×	×	*	×	$(k+1)\|\|G_1\|\| + \|\|m\|\|$	1	3 (2)	3	1 (1)
B [5]	√	√	√	*	$(k+1)\|\|G_1\|\| + \|\|m\|\|$	1	4 (3)	4	2 (2)
CYSC [9]	×	√	√	√	$k(\|\|G_1\|\| + \|\|F_p\|\|) + \|\|m\|\|$	2	2 (1)	4	1 (1)
LQ2 [14]	√	√	√	*	$(k+1)\|\|G_1\|\| + \|\|m + \delta\|\|$	1	4 (3)	4	1 (1)
This scheme	√	√	√	√	$(k+1)\|\|G_1\|\| + \|\|m\|\|$	1	4 (1)	3	1 (1)

are IND-C secure in their own models only, but they are not IND-C secure in Boyen's model and our model. It is because the private key of sender is known to the adversary in our strengthened model.

NR's scheme is not EU-CMA secure. Any adversary can forge a signcryption from any sender to recipient ID_B, where private key of ID_B is known to the adversary. Boyen's scheme has unforgeability for the signature only. It does not satisfy the unforgeability for ciphertext in our security model and also the security model of standard signcryption in [1]. LQ2 scheme is similar to Boyen's in this aspect. Our IBSC scheme avoids this controversial property of unlinkability and achieves unforgeability for ciphertext.

As we can see, our IBSC scheme is the fastest, with shortest ciphertext size and proven secure in the strongest model among the existing schemes[1].

Additional Functionalities of Our Scheme: From our new efficient IBSC scheme, we can achieve further functionalities which are useful in reality. They are the TA compatibility and forward secrecy.

TA Compatibility. In reality, sender and recipient may use different TAs. If it happens, our scheme can still be used with slight changes. Assume all TAs use same pairing e, hash functions and $P \in G_1$. Now let Alice uses TA_1 with master key s_1. and Bob uses TA_2 with master key s_2. In Encrypt, change $V = e(Q_B^r, P_{TA_2})$. In Verify, $e(P, Y) = e(P_{TA_1}{}^h X, Q_A)$. Others remain unchanged.

Forward Secrecy. Our scheme can achieve forward secrecy. It is implied by IND-CCA2. If sender and recipient uses different TAs, then it can even achieve partial

---

[1] After the completion of this research, two new pairing-based IBSC schemes are proposed in [8] and [17]. [8] has the same efficiency after pre-compute and has similar security as our IBSC scheme. [17] proposed a faster scheme with same bandwidth, but there is no security proof for it.

TA forward secrecy. If the master key of $TA1$ is compromised, then past communications with users using different TAs will not be compromised, since the adversary still cannot compute V.

7 Conclusion

In this paper, we have proposed a new BIBSC scheme and its security model. We introduce the generic group and pairing model (GGPM). We proof the BIBSC scheme is secure against plm-uf in ROM+GGPM.

For the IBSC scheme, our scheme is the fastest, with shortest ciphertext and proven secure in a stronger security model when comparing with existing schemes. We provide the flexibility for choosing linkability of ciphertext or not.

References

1. J.H. An, Y. Dodis, and T. Rabin. On the security of joint signature and encryption. In *Proc. CRYPTO 2002*, pages 83–107. Springer-Verlag, 2002. Lecture Notes in Computer Science No. 2332.
2. M. Bellare, C. Namprempre, D. Pointcheval, and M. Semanko. The one-more-RSA-inversion problem and the security of Chaum's blind signature scheme. *J. of Cryptology*, pages 185–215, 2003.
3. A. Boldyreva. Efficient threshold signature, multisignature, and blind signature schemes based on the Gap-Diffie-Hellman-group signature scheme. In *PKC'03*, pages 31–46. Springer-Verlag, 2003. Lecture Notes in Computer Science No. 567.
4. D. Boneh and M. Franklin. Identity-based encryption from the weil paring. In *Proc. CRYPTO 2001*, pages 213–229. Springer-Verlag, 2001. Lecture Notes in Computer Science No. 2139.
5. X. Boyen. Multipurpose identity-based signcryption: A swiss army knife for identity-based cryptography. In *Proc. CRYPTO 2003*, pages 382–398. Springer-Verlag, 2003. Lecture Notes in Computer Science No. 2729.
6. J.C. Cha and J.H. Cheon. An identity-based signature from gap diffie-hellman groups. In *Practice and Theory in Public Key Cryptography – PKC'2003*, pages 18–30. Springer-Verlag, 2003. Lecture Notes in Computer Science No. 2567.
7. D. Chaum. Blind signatures for untraceable payments. In *Proc. CRYPTO 82*, pages 199–203. NY, 1983. Plenum.
8. L. Chen and J. Malone-Lee. Improved identity-based signcryption. Cryptology ePrint Archive, Report 2004/114, 2004. http://eprint.iacr.org/.
9. S. Chow, S.M. Yiu, L. Hui, and K.P. Chow. Efficient forward and provably secure ID-based signcryption scheme with public verifiability and public ciphertext authenticity. In *ICISC 2003*, pages 352–369. Springer-Verlag, 2003. Lecture Notes in Computer Science No. 2971.
10. C. Cocks. Non-interactive public-key cryptography. In *Cryptography and Coding*, pages 360–363. Springer-Verlag, 2001. Lecture Notes in Computer Science No. 2260.
11. K.C. Reddy D. Nalla. Signcryption scheme for identity-based cryptosystems. Cryptology ePrint Archive, Report 2003/066, 2003. http://eprint.iacr.org/.

12. Y. Desmedt and J. Quisquater. Public-key systems based on the difficulty of tampering. In *Proc. CRYPTO 86*, pages 111–117. Springer-Verlag, 1986. Lecture Notes in Computer Science No. 263.
13. B. Libert and J.-J. Quisquater. New identity based signcryption schemes from pairings. IEEE Information Theory Workshop, Paris (France), 2003.
14. B. Libert and J.-J. Quisquater. The exact security of an identity based signature and its applications. Cryptology ePrint Archive, Report 2004/102, 2004. http://eprint.iacr.org/.
15. J. Malone-Lee. Identity-based signcryption. Cryptology ePrint Archive, Report 2002/098, 2002. http://eprint.iacr.org/.
16. U. Maurer and Y. Yacobi. Non-interactive public-key cryptography. In *Proc. CRYPTO 91*, pages 498–507. Springer-Verlag, 1991. Lecture Notes in Computer Science No. 547.
17. N. McCullagh and P. S. L. M. Barreto. Efficient and forward-secure identity-based signcryption. Cryptology ePrint Archive, Report 2004/117, 2004. http://eprint.iacr.org/.
18. V.I. Nechaev. Complexity of a determinate algorithm for the discrete logarithm. *Mathematical Notes 55*, pages 165–172, 1994.
19. C. P. Schnorr. Practical security in public-key cryptography. In *Proc. ICISC*. Springer, 2001. Lecture Notes in Computer Science.
20. C. P. Schnorr. Security of blind discrete log signatures against interactive attacks. In *Proc. ICISC*, pages 1–12. Springer-Verlag, 2001. Lecture Notes in Computer Science No. 2229.
21. A. Shamir. Identity-based cryptosystems and signature schemes. In *Proc. CRYPTO 84*, pages 47–53. Springer-Verlag, 1984. Lecture Notes in Computer Science No. 196.
22. V. Shoup. Lower bounds for discrete logarithms and related problems. In *Proc. EUROCRYPT 97*, pages 256–266. Springer-Verlag, 1997. Lecture Notes in Computer Science No. 1233.
23. S. Tsuji and T. Itoh. An ID-based cryptosystem based on the discrete logarithm problem. *IEEE Journal on Selected Areas in Communication*, 7(4):467–473, 1989.
24. F. Zhang and K. Kim. ID-Based blind signature and ring signature from pairings. In *Proc. ASIACRYPT 2002*, pages 533–547. Springer-Verlag, 2002. Lecture Notes in Computer Science No. 2501.
25. F. Zhang and K. Kim. Efficient ID-based blind signature and proxy signature from bilinear pairings. In *Proc. ACISP'03*, pages 312–323. Springer-Verlag, 2003. Lecture Notes in Computer Science No. 2727.
26. F. Zhang, R. Safavi-Naini, and W. Susilo. Efficient verifiably encrypted signature and partially blind signature from bilinear pairings. In *Proc. INDOCRYPT03*, pages 191–204. Springer-Verlag, 2003. Lecture Notes in Computer Science No. 2904.
27. Y. Zheng. Digital signcryption or how to achieve cost(signature & encryption) $\ll$ cost(signature) + cost (encryption). In *Proc. CRYPTO 97*, pages 165–179. Springer-Verlag, 1997. Lecture Notes in Computer Science No. 1294.

A Powers and Limitations of GGM and GGPM

Lemma 6 implies that co-CDH is hard. The perspective is that co-CDH constitutes collisions in GGPM. The real-world interpretation of this model-based result is roughly as follows: GGM (resp. GGPM) *bans* certain operations, in the sense that it can be assumed w.l.o.g. that the generic algorithm does not use these operations. The justification is that these operations are thought to be of no help. In GGM for discrete logarithm with parameters p, q, g, the additions (resp. subtractions) in Z_p are banned. In GGM for ECDL with parameters p, q, base point G whose order is q, arithmetics in Z_p are banned. In GGPM where we have in mind the G_1, G_2, and G_3 are all groups of elliptic curve points, the GGPM model allows point operations, arithmetics in Z_q, but bans arithmetics in Z_p on the argument that they do not help.

Based on such model assumptions, GGM has been used to prove results that often cannot be proved in other models. The GGM has been used to prove the hardness of the discrete logarithm [18, 22]. It has also been used to reduce plm-uf of Schnorr or Okamoto-Schnorr blind signature to the ROS Problem [20], or the one-more discrete logarithm problem. Note that the one-more discrete logarithm problem is proven hard in the GGM by simple applications of the methods used in [19]. Based on similar model assumptions, we use GGPM to reduce plm-uf of blind signcryption to the ROS Problem or the one-more co-CDH Problem in this paper. Note that one-more co-CDH is proven hard in GGPM.

Algorithms already exist that exploit operations *banned* from GGM. The index calculus method to compute the discrete logarithm utilizes size information in Z_p to achieve efficiency. It is outside the boundary of GGM. In ECDL, it is suspected but not yet explicitly demonstrated that arithmetics in Z_p and properties of the curve can be exploited. Therefore, GGM and GGPM are used with these elliptic curves applications in mind. If and when exploitations of Z_p arithmetics or curve properties, or other unforeseen techniques outside the model, can be exhibited, both GGM and GGPM will need to be reexamined.

Lemma 6 also implies the hardness of the one-more co-CDH Problem in the GGPM. The *one-more co-CDH Problem* is (roughly speaking): Given q_B queries to the co-CDH Oracle, compute $q_B + 1$ co-CDH Problems.

B Proofs

B.1 Proof Sketch of Theorem 2

Setting Up: Dealer D gives $(P, P^\alpha, P^\beta, Q)$ to Simulator S and wants S to compute $e(P,Q)^{\alpha\beta}$. S sends the system parameter to Forger F with $P_{TA} = P^\beta$ as in Setup. S randomly picks η_Q from $\{1, 2, ..., \mu_0\}$, where μ_0 is the number of queries to H_0.

Simulating Oracles: As regards queries to the oracles:

- Query on H_0 for identity ID is handled as follows:
  - The η_Q-th distinct query to H_0 is back patched to the value Q. The corresponding identity is denoted as ID_Q. Adds the entry $\langle ID_Q, Q \rangle$ to tape L_0, and returns the public key Q.
  - Otherwise, picks a random $\lambda \in F_p^*$, adds the entry $\langle ID, \lambda \rangle$ to the tape L_0, and return the public key $Q_{ID} = P^\lambda$.
- Queries on H_1, H_2 and H_3 are handled by producing a random element from the codomain, and adding both query and answer to tape L_1, L_2 and L_3.

- $\mathcal{KEO}$: For input identity ID_A.
  - If $ID_A = ID_Q$, then D terminates its interaction with F, having failed to guess the targeted recipient among those in L_0.
  - Otherwise, S retrieves $\langle ID_A, \lambda_A \rangle$ from L_0 and returns $S_A = (P^\beta)^{\lambda_A}$.
- $\mathcal{SO}$: For input message m, sender ID_A, and recipient ID_B.
  - If $ID_A = ID_Q$, then S randomly chooses $r, h \in F_p^*$, and lets $X = P^r(P^\beta)^{-h}$, $W = (Q)^r$. Then, S adds the tuple $\langle m, X, h \oplus ID_B \rangle$ to L_1 to force the random oracle $H_1(m, X) = h \oplus ID_B$. Finally, S uses $\langle X, W, m, r, ID_B \rangle$ to run Signcrypt to produce the desired ciphertext σ.
  - Otherwise, S retrieves $\langle ID_A, \lambda_A \rangle$ from L_0 and computes $S_A = (P^\beta)^{\lambda_A}$. Then S will run Signcrypt using S_A and get ciphertext σ.
- $\mathcal{UO}$: For input recipient ID_B and ciphertext $\sigma = \langle X, Y, Z \rangle$.
  - If $ID_B = ID_Q$, then S searches all combinations $\langle ID_A, m, X, W \rangle$ such that $\langle m, X, h_1 \rangle \in L_1$, $\langle V, h_2 \rangle \in L_2$, $\langle V, ID_A, h_3 \rangle \in L_3$, for some h_1, h_2, h_3, V, under the constraints that $h_3 \oplus Y = W$, $h_2 \oplus Z = \langle ID_A, m \rangle$ and $\mathsf{Verify}[ID_A, m, X, W, ID_B] = \top$. Pick a $\langle ID_A, m \rangle$ in one of the combinations above to return as answer. If no such tuple is found, the oracle signals that the ciphertext is invalid.
  - Otherwise, S retrieves $\langle ID_B, \lambda_B \rangle$ from L_0 and computes $S_B = (P^\beta)^{\lambda_B}$. Then S will run Unsigncrypt using S_B to get $\langle ID_A, m \rangle$ or $\bot$.

Witness Extraction: As in the IND-IBSC-CCA2 game, at some point F chooses plaintext m_1, sender ID_{A1}, and recipient ID_{B1} on which he wishes to be challenged. S responds with challenge ciphertext $\langle X, Y, Z \rangle$, where $X = P^\alpha$. Y and Z are random strings of appropriate size. All further queries by F are processed adaptively as in the oracles above.

Finally, F returns its final guess. S ignores the answer from F, randomly picks an entry $\langle V, h_2 \rangle$ in L_2, and returns V as the solution to the co-BDH problem.

If the recipient identity $ID_{A1} = ID_Q$ selected by S, to recognize the challenge ciphertext $\langle X, Y, Z \rangle$ with $X = P^\alpha$ is incorrect, F needs to query random oracle $H_2(V)$ with $V = e(X, S_Q) = e(P^\alpha, Q^\beta) = e(P, Q)^{\alpha\beta}$. It will leave an entry $\langle V, h_2 \rangle$ on L_2, from which B can extract $V = e(P, Q)^{\alpha\beta}$. □

B.2 Proof Sketch of Theorem 3

Setting up: Dealer D gives (P, P^β, Q) to Simulator S and wants S to compute Q^β. Others are same as in the proof sketch of Theorem 2.

Oracle Simulation: The signcryption oracle, unsigncryption oracle, and key extraction oracle are simulated in the same way as in the proof of Theorem 2.

Witness Extraction: Assume $\mathcal{F}$ is a PPT forger. Rewind $\mathcal{F}$ to the random oracle query whose output appears in verification of Unsigncrypt. Then we obtain $W = S_A^h Q_A^r$ and $W' = S_A^{h'} Q_A^r$ in respective forks. Combining, we can compute the co-CDH Problem if $Q_A = Q$. Then $Q^\beta = S_A = (W'/W)^{(h'-h)^{-1}}$. □

B.3 Proof Sketch of Theorem 4

To prove the blindness of BIBSC scheme, we show that given a valid ciphertext $\langle \hat{X}, \hat{Y}, \hat{Z} \rangle$ and any transcript of blind signcryption (X, h, W, V), there always exists a unique pair of blinding factors $\alpha, \beta \in Z_q^*$. Since the blinding factors are randomly chosen, the blindness of BIBSC scheme is achieved.

Given a valid ciphertext $\langle \hat{X}, \hat{Y}, \hat{Z} \rangle$, then there exists a unique $(\hat{X}, \hat{W}, \hat{V}, m)$ for this ciphertext. Then for any transcript of blind signcryption (X, h, W, V), the following equations must hold for $\alpha, \beta \in Z_q^*$:

$$\hat{X} = X^\alpha P^\beta$$
$$h = \alpha^{-1} H_1(m, \hat{X})$$
$$\hat{W} = W^\alpha Q_A^\beta$$
$$\hat{V} = V^\alpha e(P_{TA}{}^\beta, Q_B)$$

From the second equation, we see that there exists a blinding factor $\alpha = H_1(m, \hat{X})/h$. For this α, there exists a blinding factor β from the first equation and $\beta = log_P(\hat{X} X^{-\alpha})$. Therefore we have to show that these blinding factors α, β satisfy the last two equations.

Notice that there exists a S_B which is the private key for Q_B. Then:

$$\begin{aligned}\hat{V} &= e(\hat{X}, S_B) \\ &= e(X^\alpha P^\beta, S_B) \\ &= e(X, S_B)^\alpha e(P^\beta, S_B) \\ &= V^\alpha e(P_{TA}{}^\beta, Q_B)\end{aligned}$$

Furthermore, $\langle \hat{X}, \hat{W}, m \rangle$ is a valid signature. Therefore we have:

$$\begin{aligned}e(P, \hat{W}) &= e(\hat{X}, Q_A) e(P_{TA}, Q_A)^{H_1(m, \hat{X})} \\ &= e(X^\alpha P^\beta, Q_A) e(P_{TA}, Q_A)^{\alpha h} \\ &= e(X P_{TA}{}^h, Q_A)^\alpha e(P^\beta, Q_A) \\ &= e(P, W)^\alpha e(P, Q_A^\beta) \\ &= e(P, W^\alpha Q_A^\beta)\end{aligned}$$

Hence, given a valid ciphertext $\langle \hat{X}, \hat{Y}, \hat{Z} \rangle$ and any transcript of blind signcryption (X, h, W, V), there always exists a unique pair of blinding factors $\alpha, \beta \in Z_q^*$. Therefore, $Prob\{\sigma\ by\ Warden\} = Prob\{\sigma\ by\ Warden | T\}$. The blindness of BIBSC scheme is proved. □

B.4 Proof Sketch of Theorem 5

This section refers to a generic adversary $\mathcal{A}$ performing some t generic steps, including some q_B interactions $(X_1, h_1, W_1, V_1), \cdots, (X_{q_B}, h_{q_B}, W_{q_B}, V_{q_B})$ with $\mathcal{BSO}$, producing some $t'^{(u)}$ group elements in G_u. We let $r = (r_1, \cdots, r_{q_B})$ denote $\mathcal{BSO}$ random coins. Let $f_1 = P, f_2 = P_{TA}, f_3, \cdots, f_{t'^{(1)}} \in G_1$ denote the group elements of $\mathcal{A}$'s computation. The generic $\mathcal{A}$ computes $f_j = P^{a_{j,-1}} P_{TA}^{a_{j,0}} \prod_{\ell=1}^{q_B} X_\ell^{a_{j,\ell}}$ where X_ℓ are $\mathcal{BSO}$ commitments and the exponents depend arbitrarily on previously computed non-group data.

Schnorr's Lemma 2 implies DLP is hard (uncomputable by PPT generic adversary) in GGM. Similarly, it applies here. It is hard to get s from $Q_B{}^s$.

Let $\mathcal{A}$'s outputs $(\hat{X}_i, \hat{W}_i, \hat{V}_i)$ be valid for message $\hat{m}_i$, sender ID_A and recipient ID_{B_i}, $1 \leq i \leq q_B + 1$. Then we have $\hat{h}_i = H_1(\hat{X}_i, \hat{m}_i, ID_{B_i})$ for some hash query satisfying $e(\hat{X}_i P_{TA}^{\hat{h}_i}, Q_A) = e(P, \hat{W}_i)$. Let $\hat{X}_i = f_{\sigma_i}^{(1)}$.

The equation $e(P, \hat{W}_i)e(P_{TA}^{-\hat{h}_i}, Q_A) = e(f_{\sigma_i}, Q_A) = e(P^{a_{\sigma_i,-1}} P_{TA}^{a_{\sigma_i,0}} \prod_{\ell=1}^{q_B} X_\ell^{a_{\sigma_i,\ell}}, Q_A)$ and $e(X_\ell, Q_A) = e(P, W_\ell)e(P_{TA}^{-h_\ell}, Q_A)$ imply:

$$\hat{W}_i = Q_A{}^{a_{\sigma_i,-1}} \cdot \prod_{\ell=1}^{q_B} W_\ell^{a_{\sigma_i,\ell}} \cdot Q_A{}^{(a_{\sigma_i,0} - \sum_{\ell=1}^{q_B} a_{\sigma_i,\ell} h_\ell + \hat{h}_i)s}$$

If $\hat{h}_i = -a_{\sigma_i,0} + \sum_{\ell=1}^{l} a_{\sigma_i,\ell} h_\ell$, then $\mathcal{A}$ can easily compute the correct $\hat{W}_i$. Then we have $\hat{W}_i = Q_A{}^{a_{\sigma_i,-1}} \prod_{\ell=1}^{l} W_\ell^{a_{\sigma_i,\ell}}$ where $W_1, \cdots, W_l, a_{\sigma_i,-1}, \cdots, a_{\sigma_i,l}$ are known to $\mathcal{A}$.

Conversely, $\mathcal{A}$ must select $h_1, \cdots, h_l$ as to zero the coefficient involving the master secret key s. Otherwise we can recover $Q_A{}^s$ from $W_1, \cdots, W_l, a_{\sigma_i,-1}, \cdots, a_{\sigma_i,l}, \hat{h}_i, \hat{W}_i$ which are known to $\mathcal{A}$. Then it can solve the 1m-co-CDH problem, as we get q_K private keys from $\mathcal{KEO}$. The probability of solving 1m-co-CDH in GGPM is negligible. Hence $\mathcal{A}$ must solve the ROS problem. □

A Systematic Evaluation of Compact Hardware Implementations for the Rijndael S-Box

Nele Mentens*, Lejla Batina*, Bart Preneel, and Ingrid Verbauwhede

K.U. Leuven ESAT/COSIC, Kasteelpark Arenberg 10,
B-3001 Leuven-Heverlee, Belgium
{Nele.Mentens,Lejla.Batina,Bart.Preneel,Ingrid.Verbauwhede}
@esat.kuleuven.ac.be

Abstract. This work proposes a compact implementation of the AES S-box using composite field arithmetic in $GF(((2^2)^2)^2)$. It describes a systematic exploration of different choices for the irreducible polynomials that generate the extension fields. It also examines all possible transformation matrices that map one field representation to another. We show that the area of Satoh's S-box, which is the most compact to our knowledge, is at least 5% away from an optimal solution. We implemented this optimal solution and Satoh's design using a 0.18 μm standard cell library.

Keywords: AES, S-box, inversion in $GF(2^n)$, composite fields, smart card implementation

1 Introduction

After an open competition ending in 2000, the National Institute for Standard and Technology (NIST) has selected the Rijndael block cipher as the new Advanced Encryption Standard (AES) [1]. The AES algorithm, designed by Joan Daemen and Vincent Rijmen, has an SPN (Substitution Permutation Network) structure. Its use is mandatory for the encryption of sensitive but unclassified US government information; in 2003 the US government has announced that it can also be used for encrypting secret and top secret information (for the last category key lengths of at least 192 bits need to be used). AES is currently replacing the Data Encryption Standard as the worldwide standard algorithm.

Since 2000, extensive research has been performed on AES implementations. In this article we are focusing on compact hardware implementations for mobile devices and smart cards, but our results can also be applied in high-speed pipelined implementations for network security and e-commerce applications. Note that the best known software implementations achieve about 15 cycles/byte on a modern PC.

---

* Lejla Batina and Nele Mentens are funded by research grants of the Katholieke Universiteit Leuven, Belgium. This work was supported by FWO project (G.0450.04).

Design challenges for AES mainly lie in exploring all the options for the S-box design. The most common strategy to reduce the gate complexity consists of exploiting composite field arithmetic. By following that approach one still has several options to represent the finite field $GF(2^8)$. In this paper we represent $GF(2^8)$ as the composite field $GF(((2^2)^2)^2)$. In this way, we reduce the arithmetic in $GF(2^8)$ to operations in smaller fields. There are many ways to represent $GF(2^8)$ as $GF(((2^2)^2)^2)$. Choices need to be made with respect to the irreducible polynomials that are used to create the extension fields and with respect to the transformation matrices that map elements from one representation to the other. Exploring these two degrees of freedom we optimize the S-box of Satoh et al. [17], which is to our knowledge the most compact implementation today. Another area efficient implementation is the one of Wolkerstorfer et al. [20]. More precisely, according to Daemen and Rijmen [6], the number of kgates for the implementations of [17] and [20] are 5.4 and 5.7 respectively.

The remainder of this paper is organized as follows. In Sect. 2 some details on the AES algorithm are discussed. Section 3 lists previous work on hardware implementations of Rijndael. In Sect. 4 we explain our approach to minimize the area of the S-box and compare our new solution with the S-box of Satoh. Section 5 concludes the paper and outlines future work.

2 The AES Algorithm

Rijndael has a variable block and key length which can be 128, 192 or 256 bits; the AES standard includes only block lengths of 128 bits. In this implementation we focus on the 128-bit key version of AES which has 10 rounds. In this case, each round and the initial stage require a 128-bit round key. In total 10 sets of round keys are generated from the secret key by using the S-box. The input data is arranged as a table i.e., a matrix of bytes. Figure 1 outlines the basic structure of the algorithm. The round transformation consists of four different transformations: ByteSub, ShiftRow, MixColumn and AddRoundKey. They are performed in this order with the exception of the final round which is slightly different. All transformations are based on byte-oriented arithmetic and AddRoundKey is a bitwise XOR operation. The transformations operate on the intermediate result, which is called the State.

The ByteSub transformation is a non-linear byte substitution also called S-box (substitution table). It operates on bytes independently. The S-box is invertible and consists of the following two transformations:

1. Inversion in the $GF(2^8)$ field, modulo the irreducible polynomial $m(x) = x^8 + x^4 + x^3 + x + 1$.
2. Affine transformation defined with: $Y = AX^{-1} + b$, where A is a 8×8 fixed matrix and b is a 8×1 vector-matrix.

The schematic of the complete AES algorithm is shown in Fig. 1. Further details on the AES algorithm can be found in [4, 5].

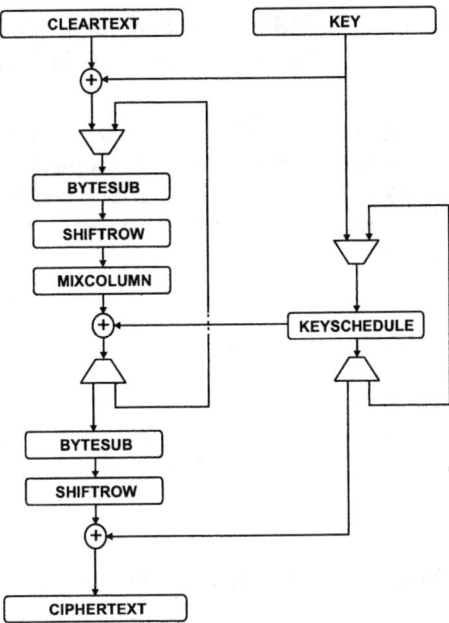

Fig. 1. Schematic of the AES encryption algorithm.

3 Previous Work

Many hardware architectures for Rijndael were proposed as either ASIC [11, 12, 19] or FPGA implementations [2, 3, 7, 8, 10, 14, 18]. Most of the known implementations, particularly the early ones, were quite simple and not small enough as they did not exploit composite field arithmetic. Among those who tried to produce a really small circuit we mention the work of Satoh *et al.* [17] and Wolkerstorfer *et al.* [21]. In [16] the use of the composite field $GF((2^4)^2)$ was also proposed but no hardware implementation was presented.

Satoh *et al.* introduced a new composite field $GF(((2^2)^2)^2)$ which resulted in an optimized S-box. More precisely, their S-box requires less than 1/4 of the size of one using a look-up table. This resulted in a compact AES implementation with a gate complexity of 5.4 kgates. To our knowledge this is the most compact architecture so far. Wolkerstorfer *et al.* used arithmetic in $GF((2^4)^2)$ to achieve an implementation with a gate count comparable to Satoh *et al.* (5.7 kgates). In the solution of Wolkerstorfer *et al.* inversion of two-term polynomials $a_h x + a_l \in GF((2^4)^2)$ involves only operations in $GF(2^4)$, which are easily computed using combinational logic. Macchetti and Bertoni [13] have described an ASIC implementation for the same composite field $GF((2^4)^2)$, but with a representation as given in [16]. The work of Chodowiec and Gaj [3] also offers a compact design that is targeting low-cost embedded applications. They used dedicated Block RAMs for the implementation of the S-boxes. Recently, the work of Wu et al. [22] gives an area and delay reduction of 1/6 and 1/4 respectively compared to [21]. The proposed approach uses dual AES in combination with a composite field.

Here we use the composite field $GF(((2^2)^2)^2)$, which was only explored by Satoh et al. [17]. By a systematic exploration of all options we show that Satoh's S-box is at least 5% away from an optimal solution. The implementation of this optimal solution and the approach we use to explore all design possibilities is explained in the next section.

4 Hardware Implementation

In this section we examine the S-box of Satoh et al. [17] and we try to optimize it for area. Section 4.1 describes the approach we used to optimize Satoh's S-box. Section 4.2 presents our new S-box. Implementation results and comparison with Satoh's S-box are given in Sect. 4.3.

4.1 Theoretical Approach to Optimize the S-Box

Let us view $GF(2^{2m})$ as a field extension of degree 2 over $GF(2^m)$. The field $GF(2^{2m})$ is generated as an extension field of $GF(2^m)$ using an irreducible polynomial say $f(x) = x^2 + \alpha x + \beta$, where $\alpha, \beta \in GF(2^m)$. Then we have $GF(2^{2m}) = GF(2^m)[\omega]$, where ω is a root of $f(x)$ and $GF(2^{2m})$ can be viewed as a two-dimensional vector space over $GF(2^m)$. Hence, an arbitrary element $\Delta \in GF(2^{2m})$ can be written as $\Delta = \delta_1 \omega + \delta_0$, where $\delta_1, \delta_0 \in GF(2^m)$. We want to calculate the inverse of Δ i.e. Δ^{-1} such that $\Delta \cdot \Delta^{-1} \equiv 1 \bmod f(x)$.

The multiplicative inverse of $\Delta \in GF(2^{2m})$ can therefore be computed as:

$$\Delta^{-1} = (\delta_1 \omega + \delta_0)^{-1} = \delta_1 (\delta_1^2 \beta + \delta_1 \delta_0 \alpha + \delta_0^2)^{-1} \omega + (\delta_0 + \delta_1 \alpha)(\delta_1^2 \beta + \delta_1 \delta_0 \alpha + \delta_0^2)^{-1} \tag{1}$$

This equation consists of operations which can be performed in the subfield $GF(2^m)$ [9].

Equation (1) can be used recursively to find the inverse in $GF(((2^2)^2)^2)$. $GF(((2^2)^2)^2)$ is a field extension of degree 2 over $GF((2^2)^2)$ constructed using the irreducible polynomial $P(x) = x^2 + p_1 x + p_0$, where $p_1, p_0 \in GF((2^2)^2)$. Let us call a root of P also x. $GF((2^2)^2)$ is a field extension of degree 2 over $GF(2^2)$ using the irreducible polynomial $Q(y) = y^2 + q_1 y + q_0$, with y a root of the polynomial and $q_1, q_0 \in GF(2^2)$. $GF(2^2)$ is a field extension of degree 2 over $GF(2)$ using the irreducible polynomial $R(z) = z^2 + z + 1$, with root z.

In Satoh et al. the following choices are made for the coefficients of the irreducible polynomials: $p_1 = 1 = \{0001\}_2$, $p_0 = \lambda = (z+1)y = \{1100\}_2$, $q_1 = 1 = \{01\}_2$ and $q_0 = \phi = z = \{10\}_2$. Equation (1) is implemented as

$$\Delta_2 = \delta_{21}x + \delta_{20} \in GF(((2^2)^2)^2):$$
$$\Delta_2^{-1} = (\delta_{21}x + (\delta_{21} + \delta_{20})) \cdot (\lambda\delta_{21}^2 + (\delta_{21} + \delta_{20})\delta_{20})^{-1}, \qquad (2)$$
$$\Delta_1 = \delta_{11}y + \delta_{10} \in GF((2^2)^2):$$
$$\Delta_1^{-1} = (\delta_{11}y + (\delta_{11} + \delta_{10})) \cdot (\phi\delta_{11}^2 + (\delta_{11} + \delta_{10})\delta_{10})^{-1}.$$

Inversion in $GF(2^2)$ requires only one addition:

$$\Delta_0 = \delta_{01}z + \delta_{00} \in GF(2^2) : \Delta_0^{-1} = \delta_{01}z + (\delta_{01} + \delta_{00}). \qquad (3)$$

The inversion in $GF(2^8)$ is finally decomposed into operations in $GF(2^2)$. Therefore a transformation is needed to transform a representation in $GF(2^8)$ to a representation in $GF(((2^2)^2)^2)$. In [15], Paar explains how a matrix can be created to perform this transformation. Different choices for the irreducible polynomials $P(x)$ and $Q(y)$ lead to different transformation matrices. For every combination of $P(x)$ and $Q(y)$ there are 8 possibilities for the transformation matrix. For hardware implementations, the most area efficient transformation matrix is the one that has the least '1' entries, because this number determines the XOR gate count for the transformation. After performing the inversion using the $GF(((2^2)^2)^2)$ representation we need to go back to the $GF(2^8)$ representation using the inverse of the transformation matrix. This matrix can be combined with the affine transformation matrix at the end of the S-box.

We stick to the choice of Satoh et al. to make $p_1 = q_1 = 1$ and $q_0 = \phi = z$. Based on (2) and the fact that the transformation matrix depends on $P(x)$ and $Q(y)$, we conclude that the hardware complexity of the circuit depends on the choice of $p_0 = \lambda$. That is why we explored all values of λ to determine the most compact solution for the S-box. There are 8 choices for λ. The two elements that determine the hardware complexity of the circuit are:

- the number of gates in the constant multiplication with λ in $GF((2^2)^2)$,
- the number of '1' entries in the transformation matrix and in the combination of the inverse transformation matrix with the affine transformation matrix.

For every λ, Table 1 gives the number of 2-input XORs for the constant multiplication and the total number of '1' entries for every option of the transformation matrix. Out of 8 possible transformation matrices for every λ, the one that gives the least total number of '1' entries is given in the last column.

The values in the table are depicted in Fig. 2.

From Table 1 and Fig. 2 we conclude that the solution of Satoh et al., which has $\lambda = (z+1)y$ uses the most area efficient constant multiplication. The transformation matrix they chose gives a total number of '1' entries equal to 61. Their implementation can be made more efficient by choosing the most compact transformation matrix which leads to a total number of '1' entries equal to 59. But the most optimal solution ("best case") would be to change the implementation even more by taking $\lambda = zy$. The constant multiplication requires only 1 XOR more than Satoh's constant multiplication, but the total number of '1' entries in the matrices is reduced by 5. On the other hand, the design with a maximized

Table 1. Comparison of the hardware complexity when using different polynomials $x^2 + x + \lambda$ for the generation of $GF(((2^2)^2)^2)$.

λ	$\{\lambda\}_2$	# XORs in $\star\lambda$	# '1' in matrices	min. # '1'
$(z+1)y + z$	$\{1110\}_2$	4	57, 58, 59, 60, 62, 63, 63, 67	57
zy	$\{1000\}_2$	4	54, 57, 59, 59, 61, 62, 63, 66	54
$zy + (z+1)$	$\{1011\}_2$	5	59, 63, 63, 64, 65, 65, 67, 71	59
$zy + z$	$\{1010\}_2$	4	56, 59, 59, 61, 61, 66, 70, 71	56
$(z+1)y$	$\{1100\}_2$	3	59, 61, 62, 63, 63, 64, 66, 69	59
$(z+1)y + 1$	$\{1101\}_2$	4	60, 61, 62, 62, 63, 64, 65, 68	60
$zy + 1$	$\{1001\}_2$	5	55, 59, 59, 61, 62, 63, 65, 67	55
$(z+1)y + (z+1)$	$\{1111\}_2$	3	59, 59, 61, 62, 64, 64, 66, 72	59

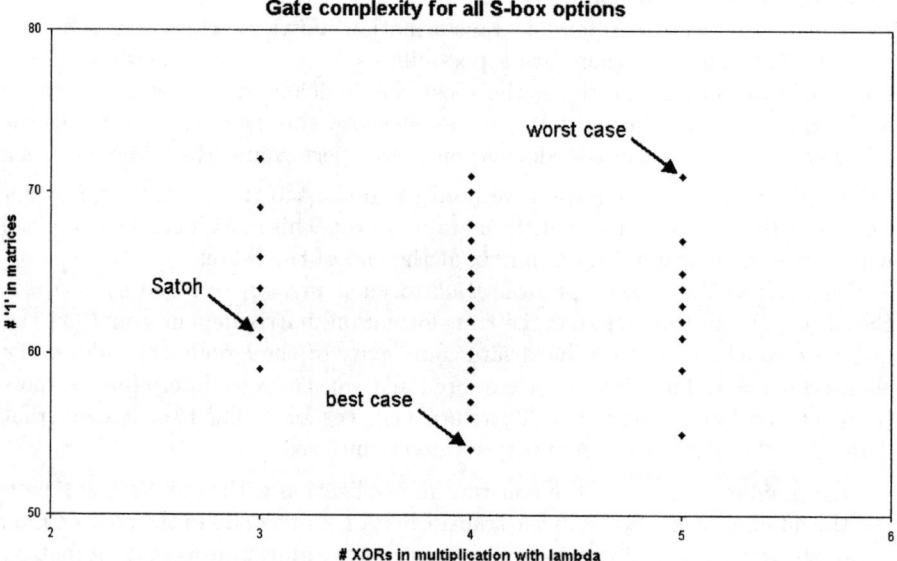

Fig. 2. Graphical representation of the values in Table 1. The arrows point at Satoh's S-box, the S-box with minimized gate count ("best case") and the S-box with maximized gate count ("worst case").

gate count ("worst case") uses 5 XOR gates for the constant multiplication and 71 '1' entries in the matrices.

The implementation of the new optimized S-box is explained in the next section. Implementation results for Satoh's S-box, the "best case" and the "worst case" are given in Sect. 4.3.

4.2 Implementation of the New Optimized S-Box

Figure 3 shows the structure of the S-box implementation.

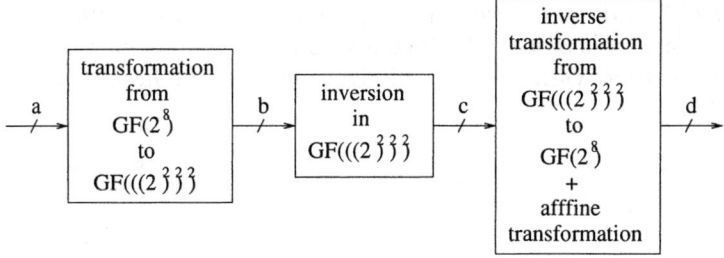

Fig. 3. Structure of the S-box implementation.

The transformation used here is

$$\begin{bmatrix} b_0 \\ b_1 \\ b_2 \\ b_3 \\ b_4 \\ b_5 \\ b_6 \\ b_7 \end{bmatrix} = \begin{bmatrix} 1\,0\,1\,0\,0\,0\,0\,0 \\ 0\,1\,0\,0\,0\,0\,1\,1 \\ 0\,0\,1\,0\,0\,1\,0\,0 \\ 0\,1\,0\,1\,0\,0\,1\,1 \\ 0\,1\,0\,0\,0\,1\,0\,1 \\ 0\,1\,0\,0\,1\,1\,1\,0 \\ 0\,1\,1\,1\,1\,1\,1\,0 \\ 0\,0\,0\,0\,0\,1\,0\,1 \end{bmatrix} \begin{bmatrix} a_0 \\ a_1 \\ a_2 \\ a_3 \\ a_4 \\ a_5 \\ a_6 \\ a_7 \end{bmatrix},$$

where $a_i, b_i \in GF(2)$ are the coefficients of $a \in GF(2^8)$, $b \in GF(((2^2)^2)^2)$ respectively.

This results in the following combination of the inverse transformation with the affine transformation

$$\begin{bmatrix} d_0 \\ d_1 \\ d_2 \\ d_3 \\ d_4 \\ d_5 \\ d_6 \\ d_7 \end{bmatrix} = \begin{bmatrix} 1\,0\,1\,0\,1\,1\,0\,0 \\ 1\,1\,1\,0\,0\,0\,0\,0 \\ 1\,1\,0\,0\,0\,0\,0\,0 \\ 1\,0\,1\,0\,1\,1\,1\,0 \\ 1\,0\,0\,1\,1\,1\,0\,0 \\ 0\,0\,1\,1\,1\,1\,0\,0 \\ 0\,0\,0\,0\,1\,0\,1\,1 \\ 0\,0\,1\,0\,1\,0\,1\,0 \end{bmatrix} \begin{bmatrix} c_0 \\ c_1 \\ c_2 \\ c_3 \\ c_4 \\ c_5 \\ c_6 \\ c_7 \end{bmatrix} + \begin{bmatrix} 1 \\ 1 \\ 0 \\ 0 \\ 0 \\ 1 \\ 1 \\ 0 \end{bmatrix},$$

where $c_i, d_i \in GF(2)$ are the coefficients of $c \in GF(((2^2)^2)^2)$, $d \in GF(2^8)$ respectively.

The total number of '1' entries in both 8×8 matrices is equal to 54. The addition of the column vector in the affine transformation is fixed and hence does not have to be considered for optimization. The number of '1' entries in the matrices in Satoh's implementation is equal to 61. Implementing the matrices in a straightforward way, the number of XORs would be equal to the number of '1' entries minus the number of rows in the matrices. This would lead to an XOR gate count of 38 and 45 for our and Satoh's S-box respectively, which results in a reduction of 7 XOR gates. By finding common terms in the XOR equations and exploring some rools of logic it is possible to reduce the number of XOR gates. We leave this to a synthesis tool and give results on the final implementations in Sect. 4.3.

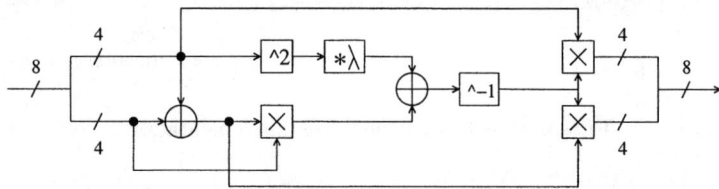

Fig. 4. Structure of the S-box implementation.

The inversion in $GF(((2^2)^2)^2)$ has many levels of hierarchy. At the highest level the architecture looks the same as Satoh's architecture (see Fig. 4). At the next level of hierarchy, the only difference with Satoh's design is the implementation of the constant multiplication with λ. Figure 5 gives the gate-level implementation of both Satoh's (top) and our (bottom) constant multiplication. As can be seen, our constant multiplication requires one extra XOR gate compared to Satoh's implementation.

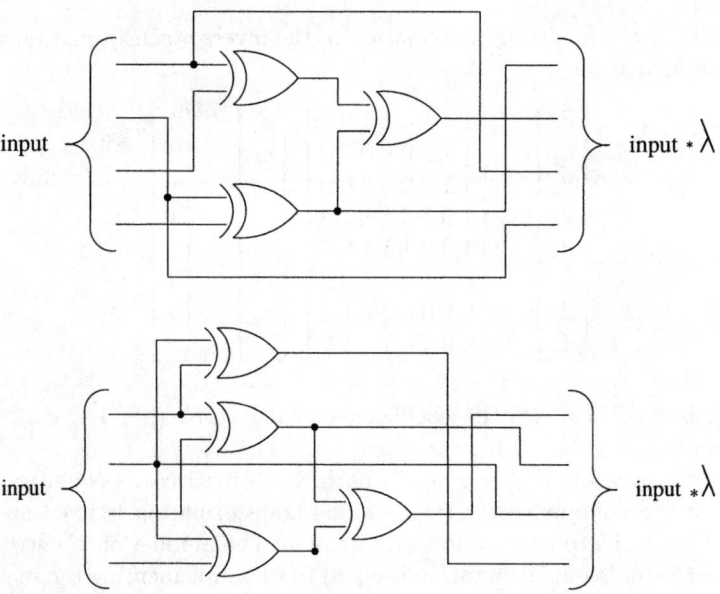

Fig. 5. Gate-level implementation of Satoh's (top) and our (bottom) constant multiplication with λ.

We can summarize this section by stating that, in a straightforward implementation, our S-box would require 6 XOR gates less than Satoh's S-box (7 less for the implementation of the matrices and 1 extra for the multiplication with λ). The implementation results of both approaches after synthesis are compared in the next section. To show the upper bound of the gate complexity, we also implemented the "worst case" S-box and included it in the comparison.

4.3 Implementation Results and Comparison

We implemented both our new optimized S-box and Satoh's S-box using a 0.18 μm CMOS standard cell library. To show that the area is sensitive to the choice of the polynomials and the transformation matrix we also implemented the "worst case" S-box (with maximized gate count). All three implementations are synthesized with a rather slow target delay of 10 ns. Table 2 gives the number of gates (in equivalent number of 2-input NAND gates) for all designs.

Table 2. Area comparison of our S-box with minimized/maximized gate count and Satoh's S-box (in equivalent number of 2-input NAND gates).

	min. area	Satoh	max. area
number of gates	272	286	297

Satoh *et al.* implemented their S-box using a 0.11 μm CMOS standard cell library, which resulted in 294 gates with a delay of 3.69 ns. This corresponds to 286 gates in 0.18 μm CMOS with a maximum delay of 10 ns. The table shows our new S-box has a 5% area reduction compared to Satoh's S-box. This is equivalent to the expected reduction of 6 XOR gates. The table also shows that a bad choice for the polynomials and the transformation matrix can lead to an area enlargement of 9%.

5 Conclusions and Future Work

We explored various options for low gate counts in the design of the AES S-box. We used the architecture of Satoh as a reference and we showed that it is 5% away from an optimal solution. Furthermore, we proved that the "worst case" S-box leads to a 9% area increase. We optimized Satoh's S-box by choosing the irreducible polynomial and transformation matrix that lead to the most compact solution.

There exists a possibility that a more compact S-box can be achieved by choosing an irreducible polynomial $P(x) = x^2 + p_1 x + p_0$ with $p_1 \neq 1$. The high level architecture of Satoh does not stay the same in this case. On the other hand, in this work we only considered the gate complexity of the S-box while encryption is done. The same strategy can be applied to the decryption operation as well.

Acknowledgements

For this work we used the Magma software package. We thank Jasper Scholten from COSIC, Katholieke Universiteit Leuven for his help.

References

1. FIPS Pub. 197: Specification for the AES, Nov. 2001. http://csrc.nist.gov/publications/fips/fips197/fips-197.pdf.
2. M. Alam, W. Badawy, and G. Jullien. A novel pipelined threads architecture for aes encryption algorithm. In M. Schulte, S. Bhattacharyya, N. Burgess, and R. Schreiber, editors, *Proceedings of the IEEE International Conference on Application-Specific Systems, Architectures, and Processors (ASAP)*, pages 296–302, San Jose, CA, USA, July 17-19 2002. IEEE Computer Society Press.
3. P. Chodowiec and K. Gaj. Very compact FPGA implementation of the AES algorithm. In C. Walter, Ç. K. Koç, and C. Paar, editors, *Proceedings of 5th International Workshop on Cryptographic Hardware and Embedded Systems (CHES)*, number 2779 in Lecture Notes in Computer Science, pages 319–333, Cologne, Germany, September 7-10 2003. Springer-Verlag.
4. J. Daemen and V. Rijmen. AES proposal: Rijndael, September 2001. http://csrc.nist.gov/CryptoToolkit/aes/rijndael/Rijndael.pdf.
5. J. Daemen and V. Rijmen. *The design of Rijndael: AES-The Advanced Encryption Standard*. Springer-Verlag, 2002.
6. J. Daemen and V. Rijmen. Security of a wide trail design. In A. Menezes and P. Sarkar, editors, *Proceedings of Third International Conference on Cryptology in India*, volume 2551 of *LNCS*, pages 1–11, Hyderabad, India, December 16-18 2002. Springer-Verlag. Invited talk.
7. V. Fischer and M. Drutarovský. Two methods of Rijndael implementation in reconfigurable hardware. In Ç. K. Koç, D. Naccache, and C. Paar, editors, *Proceedings of 3rd International Workshop on Cryptographic Hardware and Embedded Systems (CHES)*, number 2162 in Lecture Notes in Computer Science, page 77–92, Paris, France, May 13-16 2001. Springer-Verlag.
8. K. Gaj and P. Chodowiec. Fast implementation and fair comparison of the final candidates for advanced encryption standard using field programmable gate arrays. In D. Naccache, editor, *Proceedings of RSA Security Conference: Topics in Cryptology - CT-RSA*, number 2020 in Lecture Notes in Computer Science, San Francisco, CA, USA, April 8-12 2001. Springer-Verlag.
9. J. Guajardo and C. Paar. Efficient algorithms for elliptic curve cryptosystems. In B. S. Kaliski Jr., editor, *Advances in Cryptology: Proceedings of CRYPTO'97*, number 1294 in Lecture Notes in Computer Science, pages 342–356. Springer-Verlag, 1997.
10. K. Jdrvinen, M. Tommiska, and J. Skyttd. A fully pipelined memoryless 17.8 Gbps AES-128 encyptor. In *Proceedings of the 11th ACM International Symposium on Field-Programmable Gate Arrays (FPGA)*, Monterey, CA, USA, February 23-25 2003.
11. H. Kuo and I. Verbauwhede. Architectural optimization for a 1.82Gbits/sec VLSI implementation of the AES rijndael algorithm. In Ç. K. Koç, D. Naccache, and C. Paar, editors, *Proceedings of 3rd International Workshop on Cryptographic Hardware and Embedded Systems (CHES)*, number 2162 in Lecture Notes in Computer Science, pages 51–64, Paris, France, May 13-16 2001. Springer-Verlag.

12. C.-C. Lu and S.-Y. Tseng. Integrated design of AES (Advanced Encryption Standard) encrypter and decrypter. In M. Schulte, S. Bhattacharyya, N. Burgess, and R. Schreiber, editors, *Proceedings of the IEEE International Conference on Application-Specific Systems, Architectures, and Processors (ASAP)*, pages 277–285, San Jose, CA, USA, July 17-19 2002. IEEE Computer Society Press.
13. M. Macchetti and G. Bertoni. Hardware implementation of the Rijndael Sbox: A case study. *ST Journal of system research*, (0):84–91, 2002.
14. M. McLoone and J.V McCanny. High performance single-chip FPGA Rijndael algorithm implementations. In Ç. K. Koç, D. Naccache, and C. Paar, editors, *Proceedings of 3rd International Workshop on Cryptographic Hardware and Embedded Systems (CHES)*, number 2162 in Lecture Notes in Computer Science, pages 65–76, Paris, France, May 13-16 2001. Springer-Verlag.
15. C. Paar. *Efficient VLSI Architectures for Bit-Parallel Computation in Galois Fields*. PhD thesis, Institute for Experimental Mathematics, University of Essen, Germany, 1994.
16. A. Rudra, P. K. Dubey, C. S. Jutla, V. Kumar, J. R. Rao, and P. Rohatgi. Efficient Rijndael encryption implementation with composite field arithmetic. In Ç. K. Koç, D. Naccache, and C. Paar, editors, *Proceedings of 3rd International Workshop on Cryptograpic Hardware and Embedded Systems (CHES)*, number 2162 in Lecture Notes in Computer Science, pages 171–184, Paris, France, May 14-16 2001. Springer-Verlag.
17. A. Satoh, S. Morioka, K. Takano, and S. Munetoh. A compact Rijndael hardware architecture with S-Box optimization. In C. Boyd, editor, *Proceedings of Advances in Cryptology - ASIACRYPT: 7th International Conference on the Theory and Application of Cryptology and Information Security*, number 2248 in Lecture Notes in Computer Science, pages 239–254, Gold Coast, Australia, December 2001. Springer-Verlag.
18. F.-X. Standaert, G. Rouvroy, J.-J. Quisquater, and J.-D. Legat. Efficient implementation of rijndael encryption in reconfigurable hardware: Improvements and design tradeoffs. In C. Walter, Ç. K. Koç, and C. Paar, editors, *Proceedings of 5th International Workshop on Cryptograpic Hardware and Embedded Systems (CHES)*, number 2779 in Lecture Notes in Computer Science, pages 334–350, Cologne, Germany, September 7-10 2003. Springer-Verlag.
19. I. Verbauwhede, P. Schaumont, and H. Kuo. Design and performance testing of a 2.29-Gb/s Rijndael processor. *IEEE Journal of Solid-State Circuits*, 38(3):569–572, March 2003.
20. J. Wolkerstorfer. Dual-field arithmetic unit for $GF(p)$ and $GF(2^m)$. In B. S. Kaliski Jr., Ç. Koç, and C. Paar, editors, *Proceedings of 4th International Workshop on Cryptographic Hardware and Embedded Systems (CHES)*, number 2523 in Lecture Notes in Computer Science, pages 500–514, Redwood Shores, CA, USA, August 13-15 2002. Springer-Verlag.
21. J. Wolkerstorfer, E. Oswald, and M. Lamberger. An ASIC implementation of the AES S-Boxes. In B. Preneel, editor, *Proceedings of the RSA Conference - Topics in Cryptography (CT-RSA)*, number 2271 in Lecture Notes in Computer Science, pages 67–78, San Jose, USA, February 18-22 2002. Springer-Verlag.
22. S.-Y. Wu, S.-C. Lu, and C. S. Laih. Design of AES based on dual cipher and composite field. In T. Okamoto, editor, *Proceedings of RSA Cryptographers' Track*, number 2964 in Lecture Notes in Computer Science, pages 25–38, San Fransisco, USA, February 23-27 2004. Springer-Verlag.

CryptoGraphics:
Secret Key Cryptography Using Graphics Cards

Debra L. Cook[1], John Ioannidis[1], Angelos D. Keromytis[1], and Jake Luck[2]

[1] Department of Computer Science, Columbia University, New York, NY, USA
{dcook,ji,angelos}@cs.columbia.edu
[2] 10K Interactive
jake301@10k.org

Abstract. We study the feasibility of using Graphics Processing Units (GPUs) for cryptographic processing, by exploiting the ability for GPUs to simultaneously process large quantities of pixels, to offload symmetric key encryption from the main processor. We demonstrate the use of GPUs for applying the key stream when using stream ciphers. We also investigate the use of GPUs for block ciphers, discuss operations that make certain ciphers unsuitable for use with a GPU, and compare the performance of an OpenGL-based implementation of AES with implementations utilizing general CPUs. While we conclude that existing symmetric key ciphers are not suitable for implementation within a GPU given present APIs, we discuss the applicability of moving encryption and decryption into the GPU to image processing, including the handling of displays in thin-client applications and streaming video, in scenarios in which it is desired to limit exposure of the plaintext to within the GPU on untrusted clients.

Keywords: Graphics Processing Unit, Block Ciphers, Stream Ciphers, AES.

1 Introduction

We investigate the potential for utilizing Graphics Processing Units (GPUs) for symmetric key encryption. The motivation for our work is twofold. First, our initial motivation was a desire to exploit existing system resources to speed up cryptographic processing and offload system resources. Second, there is the need to avoid exposing unencrypted images and graphical displays to untrusted systems while still allowing remote viewing. While we show that moving symmetric key encryption into the GPU offers limited benefits compared to utilizing general CPUs with respect to non-graphics applications, our work provides a starting point towards achieving the second goal, by determining the feasibility of moving existing symmetric key ciphers into the GPU. Our initial intent is to determine the use of standard GPUs and configurations for cryptographic applications, as opposed to requiring enhancements to GPUs, their drivers, or other system components. Avoiding specialized requirements is necessary to provide a benefit to generalized environments. The focus of our work is on symmetric key ciphers (as opposed to asymmetric schemes) due to the general use of symmetric key ciphers for encryption of large quantities of data and the lack of basic modular arithmetic support in the standard API (OpenGL) for GPUs.

In a large-scale distributed environment such as the Internet, cryptographic protocols and mechanisms play an important role in ensuring the safety and integrity of the interconnected systems and the resources that are available through them. The fundamental building block such protocols depend on are cryptographic primitives, whose algorithmic complexity often turns them into a real or perceived performance bottleneck to the systems that employ them [4]. To address this issue, vendors have been marketing hardware cryptographic accelerators that implement such algorithms [7, 12, 14, 16, 17]. Others have experimented with taking advantage of special functions available in some CPUs, such as MMX instructions [1, 15].

While the performance improvement that can be derived from accelerators is significant [13], only a relatively small number of systems employ such dedicated hardware. Our approach is to exploit resources typically available in most systems. We observe that the large majority of systems, in particular workstations and laptops, but also servers, include a high-performance GPU, also known as a graphics accelerator. Due to intense competition and considerable demand (primarily from the gaming community) for high-performance graphics, such GPUs pack more transistors than the CPUs found in the same PC enclosure [18] at a smaller price. GPUs provide parallel processing of large quantities of data relative to what can be provided by a general CPU. Performance levels equivalent to the processing speed of 10Ghz Pentium processor have been reached, and GPUs from Nvidia and ATI are functioning as co-processors to CPUs in various graphics subsystems [18]. GPUs are already being used for non-graphics applications, but presently none are oriented towards security [11, 27].

With respect to our second goal, limiting exposure of images and graphical displays to be within a GPU, implementing ciphers within the GPU allows images to be encrypted and decrypted without writing the image temporarily as plaintext to system memory. Potential applications include thin clients, in which servers export displays to remote clients, and streaming video applications. While existing Digital Rights Management (DRM) solutions provide decryption of video within the media player and only allow authenticated media players to decrypt the images, such solutions are still utilizing the system's memory and do not readily lend themselves to generic applications exporting displays to clients [20].

Our work consists of several related experiments regarding the use of GPUs for symmetric key ciphers. First, we experiment with the use of GPUs for stream ciphers, leveraging the parallel processing to quickly apply the key stream to large segments of data. Second, we determine if AES can be implemented to utilize a GPU in a manner that allows for offloading work from other system resources (*e.g.,* the CPU). Our work illustrates why algorithms involving certain byte-level operations and substantial byte-level manipulation are unsuitable for use with GPUs given current APIs. Finally, we investigate the potential for implementing ciphers in GPUs for image processing to avoid the image being written to system memory as plaintext.

1.1 Paper Organization

The remainder of the paper is organized as follows. We provide background on the OpenGL commands and pixel processing used in our implementations in Section 2. Section 3 explains how GPUs can be utilized for the combination of a stream cipher's keystream with data in certain applications, and includes performance results. Section 4 describes the representation of AES which we implemented in OpenGL and includes a general discussion of why certain block ciphers are not suitable candidates for use with a GPU given the existing APIs. Section 5 provides an overview of our implementation of AES that utilizes a GPU and provides performance results. We discuss the potential use of GPU-embedded versions of symmetric key ciphers in image processing and thin client applications in Section 6. Our conclusions and future areas of work are covered in Section 7. Appendix A describes the experimental environments, including the minimum required specifications for the GPUs. Appendix B contains pseudo-code for our OpenGL AES encryption routine.

2 OpenGL and GPU Background

Before describing our work with symmetric key ciphers in GPUs, we give a brief overview of the OpenGL pipeline, modeled after the way modern GPUs operate, and the OpenGL commands relevant to our experiments. The two most common APIs for GPUs are OpenGL and Direct3D [19]. We use OpenGL in order to provide platform independence (in contrast to Microsoft's Direct3D). We choose to avoid higher level languages built on top of these APIs in order to ensure that specific OpenGL commands are being used and executed in the GPU when using full hardware acceleration. Examples of such languages include Cg [8] (HLSL in DirectX [19]) and, from more recent research, Brook (the BrookGPU compiler [3] uses Cg in addition to OpenGL and Direct3D). Higher level languages do not allow the developer to specify which OpenGL commands are utilized when there are multiple ways of implementing a function via OpenGL commands and do not even guarantee the operations will be transformed into OpenGL commands but instead may transform it into C code. For example, code in a higher level language that XORs two bytes will likely be transformed into code executed in the operating system rather than converted into OpenGL commands that converts the bytes to pixels and XORs pixels. We use the OpenGL Utility Toolkit (GLUT) [28] to open the display window. GLUT serves as a wrapper for window system APIs, allowing the code to be independent of the window system.

Our implementations process data as 32 bit pixels treated as floating point values, with one byte of data stored in each pixel component[1]. We do not use OpenGL's capabilities of processing pixels as color and stencil indices, and we do not use OpenGL's vertex processing (refer to [21] and [28] for a complete description). OpenGL version 1.4 was used in all experiments. Figure 1 shows the components of the OpenGL pipeline that are relevant to pixel processing when pixels are treated as floating point values. While implementations are not required to adhere to the pipeline, it serves as a general guideline for how data is processed. We also point out that OpenGL requires support for at least a front buffer (image is visible) and a back buffer (image is not visible) but does not require support for the Alpha pixel component in the back buffer. This limits us to three bytes per pixel (the Red, Green, Blue components) when performing operations

---

[1] When using 32 bit pixels, 1 byte is typically dedicated to each of the Red, Green, Blue and Alpha components. A format with 10 bits for each of the Red, Green and Blue components and 2 bits for the Alpha component may also be supported.

in the back buffer. It is worth mentioning that while a 32 bit pixel format is used, the 32 bits cannot be operated on as a single 32 bit value, but rather is interpreted in terms of pixel components. For example, it is not possible to add or multiply two 32 bit integers by representing them as pixels.

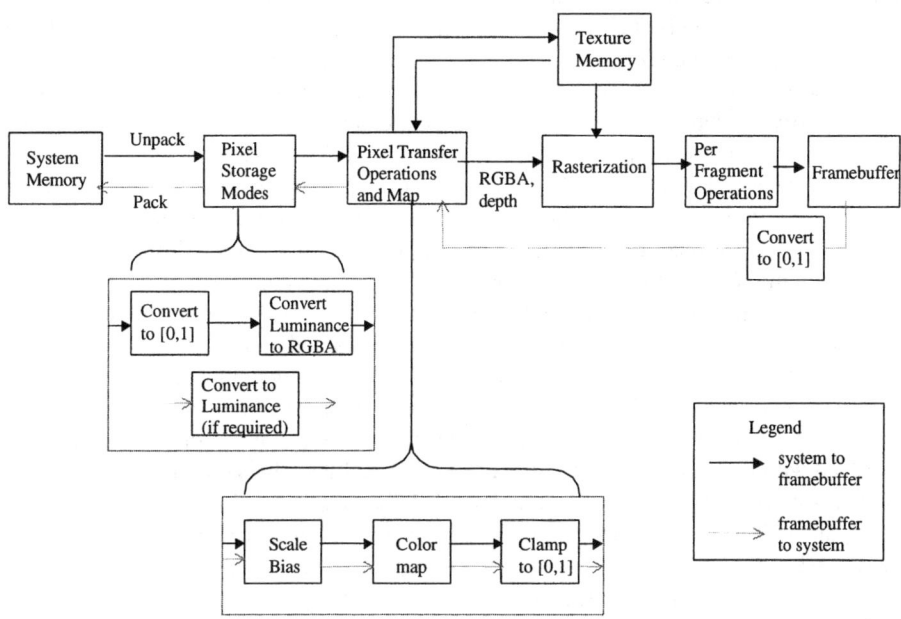

Fig. 1. OpenGL Pipeline for Pixel Processing

A data format indicating such items as number of bits per pixel and the ordering of color components specifies how the GPU interprets and packs/unpacks the bits when reading data to and from system memory. The data format may indicate that the pixels are to be treated as floating point numbers, color indices, or stencil indices. The following description concerns the floating point interpretation. When reading data from system memory, the data is unpacked and converted into floating point values in the range $[0, 1]$. Luminance, scaling and bias are applied per color component. The next step is to apply the color map, which we describe later in more detail. The values of the color components are then clamped to be within the range $[0, 1]$.

Rasterization is the conversion of data into fragments, with each fragment corresponding to one pixel in the frame buffer. In our work this step has no impact. The fragment operations relevant to pixel processing include dithering, threshold based tests, such as discarding pixels based on alpha value and stencils, and blending and logical operations that combine pixels being drawn into the frame buffer with those already in the destination area of the frame buffer. Dithering, which is enabled by default, must be turned off in our implementations in order to prevent pixels from being averaged with their neighbors.

When reading data from the frame buffer to system memory, the pixel values are mapped to the range $[0,1]$. Scaling, bias, and color maps are applied to each of the RGBA components and the result clamped to the range $[0,1]$. The components or luminance are then packed into system memory according to the format specified. When copying pixels between areas of the frame buffer, the processing occurs as if the pixels were being read back to system memory, except that the data is written to the new location in the frame buffer according to the format specified for reading pixels from system memory to the GPU.

Aside from reading the input from system memory and writing the result to system memory, the OpenGL commands in our implementations consist of copying pixels between coordinates, with color mapping and a logical operation of XOR enabled or disabled as needed. Unfortunately, the copying of pixels and color maps are two of the slowest operations to perform [28]. The logical operation of XOR produces a bitwise-XOR between the pixel being copied and the pixel currently in the destination of the copy, with the result being written to the destination of the copy.

A color map is applied to a particular component of a pixel when the pixel is copied from one coordinate to another. A color map can be enabled individually for each of the RGBA components. The color map is a static table of floating point numbers between 0 and 1. Internal to the GPU, the value of the pixel component being mapped is converted to an integer value which is used as the index into the table and the pixel component is replaced with the value from the table. For example, if the table consists of 256 entries, as in our AES implementation, and the map is being applied to the red component of a pixel, the 8 bits of the red value are treated as an integer between 0 and 255, and the red value updated with the corresponding entry from the table. In order to implement the tables of equation (III) in Section 4 as color maps, the tables must be converted to tables of floating point numbers between 0 and 1, and hard-coded in the program as constants. The table entries, which would vary from 0 to 255 if the bytes were in integer format, are converted to floating point values by dividing by 255. Because pixels are stored as floating point numbers and the values are truncated when they are converted to integers to index into a color map, 0.000001 is added to the result (except to 0 and 1) to prevent errors due to truncation.

3 Graphics Cards and Stream Ciphers

As a first step in evaluating the usefulness of GPUs for implementing cryptographic primitives, we implemented the mixing component of a stream cipher (the XOR operation) inside the GPU. GPUs have the ability to XOR many pixels simultaneously, which can be beneficial in stream cipher implementations. For applications that pre-compute segments of key streams, a segment can be stored in an array of bytes which is then read into the GPU's memory and treated as a collection of pixels. The data to be encrypted or decrypted is also stored in an array of bytes which is read into the same area of the GPU's memory as the key stream segment, with the logical operation of XOR enabled during the read. The result is then written to system memory. Overall, XORing the data with the key stream requires two reads of data into the GPU from system memory and one read from the GPU to system memory.

The number of bytes can be at most three times the number of pixels supported if the data is processed in a back buffer utilizing only RGB components. The number of bytes can be four times the number of pixels if the front buffer can be used or the back buffer supports the Alpha component. If the key stream is not computed in the GPU, the cost of computing the key stream and temporarily storing it in an array is the same as in an implementation not utilizing a GPU. At least one stream cipher, RC4 [26], can be implemented such that the key stream is generated within the GPU. However, the operations involved result in decreased performance compared to an implementation with a general CPU. Our work with AES serves to illustrate the problems with implementing byte-level operations within a GPU and thus we omit further discussion of RC4 within this paper. Others, such as SEAL [24] which requires 9-bit rotations, involve operations which make it difficult or impossible to implement in the GPU given current APIs.

Table 1. XOR Rate Using System Resources (CPU)

	CPU		
	1.8 Ghz	1.3 Ghz	800 Mhz
XOR Rate	139MB/s	93.9MB/s	56MB/s

We compared the rate at which data can be XORed with a key stream in an OpenGL implementation to that of a C implementation (Visual C++ 6.0). We conducted the tests using a PC with a 1.8Ghz Pentium IV processor and an Nvidia GeForce3 graphics card, a laptop with a 1.3Ghz Pentium Centrino Processor and a ATI Mobility Radeon graphics card, and a PC with a 800Mhz Pentium III Processor and an Nvidia TNT2 graphics card. Refer to Appendix A for additional details on the test environments. We provide the results from the C implementation in Table 1. We tested several data sizes to determine the ranges for which the OpenGL implementation would be useful. As expected, the benefit of the GPU's simultaneous processing is diminished if the processed data is too small. Table 2 indicates the average encryption rates over 10 trials of encrypting 1000 data segments of size $3Y^2$ and $4Y^2$, respectively, where the area of pixels is Y by Y. For the number of pixels involved in our images, the transfer rate to the GPU was measured to be equal to the transfer rate from the GPU, thus each read and write contributed equally to the overall time.

Notice that the encryption rate was fairly constant for all data sizes on the slowest processor with the oldest GPU (Nvidia TNT2). Possible explanations include slow memory controller, memory bus, or GPU, although we have not investigated this further. With the GeForce3 Ti200 card, the efficiency increased as more bytes were XORed simultaneously. On the laptop the peak rates were obtained with 200x200 to 400x400 square pixel areas.

When using the RGB components, the highest rate obtained by the GPUs compared to the C program is 58% for the Nvidia GeForce3 Ti200 card, 48.5% for the ATI Mobility Radeon card, and 51.4% for the Nvidia TNT2 card. With both the GeForce3 Ti200 and the ATI Radeon cards, results with the 50x50 pixel area was significantly slower than with larger areas due to the time to read data to/from system memory representing

Table 2. XOR Rate Using GPUs - RGB and RGBA Pixel Components

Area (in pixels)	Using RGB components			Using RGBA components		
	Nvidia GeForce3 Ti200	ATI Mobility Radeon 7500	Nvidia TNT2	Nvidia GeForce3 Ti200	ATI Mobility Radeon 7500	Nvidia TNT2
50x50	35.7MBps	23.5MBps	27.8MBps	49.3MBps	26.3MBps	37.0MBps
100x100	53.4MBps	38.5MBps	28.8MBps	69.2MBps	38.1MBps	38.4MBps
200x200	64.5MBps	45.5MBps	26.0MBps	86.8MBps	45.7MBps	32.0MBps
300x300	70.1MBps	45.0MBps	26.0MBps	94.8MBps	42.3MBps	32.0MBps
400x400	75.4MBps	43.0MBps	27.0MBps	95.9MBps	49.0MBps	32.8MBps
500x500	77.3MBps	38.0MBps	26.6MBps	97.5MBps	37.0MBps	32.6MBps
600x600	81.2MBps	41.7MBps	27.7MBps	105.0MBps	41.5MBps	32.8MBps

a larger portion of the total time. In both cases the rate is approximately 25% of that of the C program. When using the RGBA components, the highest rates on the Nvidia GeForce Ti200, ATI Radeon and Nvidia TNT2 cards are 75.5%, 52% and 68% of the C program, respectively.

4 Graphics Cards and Block Ciphers

We now turn our attention to the use of GPUs for implementing block ciphers. The first step in our work is to determine if AES can be represented in a manner which allows it to be implemented within a GPU. We describe the derivation of the OpenGL version of AES and its implementation in some detail, in order to illustrate the difficulties that arise when utilizing GPUs for algorithms performing byte-level operations. We also briefly comment on the suitability of using GPUs for block ciphers in general. While GPUs are advantageous in various aspects, the use of floating point arithmetic and the fact that the APIs are not designed for typical byte-level operations, as required in most block ciphers, present severe obstacles. For 128-bit blocks, the AES round function for encryption is typically described with data represented as a 4x4-byte matrix upon which the following series of steps are performed:

(I) SubBytes (S-Box applied to each entry)
ShiftRows (bytes within each row of the 4x4 matrix are shifted 0 to 3 columns)
MixColumns (a matrix multiplication; absent in last round)
AddRoundKey (the 4x4 matrix is XORed with a round key)

Ten rounds are performed, with the data XORed with key material prior to the first round and the MixColumns step omitted in the last round. The round function for decryption differs from encryption in that inverse functions for SubBytes, ShiftRows and MixColumns are used. Refer to [9] for a complete description of each function.

A faster implementation for environments with sufficient memory operates on 32-bit words and reduces the AES round function to four table lookups and four XORs. If A denotes a 4x4 matrix input to the round, $a_{i,j}$ denotes the ith row and j^{th} column of A, $j - x$ is computed modulo 4, and Tk are tables with 256 32-bit entries, the round function is reduced to the form:

(II) $A'_j = T0[a_{0,j}] \oplus T1[a_{1,j-1}] \oplus T2[a_{2,j-2}] \oplus T3[a_{3,j-3}] \oplus RoundKey$

where A'_j denotes the jth column of the round's output. Refer to pages 58–59 of [6] for a complete description. The entries in the tables in (II) are concatenations of 1, 2, and 3 times the S-Box entries. This version is due to the fact that the order of the SubBytes and ShiftRows steps can be switched and the MixColumn step can be viewed as the linear combination of four column vectors, which is actually a linear combination of the S-Box entries.

The AES round function cannot easily be implemented with OpenGL as the standard series of four steps. The SubBytes can be performed using a color map, and the ShiftRows and AddRoundKey can be performed by copying pixels to change their location or to XOR them with other pixels. However, the MixColumn step would have to be expanded to a series of color maps to perform individual multiplications and copying of pixels to perform additions due to the lack of a corresponding matrix multiplication with modular arithmetic in OpenGL. The view of AES as four table lookups and XORs also cannot be implemented in OpenGL due to the lack of a 32-bit data structure. While the RGBA format is 32 bits, it is not possible to use all 32 bits as an index into a color map or to swap values between components, both of which would be necessary to implement the version in (II). As a result, we use an intermediate step in the transformation of the standard algorithm to the version in (II). Letting A'_j and $a_{i,j}$ be defined as in (II) and letting $S[a_{i,j}]$ denote the S-Box entry corresponding to $a_{i,j}$, the encryption round function for rounds 1 to 9 is represented as:

(III) $A'_j =$

$$\begin{pmatrix} 02S[a_{0,j}] \\ 01S[a_{0,j}] \\ 01S[a_{0,j}] \\ 03S[a_{0,j}] \end{pmatrix} \oplus \begin{pmatrix} 03S[a_{1,j-1}] \\ 02S[a_{1,j-1}] \\ 01S[a_{1,j-1}] \\ 01S[a_{1,j-1}] \end{pmatrix} \oplus \begin{pmatrix} 01S[a_{2,j-2}] \\ 03S[a_{2,j-2}] \\ 02S[a_{2,j-2}] \\ 01S[a_{2,j-2}] \end{pmatrix} \oplus \begin{pmatrix} 01S[a_{3,j-3}] \\ 03S[a_{3,j-3}] \\ 02S[a_{3,j-3}] \\ 01S[a_{3,j-3}] \end{pmatrix} \oplus Roundkey$$

If three tables, representing 1, 2, and 3 times the S-Box entries are stored, (III) reduces to a series of table lookups and XORs. This allows AES to be implemented using color maps and copying of pixels. The 10th round is implemented as (III) with all the coefficients of 2 and 3 replaced by 1. Since decryption uses the inverses of the S-Box and matrix multiplication, five tables need to be stored, representing 0E, 0B, 0D, 09 and 01 times the S-Box inverse. Notice that this representation of AES processes data as individual bytes, instead of 4-byte words. However, the manner in which the pixel components are utilized in the implementation when encrypting multiple blocks allows 4 bytes to be processed simultaneously per pixel, compensating for the loss of not being able to use 32-bit words as in (II).

In general, algorithms performing certain byte and bit-level operations are not suitable for GPUs given current APIs. While simple logical operations can be performed efficiently in GPUs on large quantities of bytes, as shown in Section 3, the byte and bit-level operations typically found in symmetric key ciphers, such as shifts and rotates, are not available via the APIs to GPUs. Modular arithmetic operations are also not readily available. While some operations, such as defining masks of pixels and using multiple copy commands to perform rotations and shifts on single bytes, can be performed via combinations of OpenGL commands, other operations, such as shifts across multiple bytes and table lookups based on specific bits, prove to be more difficult. For example, there is no straightforward way to implement in OpenGL the data dependent rotations found in RC6 [23] and MARS [5]. Also consider the DES S-Boxes [10]. The index into the S-Box is based on six key bits XORed with six data bits. Two of the bits are used to select the S-Box and the remaining four are the index into the S-Box. Masks of pixels copied onto the data can be used to "extract" the desired bits, but to merely XOR the six key bits with six data bits requires copying the pixel containing the desired key bits onto the pixel containing the mask with XOR turned on, doing the same for the data pixel, then copying the two resulting pixels to the same position. Color maps are required to emulate the S-Box. Overall, to use OpenGL for the S-Box step in DES, a larger number of less efficient operations are required than in a C implementation.

5 OpenGL Version of AES

5.1 Implementation Overview

We describe an implementation of AES's encryption and decryption functions for 128-bit blocks that works with any GPU supporting 32-bit pixels and OpenGL. The key schedule is not implemented inside the GPU. While the GPU allows for parallel processing of a large number of blocks, due to the simplicity in which AES can be implemented in software as a series of table lookups and XORs, the overall encryption rate using the GPU is below the rate that can be obtained with a C implementation utilizing only system resources.

The code consisted of C, OpenGL and GLUT. The C portion of the code sets up the plaintext or ciphertext and key. The OpenGL and GLUT commands are called from within the C program. GLUT commands are used to open the display window. All of the encryption and decryption computations are performed with OpenGL functions, with data being stored and processed as pixels. To accomplish this, it is necessary to represent AES in a manner that requires only the specific transformations or functions supported by the graphics hardware. As explained in Section 4, we use a representation that can be implemented in OpenGL solely via color maps and pixel copying. The implementation allows encrypting $4*n$ blocks simultaneously, where n is the number of pixels utilized for the data being encrypted or decrypted and may be any integer less than the display's maximum pixel height supported by the GPU. The encryption of multiple blocks simultaneously from the same plaintext is useful if ECB or CTR mode are used. Alternatively, we can process one block from several messages in parallel.

Figure 2 illustrates the pixel coordinates utilized by the algorithm. The initial data blocks are read into the 16 x n area starting at the origin, indicated by "DATA" in the diagram. One byte of data is stored in each pixel component, allowing us to process $4*n$ blocks of data when all of the RGBA components are used. The ith column contains the ith byte of each block. This area is also used to store the output from each round. To maximize throughput, for each data block one copy of the expanded key is read into the area labeled "KEY" in the diagram. This area is 176 x n pixels starting at (17, 0) and the round keys are stored in order, each encompassing 16 columns. The tables are

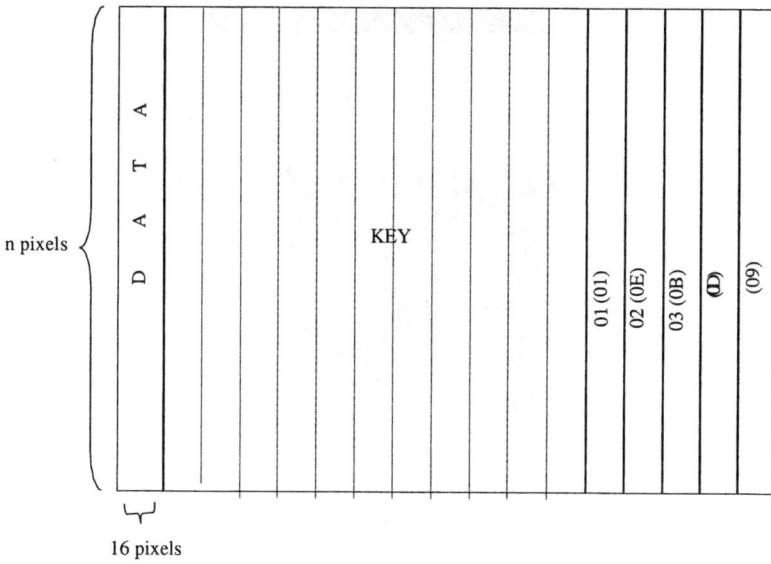

Fig. 2. Layout of Data in Pixel Coordinates used in OpenGL Version of AES

stored as color maps and do not appear in the layout. The data stored in the first 16 columns is copied 3 times for encryption and 5 times for decryption, applying a color map each time. The results are stored in the areas indicated by the hex values in the diagram and are computed per round. The values in parenthesis indicate the location of the transformations for decryption. The hex value indicates the value by which the S-Box (or inverse S-Box, when decrypting) entries are multiplied. See Appendix B for pseudo-code of the GPU AES encryption process. Figure 3 shows an example of the resulting display when the front buffer and RGB components are used to encrypt 300 identical data blocks simultaneously.

Two C implementations of AES are used for comparison. The first is the AES representation corresponding to variant (I) in Section 4, with the multiplication steps performed via table lookups, and reflects environments in which system resources for storing the tables required by variant (II) are not available. The second is a C implementation of variant (II), which offers increased encryption and decryption rates over (I) at the cost of requiring additional memory for tables. The code for (II) is a subset of [22].

5.2 Experiments

We compare the rate of encryption provided with the GPU to that provided by the C implementation running on the system CPU. Tests were conducted using the same three environments used for the stream cipher experiments. When describing the results, AES-GL indicates the implementation using OpenGL and AES-C indicates the C implementations, with the specific variant from Section 4 indicated by I and II. The AES-C programs have a hard-coded key and single 128-bit block of data. The programs expand the key then loop through encrypting a single block of data, with the

Fig. 3. Encryption of 300 Identical Blocks in RGB Components

output from the previous iteration being encrypted each time. No data is written to files and the measurements exclude the key setup (which is common for all variants). The AES-GL program uses a hard-coded expanded key and one or four blocks of data in the cases when the red or RGBA pixel components are used, respectively. Both the key and data are read in n times to provide n copies. Similar to the AES-C programs, the AES-GL program loops through encrypting blocks of data, with the output from the previous iteration being encrypted each time. The times exclude reading in the initial data and key, and no data is read from or written to system memory during the loop. Trials were conducted with the values of n ranging from 100 to 600 in increments of 100. The rates for values of $n \geq 300$ varied by less than 2% and the rates across all values of n varied by at most 8%. The results for AES-GL in Table 3 are the averages over $n \geq 300$ when a single pixel component and all of the RGBA pixel components are utilized. The corresponding decryption rates for the C and OpenGL implementations will be slightly lower than the encryption rates due to a small difference in the number of operations in the decryption function compared to that of the encryption function.

The layout of the pixels was chosen to simplify indexing while allowing for a few thousand blocks to be encrypted simultaneously. Since the layout does not utilize all of the available pixels, the number of blocks encrypted at once can be increased if the

Table 3. Encryption Rates for AES

	AES Version			
PC and GPU	AES-GL R	AES-GL RGBA	AES-C (I)	AES-C (II)
800Mhz Nvidia TNT2	184Kbps	732Kbps	1.68Mbps	30Mbps
1.3Ghz ATI Mobility Radeon	55Kbps	278.3Kbps	2.52Mbps	45Mbps
1.8Ghz Nvidia GeForce3	380Kbps	1.53Mbps	3.5Mbps	64Mbps

display area is utilized differently. For example, if the number of blocks is n^2, the layout can be altered such that the various segments are laid out in $n \times n$ areas instead of as columns. Performance recommendations for OpenGL include processing square regions of pixels as opposed to processing narrower rectangles [28]. We tried a modification of the program, which performed the same number of steps on square regions instead of the configuration shown in Figure 2. There was no change in the encryption rate, most likely because the program appears to be CPU-bound as we discuss in the next section. Furthermore, using square areas makes indexing more difficult and requires the number of blocks to be a perfect square for optimal utilization of the available pixels.

5.3 Performance Analysis

With the two Nvidia graphics cards, AES-GL's encryption rate was just under 50% that of AES-C (I). However, when compared to AES-C (II), the AES-GL rate was 2.4% of the AES-C version. The ratio was lower in both cases when using ATI Mobility Radeon graphics card, with the AES-GL encryption rate being 11% of AES-C (I)'s rate and less than 1% of AES-C (II)'s rate.

To determine the factors affecting AES-GL's performance, additional tests were performed in which AES-GL and AES-C were run while monitoring system resources. When we use either AES-C or AES-GL, the CPU utilization is 100% for the duration of the program. While we expect high CPU utilization for AES-C, the result is somewhat counter-intuitive for AES-GL. We believe that this happens because of the rate at which commands are being issued to the graphics card driver. Due to the simplicity in which AES is represented, a single OpenGL command resulted in one operation from AES being performed: either the table lookup or the XORing of bytes.

We do not consider the difference between the AES representations used by AES-GL and AES-C to be a factor. While the representation of AES used in AES-GL processes data as individual bytes instead of as the 32-bit words used in AES-C (II), even when excluding the processing of n pixels simultaneously the use of the RGBA components allows 4 bytes to be processed simultaneously per pixel, compensating for the loss of not being able to use 32-bit words. We also reiterate that the actions performed upon the pixels (color maps and copying) are two of the slowest GPU operations.

6 Decryption of Images Inside the GPU

The fact that symmetric key ciphers can be implemented within a GPU implies it is possible to encrypt and decrypt images in a manner that does not require the image to ever be present outside the GPU in unencrypted format. If the decrypted image is only available in the GPU, an adversary must be able to execute reads of the GPU's memory for the area utilized by the window containing the image while the image is being displayed. As a proof of concept, we use the AES-GL implementation with the image read into the card's memory in an area not utilized by AES. The data area for AES is populated by copying the image pixels into the area *in lieu* of reading data from system memory. Trivially, the image can have a stream cipher's key stream applied to it in the GPU by XORing the image with the pixel representation of the key stream.

One potential application is encrypted streaming video in which the video frames are decrypted within the back buffer of the GPU prior to being displayed, as opposed to decrypting within the system when the data is received. Typical media player screens vary from 320 x 200 pixels to 1280x1024 pixels. For low-end video, 10 frames per second (fps) is sufficient, while full-motion video requires 15 to 30 fps, with minimal perceived difference between the two rates. Assuming 8 bits per RGB component, the decryption rate must be 1.92 MBps to support 10 fps and 2.88 MBps to support 15 fps when displaying video to a 320 x 200 pixel window, rates within the limits supported by the GPUs when using stream ciphers but which exceed the rate currently obtained with AES-GL. The AES C (I) implementation also does not support these rates. For a 1280x1024 screen, 39.25 MBps support is required for 10 fps, a rate which is supported when using a stream cipher in two of the three GPUs. At 15 fps, 58.9MBps must be supported, which can only be achieved with the Nvidia GeForce3 Ti200.

A second application, less intensive than streaming video, concerns processing of displays in thin client applications. In such applications, a server sends only the updated portion of a display to the client. For example, when a user places the mouse over a link on a web page, the server may send an update that results in the link being underlined or changing color by sending an update only for the display area containing the link, thus requiring only a small amount of data to be decrypted.

When encrypting and decrypting images within the GPU, a few issues need to be resolved, such as image compression. If an image is encrypted prior to compression, ideally no compression should be possible; therefore, when encrypting and decrypting images in the GPU compression and decompression will also need to be migrated to the GPU. Second, as mentioned previously, dithering needs to be turned off. This may produce a visible side affect if the algorithm is used on large images. However, on small images, typical of a media player when not set to full screen, the lack of dithering is not likely to be noticeable. An option would be to decrypt the image in the back buffer then have dithering on when transferring the image to the front buffer, allowing decrypted images and video to be displayed with dithering.

The current AES-GL implementation reads the expanded key from the system. Alternate methods of storing the key or conveying the key to the GPU must be considered to make the key storage secure as well. We are currently experimenting with the use of remotely-keyed encryption [2] in which a smartcard or external system conveys an encrypted secret key that is decrypted within the GPU. Our current implementation requires the ability to store a certificate within the GPU and utilizes RSA [25] to convey the secret key to the GPU, with limitations on the size of the RSA private key and modulus in order to contain decryption of the secret key within the GPU.

7 Conclusions

While symmetric key encryption is possible in GPUs, the lack of support via APIs for certain operations results in poor performance overall when using existing ciphers. Furthermore, current APIs do not permit some ciphers to be implemented within the GPU. The AES experiments prove it is possible to implement AES in a manner that utilizes a GPU to perform the computation while illustrating the difficulty in moving existing block ciphers into the GPU. The lessons learned from developing the OpenGL version of AES indicate GPUs are not suitable, given current APIs, for ciphers involving certain types of byte-level operations. GPUs can be used to offload a shared system CPU in applications using stream ciphers and which allow large segments of data to be combined with the key stream simultaneously.

Encryption and decryption of graphical displays and images may be moved into the GPU to avoid temporarily storing an image as plaintext in system memory. As GPU processing power and capabilities continue to increase, the potential uses will also increase. Our plans for future work include deriving a mechanism by which the key is not exposed outside the GPU and continuation of the work on remote keying of GPUs. We are also continuing work on the applicability to thin client and streaming video applications, such as video conferencing, and are designing a new cipher that can better exploit the capabilities of modern GPUs for use in these applications.

References

1. E. Biham, A Fast New DES Implementation in Software, *Workshop on Fast Software Encryption (FSE '97)*, LNCS 1267, Springer-Verlag, pages 260-272, 1997.
2. M. Blaze, J. Feigenbaum and M. Naor, A Formal Treatment of Remotely Keyed Encryption, *Proceedings of EUROCRYPT '98*, LNCS 1403, Springer-Verlag, pages 251-265, 1998.
3. I. Buck, T. Foley, D. Horn, J. Sugerman, K. Fatahalian, M. Houston and P. Hanrahan, Brook for GPUs: Stream Computing on Graphics Hardware *Proceedings of SIGGRAPH*, 2004.
4. A. G. Broscius and J. M. Smith, Exploiting Parallelism in Hardware Implementation of the DES, *Proceedings of CRYPTO '91*, LNCS 576, Springer-Verlag, pages 367-376, 1991.
5. D. Coppersmith, et.al., The MARS Cipher, http://www.research.ibm.com/security/mars.html, 1999.
6. J. Daemon and V. Rijmen, *The Design of Rijndael: AES the Advanced Encryption Standard*, Springer-Verlag, Berlin, 2002.
7. W. Feghali, B. Burres, G. Wolrich and D. Carrigan, Security: Adding Protection to the Network via the Network Processor, *Intel Technology Journal*, 6(3), August 2002.
8. R. Fernando and M. Kilgard, *The Cg Tutorial*, Addison-Wesley, New York, 2003.
9. FIPS 197 Advanced Encryption Standard (AES), 2001.
10. FIPS 46-3 Data Encryption Standard (DES), 1999.
11. General Purpose Computation Using Graphics Hardware, http://www.gpgpu.org.
12. Helion Technology Limited, High Performance Solutions in Silicon, AES (Rijndael) Core, http://www.heliontech.com/core2.htm, 2003.
13. A. D. Keromytis, J. L. Wright and T. de Raadt, The Design of the OpenBSD Cryptographic Framework, *Proceedings of the USENIX Annual Technical Conference*, pages 181-196, 2003.
14. H. Kuo and I. Verbauwhede, Architectual Optimization for 1.82 Gbits/sec VLSI Implementation of Rijndael Algorithm, *Proceedings of CHES*, LNCS 2162, Springer-Link, pages 51-64, 2001.
15. Helger Lipmaa, IDEA: A Cipher for Multimedia Architectures? *Selected Areas in Cryptography '98*, LNCS 1556, Springer-Verlag, pages 248–263, 1998.
16. A. Lutz, J. Treichler, F.K. Gurkeynak, H. Kaeslin, G. Bosler, A. Erni, S. Reichmuth, P. Rommens, S. Oetiker and W. Fichtner, 2G bits/s Hardware Realizations of Rijndael and Serpent: A Comparative Analysis, *Proceedings of CHES*, LNCS 2523, Springer-Link, pages 144-158, 2002.

17. M. McLoone and J. McConny, High Performance Single Chip FPGA Rijndael Algorithms Implementations, *Proceedings of CHES*, LNCS 2162, Springer-Link, pages 65-76, 2001.
18. M. Macedonia, The GPU Enters Computing's Mainstream, *IEEE Computer Magazine*, pages 106-108, October 2003.
19. Microsoft DirectX, http://www.microsoft.com/windows/directx.
20. Microsoft Windows 9 Media Series Digital Rights Management, http://www.microsoft.com/windows/windowsmedia/drm.aspx, 2004.
21. OpenGL Organization, http://www.opengl.org.
22. V. Rijmen, A. Bosselaers and P. Barreto, AES Optimized ANSI C Code, http://www.esat.kuleuven.ac.be/ rijmen/rijndael/rijndael-fst-3.0.zip.
23. Rivest, Robshaw, Sidney and Yin, RC6 Block Cipher, http://www.rsa.security.com/rsalabs/rc6, 1998.
24. P. Rogaway and D. Coppersmith, A Software Optimized Encryption Algorithm, *Journal of Cryptology*, vol. 11, pages 273-287, 1998.
25. RSA Laboratories, PKCS1 RSA Encryption Standard, Version 1.5, 1993.
26. B. Schneier, *Applied Cryptography*, 2nd edition, John Wiley and Sons, New York, 1996.
27. C. Thompson, S. Hahn and M. Oskin, Using Modern Graphics Architectures for General-Purpose Computing: A Framework and Analysis, *35th Annual IEEE/ACM International Symposium on Micro Architecture (MICRO-35)*, pages 306-317, 2002.
28. M. Woo, J. Neider, T. Davis and D. Shreiner, *The OpenGL Programming Guide*, 3rd edition, Addison-Wesley, Reading, MA, 1999.

Appendix A: Environments

GPU Requirements

For our implementations, we use OpenGL as the API to the graphics card driver. All of our programs use basic OpenGL commands and have been tested with OpenGL 1.4.0. No vendor-specific extensions are used, allowing the program to be independent of the GPU. The GPU must support 32-bit "true color" mode, because 8-bit color components are required for placing the data in pixels. At a minimum, one color component and at a maximum all four of the RGBA components are utilized by our programs. The implementations of AES and stream ciphers can be set to work with one to four pixel components. To avoid displaying the pixels to the window as the encryption is occurring, the display mode can be set to use a front and back buffer, with the rendering performed in the back buffer and the results read directly from the back buffer to system memory and never displayed on the screen. The support for the Alpha component in the back buffer is optional in OpenGL; therefore, it may be necessary to perform rendering in the front buffer and display the pixels to the screen when utilizing all of the RGBA components.

Processors

All tests were performed in three different environments, then a subset of the tests were run in other environments to verify the correctness of the implementations with additional GPUs. The environments were selected to represent a fairly current computing environment, a laptop and a low-end PC. Both Nvidia and ATI cards were used to illustrate our implementations worked with different brands of cards, but not to compare the performance of the different graphics cards. The three environments used for all tests are:

1. A Pentium IV 1.8 Ghz PC with 256KB RAM and an Nvidia GeForce3 Ti200 graphics card with 64MB of memory. The operating system is MS Windows XP.
2. A Pentium Centrino 1.3 Ghz laptop with 256KB RAM and an ATI Mobility Radeon 7500 graphics card with 32MB of memory. The operating system is MS Windows XP.
3. A Pentium III 800 Mhz PC with 256KB RAM and an Nvidia TNT32 M64 graphics card with 32MB of memory. The operating system is MS Windows 98.

In all cases, the display was set to use 32-bit true color and full hardware acceleration. Aside from MS Windows and, in some cases a CPU monitor, no programs other than that required for the experiment were running. The CPU usage averages around 8% in each environment with only the OS and CPU monitor running. All code was compiled with Visual C++ Version 6.0. Our implementations required opening a display window, though computations may be performed in a buffer that is not visible on the screen. The window opened by the program is positioned such that it does overlap with the window from which the program was executed and to which the output of the program is written. The reason for this positioning is that movement of the display window or overlap with another active window may result in a slight decrease in performance and can interfere with the results. GLUT commands were used to open the display window.

The other GPUs we tested our programs with included an Intel© 82845G/GL Graphics Controller on a 2.3 Ghz Pentium IV processor running MS Windows XP, and a Nvidia GeForce4 Ti 4200 on a Pentium III 1.4 Ghz processor running MS Windows 2000. The AES implementation was also tested using a GeForce3 Ti200 graphics card with 64MB of memory with X11 and Redhat Linux 7.3.

Configuration Factors

In order to determine configuration factors impacting performance, we ran a series of initial tests with the OpenGL implementations of AES and the stream cipher while holding the number of bytes encrypted constant. First, since the implementation required a GPU that was also being utilized by the display, we varied the refresh rate for the display, but that did not affect performance. Second, we varied the screen area (not the number of pixels utilized for the cipher) from 800x600 to 1600x1200. This also did not affect performance, and in the results cited for AES, we set the screen area to the minimum of 800x600 and the dimension that accommodated the number of pixels required by the test. Third, we tested the use of a single buffer with the pixels displayed to the screen versus a front and back buffer with all work performed in the back buffer and not displayed to the screen. Again, there was no change in the encryption rate. A fourth test was run to determine if there was any decrease in performance by using the GLUT or GLX libraries to handle the display. GLX is the X Window System extension to support OpenGL. In the test, we executed two versions of the program, one using GLUT and one using GLX with direct rendering, from a server with a Pentium III running Redhat Linux 7.3. There was no noticeable difference between the rates from the GLUT and GLX versions of the program.

Appendix B: AES Encryption Using OpenGL

In our OpenGL version of AES, encryption was implemented as the following steps:
Define static color maps corresponding to 1, 2, 3 times the S-Box entries.

main {
    Load the data into the DATA area.
    Load the expanded key into the KEY area.
    Turn the logical operation of XOR on.
    Copy the first key from the KEY area to the DATA area.
    Turn the logical operation XOR off.
    for (i=0; i < 9; ++i) {
        Copy the DATA area:
        to the 01 area with the color map corresponding to 1*S-Box turned on
        to the 02 area with the color map corresponding to 2*S-Box turned on
        to the 03 area with the color map corresponding to 1*S-Box turned on
        Turn color mapping off
        Copy the pixels from areas 01,02,03 corresponding to the first term on
            the right hand side of (III) to the DATA area.
        Turn the logical operation of XOR on.
        Copy the pixels from areas 01,02,03 corresponding to the $2^{nd}, 3^{rd}$ and
            4th terms on the right hand side of (III) to the DATA area.
        Copy the ith round key from the KEY area to the DATA area.
        Turn the logical operation XOR off.
    }
    Copy the DATA area to the 01 area with the color map corresponding to
        1*S-Box turned on.
    Turn color mapping off.
    Copy the pixels from the 01 area back to the DATA area in the order
        corresponding to ShiftRows.
    Turn the logical operation of XOR on.
    Copy the last round key from the KEY area to the DATA area.
    Turn the logical operation XOR off.
    Read the DATA area to system memory.
}

Side-Channel Leakage of Masked CMOS Gates*

Stefan Mangard[1], Thomas Popp[1], and Berndt M. Gammel[2]

[1] Institute for Applied Information Processing and Communications (IAIK)
Graz University of Technology, Inffeldgasse 16a, 8010 Graz, Austria
{Stefan.Mangard,Thomas.Popp}@iaik.at
[2] Infineon Technolgies AG
St.-Martin-Straße 76, 81541 Munich, Germany
Berndt.Gammel@infineon.com

Abstract. There are many articles and patents on the masking of logic gates. However, the existing publications assume that a masked logic gate switches its output no more than once per clock cycle. Unfortunately, this assumption usually does not hold true in practice.

In this article, we show that glitches occurring in circuits of masked gates make these circuits susceptible to classical first-order DPA attacks. Besides a thorough theoretical analysis of the DPA-resistance of masked gates in the presence of glitches, we also provide simulation results that confirm the theoretical elaborations. Glitches occur in every CMOS circuit. Consequently, the currently known masking schemes for CMOS gates do not prevent DPA attacks.

Keywords: Power Analysis, DPA, Masking, Masked Digital Circuits, Masked Logic Gates

1 Introduction

During the last years, a lot of research has been conducted on differential power-analysis (DPA) attacks [11] and on corresponding countermeasures. DPA attacks exploit the fact that the power consumption of a device executing a cryptographic algorithm is correlated to intermediate results of the algorithm. This correlation between the intermediate results and the power consumption allows an attacker to reveal the secret key that is used by a device (see [11]).

Hence, the goal of countermeasures against DPA attacks is to completely remove or at least to reduce this correlation. Essentially, there exist two approaches to achieve this goal.

The first approach is to try to make the power consumption of a device independent of the data that is processed by the device. The countermeasures that are based on this approach are usually called hardware countermeasures. Typical examples of such countermeasures are detached power supplies [19], logic styles with a data-independent power consumption [20, 21], noise generators and

---

* This work has been supported by the Austrian Science Fund (FWF Project Number P16110-N04) and by the European Commission under the Sixth Framework Programme (Project SCARD, Contract Number IST-2002-507270).

the insertion of random delays [4,12]. Each of these hardware countermeasures reduces the correlation between the data that is processed by the device and the power consumption. In practice, hardware countermeasures are typically combined. This can reduce the correlation down to a level that makes DPA attacks almost impossible in practice.

The second approach to counteract DPA attacks is to randomize the intermediate results occurring in a cryptographic algorithm. The motivation behind this approach is that the power consumption of a device processing randomized intermediate results is uncorrelated to the actual intermediate results. The randomization of intermediate results is usually called masking. Masking can be applied either at the algorithm level or at the gate level.

Applying masking at the algorithm level means that an algorithm is rewritten such that all intermediate results are randomized, while the input and the output of the algorithm are identical to those of the unmasked version. There are several publications that discuss how symmetric [1,7,8,24] and asymmetric ciphers [5,15] can be rewritten this way.

The alternative to masking at the algorithm level is the usage of masked logic gates for implementations of cryptographic algorithms. This leads to circuits where no wire stores a value that is correlated to an intermediate result of the algorithm. Clearly this approach is more generic. Masking at the gate level is independent of the implemented algorithm and in principle it can even be done completely automatically, *i.e.* a program can be used to convert a digital circuit into a circuit of masked gates. Throughout this article, we refer to such circuits as masked circuits.

The theory of masking at the gate level has been analyzed recently in [9]. An implementation of an AES co-processor that is based on masking at the gate level has been presented by Trichina and Korkishko in [22,23]. Additionally, there exist several patents on masking at the gate level (see for example [10], [13] and [14]).

However, an important issue of masking at the gate level has not been considered until now. The security analyses that have been conducted so far assume that each gate in a masked circuit switches no more than once per clock cycle. However, this assumption does not hold true in general. The input signals of a gate in a digital circuit usually do not arrive at the same time. Therefore, the output of a gate potentially switches several times during one clock cycle.

The transitions at the output of a gate that occur before the gate switches to the correct output are called *glitches*. The fact that glitches occur in digital circuits is well known and it is extensively discussed in the literature on VLSI design (see for example [17]). Glitches contribute significantly to the power consumption of CMOS circuits and hence, they are very relevant for the DPA-resistance of these circuits.

In this article, we analyze the effect of glitches on the DPA-resistance of masked gates. In fact we show that several masked CMOS implementations of nonlinear gates, such as AND and OR gates, are not resistant to DPA attacks. These implementations are susceptible to classical first-order DPA attacks. We

 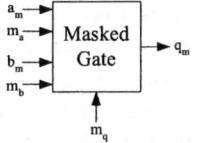

Fig. 1. The inputs and the output of a normal gate.

Fig. 2. The inputs and the output of a masked gate.

show this fact theoretically and we also provide attack results based on SPICE simulations.

This article is organized as follows: Section 2 introduces the concept of masking at the gate level and discusses existing publications and patents on this countermeasure. In Section 3, we perform a thorough theoretical analysis of the DPA-resistance of masked gates in the presence of glitches. Furthermore, we discuss the causes of glitches and elaborate on the effort that is necessary to prevent them. Section 4 presents simulation results of DPA attacks that have been conducted on implementations of masked gates as they have been proposed in [14] and [22, 23]. We show that both approaches lead to gates that are susceptible to DPA attacks in practice. The conclusions of our findings are presented in Section 5.

2 Masking at the Gate Level

The basic idea of masking at the gate level is to represent each value a occurring in a circuit by two values a_m and m_a. m_a is a random mask that is statistically independent of a and uniformly distributed. The masked value a_m is calculated by adding a and m_a modulo two: $a_m = a \oplus m_a$.

In a masked digital circuit, logic gates take the tuple (a_m, m_a) instead of a as input. In fact, all inputs and the output of every logic gate are masked. The inputs and the output of a gate in a normal digital circuit are shown in Figure 1, while the inputs and the output of a gate in a masked digital circuit are shown in Figure 2.

In a normal digital circuit, a gate with two inputs calculates the output q based on the inputs a and b: $q = f(a,b)$. In a masked circuit, the inputs as well as the output are masked. This means that $a_m = a \oplus m_a$, $b_m = b \oplus m_b$ and $q_m = q \oplus m_q$, where m_a, m_b and m_q are randomly generated masks. The masked gate calculates the output q_m of the gate based on the inputs a_m, m_a, b_m, m_b and m_q: $q_m = \tilde{f}(a_m, m_a, b_m, m_b, m_q)$.

For the sake of readability, we only discuss gates with two masked inputs and one masked output. However, this restriction can be done without loss of generality. Our results also hold true for more complex gates. Another restriction we make in this article is that we only analyze masked circuits where one data bit is masked with one mask bit. We do not consider the general case where a value a is masked with several masks: $a = a_m \oplus m_1 \oplus m_2 \oplus \ldots \oplus m_n$.

Using more than one mask bit for one data bit, as for example proposed in [9], is not very practical. Already in the case where only one mask bit is required for each data bit, the generation and the distribution of the mask bits are challenging tasks for the designers of a circuit.

In practical (commercial) applications, area and power restrictions usually rule out the generation of a fresh mask for every data bit in every clock cycle. This approach would essentially mean that for every data bit, one (pseudo-) random number generator would be required. In practical applications, designers have to reuse the same mask for several data signals or they have to use the same mask for several clock cycles.

However, in the context of this article we do not elaborate on the issue of how masks can be generated or distributed. We simply use the best-case assumption concerning the generation and distribution of masks, *i.e.* fresh masks m_a, m_b and m_q can be generated for every gate in every clock cycle. We show that even using this ideal assumption, glitches in masked digital circuits make these circuits susceptible to DPA attacks.

2.1 The Theory Behind Masked Gates

So far, the masking of algorithms has received more attention than the masking of logic gates. For example, there are several publications on how to mask DES [1, 8] and AES [1, 2, 7, 24]. However, there also exist two publications [3, 9] that discuss masking in a more generic way. In particular, [9] discusses the theory of masked gates. In this article, masked circuits are referred to as private circuits. The goal of these circuits is to provide protection against an attacker that can probe a certain number of wires in a circuit. Power-analysis attacks are modelled as probing attacks because they allow the attacker to determine the value of a particular wire.

An important assumption that is implicitly made in [9] is that every wire changes its voltage level no more than once per clock cycle. A digital circuit is modelled as a graph, where the nodes correspond to gates and the connections correspond to wires. The propagation delay of the gates is not considered and therefore, no glitches occur in this model. However, glitches occur in digital circuits in practice and they have a significant impact on the power consumption of a circuit. Therefore, the model proposed in [9] needs to be updated in order to be applicable for circuits as they are used in practice.

The model used in [3] to analyze the security of masking does also not consider the effect of glitches. Hence, also this model needs to be extended accordingly.

2.2 Building Masked Gates Based on Multiplexors

One of the first patents on masking at the gate level has been issued to Messerges, Dabbish, and Puhl in 2001 [14]. This patent describes how an arbitrary logical function can be masked based on multiplexors and crossbar switches. All inputs of the logical function as well as the output are masked. Therefore, the interfaces of masked gates implemented according to [14] correspond to the one shown in Figure 2.

Implementations of masked gates using this approach are relatively big in practice. For example, a 2-input gate consists of 3 multiplexors, 3 crossbar-switches and 4 XOR gates. Nevertheless, in [6] it has been proposed to use this approach to secure a data scrambling technique against power-analysis attacks.

In the current article, we show that masked gates based on multiplexors do not prevent DPA attacks, if glitches occur in the masked circuit.

2.3 Building Masked Gates Based on Correction Terms

In [22] and [23], an alternative approach for the implementation of masked logic gates has been proposed. The basic idea of this approach is to build masked gates based on normal (unmasked) gates.

For example, the masked AND gate that is used in [22] and [23] to implement a masked AES co-processor consists of 4 AND gates and 4 XOR gates. The interface of this AND gate also corresponds to the one shown in Figure 2.

A similar approach as the one presented by Trichina and Korkishko has been patented by Klug, Kniffler, and Gammel in [10]. The main difference between these two approaches is that in the patent, the same mask is used for the inputs and the output, *i.e.* $m_a = m_b = m_q$. This leads to significantly smaller implementations of masked gates.

However, all these approaches are vulnerable to DPA attacks in theory and practice, if glitches occur in the masked circuit.

3 Theoretical Security Analysis of Masked Gates

In digital circuits, logical values are usually represented by voltage levels of wires. The power consumption of a digital circuit is data-dependent because keeping a wire at a certain voltage level requires almost no energy, while the switching of a voltage level requires a significant amount of energy.

We denote the energy that is needed to switch a wire from the voltage representing the value 0 to the voltage representing the value 1 as $E_{0 \rightarrow 1}$. Accordingly, we denote the energy that is needed to perform a $(1 \rightarrow 0)$ transition as $E_{1 \rightarrow 0}$. In practice, these energies are usually different, *i.e.* $E_{0 \rightarrow 1} \neq E_{1 \rightarrow 0}$. Although keeping a wire at a certain voltage level requires almost no energy, we also introduce a notation for these energies. We refer to these energies as $E_{0 \rightarrow 0}$ and $E_{1 \rightarrow 1}$, respectively.

Besides a notation for the energy consumption, also certain assumptions about the data inputs of masked gates are required in order to perform an analysis of the DPA-resistance. In this article, we use the common assumption that the inputs of a gate in a digital circuit are statistically independent and uniformly distributed. Based on this assumption and the notation for the energy consumption, we analyze the DPA-resistance of different logic gates in the following subsections.

Table 1. The transitions a normal AND gate can perform during one clock cycle.

a	b	q	Energy	a	b	q	Energy
$0 \to 0$	$0 \to 0$	$0 \to 0$	$E_{0 \to 0}$	$1 \to 0$	$0 \to 0$	$0 \to 0$	$E_{0 \to 0}$
$0 \to 0$	$0 \to 1$	$0 \to 0$	$E_{0 \to 0}$	$1 \to 0$	$0 \to 1$	$0 \to 0$	$E_{0 \to 0}$
$0 \to 0$	$1 \to 0$	$0 \to 0$	$E_{0 \to 0}$	$1 \to 0$	$1 \to 0$	$1 \to 0$	$E_{1 \to 0}$
$0 \to 0$	$1 \to 1$	$0 \to 0$	$E_{0 \to 0}$	$1 \to 0$	$1 \to 1$	$1 \to 0$	$E_{1 \to 0}$
$0 \to 1$	$0 \to 0$	$0 \to 0$	$E_{0 \to 0}$	$1 \to 1$	$0 \to 0$	$0 \to 0$	$E_{0 \to 0}$
$0 \to 1$	$0 \to 1$	$0 \to 1$	$E_{0 \to 1}$	$1 \to 1$	$0 \to 1$	$0 \to 1$	$E_{0 \to 1}$
$0 \to 1$	$1 \to 0$	$0 \to 0$	$E_{0 \to 0}$	$1 \to 1$	$1 \to 0$	$1 \to 0$	$E_{1 \to 0}$
$0 \to 1$	$1 \to 1$	$0 \to 1$	$E_{0 \to 1}$	$1 \to 1$	$1 \to 1$	$1 \to 1$	$E_{1 \to 1}$

First, we analyze the DPA-resistance of normal (unmasked) gates in Subsection 3.1. This analysis is presented in order to provide a reference for the analysis of masked gates. Subsection 3.2 discusses why masked gates provide DPA-resistance, if no glitches occur in a digital circuit. This is essentially a short summary of the arguments that have been used so far to promote masked gates as a countermeasure against DPA attacks.

In Subsection 3.3, we argue why the assumption that there are no glitches in a digital circuit is typically wrong in practice. This subsection in particular also discusses the effort that is necessary to avoid glitches in digital circuits.

Finally, in Subsection 3.4 we show why masked CMOS gates do not prevent DPA attacks, if glitches occur in a digital circuit.

3.1 Analyzing the DPA-Resistance of Normal Gates

A 2-input AND gate takes the two values a and b as input to calculate $q = a \wedge b$. For our analysis, we assume that the inputs arrive at the same time and that they change their values no more than once per clock cycle. We do not need to consider glitches for our analysis of normal gates because these gates are susceptible to DPA-attacks even if no glitches occur.

Each input of the AND gate can perform one out of four transitions ($0 \to 0$, $0 \to 1$, $1 \to 0$, or $1 \to 1$) during a given clock cycle. Hence, in total there exist $4^2 = 16$ possible combinations of input transitions that can occur. These combinations of input transitions are listed in Table 1. In addition to the input transitions, Table 1 also shows the corresponding output transitions and the energy that is needed to perform these transitions. All 16 cases shown in this table have the same probability of occurrence because the inputs a and b are statistically independent and uniformly distributed.

In a DPA attack on an AND gate that is part of a digital circuit, the power consumption of the circuit is first recorded several times while the circuit performs a cryptographic operation with different inputs. Subsequently, the power measurements are split into two groups. The first group contains all measurements, where $q = 0$ at the end of the clock cycle and the second group contains all measurements, where $q = 1$.

Using the notation introduced in this section, this means that the first group contains the cases where the output performs a $(0 \rightarrow 0)$ or a $(1 \rightarrow 0)$ transition, while the second group contains the remaining cases. The attacker calculates the means of the energies of both groups and subtracts them from each other.

$$\frac{3E_{0\rightarrow 1} + E_{1\rightarrow 1}}{4} \neq \frac{3E_{1\rightarrow 0} + 9E_{0\rightarrow 0}}{12} \qquad (1)$$

The expected values of these two means are in general not equal and hence, there is a leakage of side-channel information. The processing of $q = 0$ requires a different amount of energy than the processing of $q = 1$. In practice, the number of samples that is needed to exploit this energy difference essentially depends on the background noise, e.g. due to other circuit parts, and on the values $E_{0\rightarrow 0}$, $E_{0\rightarrow 1}$, $E_{1\rightarrow 1}$, and $E_{1\rightarrow 1}$.

The corresponding analysis can also easily be carried out for other logic gates, such as OR and XOR. All these gates are susceptible to DPA attacks.

Throughout this article we focus on the correlation between the power consumption of logic gates and the data that is processed by the gates. This correlation determines the number of samples that are needed in DPA attacks in practice (see [12] and [16]).

3.2 Analyzing the DPA-Resistance of Masked Gates in Circuits Without Glitches

Assuming that no glitches occur in a digital circuit, it is relatively easy to proof that masked gates are resistant to DPA attacks. We present the basic idea of these proofs based on a masked 2-input AND gate.

A masked 2-input AND gate takes five signals as input $(a_m, m_a, b_m, m_b, m_q)$ and calculates the output $q_m = ((a_m \oplus m_a) \wedge (b_m \oplus m_b)) \oplus m_q$. The assumption that there are no glitches in a digital circuit means that every input and output signal switches only once per clock cycle. Every input can perform one out of four transitions during a given clock cycle. Hence, there are $4^5 = 1024$ possible combinations of input transitions that can occur.

Like in the previous subsection, we have created a table containing all possible input transitions, the corresponding output transitions and the energies consumed by these output transitions. Based on this table it is possible to determine whether the processing of $q = 0$ and the processing of $q = 1$ require different amounts of energy or not.

In fact, it turns out that the expected value of the energy that is needed to process $q = 0$ and the corresponding expected value for the processing of $q = 1$ are identical. Furthermore, the table can be used to show that also DPA attacks on the inputs a and b are not possible. Assuming that there are no glitches in a digital circuit, the energy dissipation of a masked AND gate is indeed independent of the unmasked inputs and the unmasked output. Accordingly, it can be shown that implementations of other masked gates (OR, XOR, ...), as described in [10], [14], [22], and [23] are also resistant against first-order DPA attacks.

This fact has been used in the past to promote masked gates. However, in the following subsection, we discuss why the assumption that there occur no glitches in digital circuits usually does not hold true in practice.

3.3 Timing and Switching Characteristics of Digital Circuits

In practice, digital circuits are usually implemented based on CMOS (see [17]). Logical functions are realized by connecting multiple CMOS gates to each other. An important property of these gates is that they have a certain propagation delay, *i.e.* it takes a certain amount of time until the output of a gate reacts to a change at an input of the gate.

This property has a significant impact on the switching activity of a digital circuit. In such a circuit, the input signals of a gate are the outputs of different combinational paths. These paths do not necessarily have the same length. For example, it can happen that the input a of a gate always arrives earlier than the input b. The consequence of such a delay between the input signals is that the gate switches its output more than once per clock cycle. The output switches when the input a performs a transition and it switches again when the input b performs a transition. It is important to note that in the time span between the arrival of the two input signals, the output of the gate is switched to a "wrong" value. This "wrong" value is potentially the input of another logic gate. Of course, such a gate reacts to this transition at its input and changes its output based on the "wrong" input value. In this way, "wrong" values propagate through the circuit.

The consequence of all this is that a lot of unintended switching activity takes place before every wire in a combinational circuit settles to the final value. In practice, glitches account for a significant amount of the power consumption of a circuit. Hence, glitches cannot be neglected in a thorough analysis of the DPA-resistance of masked gates. In the following subsection, we show that glitches make masked gates susceptible to DPA attacks.

Glitches occur in classical CMOS circuits and of course they also occur in masked circuits that are based on CMOS. However, besides CMOS there are many other logic styles that can be used to implement digital circuits. Among them, there are actually some that prevent glitches.

Glitches do not occur in so-called domino logic styles, such as for example pre-charged NMOS [17], DCVSL [17] or SABL [20]. However, pre-charged circuits are usually bigger than corresponding CMOS circuits. Another major disadvantage of these logic styles compared to CMOS is the lack of automated off-the-shelf circuit synthesis tools.

The papers and patents that have been published so far on masking at the gate level do not address the problem of glitches. Therefore, readers of these publications might implicitly assume that masking can be implemented based on CMOS. However, as we point out in the following subsection, this is not the case. In order to be sure that masked circuits are DPA-resistant, a logic style that prevents glitches needs to be used. This significantly increases the implementation costs of masked circuits.

3.4 Analyzing the Effect of Glitches on the DPA-Resistance of Masked Gates

In digital circuits that are based on CMOS, the input signals of logic gates can arrive at different moments of time. Furthermore, these signals switch potentially several times during one clock cycle. We now analyze the impact of these facts on an implementation of a masked 2-input AND gate.

In order to simplify the analysis, we make certain assumptions about how the delays between the input signals look like and about how often the input signals switch per clock cycle. However, these assumptions do not mean a loss of generality.

We assume that each input signal switches once per clock cycle and that at least one of the five input signals arrives at a different time than the other signals. Furthermore, if there is a difference between the arrival time of two signals, this difference is always assumed to be bigger than the propagation delay of the masked gate.

For the analysis of the susceptibility of the masked AND gate, we have used the same technique as in the previous subsections. We have created tables with the input transitions, the output transitions and the energy that is needed to perform the transitions.

First we have looked at the scenarios where only one of the five inputs arrives at a different moment of time than the remaining four inputs. There exist ten such scenarios. There are five input signals and each one of them can arrive either before or after the four other ones. One scenario is for example that m_q arrives first and that a_m, m_a, b_m and m_b arrive later.

Like in Subsection 3.2, in every scenario there exist $4^5 = 1024$ possible combinations of transitions that can occur at the inputs. However, in the ten scenarios where the inputs arrive at two different moments of time, the output of the masked AND gate performs two transitions instead of one. One transition is performed when the single input performs a transition and another one is performed when the other four input signals perform a transition.

We have analyzed whether the energy dissipation that is needed to perform these two transitions is correlated to $q = q_m \oplus m_q$ or not. This was done by calculating the expected value for the energy needed to process $q = 0$ and the corresponding expected value for $q = 1$ (see Subsection 3.1). The same has also been done for the unmasked inputs a and b. A masked gate is only resistant to DPA-attacks if the energy dissipation of the gate is uncorrelated to all unmasked inputs and outputs.

Unfortunately, it has turned out that in all ten scenarios the energy that is needed to perform the two transitions of the output is correlated to a, b or q. We have also investigated all remaining scenarios. These are for example the scenarios where two inputs arrive at separate moments of time either before or after the remaining three arrive. However, in the analysis of all scenarios, starting from the one where only one signal arrives at a different time to the scenario where all inputs arrive at separate moments of times, there has always been a correlation to a, b or q.

In practice, different arrival times are very common. In case of a masked gate, it is in particular very likely that the masks m_a, m_b and m_q arrive at different moments of time than the inputs a_m and b_m. The reason for this is that the masks are generated by a completely different part of the digital circuit.

Based on the scenarios we have analyzed in this section, we have to conclude that there exists no implementation of a masked AND gate based on CMOS that is resistant to DPA attacks. We have performed the same analysis as for the masked AND gate also for masked NAND, OR, NOR, XOR and XNOR gates.

It has turned out that masked nonlinear gates, such as AND, NAND, OR and NOR gates, are susceptible to DPA-attacks, while masked linear gates, such as XOR and XNOR gates, are resistant to DPA attacks. However, for implementations of operations like the AES S-Box, nonlinear operations are crucial (see [18] and [25]).

Therefore, the conclusion of our theoretical analysis is that all published gate-level masking schemes need to be implemented based on a logic style that prevents glitches.

4 The DPA-Resistance of Masked Gates in Practice

In order to empirically verify the results of the theoretical analysis presented in the previous section, we have performed simulated DPA attacks on implementations of masked 2-input AND gates. For this purpose, we have implemented the masked AND gate presented in [14] and a masked AND gate based on the approach described in [22, 23]. Both gates have been implemented using a CMOS standard cell library based on a 0.35 μm technology.

We have performed SPICE simulations of these gates for two scenarios. In the first scenario, all five inputs of the masked gates have arrived at the same time. In the second scenario, the output mask m_q has arrived first and the remaining inputs have arrived one nanosecond later. Like in the theoretical analysis, each input signal has only performed one transition per clock cycle.

We have simulated one power trace for each of the $4^5 = 1024$ combinations of input transitions that can occur. Subsequently, a DPA attack on q has been performed. The goal was to check whether the mean power consumption for $q = 0$ and the mean power consumption for $q = 1$ are indeed different or not. It is important to point out that the AND gates have been implemented exactly as described in [14] and [22, 23], respectively. Hence, there was no wire in the circuit that stored q directly. However, the glitches in the gates have lead to the fact that the power consumptions of the masked gates were correlated to q.

In order to provide a reference for the detected correlations, we have also performed a DPA attack based on simulated power traces on a normal AND gate. The results of all attacks are shown in Figure 3. The first three plots are the result of attacks that are based on simulations where the inputs have arrived at the same time. Even in this scenario, the power consumptions of the masked gates are correlated to q. In fact, this is not surprising. The masked gates consist

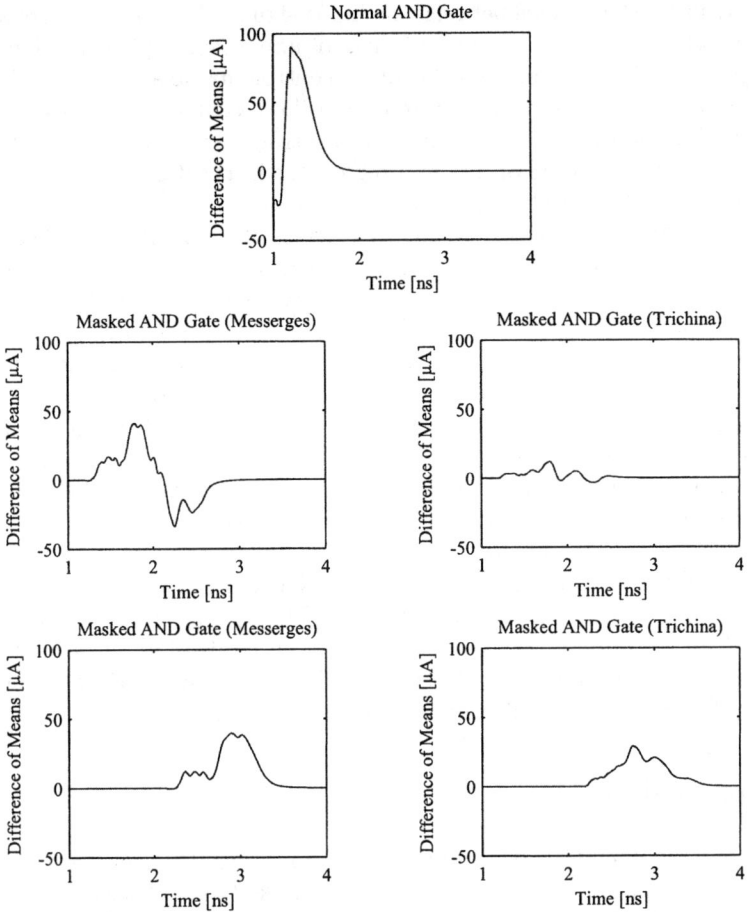

Fig. 3. The results of DPA attacks on a normal AND gate and of attacks on masked AND gates implemented according to [14] and [22, 23].

of unmasked CMOS gates. Consequently, even if the inputs arrive at the same time, glitches occur in the masked AND gates.

The last two plots show the results of attacks on implementations where m_q arrives one nanosecond before the other inputs. This time difference affects in particular the implementation according to [22, 23]. The time difference leads to a significant increase of the maximum of the DPA peak that occurs in the attack.

In the two scenarios we have analyzed for the masked gates, the DPA peaks that occur are obviously smaller than the peak that occurs in an attack on the normal AND gate. However, the two scenarios are just examples of attacks on the output q. We have also performed attacks on a and b and we have also looked at scenarios with other delays between the input signals. In fact, there are actually scenarios where peaks in the range of those of unmasked implementations occur.

In practice, it is extremely difficult to control the delay between the input signals of a gate. In the semi-custom design flows that are usually used to implement ICs, the designer has almost no control over these delays. Therefore, almost any delay scenario occurs in a big circuit in practice.

The goal of this article is to show that glitches are a problem for the DPA-resistance of masked CMOS circuits. We have not explicitly searched for the scenario that maximizes the DPA peak occurring in an attack on a particular implementation. Instead, we have presented two simple scenarios that should make our point clear. Already in these simple scenarios, the maxima of the DPA peaks are only a little bit more than halved by masking the gate. This is definitely less than one would expect from this countermeasure. A reduction of the DPA peak in this range can also be achieved by more inexpensive countermeasures such as the generation of noise [4, 12].

A last point that is important to mention is that the results of the simulated attacks presented in this section can not be compared directly with our theoretical analysis conducted before. The reason for this is the fact that the masked gates are built with unmasked CMOS gates. Hence, glitches occur not only outside the gates, but also inside the gates. The DPA peaks shown in Figure 3 are the result of the superposition of the effect of all kinds of glitches. However, as discussed in the theoretical analysis, masked gates are also susceptible to DPA attacks, if glitches occur only outside the masked gates.

5 Conclusions

There are several publications and patents on masking at the gate level. We have shown that all proposed implementations of masked gates based on CMOS are susceptible to DPA attacks because of glitches. Glitches have been completely ignored in previous analyses of masking at the gate level.

In this article, we have performed a theoretical analysis of the effect of glitches on masked gates. Furthermore, we have presented results of DPA attacks based on SPICE simulations of masked gates as they have been proposed in [14] and [22, 23]. Both approaches have turned out to be susceptible to DPA attacks.

Glitches in digital circuits can be prevented by using domino logic styles. However, implementations based on such logic styles are usually bigger than implementations based on CMOS. Also the design effort for circuits using domino logic styles is significantly higher than the one for corresponding CMOS circuits. This is a consequence of the fact that commercial synthesis tools for domino logic styles are currently not available. Hence, the protection of digital circuits against DPA attacks based on masked logic gates is very expensive in practice.

References

1. Mehdi-Laurent Akkar and Christophe Giraud. An Implementation of DES and AES, Secure against Some Attacks. In Çetin Kaya Koç, David Naccache, and Christof Paar, editors, *Cryptographic Hardware and Embedded Systems - CHES 2001, Third International Workshop, Paris, France, May 14-16, 2001, Proceedings*, volume 2162 of *Lecture Notes in Computer Science*, pages 309–318. Springer, 2001.
2. Johannes Blömer, Jorge Guajardo Merchan, and Volker Krummel. Provably Secure Masking of AES. Cryptology ePrint Archive (http://eprint.iacr.org/), Report 2004/101, 2004.
3. Suresh Chari, Charanjit S. Jutla, Josyula R. Rao, and Pankaj Rohatgi. Towards Sound Approaches to Counteract Power-Analysis Attacks. In Michael J. Wiener, editor, *Advances in Cryptology - CRYPTO '99, 19th Annual International Cryptology Conference, Santa Barbara, California, USA, August 15-19, 1999, Proceedings*, volume 1666 of *Lecture Notes in Computer Science*, pages 398–412. Springer, 1999.
4. Christophe Clavier, Jean-Sébastien Coron, and Nora Dabbous. Differential Power Analysis in the Presence of Hardware Countermeasures. In Çetin Kaya Koç and Christof Paar, editors, *Cryptographic Hardware and Embedded Systems - CHES 2000, Second International Workshop, Worcester, MA, USA, August 17-18, 2000, Proceedings*, volume 1965 of *Lecture Notes in Computer Science*, pages 252–263. Springer, 2000.
5. Jean-Sébastien Coron. Resistance against Differential Power Analysis for Elliptic Curve Cryptosystems. In Çetin Kaya Koç and Christof Paar, editors, *Cryptographic Hardware and Embedded Systems, First International Workshop, CHES'99, Worcester, MA, USA, August 12-13, 1999, Proceedings*, volume 1717 of *Lecture Notes in Computer Science*, pages 292–302. Springer, 1999.
6. Jovan D. Golić. DeKaRT: A New Paradigm for Key-Dependent Reversible Circuits. In Colin D. Walter, Çetin Kaya Koç, and Christof Paar, editors, *Cryptographic Hardware and Embedded Systems - CHES 2003, 5th International Workshop, Cologne, Germany, September 8-10, 2003, Proceedings*, volume 2779 of *Lecture Notes in Computer Science*, pages 98–112. Springer, 2003.
7. Jovan D. Golić and Christophe Tymen. Multiplicative Masking and Power Analysis of AES. In Burton S. Kaliski Jr., Çetin Kaya Koç, and Christof Paar, editors, *Cryptographic Hardware and Embedded Systems - CHES 2002, 4th International Workshop, Redwood Shores, CA, USA, August 13-15, 2002, Revised Papers*, volume 2535 of *Lecture Notes in Computer Science*, pages 198–212. Springer, 2003.
8. Louis Goubin and Jacques Patarin. DES and Differential Power Analysis - The Duplication Method. In Çetin Kaya Koç and Christof Paar, editors, *Cryptographic Hardware and Embedded Systems, First International Workshop, CHES'99, Worcester, MA, USA, August 12-13, 1999, Proceedings*, volume 1717 of *Lecture Notes in Computer Science*, pages 158–172. Springer, 1999.
9. Yuval Ishai, Amit Sahai, and David Wagner. Private Circuits: Securing Hardware against Probing Attacks. In Dan Boneh, editor, *Advances in Cryptology - CRYPTO 2003, 23rd Annual International Cryptology Conference, Santa Barbara, California, USA, August 17-21, 2003, Proceedings*, volume 2729 of *Lecture Notes in Computer Science*, pages 463–481. Springer, 2003.
10. Franz Klug, Oliver Kniffler, and Berndt Gammel. Rechenwerk, Verfahren zum Ausführen einer Operation mit einem verschlüsselten Operanden, Carry-Select-Addierer und Kryptographieprozessor. German Patent DE 10201449 C1, January 2002.

11. Paul C. Kocher, Joshua Jaffe, and Benjamin Jun. Differential Power Analysis. In Michael Wiener, editor, *Advances in Cryptology - CRYPTO '99, 19th Annual International Cryptology Conference, Santa Barbara, California, USA, August 15-19, 1999, Proceedings*, volume 1666 of *Lecture Notes in Computer Science*, pages 388–397. Springer, 1999.
12. Stefan Mangard. Hardware Countermeasures against DPA – A Statistical Analysis of Their Effectiveness. In Tatsuaki Okamoto, editor, *Topics in Cryptology - CT-RSA 2004, The Cryptographers' Track at the RSA Conference 2004, San Francisco, CA, USA, February 23-27, 2004, Proceedings*, volume 2964 of *Lecture Notes in Computer Science*, pages 222–235. Springer, 2004.
13. Renato Menicocci and Johan Pascal. Elaborazione Crittografica di Dati Digitali Mascherati. Italian Patent IT MI0020031375A, July 2003.
14. Thomas S. Messerges, Ezzy A. Dabbish, and Larry Puhl. Method and Apparatus for Preventing Information Leakage Attacks on a Microelectronic Assembly. US Patent 6,295,606, September 2001. Available online at http://www.uspto.gov/.
15. Thomas S. Messerges, Ezzy A. Dabbish, and Robert H. Sloan. Power Analysis Attacks of Modular Exponentiation in Smartcards. In Çetin Kaya Koç and Christof Paar, editors, *Cryptographic Hardware and Embedded Systems, First International Workshop, CHES'99, Worcester, MA, USA, August 12-13, 1999, Proceedings*, volume 1717 of *Lecture Notes in Computer Science*, pages 144–157. Springer, 1999.
16. Thomas S. Messerges, Ezzy A. Dabbish, and Robert H. Sloan. Examining Smart-Card Security under the Threat of Power Analysis Attacks. *IEEE Transactions on Computers*, 51(5):541–552, January 2002.
17. Jan M. Rabaey. *Digital Integrated Circuits*. Prentice Hall, 1996. ISBN 0-13-178609-1.
18. Akashi Satoh, Sumio Morioka, Kohji Takano, and Seiji Munetoh. A Compact Rijndael Hardware Architecture with S-Box Optimization. In Colin Boyd, editor, *Advances in Cryptology - ASIACRYPT 2001, 7th International Conference on the Theory and Application of Cryptology and Information Security, Gold Coast, Australia, December 9-13, 2001, Proceedings*, volume 2248 of *Lecture Notes in Computer Science*, pages 239–254. Springer, 2001.
19. Adi Shamir. Protecting Smart Cards from Passive Power Analysis with Detached Power Supplies. In Çetin Kaya Koç and Christof Paar, editors, *Cryptographic Hardware and Embedded Systems – CHES 2000, Second International Workshop, Worcester, MA, USA, August 17-18, 2000, Proceedings*, volume 1965 of *Lecture Notes in Computer Science*, pages 71–77. Springer, 2000.
20. Kris Tiri and Ingrid Verbauwhede. Securing Encryption Algorithms against DPA at the Logic Level: Next Generation Smart Card Technology. In Colin D. Walter, Çetin Kaya Koç, and Christof Paar, editors, *Cryptographic Hardware and Embedded Systems - CHES 2003, 5th International Workshop, Cologne, Germany, September 8-10, 2003, Proceedings*, volume 2779 of *Lecture Notes in Computer Science*, pages 137–151. Springer, 2003.
21. Kris Tiri and Ingrid Verbauwhede. A Logic Level Design Methodology for a Secure DPA Resistant ASIC or FPGA Implementation. In *2004 Design, Automation and Test in Europe Conference and Exposition (DATE 2004), 16-20 February 2004, Paris, France*, pages 246–251. IEEE Computer Society, 2004.
22. Elena Trichina. Combinational Logic Design for AES SubByte Transformation on Masked Data. Cryptology ePrint Archive (http://eprint.iacr.org/), Report 2003/236, 2003.

23. Elena Trichina and Tymur Korkishko. Small Size, Low Power, Side Channel-Immune AES Coprocessor: Design and Synthesis Results. In *Proceedings of the Fourth Conference on the Advanced Encryption Standard (AES)*, 2004.
24. Elena Trichina, Domenico De Seta, and Lucia Germani. Simplified Adaptive Multiplicative Masking for AES. In Burton S. Kaliski Jr., Çetin Kaya Koç, and Christof Paar, editors, *Cryptographic Hardware and Embedded Systems – CHES 2002, 4th International Workshop, Redwood Shores, CA, USA, August 13-15, 2002, Revised Papers*, volume 2535 of *Lecture Notes in Computer Science*, pages 187–197. Springer, 2003.
25. Johannes Wolkerstorfer, Elisabeth Oswald, and Mario Lamberger. An ASIC implementation of the AES SBoxes. In Bart Preneel, editor, *Topics in Cryptology - CT-RSA 2002, The Cryptographer's Track at the RSA Conference, 2002, San Jose, CA, USA, February 18-22, 2002*, volume 2271 of *Lecture Notes in Computer Science*, pages 67–78. Springer, 2002.

New Minimal Weight Representations for Left-to-Right Window Methods

James A. Muir[1] and Douglas R. Stinson[2],*

[1] Department of Combinatorics and Optimization, University of Waterloo
Waterloo, Ontario, Canada N2L 3G1
jamuir@uwaterloo.ca

[2] School of Computer Science, University of Waterloo
Waterloo, Ontario, Canada N2L 3G1
dstinson@uwaterloo.ca

Abstract. For an integer $w \geq 2$, a radix 2 representation is called a *width-w nonadjacent form* (w-NAF, for short) if each nonzero digit is an odd integer with absolute value less than 2^{w-1}, and, of any w consecutive digits, at most one is nonzero. In elliptic curve cryptography, the w-NAF *window method* is used to efficiently compute nP where n is an integer and P is an elliptic curve point. We introduce a new family of radix 2 representations which use the same digits as the w-NAF but have the advantage that they result in a window method which uses less memory. This memory savings results from the fact that these new representations can be deduced using a very simple *left-to-right* algorithm. Further, we show that like the w-NAF, these new representations have a minimal number of nonzero digits.

1 Window Methods

An operation fundamental to elliptic curve cryptography is *scalar multiplication*; that is, computing nP for an integer, n, and an elliptic curve point, P. A number of different algorithms have been proposed to perform this operation efficiently (see Ch. 3 of [7] for a recent survey). A variety of these algorithms, known as *window methods*, use the approach described in Algorithm 1.1.

For example, suppose $D = \{0, \pm 1, \pm 3\}$. Then Algorithm 1.1 first computes and stores P and $3P$. After a D-radix 2 representation of n is computed its digits are read from left to right by the "for" loop and nP is computed using doubling, addition and subtraction operations. Including negative digits in D takes advantage of the fact that subtracting an elliptic curve point can be done just as efficiently as adding it. A D-radix 2 representation of n can be computed by sliding a window of width 3 from right to left across the $\{0, 1\}$-radix 2 representation of n (see Section 3).

---

* Supported by NSERC grant RGPIN 203114-02.

Algorithm 1.1: RADIX-2-WINDOW-METHOD(n, P)

fix a set of digits, $D \subset \mathbb{Z}$.
for each $d \in D$ with $d > 0$
  do $P_d \leftarrow dP$
compute and store a representation $(a_{\ell-1} \ldots a_1 a_0)_2 = n$ with $a_i \in D$.
$Q \leftarrow \infty$
for $i = \ell - 1 \ldots 0$
$$\text{do} \begin{cases} Q \leftarrow 2Q \\ \text{if } a_i \neq 0 \\ \quad \text{then} \begin{cases} \text{if } a_i > 0 \\ \quad \text{then } Q \leftarrow Q + P_{a_i} \\ \quad \text{else } Q \leftarrow Q - P_{-a_i} \end{cases} \end{cases}$$
return Q

Blake, Seroussi and Smart [2], Cohen, Miyaji and Ono [4] and Solinas [14] independently suggested a specialization of Algorithm 1.1 called the *width-w nonadjacent form window method* (this terminology is due to Solinas). We introduce it now.

For an integer $w \geq 2$, a radix 2 representation is called a *width-w nonadjacent form* (*w*-NAF, for short) if it satisfies the following conditions:

1. Each nonzero digit is an odd integer with absolute value less than 2^{w-1}.
2. Of any w consecutive digits, at most one is nonzero.

For example, a 3-NAF of 42 is given by $(300\bar{3}0)_2$ (note that $\bar{1}$ denotes -1, $\bar{3}$ denotes -3, etc.) as it satisfies conditions 1. and 2., and

$$(300\bar{3}0)_2 = 3 \cdot 2^4 + 0 \cdot 2^3 + 0 \cdot 2^2 - 3 \cdot 2^1 + 0 \cdot 2^0 = 42.$$

When $w = 2$, the w-NAF coincides with the well known *nonadjacent form* [5]. Because of this, the w-NAF may be regarded as a generalization of the ordinary NAF. As with the ordinary NAF, an integer n has exactly one w-NAF, and it can be efficiently computed; hence, we refer to *the* w-NAF of n.

Let D_w be the set of w-NAF digits; that is,

$$D_w := \{0\} \cup \{d \in \mathbb{Z} : d \text{ odd}, |d| < 2^{w-1}\}.$$

If, in Algorithm 1.1, the digit set D_w is used and the representation $(a_{\ell-1} \ldots a_1 a_0)_2$ is always chosen to be a w-NAF, then this is the w-NAF window method.

One advantage of using the w-NAF of an integer is that it has a *minimal number of nonzero digits* [1, 11]. A nonzero integer has an infinite number of D_w-radix 2 representations and any of these representation could be used in Algorithm 1.1. However, the choice of representation affects the performance of the algorithm. In the "for" loop, an addition/subtraction operation is performed for every nonzero digit of $(a_{\ell-1} \ldots a_1 a_0)_2$. It is thus desirable to use a D_w-radix

2 representation of n with as few nonzero digits as possible. No other D_w-radix 2 representation of an integer has fewer nonzero digits than its w-NAF[1].

The w-NAF of an integer is computed by sliding a window of width w from right to left across the $\{0,1\}$-radix 2 representation of n. This procedure deduces the digits of the w-NAF from right to left; however, the "for" loop of Algorithm 1.1 reads these digits from left to right. This means that the w-NAF of n must be computed and stored in its entirety before computations inside the "for" loop can begin.

This problem of opposing directions occurs in many window methods and has been lamented by both Müller [12] and Solinas [14]. If the algorithm which computes the D_w-radix 2 representation of n worked in the same direction as the "for" loop, Algorithm 1.1 could be modified so that it uses less memory. In that case, it would be unnecessary to store the representation $(a_{\ell-1}\ldots a_1 a_0)_2$ since its digits could be computed inside the "for" loop as they are needed. This savings is most relevant for memory constrained devices like smartcards.

We propose a new family of D_w-radix 2 representations and prove that, like the w-NAFs, these representations have a minimal number of nonzero digits. The digits of these representations can be deduced from left to right and thus can be used to reduce the memory requirements of Algorithm 1.1.

Joye and Yen [9] give a very simple left-to-right algorithm for computing the digits of a $\{0, \pm 1\}$-radix 2 representation of an integer. They also prove that the representations constructed by this algorithm have a minimal number of nonzero digits. Their results apply to the digit set D_2, whereas ours apply to arbitrary D_w with $w \geq 2$.

The outline of the paper is as follows. In Section 2, we introduce the algorithm used to construct our representations and then, in Section 3, describe how it can be efficiently implemented. In Section 4, we prove minimality. We end with a discussion of related work and some remarks.

2 The Algorithm

We introduce an algorithm for computing the digits of a D_w-radix 2 representation from left to right. Our discussion is mainly intended to illustrate the idea behind algorithm. Details about how the algorithm can be efficiently implemented are postponed until Section 3.

In order to motivate our algorithm, we begin by describing a simple method of computing the $\{0,1\}$-radix 2 representation of a positive integer. This could be used if an integer were represented in some other way; for example, if we wanted to convert from radix 10 to radix 2.

Suppose we want to deduce the digits of the $\{0,1\}$-radix 2 representation of 233 from left to right. This is easily done by subtracting powers of 2. The number $n = 233$ is small enough so that we can quickly determine $2^{\lfloor \lg n \rfloor}$; this is the power of 2 closest to, but not larger than, n. Once we determine $2^{\lfloor \lg n \rfloor}$, we

---

[1] Cohen [3] presents a detailed average case analysis of the w-NAF window method.

replace n with $n - 2^{\lfloor \lg n \rfloor}$ and then repeat these steps until we reach 0. Doing so gives us

n	$2^{\lfloor \lg n \rfloor}$
233	$128 = 2^7$
105	$64 = 2^6$
41	$32 = 2^5$
9	$8 = 2^3$
1	$1 = 2^0$.

Thus, we see that $233 = 2^7 + 2^6 + 2^5 + 2^3 + 2^0 = (11101001)_2$.

We can modify this process so that it returns a $\{0,1\}$-string. We begin with a string, α, which is initially empty. In each step, we append to α a (possibly empty) run of 0's followed by a single 1. Doing so gives us

n	$2^{\lfloor \lg n \rfloor}$	α
233	$128 = 2^7$	$\alpha \parallel 1$
105	$64 = 2^6$	$\alpha \parallel 1$
41	$32 = 2^5$	$\alpha \parallel 1$
9	$8 = 2^3$	$\alpha \parallel 01$
1	$1 = 2^0$	$\alpha \parallel 001$.

Note that the symbol $\parallel$ denotes concatenation. When n reaches 0, α is equal to 11101001 and we see that $(\alpha)_2 = 233$.

For an arbitrary nonnegative integer, we can describe this process in pseudocode. Let $D = \{0, 1\}$ and define

$$\mathcal{C} := \{d \cdot 2^i : d \in D \setminus \{0\}, i \in \mathbb{Z}, i \geq 0\}.$$

The set $\mathcal{C}$ consists of all the positive powers of 2. Here is a description of the procedure:

$\alpha \leftarrow \epsilon$
while $n \neq 0$
$\quad$ **do** $\begin{cases} c \leftarrow \text{an element in } \mathcal{C} \text{ closest to, but not larger than, } n \\ \text{append digits to } \alpha \text{ according to the value of } c \\ n \leftarrow n - c \end{cases}$
return α

Note that ϵ denotes the empty string. The set $\mathcal{C}$ is infinite, however this is not a concern since we do not need to store $\mathcal{C}$. To choose an element in $\mathcal{C}$ closest to, but not larger than, n, we simply compute $2^{\lfloor \lg n \rfloor}$.

Consider now the digit set $D_2 = \{0, \pm 1\}$. We would like to somehow deduce the digits of a D_2-radix 2 representation of an integer from left to right. We can do this by modifying our previous procedure slightly. We first define

$$\mathcal{C}_2 := \{d \cdot 2^i : d \in D_2 \setminus \{0\}, i \in \mathbb{Z}, i \geq 0\}.$$

Note that $\mathcal{C}_2$ consists of the positive and negative powers of 2. Now consider the following procedure:

$$\alpha \leftarrow \epsilon$$
while $n \neq 0$
do $\begin{cases} c \leftarrow \text{an element in } \mathcal{C}_2 \text{ closest to } n \\ \text{append digits to } \alpha \text{ according to the value of } c \\ n \leftarrow n - c \end{cases}$
return α

The only change above is that the condition "closest to, but not larger than, n" is now simply "closest to n". If we apply this procedure to $n = 233$ we get

n	c	α
233	$256 = 2^8$	$\alpha \parallel 1$
-23	$-16 = -2^4$	$\alpha \parallel 000\bar{1}$
-7	$-8 = -2^3$	$\alpha \parallel \bar{1}$
1	$1 = 2^0$	$\alpha \parallel 001$.

When n reaches 0, α is equal to $1000\bar{1}\bar{1}001$ and we see that

$$(\alpha)_2 = (1000\bar{1}\bar{1}001)_2 = 2^8 - 2^4 - 2^3 + 2^0 = 233.$$

This same example is worked by Joye and Yen [9] and our representation is identical to theirs. Note also that, as in the previous case, the set $\mathcal{C}_2$ does not need to be stored. To choose a closest element from $\mathcal{C}_2$, we compute $2^{\lfloor \lg |n| \rfloor}$ and then compare $|n|$ to $2^{\lfloor \lg |n| \rfloor}$ and $2^{\lfloor \lg |n| \rfloor + 1}$.

In the general case, we would like to construct D_w-radix 2 representations from left to right for arbitrary $w \geq 2$. Here is a procedure which does so:

Algorithm 2.1: MSF$_w(n)$

comment: $w \geq 2$, $D_w = \{0\} \cup \{d \in \mathbb{Z} : d \text{ odd}, |d| < 2^{w-1}\}$, and
$\mathcal{C}_w = \{d \cdot 2^i : d \in D_w \setminus \{0\}, i \in \mathbb{Z}, i \geq 0\}$

$\alpha \leftarrow \epsilon$
while $n \neq 0$
do $\begin{cases} c \leftarrow \text{an element in } \mathcal{C}_w \text{ closest to } n \\ \text{append digits to } \alpha \text{ according to the value of } c \\ n \leftarrow n - c \end{cases}$
return α

As before, the set $\mathcal{C}_w$ does not need to be stored. We will see in Section 3 that this procedure can be implemented efficiently by sliding a window of width $w+1$ from left to right across the $\{0, 1\}$-radix 2 representation of n. A description of how digits are appended to α will be provided shortly. We call this procedure MSF$_w(n)$.

We have given this procedure the title "Algorithm". To justify this we must show that MSF$_w(n)$ terminates for all $n \in \mathbb{Z}$. If $n = 0$, then MSF$_w(n)$ clearly terminates, so we need only consider $n \neq 0$. To finish the argument we need a Lemma.

Lemma 1 ([10]). *Let n be a nonzero integer. If c is an element in C_w closest to n, then*

$$|n - c| \leq \frac{2^{\lfloor \lg |n| \rfloor}}{2^{w-1}}.$$

A proof of Lemma 1, as well as proofs of some other Lemmas we require later on, can be found in the extended version of this paper [10].

To show that $\text{MSF}_w(n)$ terminates for $n \neq 0$, it suffices to show that $|n| > |n - c|$. Suppose to the contrary that $|n| \leq |n - c|$. Then

$$|n| \leq |n - c|$$
$$\implies |n| \leq 2^{\lfloor \lg |n| \rfloor}/2^{w-1} \quad \text{(by Lemma 1)}$$
$$\implies 2^{\lfloor \lg |n| \rfloor} \leq 2^{\lfloor \lg |n| \rfloor}/2^{w-1}$$
$$\implies 1 \leq 1/2^{w-1}$$
$$\implies w \leq 1.$$

However, this is a contradiction because $w \geq 2$. So, the sequence formed by taking the absolute value of the variable n during the execution of $\text{MSF}_w(n)$ is strictly decreasing. Thus, the variable n must reach 0, and so $\text{MSF}_w(n)$ terminates for all $n \in \mathbb{Z}$.

The string α returned by $\text{MSF}_w(n)$ has been defined somewhat informally. We present a more rigorous definition based on the values that the variable c assumes during the execution of $\text{MSF}_w(n)$. For an input, n, we define $\alpha = \ldots a_2 a_1 a_0$ to be the string such that

$$a_i := \begin{cases} d & \text{if } c \text{ assumes the value } d \cdot 2^i \text{ at some point in the algorithm,} \\ 0 & \text{otherwise.} \end{cases} \quad (1)$$

Clearly, each $a_i \in D_w$, and so α is a D_w-string. There is, however, one possible problem with this definition. Suppose c assumes the two distinct values $d_0 \cdot 2^i$ and $d_1 \cdot 2^i$ which share the same power of 2. In that case, the value of a_i is undefined. Fortunately, this problem never occurs, as is shown in the following Lemma.

Lemma 2 ([10]). *Let c_0, c_1 and n be nonzero integers such that c_0 is an element in C_w closest to n and c_1 is an element in C_w closest to $n - c_0$. If $c_0 = d_0 2^{i_0}$ and $c_1 = d_1 2^{i_1}$ with $d_0, d_1 \in D_w$, then $i_0 > i_1$.*

By Lemma 2, the string α is well defined. Moreover, as we saw in our earlier examples, Lemma 2 tells us that α can be constructed using operations of the form

$$\alpha \leftarrow \alpha \parallel 0^t d \quad \text{where} \quad t \geq 0, \ d \in D_w, \ d > 0.$$

Actually, we need to be a bit more precise here. If n is odd, then α can be constructed using only operations like the one above; however, if n is even, then α will need to have a run of zeros appended to it before it is returned.

From the definition given in (1) we can now show that the string returned by $\mathrm{MSF}_w(n)$ is in fact a D_w-radix 2 representation of n (i.e., the algorithm is correct). Let S be the set of values that the variable c assumes during the execution of $\mathrm{MSF}_w(n)$. For $\alpha = \ldots a_1 a_0$ we have

$$(\alpha)_2 = \sum_{i \geq 0} a_i 2^i = \sum_{\substack{i \\ a_i \neq 0}} a_i 2^i = \sum_{c \in S} c = n.$$

The representation returned by $\mathrm{MSF}_w(n)$ for a given value of n is not necessarily unique. For example, when $w = 3$, $D_3 = \{0, \pm 1, \pm 3\}$ and for $n = 5$ we see that both $4 = 2^2$ and $6 = 3 \cdot 2^1$ are elements in C_3 closest to 5. Thus, $\mathrm{MSF}_3(5)$ will return one of the representations

$$(101)_2 \quad \text{or} \quad (3\bar{1})_2.$$

From the description of Algorithm 2.1, it is apparent that $\mathrm{MSF}_w(n)$ will have more than one possible output only when some value of the variable n has more than one closest element in C_w. This occurs only when a value of the variable n is the midpoint between neighbouring elements of C_w.

We argue that there are at most two distinct outputs of $\mathrm{MSF}_w(n)$ for any $n \in \mathbb{Z}$. Imagine a list of outputs of $\mathrm{MSF}_w(n)$. Let i_0 be the largest value of i such that two outputs differ at digit i. If i_0 does not exist then all the outputs are the same; otherwise, let $\alpha = \ldots a_2 a_1 a_0$ and $\beta = \ldots b_2 b_1 b_0$ be two outputs with $a_{i_0} \neq b_{i_0}$. We have

$$n = (\ldots a_{i_0} \ldots a_1 a_0)_2 = (\ldots b_{i_0} \ldots b_1 b_0)_2.$$

Let

$$n' = (a_{i_0} \ldots a_1 a_0)_2 = (b_{i_0} \ldots b_1 b_0)_2.$$

At least one of a_{i_0} and b_{i_0} is nonzero. We assume, without loss of generality, that $a_{i_0} \neq 0$. Let $c_0 = a_{i_0} 2^{i_0}$. The value c_0 is an element in C_w is closest to n'. Since $a_{i_0} \neq b_{i_0}$ there must be another value, say c_1, closest to n', and this value must be encoded as the most significant nonzero digit of $(b_{i_0} \ldots b_1 b_0)_2$. Since both c_0 and c_1 are closest to n', n' must be the midpoint between c_0 and c_1. Thus, $|n' - c_0|$ is as large as possible, so by Lemma 1 we have

$$|n' - c_0| = |n' - c_1| = 2^{\lfloor \lg |n'| \rfloor - w + 1}.$$

Let $t = \lfloor \lg |n'| \rfloor - w + 1$. Since $n' - c_0 = \pm 2^t$, $n' - c_0 \in C_w$ and $n' - c_0$ is the unique element in C_w closest to $n' - c_0$. Similarly, $n' - c_1$ is the unique element in C_w closest to $n' - c_1$. Thus, the least significant nonzero digits of α and β correspond to the values $n' - c_0$ and $n' - c_1$; that is, the least significant nonzero digits of α and β are a_t and b_t where one of a_t, b_t is 1 and the other -1, so

$$n' = c_0 + 2^t = c_1 - 2^t \quad \text{or} \quad n' = c_0 - 2^t = c_1 + 2^t.$$

Note that the example above demonstrates this property. Thus, there are just two kinds of outputs: ones that encode c_0 and ones that encode c_1.

From the preceding discussion, we can derive the following Lemma:

Lemma 3. *Let α and β be two outputs of $MSF_w(n)$. Then α and β have the same number of nonzero digits.*

Proof. If $\alpha = \beta$ then there is nothing to prove, so we can assume $\alpha \neq \beta$. Let

$$\alpha = \ldots a_{i_0} \ldots a_1 a_0 \quad \text{and} \quad \beta = \ldots b_{i_0} \ldots b_1 b_0,$$

where i_0 is the largest value of i such that $a_i \neq b_i$. From our discussion above, the strings $a_{i_0} \ldots a_1 a_0$ and $b_{i_0} \ldots b_1 b_0$ each contain exactly two nonzero digits. Thus α and β have the same number of nonzero digits. □

It is possible to implement Algorithm 2.1 in such a way that it returns a unique representation for every $n \in \mathbb{Z}$. For example, if c_0 and c_1 are both closest to n then we might resolve this ambiguity by choosing the larger one. Because of Lemma 3, we know that, however Algorithm 2.1 is implemented, all outputs will have the same number of nonzero digits (for a given input). In fact, any representation of n constructed by Algorithm 2.1 will have a *minimal number of nonzero digits*, and we will prove this in Section 4. In the next section, we describe how to implement Algorithm 2.1 efficiently.

3 Implementations

We first review a known right-to-left sliding window method for constructing D_w-radix 2 representations. Then, we describe how Algorithm 2.1 can implemented using a left-to-right sliding window method. We also give a new implementation of Algorithm 1.1 which incorporates our left-to-right representations.

3.1 Right-to-Left

Suppose that we want to deduce a radix 2 representation of the integer 379 using the digits $D_3 = \{0, \pm 1, \pm 3\}$. If we know the $\{0, 1\}$-radix 2 representation of 379 then this is easily done. Consider the following table

β	c	β'	c'
001	0	001	0
011	0	003	0
101	0	00$\bar{3}$	1
111	0	00$\bar{1}$	1
000	1	001	0
010	1	003	0
100	1	00$\bar{3}$	1
110	1	00$\bar{1}$	1

This table describes a map, $(\beta, c) \mapsto (\beta', c')$, between ordered pairs. The ordered pairs consist of a 3-digit string and a *carry*, c. Notice that for each row of the table, the string β' corresponds to $((\beta)_2 + c)$ mods 2^3. After initializing the carry to 0, we can apply these transformations by sliding a 3-digit *window* from right to left across the $\{0,1\}$-radix 2 representation:

$$
\begin{array}{c|c}
379 = (010111 1 \overset{0}{0 1 1})_2 & 00\bar{1} \\
(0101 \overset{0}{1 1 1} 011)_2 & 00\bar{1}003 \\
(01 \overset{1}{0 1} 1111011)_2 & 000\bar{1}003 \\
(0 \overset{1}{1 0 1} 1111011)_2 & 003000\bar{1}003
\end{array}
$$

Each time the contents of the window and the value of the carry match an entry in the left hand column of the table we output the corresponding 3-digit string, update the carry and then advance the window 3 digits to the left. Otherwise, we output a single 0, leave the carry unchanged and advance the window 1 digit to the left.

This process constructs an integer's 3-NAF, and it does so using only a lookup table. If we allow simple bit operations, like xor, the number of rows in the table can be halved.

3.2 Left-to-Right

When $w = 3$ we have $C_3 = \{d \cdot 2^i : d \in D_3 \setminus \{0\}, i \in \mathbb{Z}, i \geq 0\}$. The first few positive elements of C_3 are

$$1, \underline{2}, 3, \underline{4}, 6, \underline{8}, 12, \underline{16}, 24, \underline{32}, 48, \underline{64}, 96, \underline{128}, 192, \underline{256}, 384, \underline{512} \ldots$$

Notice that for $i \geq 2$, the intervals $[2^{i-1}, 2^i]$ (which are underlined) each contain exactly 3 elements of C_3. Consider the integer 379. From the list of values above, we see that 384 is the element in C_3 closest to 379, however this can also be determined from the $\{0,1\}$-radix 2 representation of 379.

We first determine two neighbouring elements of C_3, call them c' and c'', such that $379 \in [c', c'']$. The value c' is the unique element in C_3 closest to, but not larger than, 379. If $379 = (0b_{\ell-1}b_{\ell-2}\ldots b_1 b_0)_2$ with $b_i \in \{0,1\}$ and $b_{\ell-1} = 1$, we can determine the value of c' by simply reading the 3 digit prefix of this representation (i.e., $0b_{\ell-1}b_{\ell-2}$). If the prefix is 010, then $c' = (010)_2 \cdot 2^{\ell-2} = 2^{\ell-1}$. If the prefix is 011, then $c' = (011)_2 \cdot 2^{\ell-2} = 3 \cdot 2^{\ell-2}$. Since $379 = (0101111011)_2$, we see that $c' = (010)_2 \cdot 2^7 = 256$.

The most significant nonzero digit of the representation $379 = (0101111011)_2$ tells us that $2^8 \leq 379 < 2^9$. Since 2^8 and 2^9 are both in C_3, it must be that $[c', c''] \subseteq [2^8, 2^9]$. The interval $[2^8, 2^9]$ has length 2^8 and contains exactly three elements of C_3, thus the interval $[c', c'']$ must have length $2^8/2 = 2^7$. So, we see that $c'' = c' + 2^7 = 256 + 128 = 384$.

We have deduced that $379 \in [256, 384]$ where 256 and 384 are neighbouring elements of C_3. Now, the question is, which of $256, 384$ is closer to 379. This is determined by the digit immediately to the right of the 3 digit prefix we considered above. If this digit is 0, 256 is closest, otherwise 384 is closest. Since $379 = (0101111011)_2$ we see that 384 is the element in C_3 closest to 379.

To continue building a representation of 379 using Algorithm 2.1, we must now determine an element closest to $379 - 384 = -5$. Again, we can use the $\{0,1\}$-radix 2 representation of 379 to make this determination.

Suppose we have $-5 = (\bar{1}b_{\ell-1}b_{\ell-2}\ldots b_1b_0)_2$ with $b_i \in \{0,1\}$ and $b_{\ell-1} = 0$. Then, as before, we can determine neighbouring elements of $\mathcal{C}_3$, c' and c'', such that $-5 \in [c', c'']$. The value c' is determined by simply reading the 3 digit prefix of this representation (i.e., $\bar{1}b_{\ell-1}b_{\ell-2}$). If the prefix is $\bar{1}00$, then $c' = (\bar{1}00)_2 \cdot 2^{\ell-2} = -2^\ell$. If the prefix is $\bar{1}01$, then $c' = (\bar{1}01)_2 \cdot 2^{\ell-2} = -3 \cdot 2^{\ell-2}$.

It is not difficult to construct such a representation of -5. Observe

$$379 = (0101111011)_2$$
$$384 = (0110000000)_2$$
$$\Longrightarrow 379 - 384 = \;\;(\bar{1}1111011)_2.$$

Since the digits $\bar{1}1$ can be replaced by $0\bar{1}$, we have that $-5 = (\bar{1}011)_2$. Now we see that $c' = (\bar{1}01)_2 \cdot 2 = -3 \cdot 2 = -6$.

The most significant nonzero digit of the representation $-5 = (\bar{1}011)_2$ stands for -2^3. Because the following digit is 0, we have that $-2^3 \leq -5 < -2^2$. Since -2^3 and -2^2 are both in $\mathcal{C}_3$, it must be that $[c', c''] \subseteq [-2^3, -2^2]$. The interval $[-2^3, -2^2]$ has length 2^2 and contains exactly three elements of $\mathcal{C}_3$, thus the interval $[c', c'']$ must have length $2^2/2 = 2$. So, we see that $c'' = c' + 2 = -6 + 2 = -4$.

Now we must decide which of c', c'' is closest to -5. As before, we can determine this by reading the the digit immediately to the right of the 3 digit prefix. If this digit is 0, c' is closest. If this digit is 1, c'' is closest. In this case, both $c' = -6$ and $c'' = -4$ are closest to -5, however, this rule simply distinguishes one of them. Since $-5 = (\bar{1}01\underline{1})_2$ we see that -4 is an element in $\mathcal{C}_3$ closest to -5.

To finish building our representation of 379 we must now determine an element closest to $-5 - (-4) = -1$. Clearly, -1 is the element in $\mathcal{C}_3$ closest to -1, however, we can also make this determination be applying our previous arguments to the representation $-1 = (\bar{1}.000)_2$. We can always examine a 3 digit prefix of a representation by taking zeros from the right of the radix point when necessary.

The techniques we have described for determining closest elements in $\mathcal{C}_3$ can be implemented as a 4-digit window slides left to right across a $\{0,1\}$-radix 2 representation. Consider the following table

β	β'
0100	010
0101	003
0110	003
0111	100
1000	$\bar{1}00$
1001	$00\bar{3}$
1010	$00\bar{3}$
1011	$0\bar{1}0$

This table describes a map, $\beta \mapsto \beta'$, between strings. The relation between these strings is based on choosing closest elements in $\mathcal{C}_3$.

The first four rows of the table are filled in by determining closest elements to integers represented as $(01b_{\ell-2}b_{\ell-3}\ldots)_2$. The last four rows of the table are filled in by determining closest elements to integers represented as $(\overline{1}0b_{\ell-2}b_{\ell-3}\ldots)_2$. It can be shown that if $n = (\overline{b_\ell}b_{\ell-1}b_{\ell-2}b_{\ell-3}\ldots b_0)_2$ with $b_i \in \{0,1\}$ and $b_\ell \neq b_{\ell-1}$, then, for the element c closest to n that we choose, we have $n - c = (\overline{b_{\ell-3}}b_{\ell-4}\ldots b_0)_2$.

Returning to our example, $n = 379$, we have

$$379 = \begin{array}{r|l} (0\underline{101}1111011.000)_2 & 003 \\ (010\underline{1111}011.000)_2 & 0030 \\ (0101\underline{111}1011.000)_2 & 00300 \\ (01011\underline{1011}.000)_2 & 003000 \\ (010111\underline{1011}.000)_2 & 00300 00\overline{1}0 \\ (0101111\underline{1011.000})_2 & 0030000\overline{1}0\overline{1}.000 \end{array}$$

As with the construction of the 3-NAF of 379, each time the contents of the window match an entry in the left hand column of the table we output the corresponding 3-digit string and then advance the window 3 digits to the right. Otherwise, we output a single 0 and advance the window 1 digit to the right.

If we work from the description of Algorithm 2.1 in Section 2, we might construct a different representation of 379 than the one above. Since $379 - 3 \cdot 2^7 = -5$ and -5 has two closest elements in $\mathcal{C}_3$, Algorithm 2.1 might also return $379 = (300000\overline{3}1)_2$ (note that this example demonstrates that, unlike the 3-NAFs, the representations constructed by Algorithm 2.1 can have adjacent nonzero digits). The implementation we have described is deterministic, thus it must somehow distinguish one of two closest elements in $\mathcal{C}_w$. It does so by always selecting a *largest* closest element.

For *general* $w \geq 2$, Algorithm 2.1 can be implemented by sliding a window of width $w + 1$ from left to right across the $\{0,1\}$-radix 2 representation of n. This implementation is based on the following facts. If

$$n = (\overline{b_\ell}b_{\ell-1}\ldots b_1 b_0)_2 \text{ with } b_i \in \{0,1\} \text{ and } b_\ell \neq b_{\ell-1}, \tag{2}$$

then we can determine $c \in \mathcal{C}_w$ closest to n from the $w + 1$ digit string $b_\ell b_{\ell-1} \ldots b_{\ell-w}$. For this value c closest to n, we have

$$n - c = (\overline{b_{\ell-w}}b_{\ell-w-1}\ldots b_1 b_0)_2. \tag{3}$$

The resulting look-up table will contain 2^w rows and describes a map from $w + 1$ digit strings to w digit strings. Due to the symmetry in the table, if we allow simple bit operations, like xor, the second half of the table does not need to be stored.

3.3 A New Window Method

In our example implementation for $w = 3$, our window slides either 3 digits to the right (after the window matches an entry in the table) or one digit to the right (otherwise). This is because the strings output in these cases have length 3 (a string β') or length 1 (a single 0). However, it is not necessary for the strings β' to all have the same length.

If we take our previous table and delete the trailing zeros from each string β' then we get

β	β'
0100	01
0101	003
0110	003
0111	1
1000	$\bar{1}$
1001	$00\bar{3}$
1010	$00\bar{3}$
1011	$0\bar{1}$

β	j	d
0100	2	1
0101	3	3
0110	3	3
0111	1	1
1000	1	$\bar{1}$
1001	3	$\bar{3}$
1010	3	$\bar{3}$
1011	2	$\bar{1}$

In the left table, the strings β' are all of the form $0^{j-1}d$ where d is a nonzero digit in D_3. The right table is just an encoding of the left table. The left table can be used to construct D_3-radix representations similar to the way we described in the previous section. The only difference is the window slides right 1, 2 or 3 digits at a time; the number being equal to the length of the output string (either β' or a single 0).

This implementation of Algorithm 2.1 can be incorporated easily with Algorithm 1.1. From the right table, we can define a function T_3 which maps strings in $\{0,1\}^4$ to ordered pairs, (j,d), with the additional condition that if $\beta \in \{0,1\}^4$ does not appear in the table then $T_3(\beta) = (1,0)$. Here is the resulting algorithm for scalar multiplication, which works for an arbitrary value of $w \geq 2$:

Algorithm 3.1: w-MSF-WINDOW-METHOD(n, P)

comment: $w \geq 2$, $D_w = \{0\} \cup \{d \in \mathbb{Z} : d \text{ odd}, |d| < 2^{w-1}\}$
$n = (0b_{\ell-1} \ldots b_1 b_0)_2$, where $b_i \in \{0,1\}$
$\beta_i = b_i b_{i-1} \ldots b_{i-w}$

external $T_w : \beta_i \mapsto (j, d)$

for each $d \in D_w$ with $d > 0$
  do $P_d \leftarrow dP$
$Q \leftarrow \infty$, $i \leftarrow \ell$
while $i \geq 0$
do $\begin{cases} (j,d) \leftarrow T_w(\beta_i) \\ Q \leftarrow 2^j Q \\ \text{if } d \neq 0 \\ \quad \text{then } \begin{cases} \text{if } d > 0 \\ \quad \text{then } Q \leftarrow Q + P_d \\ \quad \text{else } Q \leftarrow Q - P_{-d} \end{cases} \\ i \leftarrow i - j \end{cases}$
return Q

Constructing the function $T_w : \{0,1\}^{w+1} \to \{1,2,\ldots,w\} \times D_w$ for arbitrary $w \geq 2$ is straightforward, provided we already have the width-w look-up table described in Section 3.2; we simply delete the trailing zeros on the output strings and then encode the output strings as ordered pairs.

Building the width-w look-up table all at once is not difficult, however, it is also possible to build the table on the fly. An algebraic expression for the element $c \in \mathcal{C}_w$ closest to n can be obtained by subtracting the representation for $n - c$ in (3) from that of n in (2). The resulting expression for c is

$$c = \left((\overline{b_\ell} b_{\ell-1} \ldots b_{\ell-w+1})_2 + b_{\ell-w}\right) \cdot 2^{\ell-w+1} \tag{4}$$

which is a function of the $w+1$ digits $b_\ell b_{\ell-1} \ldots b_{\ell-w}$.

4 Minimality

If α is a string of digits, we denote the number of nonzero digits in α by $\mathsf{wt}(\alpha)$. We will refer to $\mathsf{wt}(\alpha)$ as the *weight* of the string α. The set of all strings composed of digits in D_w is denoted by D_w^*. For an integer n, we define

$$\mathsf{wt}^*(n) := \min\{\mathsf{wt}(\alpha) : \alpha \in D_w^*, (\alpha)_2 = n\}.$$

So, $\mathsf{wt}^*(n)$ is the minimum number of nonzero digits required to represent n using a D_w-radix 2 representation. If $\alpha \in D_w^*$ and $(\alpha)_2 = n$ then it must be that $\mathsf{wt}(\alpha) \geq \mathsf{wt}^*(n)$; if $\mathsf{wt}(\alpha) = \mathsf{wt}^*(n)$ we say that α has *minimal weight*.

In this section, we will prove the following Theorem:

Theorem 1. *Let $w \geq 2$ be an integer. For any $n \in \mathbb{Z}$, the representation returned by $\mathrm{MSF}_w(n)$ has a minimal number of nonzero digits.*

It will be convenient to let $\mathrm{MSF}_w(n)$ denote a string returned by the algorithm on input n. To prove Theorem 1 we will show that for any $n \in \mathbb{Z}$, $\mathsf{wt}(\mathrm{MSF}_w(n)) = \mathsf{wt}^*(n)$. In doing so, we will make use of a number of short Lemmas concerning the functions $\mathsf{wt}^*(n)$ and $\mathsf{wt}(\mathrm{MSF}_w(n))$.

Lemma 4. *If n is even then $\mathsf{wt}^*(n) = \mathsf{wt}^*(n/2)$.*

Proof. Let $(\ldots a_2 a_1 a_0)_2$ be a minimal weight representation of n. Since n is even, $a_0 = 0$ and so $(\ldots a_2 a_1)_2 = n/2$. Thus, $\mathsf{wt}^*(n/2) \leq \mathsf{wt}^*(n)$. Let $(\ldots b_2 b_1 b_0)_2$ be a minimal weight representation of $n/2$. Then $(\ldots b_2 b_1 b_0 0)_2 = n$ and so $\mathsf{wt}^*(n) \leq \mathsf{wt}^*(n/2)$. □

For any $n \in \mathbb{Z}$, there exists a unique pair of integers, q and r, such that

$$n = q \cdot 2^w + r \quad \text{where} \quad -2^{w-1} < r \leq 2^{w-1}.$$

We will denote this value of r by "n mods 2^w". For example, if $w = 3$ then 13 mods $2^3 = -3$. Note that if n is odd then so is n mods 2^w. As well, when $n > 0$ it must be that $q \geq 0$, and similarly, when $n < 0$, $q \leq 0$. So, for $n \neq 0$, we have $q/n \geq 0$.

If we write $n = q \cdot 2^w + r$ with $r = n$ mods 2^w then $q \cdot 2^w$ is a *multiple of 2^w closest* to n. We will make use of this fact later on.

The w-NAF of an integer has minimal weight [1,11]. If n is odd then the least significant digit of its w-NAF is equal to n mods 2^w. From this fact, we can deduce the following Lemma:

Lemma 5. *If n is odd and $r = n$ mods 2^w, then $\mathsf{wt}^*(n) = 1 + \mathsf{wt}^*((n-r)/2)$.*

Lemma 5 can proved in the same way as Lemma 4.

To show that $\mathsf{wt}(\mathrm{MSF}_w(n)) = \mathsf{wt}^*(n)$, we will argue by induction on $|n|$. For n odd, it is thus useful to establish that $|(n-r)/2| < |n|$.

Lemma 6 ([10]). *Let n be an odd integer and let $r = n$ mods 2^w. Then $|(n-r)/2| < |n|$.*

We now give two Lemmas which involve the function $\mathsf{wt}(\mathrm{MSF}_w(n))$.

Lemma 7. *If n is an even integer then $\mathsf{wt}(\mathrm{MSF}_w(n)) = \mathsf{wt}(\mathrm{MSF}_w(n/2))$.*

Proof. If $n = 0$ then the result is clearly true, so we can assume $n \neq 0$. Let $\alpha = a_{\ell-1} \ldots a_2 a_1 a_0$ be an output of $\mathrm{MSF}_w(n)$ with $a_{\ell-1} \neq 0$. Since n is even and $n = (\alpha)_2$ it must be that $a_0 = 0$. Thus, the strings α and $\alpha' = a_{\ell-1} \ldots a_2 a_1$ have the same weight. We show α' is an output of $\mathrm{MSF}_w(n/2)$, and then the result follows from Lemma 3.

Let $c = a_{\ell-1} 2^{\ell-1}$; c is an element in $\mathcal{C}_w$ closest to n. Since $a_0 = 0$ and $a_{\ell-1} \neq 0$ it must be that $\ell - 1 \geq 1$, and so c is even. Thus, $c/2 \in \mathcal{C}_w$. Now,

$$c \text{ is closest to } n \implies c/2 \text{ is closest to } n/2,$$

so there is an output of $\mathrm{MSF}_w(n/2)$ where the most significant nonzero digit encodes $c/2 = a_{\ell-1} 2^{\ell-2}$. By repeating this argument, we see that $\alpha' = a_{\ell-1} \ldots a_2 a_1$ is indeed an output of $\mathrm{MSF}_w(n/2)$. This proves the result. □

Lemma 8. *If c is an element of $\mathcal{C}_w$ closest to n, then $\mathsf{wt}(\mathrm{MSF}_w(n)) = 1 + \mathsf{wt}(\mathrm{MSF}_w(n-c))$.*

Lemma 8 follows from the description of Algorithm 2.1.

Now we have everything we need to prove our main result.

Proof (of Theorem 1). We show that for any $n \in \mathbb{Z}$,

$$\mathsf{wt}(\mathrm{MSF}_w(n)) = \mathsf{wt}^*(n). \tag{5}$$

When $n = 0$, $\mathrm{MSF}_w(n)$ returns the empty string; thus

$$\mathsf{wt}(\mathrm{MSF}_w(0)) = 0 = \mathsf{wt}^*(0). \tag{6}$$

Also, if n is even then from Lemmas 4 and 7 we have

$$\mathsf{wt}(\mathrm{MSF}_w(n)) = \mathsf{wt}^*(n) \iff \mathsf{wt}(\mathrm{MSF}_w(n/2)) = \mathsf{wt}^*(n/2). \tag{7}$$

Thus, if we can show that (5) holds for all n with $|n| \geq 1$ and n odd, then by (6) and (7), it holds for all n.

Let n be an odd nonzero integer. We argue by induction on $|n|$. For our base cases, we consider the values of n that satisfy $1 \leq |n| < 2^{2w-1}$. We deal with this interval in two parts.

First, suppose $1 \leq |n| < 2^{w-1}$. Then $n \in D_w$ (because n is odd) and thus $\text{wt}(\text{MSF}_w(n)) = 1$. Any odd integer n has $\text{wt}^*(n) \geq 1$, thus we see that

$$\text{wt}(\text{MSF}_w(n)) = 1 = \text{wt}^*(n).$$

Next, suppose $2^{w-1} \leq |n| < 2^{2w-1}$. Note that $\lfloor \lg |n| \rfloor \leq 2w - 2$. Let c be an element in $\mathcal{C}_w$ closest to n. Note that c must be even since $|n| \geq 2^{w-1}$. By Lemma 1, we have

$$|n - c| \leq 2^{\lfloor \lg |n| \rfloor - w + 1}. \tag{8}$$

However,

$$\lfloor \lg |n| \rfloor \leq 2w - 2 \implies \lfloor \lg |n| \rfloor - w + 1 \leq w - 1,$$

and so

$$|n - c| \leq 2^{w-1}.$$

Since n is odd and c is even, $n - c$ is odd and thus, $n - c \in D_w$. So Algorithm 2.1 uses just two elements of $\mathcal{C}_w$ to represent n (namely, c and $n - c$); thus $\text{wt}(\text{MSF}_w(n)) = 2$. Any odd integer n with $|n| > 2^{w-1}$ (i.e., $n \notin D_w$) has $\text{wt}^*(n) \geq 2$, and from this we see that

$$\text{wt}(\text{MSF}_w(n)) = 2 = \text{wt}^*(n).$$

With our base cases established, we now consider n odd with $|n| \geq 2^{2w-1}$. Note that $\lfloor \lg |n| \rfloor \geq 2w - 1$. Let c be an element in $\mathcal{C}_w$ closest to n and let $r = n \bmod s\, 2^w$. We claim that c is also closest to $n - r$. To see this, first note that n lies in one of the intervals

$$[2^{\lfloor \lg |n| \rfloor}, 2^{\lfloor \lg |n| \rfloor + 1}] \quad \text{or} \quad [-2^{\lfloor \lg |n| \rfloor}, -2^{\lfloor \lg |n| \rfloor + 1}].$$

From the proof of Lemma 2, we know that all elements of $\mathcal{C}_w$ in these intervals have the form $d \cdot 2^i$ with $d \in D_w$ and

$$i \in \{\lfloor \lg |n| \rfloor - w + 2, \ldots, \lfloor \lg |n| \rfloor, \lfloor \lg |n| \rfloor + 1\}.$$

Thus,

$$i \geq \lfloor \lg |n| \rfloor - w + 2 \geq 2w - 1 - w + 2 = w + 1,$$

and so 2^{w+1} divides c. There are two neighbouring elements of $\mathcal{C}_w$, say c_0 and c_1, such that $n \in [c_0, c_1]$. Let m be the midpoint of $[c_0, c_1]$. We have

$$2^{w+1} | c_0 \text{ and } 2^{w+1} | c_1 \implies 2^{w+1} | (c_0 + c_1)$$
$$\implies 2^w \Big| \frac{c_0 + c_1}{2}$$
$$\implies 2^w | m.$$

So c_0, c_1 and m are all multiples of 2^w. One of c_0 or c_1 is equal to c. If $c = c_0$, then $n \in [c, m]$; whereas, if $c = c_1$, then $n \in [m, c]$. In either case, it can be shown that $n - r$ is an element in the same closed interval (this follows because $n - r$ is the multiple of 2^w closest to n). Thus, we see that c is closest to $n - r$. Further, since both c and $n - r$ are even, we have that

$$c/2 \text{ is closest to } (n-r)/2. \tag{9}$$

Now we are ready to finish the proof. Notice that, because $2^w | c$, we have

$$n - c \text{ mods } 2^w = n \text{ mods } 2^w = r. \tag{10}$$

By induction, we have that $\text{wt}(\text{MSF}_w(n')) = \text{wt}^*(n')$ for all n' with $|n'| < |n|$. Using this and our Lemmas, we find that

$$\begin{aligned}
\text{wt}(\text{MSF}_w(n)) &= 1 + \text{wt}(\text{MSF}_w(n - c)) && \text{(by Lemma 8)} \\
&= 1 + \text{wt}^*(n - c) && \text{(by induction)} \\
&= 1 + 1 + \text{wt}^*\left(\frac{(n-c) - r}{2}\right) && \text{(by (10) and Lemma 5)} \\
&= 1 + 1 + \text{wt}\left(\text{MSF}_w\left(\frac{(n-c) - r}{2}\right)\right) && \text{(by induction)} \\
&= 1 + 1 + \text{wt}\left(\text{MSF}_w\left(\frac{n-r}{2} - \frac{c}{2}\right)\right) \\
&= 1 + \text{wt}\left(\text{MSF}_w\left(\frac{n-r}{2}\right)\right) && \text{(by (9) and Lemma 8)} \\
&= 1 + \text{wt}^*\left(\frac{n-r}{2}\right) && \text{(by induction)} \\
&= \text{wt}^*(n) && \text{(by Lemma 5).}
\end{aligned}$$

Each of the inductive steps above is justified by either Lemma 6 or the fact that $|n - c| < |n|$. This concludes the proof. □

5 Related Work

Avanzi [1] independently obtained similar results which were presented at SAC 2004. In particular, Avanzi describes a deterministic algorithm which constructs D_w-radix 2 representations by scanning the binary representation of an integer from left to right. He also proves that these representation have minimal weight. As the input is scanned, Avanzi's algorithm works by applying arithmetic operations to windows of $w+1$ digits; his algorithm does not require a stored table. By comparing the expression for c in equation (4) to Avanzi's algorithm, it can be shown that Avanzi's algorithm is a deterministic implementation of Algorithm 2.1; that is, Avanzi's algorithm works by choosing closest elements from the set $\mathcal{C}_w$.

At CRYPTO 2004, Okeya, Schmidt-Samoa, Spahn and Takagi [13] presented a very simple technique that allows D_w-radix 2 representations to be constructed from either right to left or left to right. They consider a canonical $\{0, \pm 1\}$-radix 2 representation of an integer, n, constructed by the digit-wise subtraction of the binary representation of n from that of $2n$. Once this representation is constructed, D_w-radix 2 representations can be obtained by sliding windows of width-w across it. Sliding the window right to left gives the w-NAF, and sliding the window left to right gives the same representation constructed by Avanzi's algorithm. Okeya et al. show that the average density of nonzero digits in their left-to-right representations is asymptotically $1/(w+1)$, however they do not prove minimality.

The same canonical $\{0, \pm 1\}$-radix 2 representation defined by Okeya et al. can be found in work by Grabner, Heuberger, Prodinger and Thuswaldner [6]. Grabner et al. use these representations to construct minimal weight joint $\{0, \pm 1\}$-radix 2 representations of pairs of integers from left to right. Heuberger, Katti, Prodinger and Ruan [8] also use the canonical representations. They generalize the results of Grabner et al. to joint representations of $d \geq 2$ integers. As well, Heuberger et al. show how Avanzi's left-to-right algorithm can be obtained from the canonical representation.

What is unique to our work is the idea of choosing closest elements in the set $\mathcal{C}_w$, our simple nondeterministic algorithm, our technique for proving minimality and our method of incorporating our left-to-right representations into the algorithm for scalar multiplication.

6 Remarks

In proving that our new representations have a minimal number of nonzero digits, we essentially dealt with the following two statements concerning odd integers:

$$\text{wt}^*(n) = 1 + \text{wt}^*((n-r)/2) \quad \text{where } r = n \text{ mods } 2^w \tag{11}$$

$$\text{wt}^*(n) = 1 + \text{wt}^*(n-c) \quad \text{where } c \in \mathcal{C}_w \text{ is closest to } n. \tag{12}$$

In our proof, we noted that (11) is true (by the minimality of the w-NAF) and then showed that (11) implies (12). The same arguments can be used to show that (12) implies (11). Thus, (11) and (12) are logically equivalent, which is perhaps surprising.

The w-NAF has a very simple combinatorial description. From this description, it is very easy to look at a representation and quickly decide whether or not it is a w-NAF. For our new representations, this does not appear to be quite so easy.

Acknowledgments

The authors thank Roberto Avanzi, Bodo Möller and the anonymous reviewers for helpful comments on earlier versions of this manuscript.

References

1. R. M. Avanzi A Note on the Signed Sliding Window Integer Recoding and its Left-to-Right Analogue, in "Selected Areas in Cryptography 2004". To appear in *Lecture Notes in Computer Science*.
2. I. F. Blake, G. Seroussi and N. P. Smart. *Elliptic Curves in Cryptography*, Cambridge University Press, 1999.
3. H. Cohen. Analysis of the Flexible Window Powering Algorithm. To appear in *Journal of Cryptology*.
   Available from http://www.math.u-bordeaux.fr/~cohen/window.dvi.
4. H. Cohen, A. Miyaji and T. Ono. Efficient Elliptic Curve Exponentiation Using Mixed Coordinates, in "Advances in Cryptology – ASIACRYPT '98", *Lecture Notes in Computer Science* **1514** (1998), 51–65.
5. D. M. Gordon. A Survey of Fast Exponentiation Methods, *Journal of Algorithms* **27** (1998), 129–146.
6. P. Grabner, C. Heuberger, H. Prodinger and J. Thuswaldner. Analysis of Linear Combination Algorithms in Cryptography. Submitted.
   Available from http://www.opt.math.tu-graz.ac.at/~cheub/publications/.
7. D. Hankerson, A. Menezes and S. Vanstone. *Guide to Elliptic Curve Cryptography*, Springer, 2004.
8. C. Heuberger, R. Katti, H. Prodinger, and X. Ruan. The Alternating Greedy Expansion and Applications to Left-To-Right Algorithms in Cryptography. Submitted.
   Available from http://www.opt.math.tu-graz.ac.at/~cheub/publications/.
9. M. Joye and S. Yen. Optimal Left-to-Right Binary Signed-Digit Recoding, *IEEE Transactions on Computers* **49** (2000), 740–748.
10. J. A. Muir and D. R. Stinson. New Minimal Weight Representations for Left-to-Right Window Methods (Extended Version).
    Available from http://www.math.uwaterloo.ca/~jamuir/papers.htm.
11. J. A. Muir and D. R. Stinson. Minimality and Other Properties of the Width-w Nonadjacent Form. To appear in *Mathematics of Computation*.
    Available from http://www.math.uwaterloo.ca/~jamuir/papers.htm.
12. V. Müller. Fast Multiplication on Elliptic Curves over Small Fields of Characteristic Two. *Journal of Cryptology* **11** (1998), 219–234.
13. K. Okeya, K. Schmidt-Samoa, C. Spahn and T. Takagi. Signed Binary Representations Revisited, in "Advances in Cryptology – CRYPTO 2004", *Lecture Notes in Computer Science* **3152** (2004), 123–139.
14. J. A. Solinas. Efficient arithmetic on Koblitz curves. *Designs, Codes and Cryptography* **19** (2000), 195–249.

Author Index

Abdalla, Michel   191

Bao, Feng   72
Barreto, Paulo S.L.M.   262
Batina, Lejla   323
Bellare, Mihir   136
Billet, Olivier   19
Boneh, Dan   87
Braeken, An   29

Cook, Debra L.   334
Cui, Yang   104

Dwork, Cynthia   1

Feng, Dengguo   72

Gammel, Berndt M.   351
Gilbert, Henri   19
Gutterman, Zvi   44

Herzberg, Amir   172
Howgrave-Graham, Nick   118

Imai, Hideki   104
Ioannidis, John   334

Kaliski, Burton   227
Katz, Jonathan   87
Keromytis, Angelos D.   334
Kobara, Kazukuni   104

Laguillaumie, Fabien   154
Luck, Jake   334

Mackenzie, Philip   209
Malkhi, Dahlia   44
Mangard, Stefan   351

McCullagh, Noel   262
Mentens, Nele   323
Muir, James A.   366

Nguyen, Lan   275

Oswald, Elisabeth   58

Patel, Sarvar   209
Pointcheval, David   191
Popp, Thomas   351
Preneel, Bart   29, 323

Rijmen, Vincent   58

Scott, Michael   293
Shi, Haixia   136
Silverman, Joseph H.   118
Stinson, Douglas R.   366
Szydlo, Michael   227

Verbauwhede, Ingrid   323
Vergnaud, Damien   154

Wei, Victor K.   305
Whyte, William   118
Wolf, Christopher   29
Wu, Hongjun   72

Yao, Frances F.   245
Yin, Yiqun Lisa   245
Young, Adam   7
Yuen, Tsz Hon   305
Yung, Moti   7

Zhang, Bin   72
Zhang, Chong   136

Lecture Notes in Computer Science

For information about Vols. 1–3262

please contact your bookseller or Springer

Vol. 3385: R. Cousot (Ed.), Verification, Model Checking, and Abstract Interpretation. XII, 483 pages. 2004.

Vol. 3382: J. Odell, P. Giorgini, J.P. Müller (Eds.), Agent-Oriented Software Engineering V. X, 239 pages. 2004.

Vol. 3381: M. Bieliková, B. Charon-Bost, O. Sýkora, P. Vojtáš (Eds.), SOFSEM 2005: Theory and Practice of Computer Science. XV, 428 pages. 2004.

Vol. 3376: A. Menezes (Ed.), Topics in Cryptology – CT-RSA 2005. X, 385 pages. 2004.

Vol. 3363: T. Eiter, L. Libkin (Eds.), Database Theory - ICDT 2005. XI, 413 pages. 2004.

Vol. 3362: G. Barthe, L. Burdy, M. Huisman, J.-L. Lanet, T. Muntean (Eds.), Construction and Analysis of Safe, Secure, and Interoperable Smart Devices. IX, 257 pages. 2004.

Vol. 3360: S. Spaccapietra, E. Bertino, S. Jajodia, R. King, D. McLeod, M.E. Orlowska, L. Strous (Eds.), Journal on Data Semantics II. XI, 233 pages. 2004.

Vol. 3358: J. Cao, L.T. Yang, M. Guo, F. Lau (Eds.), Parallel and Distributed Processing and Applications. XXIV, 1058 pages. 2004.

Vol. 3357: H. Handschuh, M.A. Hasan (Eds.), Selected Areas in Cryptography. XI, 355 pages. 2004.

Vol. 3356: G. Das, V.P. Gulati (Eds.), Intelligent Information Technology. XII, 428 pages. 2004.

Vol. 3353: J. Hromkovič, M. Nagl, B. Westfechtel (Eds.), Graph-Theoretic Concepts in Computer Science. XI, 404 pages. 2004.

Vol. 3350: M. Hermenegildo, D. Cabeza (Eds.), Practical Aspects of Declarative Languages. VIII, 269 pages. 2004.

Vol. 3348: A. Canteaut, K. Viswanathan (Eds.), Progress in Cryptology - INDOCRYPT 2004. XIV, 431 pages. 2004.

Vol. 3347: R.K. Ghosh, H. Mohanty (Eds.), Distributed Computing and Internet Technology. XX, 472 pages. 2004.

Vol. 3344: J. Malenfant, B.M. Østvold (Eds.), Object-Oriented Technology. ECOOP 2004 Workshop Reader. VIII, 215 pages. 2004.

Vol. 3342: E. Şahin, W.M. Spears (Eds.), Swarm Robotics. X, 175 pages. 2004.

Vol. 3341: R. Fleischer, G. Trippen (Eds.), Algorithms and Computation. XVII, 935 pages. 2004.

Vol. 3340: C.S. Calude, E. Calude, M.J. Dinneen (Eds.), Developments in Language Theory. XI, 431 pages. 2004.

Vol. 3339: G.I. Webb, X. Yu (Eds.), AI 2004: Advances in Artificial Intelligence. XXII, 1272 pages. 2004. (Subseries LNAI).

Vol. 3338: S.Z. Li, J. Lai, T. Tan, G. Feng, Y. Wang (Eds.), Advances in Biometric Person Authentication. XVIII, 699 pages. 2004.

Vol. 3337: J.M. Barreiro, F. Martin-Sanchez, V. Maojo, F. Sanz (Eds.), Biological and Medical Data Analysis. XI, 508 pages. 2004.

Vol. 3336: D. Karagiannis, U. Reimer (Eds.), Practical Aspects of Knowledge Management. X, 523 pages. 2004. (Subseries LNAI).

Vol. 3334: Z. Chen, H. Chen, Q. Miao, Y. Fu, E. Fox, E.-p. Lim (Eds.), Digital Libraries: International Collaboration and Cross-Fertilization. XX, 690 pages. 2004.

Vol. 3333: K. Aizawa, Y. Nakamura, S. Satoh (Eds.), Advances in Multimedia Information Processing - PCM 2004, Part III. XXXV, 785 pages. 2004.

Vol. 3332: K. Aizawa, Y. Nakamura, S. Satoh (Eds.), Advances in Multimedia Information Processing - PCM 2004, Part II. XXXVI, 1051 pages. 2004.

Vol. 3331: K. Aizawa, Y. Nakamura, S. Satoh (Eds.), Advances in Multimedia Information Processing - PCM 2004, Part I. XXXVI, 667 pages. 2004.

Vol. 3329: P.J. Lee (Ed.), Advances in Cryptology - ASIACRYPT 2004. XVI, 546 pages. 2004.

Vol. 3328: K. Lodaya, M. Mahajan (Eds.), FSTTCS 2004: Foundations of Software Technology and Theoretical Computer Science. XVI, 532 pages. 2004.

Vol. 3326: A. Sen, N. Das, S.K. Das, B.P. Sinha (Eds.), Distributed Computing - IWDC 2004. XIX, 546 pages. 2004.

Vol. 3323: G. Antoniou, H. Boley (Eds.), Rules and Rule Markup Languages for the Semantic Web. X, 215 pages. 2004.

Vol. 3322: R. Klette, J. Žunić (Eds.), Combinatorial Image Analysis. XII, 760 pages. 2004.

Vol. 3321: M.J. Maher (Ed.), Advances in Computer Science - ASIAN 2004. XII, 510 pages. 2004.

Vol. 3320: K.-M. Liew, H. Shen, S. See, W. Cai (Eds.), Parallel and Distributed Computing: Applications and Technologies. XXIV, 891 pages. 2004.

Vol. 3317: M. Domaratzki, A. Okhotin, K. Salomaa, S. Yu (Eds.), Implementation and Application of Automata. XII, 336 pages. 2004.

Vol. 3316: N.R. Pal, N.K. Kasabov, R.K. Mudi, S. Pal, S.K. Parui (Eds.), Neural Information Processing. XXX, 1368 pages. 2004.

Vol. 3315: C. Lemaître, C.A. Reyes, J.A. González (Eds.), Advances in Artificial Intelligence – IBERAMIA 2004. XX, 987 pages. 2004. (Subseries LNAI).

Vol. 3314: J. Zhang, J.-H. He, Y. Fu (Eds.), Computational and Information Science. XXIV, 1259 pages. 2004.

Vol. 3312: A.J. Hu, A.K. Martin (Eds.), Formal Methods in Computer-Aided Design. XI, 445 pages. 2004.

Vol. 3311: V. Roca, F. Rousseau (Eds.), Interactive Multimedia and Next Generation Networks. XIII, 287 pages. 2004.

Vol. 3309: C.-H. Chi, K.-Y. Lam (Eds.), Content Computing. XII, 510 pages. 2004.

Vol. 3308: J. Davies, W. Schulte, M. Barnett (Eds.), Formal Methods and Software Engineering. XIII, 500 pages. 2004.

Vol. 3307: C. Bussler, S.-k. Hong, W. Jun, R. Kaschek, D.. Kinshuk, S. Krishnaswamy, S.W. Loke, D. Oberle, D. Richards, A. Sharma, Y. Sure, B. Thalheim (Eds.), Web Information Systems – WISE 2004 Workshops. XV, 277 pages. 2004.

Vol. 3306: X. Zhou, S. Su, M.P. Papazoglou, M.E. Orlowska, K.G. Jeffery (Eds.), Web Information Systems – WISE 2004. XVII, 745 pages. 2004.

Vol. 3305: P.M.A. Sloot, B. Chopard, A.G. Hoekstra (Eds.), Cellular Automata. XV, 883 pages. 2004.

Vol. 3303: J.A. López, E. Benfenati, W. Dubitzky (Eds.), Knowledge Exploration in Life Science Informatics. X, 249 pages. 2004. (Subseries LNAI).

Vol. 3302: W.-N. Chin (Ed.), Programming Languages and Systems. XIII, 453 pages. 2004.

Vol. 3300: L. Bertossi, A. Hunter, T. Schaub (Eds.), Inconsistency Tolerance. VII, 295 pages. 2004.

Vol. 3299: F. Wang (Ed.), Automated Technology for Verification and Analysis. XII, 506 pages. 2004.

Vol. 3298: S.A. McIlraith, D. Plexousakis, F. van Harmelen (Eds.), The Semantic Web – ISWC 2004. XXI, 841 pages. 2004.

Vol. 3296: L. Bougé, V.K. Prasanna (Eds.), High Performance Computing - HiPC 2004. XXV, 530 pages. 2004.

Vol. 3295: P. Markopoulos, B. Eggen, E. Aarts, J.L. Crowley (Eds.), Ambient Intelligence. XIII, 388 pages. 2004.

Vol. 3294: C.N. Dean, R.T. Boute (Eds.), Teaching Formal Methods. X, 249 pages. 2004.

Vol. 3293: C.-H. Chi, M. van Steen, C. Wills (Eds.), Web Content Caching and Distribution. IX, 283 pages. 2004.

Vol. 3292: R. Meersman, Z. Tari, A. Corsaro (Eds.), On the Move to Meaningful Internet Systems 2004: OTM 2004 Workshops. XXIII, 885 pages. 2004.

Vol. 3291: R. Meersman, Z. Tari (Eds.), On the Move to Meaningful Internet Systems 2004: CoopIS, DOA, and ODBASE, Part II. XXV, 824 pages. 2004.

Vol. 3290: R. Meersman, Z. Tari (Eds.), On the Move to Meaningful Internet Systems 2004: CoopIS, DOA, and ODBASE, Part I. XXV, 823 pages. 2004.

Vol. 3289: S. Wang, K. Tanaka, S. Zhou, T.W. Ling, J. Guan, D. Yang, F. Grandi, E. Mangina, I.-Y. Song, H.C. Mayr (Eds.), Conceptual Modeling for Advanced Application Domains. XXII, 692 pages. 2004.

Vol. 3288: P. Atzeni, W. Chu, H. Lu, S. Zhou, T.W. Ling (Eds.), Conceptual Modeling – ER 2004. XXI, 869 pages. 2004.

Vol. 3287: A. Sanfeliu, J.F. Martínez Trinidad, J.A. Carrasco Ochoa (Eds.), Progress in Pattern Recognition, Image Analysis and Applications. XVII, 703 pages. 2004.

Vol. 3286: G. Karsai, E. Visser (Eds.), Generative Programming and Component Engineering. XIII, 491 pages. 2004.

Vol. 3285: S. Manandhar, J. Austin, U.B. Desai, Y. Oyanagi, A. Talukder (Eds.), Applied Computing. XII, 334 pages. 2004.

Vol. 3284: A. Karmouch, L. Korba, E.R.M. Madeira (Eds.), Mobility Aware Technologies and Applications. XII, 382 pages. 2004.

Vol. 3283: F.A. Aagesen, C. Anutariya, V. Wuwongse (Eds.), Intelligence in Communication Systems. XIII, 327 pages. 2004.

Vol. 3282: V. Guruswami, List Decoding of Error-Correcting Codes. XIX, 350 pages. 2004.

Vol. 3281: T. Dingsøyr (Ed.), Software Process Improvement. X, 207 pages. 2004.

Vol. 3280: C. Aykanat, T. Dayar, İ. Körpeoğlu (Eds.), Computer and Information Sciences - ISCIS 2004. XVIII, 1009 pages. 2004.

Vol. 3279: G.M. Voelker, S. Shenker (Eds.), Peer-to-Peer Systems III. XI, 300 pages. 2004.

Vol. 3278: A. Sahai, F. Wu (Eds.), Utility Computing. XI, 272 pages. 2004.

Vol. 3275: P. Perner (Ed.), Advances in Data Mining. VIII, 173 pages. 2004. (Subseries LNAI).

Vol. 3274: R. Guerraoui (Ed.), Distributed Computing. XIII, 465 pages. 2004.

Vol. 3273: T. Baar, A. Strohmeier, A. Moreira, S.J. Mellor (Eds.), <<UML>> 2004 - The Unified Modelling Language. XIII, 454 pages. 2004.

Vol. 3272: L. Baresi, S. Dustdar, H. Gall, M. Matera (Eds.), Ubiquitous Mobile Information and Collaboration Systems. VIII, 197 pages. 2004.

Vol. 3271: J. Vicente, D. Hutchison (Eds.), Management of Multimedia Networks and Services. XIII, 335 pages. 2004.

Vol. 3270: M. Jeckle, R. Kowalczyk, P. Braun (Eds.), Grid Services Engineering and Management. X, 165 pages. 2004.

Vol. 3269: J. Lopez, S. Qing, E. Okamoto (Eds.), Information and Communications Security. XI, 564 pages. 2004.

Vol. 3268: W. Lindner, M. Mesiti, C. Türker, Y. Tzitzikas, A. Vakali (Eds.), Current Trends in Database Technology - EDBT 2004 Workshops. XVIII, 608 pages. 2004.

Vol. 3267: C. Priami, P. Quaglia (Eds.), Global Computing. VIII, 377 pages. 2004.

Vol. 3266: J. Solé-Pareta, M. Smirnov, P.V. Mieghem, J. Domingo-Pascual, E. Monteiro, P. Reichl, B. Stiller, R.J. Gibbens (Eds.), Quality of Service in the Emerging Networking Panorama. XVI, 390 pages. 2004.

Vol. 3265: R.E. Frederking, K.B. Taylor (Eds.), Machine Translation: From Real Users to Research. XI, 392 pages. 2004. (Subseries LNAI).

Vol. 3264: G. Paliouras, Y. Sakakibara (Eds.), Grammatical Inference: Algorithms and Applications. XI, 291 pages. 2004. (Subseries LNAI).

Vol. 3263: M. Weske, P. Liggesmeyer (Eds.), Object-Oriented and Internet-Based Technologies. XII, 239 pages. 2004.